CURVES AND SYMMETRY

VOLUME 1

To Eva Fabé

CURVES
AND
SYMMETRY
VOLUME 1

J. Lee Kavanau
University of California, Los Angeles

Science Software Systems, Inc.
Los Angeles, California

ISBN #0-937292-01-x

Library of Congress Catalogue Card Number 81-85228

Published in the United States of America by Science Software Systems, Inc.

PREFACE

The present work is the first of two volumes to illustrate and elaborate upon the topics of *Symmetry, An Analytical Treatment* (Kavanau, 1980) in which only the first 5 of 14 chapters were accompanied by figures. This volume illustrates and amplifies the material of Chapters VI-VIII. Among the topics treated are: circumpolar intercept transforms of conic sections; inversions of quadratics, cubics, quartics, quintics, and sextics; *central Cartesians* and their inversions--including serpentine circular cubics, bicircular quartics, and corresponding monovular representatives; *curves of demarcation*; *non-focal self-inversion loci*; some general and specific mapping properties of the inversion transformation; and *Inversion Taxonomy*--the classification of curves upon the basis of their relatedness by inversion.

The emphasis of the contents upon inversion reflects the fact that the inversion transformation occupies a unique position in relation to classifying curves and organizing curves for systematic symmetry analyses. This is a consequence of its recently discovered fundamental property of *immediate closure* (Kavanau, 1980). If B' is the ensemble of the inverse loci of a basis curve B about all points in the extended plane, then the inversion of any member of B' about any point in the extended plane also is a member of B'. Further, if D and E are any two distinct curves of B', then poles exist about which each inverts to the other. It follows that if C' is the ensemble of all inversions of a curve C, then either B' and C' coincide or are disjoint. This property of the inversion transformation makes possible both the classification of curves upon the basis of their membership in narrowly circumscribed ensembles and the ready identification of every member of the ensembles. No other non-trivial point transformation is known to be characterized by *immediate closure*. Accordingly, no other point transformation delineates identical closed ensembles of curves upon repeated application, the feature that lies at the heart of the utility of inversion for classifying and organizing curves.

Although the inversion transformation was discovered independently several times in the early years of the 19th century and has been studied intensively for over 140 years, its most fertile and intriguing properties, namely, the modes of mapping of *equivalent inversion poles* (points in the plane about which inversions yield similar curves), have heretofore escaped notice. The oversight would appear to be related to the fact that inversions of lines and circles, and ensembles thereof (or planes and the surfaces of spheres in the solid domain), are the most readily visualized of all inversion relationships and lead to very useful theorems. Because of the utility of these theorems in geometrical proofs and constructions, they have dominated completely almost all previous treatments of inversion. But if one deals exclusively with the inverse loci of lines and circles, one encounters solely lines and circles, which are of practically no heuristic value in relation to inversion-mapping properties.

Inverse loci of other curves are mentioned and tabulated routinely in the classical literature, but these are few in number and consist almost entirely of inversions about exceptional points on the lines of symmetry of well-known curves and about the poles of spirals. In the few instances in which the inverse loci about non-axial points of curves were studied in depth, the chief concern was with such topics as the relationships between the points of contact or intersection of tangents and bitangents (for example, whether these points lay on a conic), the envelopes of circles with centers related in a certain manner, etc., rather than with questions related to the fundamental and much more fruitful topic of the mapping properties of the inversion transformation.

If one deviates from the conventional approaches mentioned above, one is confronted forthwith by a plethora of provocative questions. In consequence, my original intention to present all of the illustrations for Chapters VI-XIV in a single volume was short lived. Barely having begun the figures for Chapter VI, basic questions concerning the inversion transformation intruded. Answering these questions led to the discovery of the fundamental mapping properties of equivalent inversion poles, which comprise universal and remarkably simple relationships. In keeping with the unique suitability of the transformation for classifying and organizing curves, the answer to virtually every inversion-mapping query is one of elegant simplicity. For example:

Question...what is the inverse locus of a given curve about a point of the ideal line at infinity (for an appropriate infinite value of the unit of linear dimension)?
Answer.....a reflection of itself in a line.

Question...what locus is obtained by inverting an inverse curve about a given point in its plane?
Answer.....the same locus obtained by inverting its basis curve about the point inverse to the given point.

A notable consequence of the above-mentioned discovery of the mapping properties of the inversion transformation relates to the continuum of ideal points of the line at infinity. One cannot regard these points as equivalent to one another from the point of view of inversion geometry, and unify them under the designation "the point at infinity." Not only do the inverse loci of a curve about different points of the ideal line at infinity differ in their orientations, the points of this line may differ in their circumpolar focal status. Thus, each ideal point of the line at infinity in a direction orthogonal to a line of symmetry of a given basis curve is a focus of self-inversion, whereas the other points are merely poles of *congruent-inversion* (see below).

In the same vein, inversion mapping identifies the line at infinity as an asymptote of certain curves, whereas classical approaches generally exclude this line from consideration. The asymptote of an inverse curve obtained by inverting a given basis curve about an incident point is the inverse of the circle of curvature of the basis curve at that point. As the curvature of such circles becomes infinite, the corresponding asymptotes approach the line at infinity. Thus, from the point of view of inversion geometry, there is no more reason to exclude the ideal line at infinity as an asymptote of an inverse curve than to exclude a point of infinite curvature from a basis curve.

Perhaps the most significant new property to emerge from inversion-mapping studies is that of *congruent-inversion*, i.e., the existence of previously unknown poles about which certain curves invert to congruent loci differing from the basis curves only in their orientations and locations. Non-axial inverse loci of central conics, for example, possess one pole of congruent-inversion in the finite plane, while the corresponding inverse loci of biovular *central Cartesians* possess 3 such poles. Thus, foci of self-inversion can be regarded merely as special cases of poles of congruent-inversion; about the former poles the inverse loci are both congruent and coincident, as opposed to being merely congruent about the latter poles.

The term *identical subspecific-inversion* denotes both self-inversion and congruent-inversion (any similar inverse locus qualifies as a congruent-inversion, since the unit of linear dimension is at one's disposal). The inverse of a curve with 4 lines of symmetry may have as many as 15 poles of identical subspecific-inversion, not more than 6 of which can be foci of self-inversion. On the other hand, the inverse loci of some periodic functions have an infinite number of poles of identical subspecific-inversion, including infinite numbers of both poles of congruent-inversion and foci of self-inversion; the inverse loci of other periodic functions have no foci of self-inversion but an infinite number of poles of congruent-inversion.

Since topics relating to inversion are dealt with in both Chapters VI and VIII of *Symmetry, An Analytical Treatment*, the length of the present volume was much increased on that account. A second principal factor that led to its expansion was the treatment of *non-focal self-inversion loci* in Chapter VIII. In the original text, only the non-focal self-inversion loci of conic sections were derived and described. A very limited extension of the treatment to limacons and *parabolic Cartesians* revealed so rich and compelling a new area, that the 15 pages of figures devoted to this topic amount to no more than an exploratory treatment.

Both the topic of non-focal self-inversion loci and the topic of *curves of demarcation* of Chapter VIII lead to considerations of *Cartesian self-inverters* (including curves known classically as "Cartesians"). When the affinities of these curves are viewed in the perspective of Inversion Taxonomy, one is led to a consideration of the central basis curves, *central Cartesians*. The latter loci possess the distinction of being the quartic equivalents of central conics outside the *quadratic-based inversion superfamily* (*QBI superfamily*). In fact, central Cartesians, central conics, and *central quartics*--the inverse loci of central conics about their centers--are the only central curves whose point-in-the-plane inverse loci have rectangular equations that do not exceed a degree of 4. Thus, these 3 groups are classified together under the designation *central curves with 4th-degree-maximum inverse loci*.

The fact, alone, that central Cartesians are the quartic non-QBI equivalents of central conics would have justified analyses of their circumpolar symmetry and inversion properties. When it also became evident that the intriguing geometrical properties of central conics and their inverse loci are at least matched in interest by corresponding properties of central Cartesians and their inversion derivatives, extensive further lengthening of the projected treatment was inevitable; 19 pages of illustrations for Chapter VIII are devoted to the latter curves.

These expansions of the volume do not refer to illustrations alone; the new findings are accompanied by analytical treatments, to the extent of 294 equations, 14 tables, and 312 pages of text. In addition, 114 new *Circumpolar Maxims* are presented, more than twice the number to be found in the original work. All but two of these deal with properties of the inversion transformation. As a result of these amplifications of the material of *Symmetry, An Analytical Treatment*, more than 80% of the present work consists of entirely new material, as opposed to legends for illustrations of curves and relationships described in the original volume.

Circumpolar Maxims are based on current knowledge, and are summarizing statements of the properties of curves in relation to point transformations. Nearly all of the 114 new Maxims are firmly established or readily susceptible to proof but, in some cases, they are subject to restrictions, extension, or other modifications as further information is obtained. Only a small fraction of the material presented in these Maxims previously was known.

It is abundantly clear from the present studies that findings concerning the circumpolar symmetry and inversion properties of QBI curves (quadratics and their inverse loci), including conservation of circumpolar focal loci under inversion, provide poor foundations for generalization to members of other inversion superfamilies. Accordingly, some of the Maxims of the original work, based almost entirely upon experience with QBI curves, have to be modified or restricted to that domain (see particularly the legend to Fig. 8-20).

The approach to the presentation of curves and relationships already dealt with but unillustrated in the basic text has been to outline the background material briefly, repeating equations where desirable, in order to make the figures comprehensible without absolute need to consult the original volume. In the same vein, the *Glossary of Terms* of the original work has been included in expanded form and many of the entries have been amplified. Figs. 1-1 and 5-1 of the basic text are duplicated in the present volume, since they are key illustrations of geometrical constructions of intercept transforms. Fig. 3-7, a treatment of *polar-linear conchoids*, supplements the illustrations for Chapter III and is important in its relationship to the *LCQ sextics* of Figs. 8-28 and 8-42. The other 84 pages of illustrations are for Chapters VI-VIII, for which there are no figures in the original work.

Unless qualified, the description of a curve or group of curves as being *highly symmetrical* refers to the *circumpolar symmetry* (symmetry with respect to a point) of the locus or loci in question. When not associated with roots or otherwise specified, upper alternate signs apply to derivatives of basis hyperbolas and lower alternate signs to those of ellipses. Dimensional balance is maintained in all equations. In addition to its didactic advantages, this greatly facilitates dealing by hand with the complicated equations encountered in symmetry and inversion analyses.

References here, to chapters, equations, tables, and appendices, to be found in *Symmetry, An Analytical Treatment*, employ Roman numerals for the chapter and appendix number; for example, page XI-3, equation XII-16, Table X-1, and Appendix II. Maxims in the original work (numbered 1 to 49) are referred to by Arabic numbers, for example, Maxim 18. References to material within the present work employ Arabic numbers for the chapters, followed by a dash and the figure number, followed by a dash and the number of the equation, table, or Maxim, as, for example, equation 8-30-2, Table 8-20-1, and Maxim 6-4-2. The appendices are referred to as Appendix 1 or 2.

As in the case of the original volume, I have sought to minimize errors by typing the manuscript and preparing the figures myself for direct reproduction. Points of the curves were first plotted by pencil, after which the loci were drawn by hand by inking successive short segments with the aid of French curves and other templates. To gain a proper perspective of the illustrations it often is necessary to view them from several angles. As an expediency, the cedilla has been omitted in typing of the name *limaçon*. Abbreviations have been held to a bare minimum--*CW, CCW, DP, QBI, LR,* and *eqs.,* for *clockwise, counter-clockwise, double point, quadratic-based inversion, latus rectum,* and *equations.*

To avoid confusion, the degree sign, °, always is employed for numerical angular measure. Terms, phrases, and sentences are italicized either for emphasis, in statements of Maxims, to introduce new terminology, as headings or titles, or as taxonomic categories. Italics are not employed for existing unusual or little-known terminology or for the unconventional usage of common terms. In the latter cases, quotation marks are used. Footnotes usually are enclosed in brackets, [ ], within the text.

As in the original work, new terms have been made as descriptive as practical and are defined in Appendix 2, the *Glossary of Terms.* Existing terminology is adhered to fairly closely, despite not infrequent disadvantages. For example, an *inversion superfamily* has been designated as the *Cassinian-based inversion superfamily*, out of deference to the fact that Cassinians are known curves; the designation *central parabolic Cartesian-based inversion superfamily* would have been more appropriate, but the name *central parabolic Cartesian* is newly assigned. The term "vertex" is employed in a broader than conventional sense and the term "oval" is used with the broadest classical meaning.

In the interests of paralleling the long-established nomenclature of biological classification, the taxons--*subspecies, species, genus, supergenus, subfamily, family, superfamily, suborder, order, class,* etc.--have been employed, though other mathematical applications of some of these terms have priority.

I wish to thank my son Warren for proofreading the final draft of the manuscript. I am indebted to my son Christopher for carrying out a number of derivations for me and for developing a useful parametrization technique for plotting inverse loci and intercept transforms of high rectangular degree. Christopher also was the first to prove Maxim 6-7-5 and he originated a fruitful method for locating the asymptotes of inverse curves. Thanks also are due to my student Roozbeh Sadeghiyan for elegant geometrical proofs of Maxims 6-4-2, 6-4-4, and 6-7-5 by procedures that also can be employed to prove many other inversion-mapping Maxims, and for calling my attention to the inversions of periodic functions with an infinite number of (parallel) lines of symmetry.

I owe a great debt of gratitude to Aaron E. Whitehorn for a meticulous reading and correcting of the penultimate draft of the manuscript and for numerous improvements and incisive comments. Technical defects that remain are my responsibility; Mr. Whitehorn's assistance infrequently included confirmations of my analyses. Lastly, I wish to thank Florence Kaplan for her valuable assistance.

J. Lee Kavanau
Los Angeles, California
January, 1982

ART IN MATHEMATICAL DESCRIPTION

When you are at a museum, notice that many visitors are eager to examine the label of an art work. They read the title, author, and date; they refer back to their catalogues and then stand back and examine the work. A description enhances our appreciation of art, as thirst for the intellectual and the visual go hand in hand. When Michelangelo was asked how he drew, he replied, "with my mind." Perhaps the author of this work would say, "my drawings guided my thoughts."

When the art of mathematics is neglected the appreciator is not fully requited. This treatment is an exception; in the following pages we have the benefit of new geometrical concepts, their descriptions, and aesthetic graphics, in a work that can be enjoyed on the merits of its art alone.

Frederick I. Sauls

THE COVER ILLUSTRATION

The cover illustration is a reproduction of Fig. 7-10 depicting segments of 8 circumpolar intercept transforms of the equilateral hyperbola about an axial pole at $h = 3a$ for angles of transformation from 30° (lower left) to 150° (upper right).

TABLE OF CONTENTS

LIST OF TABLES

CHAPTER I

Fig. 1-1. Symmetry relative to single reference elements. The symmetry of curves relative to single reference elements is assessed, in general, by means of two radius vectors emanating from points on the reference elements to the curves. When the reference element is a point, only a single angle is associated with a given characterization of the symmetry, namely the angle, α, between the two radius vectors. In the broadest sense, the symmetry of a basis curve about a given point-pole at a given angle--α-*circumpolar symmetry* --is characterized by a specific transform curve--the α-*circumpolar intercept transform*.

This transform is obtained from a rectangular plot of the lengths (intercepts) of two radius vectors, x and y, that emanate from the reference pole to the curve at the fixed angle, α, to one another, as these vectors rotate about the pole and scan the basis curve. Fig. a illustrates the manner in which the 90°-circumpolar intercept transform of an ellipse about a non-axial interior pole is obtained (see also Fig. 7-11). Successive lengths of the 4 pairs of orthogonal vectors at the left are plotted at the right along rectangular axes. [A more complete illustration of the manner of derivation of circumpolar intercept transforms (for 180°) is given in Fig. 5-1.]

The manner of derivation of an intercept transform curve that characterizes the α-θ-*circumlinear symmetry* of a basis curve is illustrated in Fig. b. In this case the symmetry is assessed relative to two reference angles, the angle α between the two radius vectors, and the angle θ between the x radius vector and the x-axis; both angles are taken to be 45° in the illustration of Fig. b. In the case of circumlinear symmetry, the radius vectors do not rotate about a pole but, rather, translate from point to point along the reference line, always maintaining the same orientation to one another and the line. The x and y intercepts obtained in this manner at 3 positions of a pole on a reference line consisting of the major axis of the basis ellipse, at the left in Fig. b, are plotted on rectangular axes at the right.

The characterization of the α-θ-*circumcurvilinear symmetry* of a basis curve is accomplished in a parallel fashion in Fig. c. The intercept transform curve that characterizes the 90°-0°-circumelliptical symmetry of a basis circle--the major auxiliary circle of the reference ellipse--is derived by plotting at the right the intercepts obtained at the left along the reference ellipse. The plots are illustrated for 3 pairs of radius vectors.

Symmetry Relative to Single Reference Elements

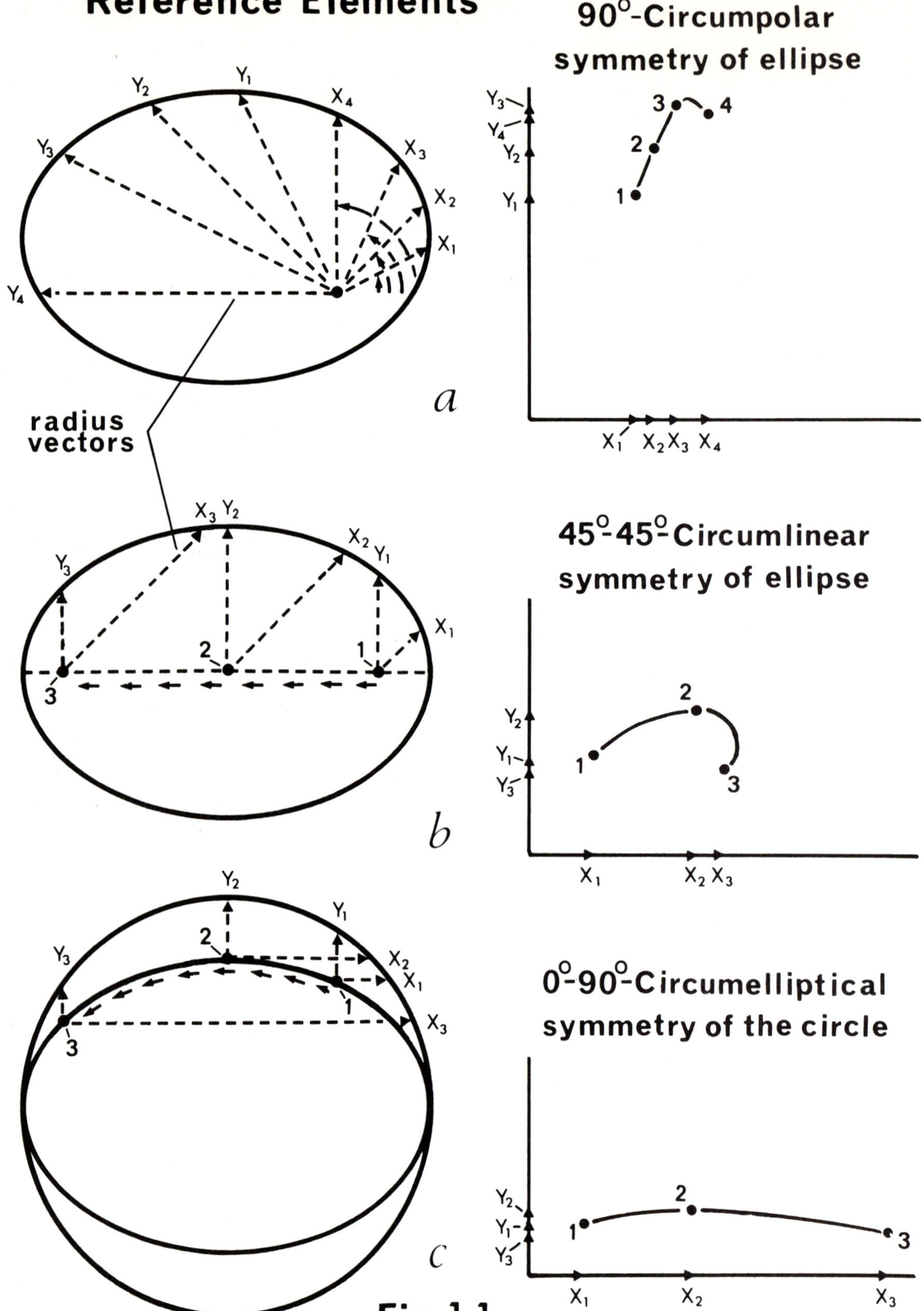

Fig. 1-1

CHAPTER III

Fig. 3-7. Polar-linear conchoids. The classical conchoid construction is a *non-orthogonal polar-curvilinear* construction in which equal distances along a line are marked off from its point of intersection with a curve for various positions of the line as it rotates about a point. The conchoid is the locus of the endpoints of the line segments obtained in this manner. Accordingly, classical conchoid constructions are not *extremal distance* constructions relative to all reference elements, which are the sole type dealt with in Chapter II-IV, and upon which the discussions of symmetry therein are based.

However, the general definition of the most symmetrical curve relative to two or more fixed reference elements (page I-4) may be applied to any type of construction. [In the case of the classical conchoid construction, the distinction regarding undirected distances is of no significance, since the distances are uniquely determined.] Since the eq. of the conchoid of Nicomedes (employed by Nicomedes for the trisection of an angle) in the classical construction is $u = v$, that curve is the most symmetrical locus within that constructional framework. Similarly, the cardioid, $u = v$, is the most symmetrical *incident polar-circular* classical conchoid construction.

On the other hand, within the framework of the polar-linear coordinate system of Chapter III, the conchoid of Nicomedes does not number among the most symmetrical polar-linear loci. In fact, its eq. (Fig. a) in this system is a quartic, although each arm can be represented individually by the quadratic eqs. $uv = a(d+u)$ and $uv = a(d-u)$. The arm of the curve on the near side of the line possesses a cusp at the point-pole when $a = d$ (i.e., when the length of the line segment is equal to the distance of the point-pole from the line), it possesses a DP when $a > d$, but it possesses no exceptional point when $a < d$.

If the complete locus of the conchoid of Nicomedes is compared with other two-arm conchoid-like polar-linear constructions (Figs. b and c), the comparison leads to the conclusion that the conchoid of Nicomedes is not one of the most symmetrical of the polar-linear conchoid-like curves. Thus, the equilateral hyperbolic and parabolic polar-linear conchoids (Figs. b and c, respectively) are examples of more symmetrical constructions.

The closer relationship of the conchoid of Nicomedes to equilateral hyperbolic conchoids than to parabolic polar-linear conchoids is suggested by the

Polar-Linear Conchoids

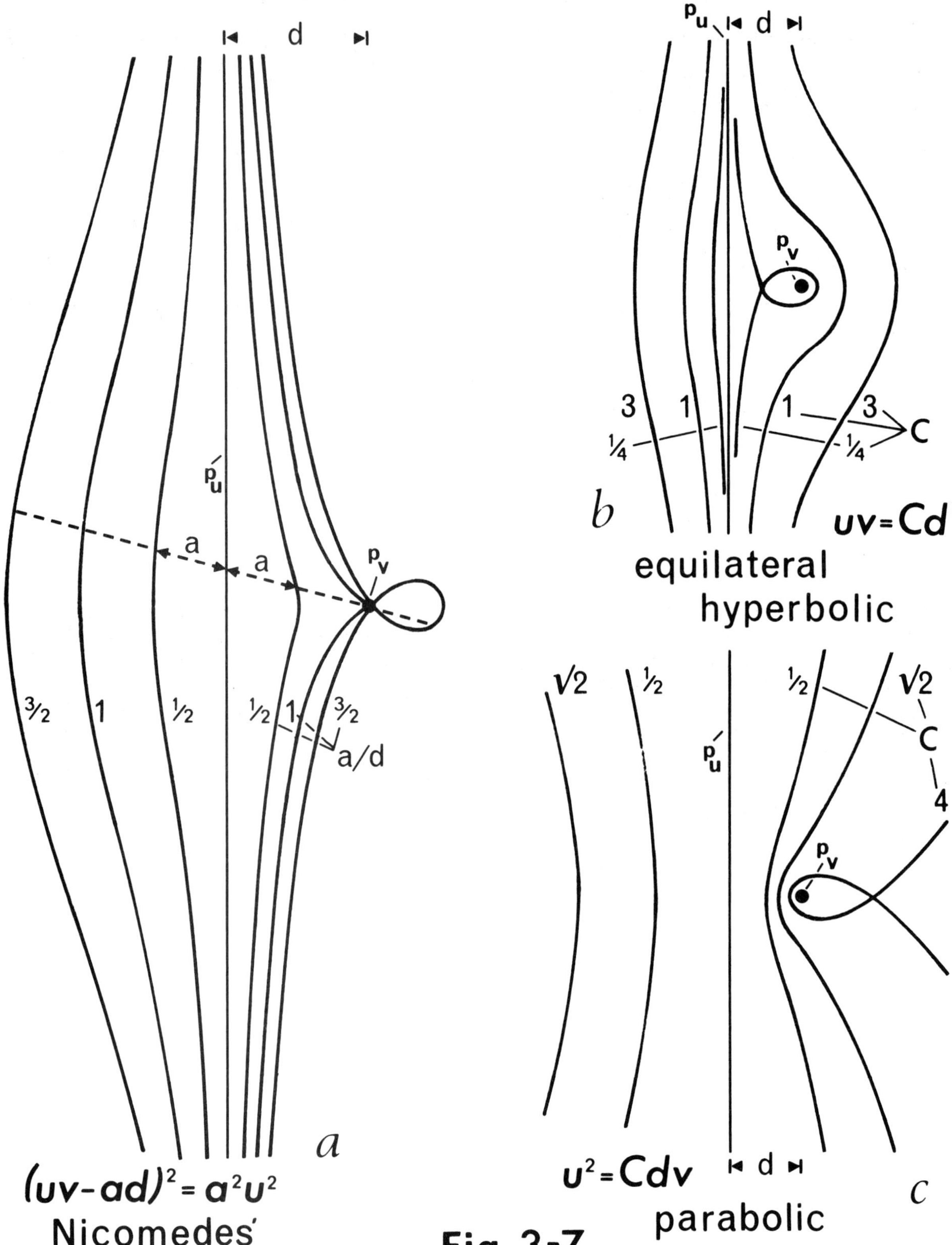

Fig. 3-7

fact that the eqs. of the single arms of the conchoid of Nicomedes also have equilateral hyperbolic form (see above eqs.). This affinity also is supported by the properties of the rectangular eqs., which differ in that a term C^2d^4 of the equilateral hyperbolic eqs. is replaced by a term $a^2(x-d)^2$ in the eqs. of the conchoid of Nicomedes [$x^2y^2 = C^2d^4 - x^2(x-d)^2$ versus $x^2y^2 = a^2(x-d)^2 - x^2(x-d)^2$].

Polar-linear equilateral hyperbolic conchoids (Fig. b) possess the distinction of being the basis curves for the inversion superfamily (see Fig. 8-42) to which belongs a sextic genus of high circumpolar symmetry--the LCQ sextics (see Fig. 8-28). The simple intercept format of the latter curves cast about the pole inverse to the polar-linear point-pole, p_v, is eq. 1a.

(a) $\cos\theta = \dfrac{x(x-a)}{b^2}$ simple intercept format of LCQ sextics cast about the tangent vertex of the small and medium-sized ovals (3-7-1)

(b) $\cos\theta = \dfrac{d(Cd-x)}{x^2}$ simple intercept format of equilateral hyperbolic polar-linear conchoids cast about the point-pole, p_v

[These curves are named for the fact that their 0° circumpolar intercept transforms are linear, their 180° transforms are circular, and their 90° transforms are quartics (these designations refer to the transforms derived from eq. 1a, rather than to the algebraically complete transforms; see Chapter VII, *Negative Intercepts, Trivial Transforms, The Degree of Transforms, and The Duality of 0° and 180° Transforms.*]

The corresponding simple intercept format of equilateral hyperbolic polar-linear conchoids cast about the point-pole is eq. 1b.

CHAPTER V

Fig. 5-1. The 180°-circumpolar intercept transform about the traditional foci of central conics. This transform is illustrated with an ellipse and hyperbola (labelled "basis ellipse" and "basis hyperbola") that have identical parameters a and b and are monoconfocal (see page V-12). The 180° transforms of these curves about the traditional foci consist of congruent and coincident equilateral hyperbolas, $2axy = b^2(x+y)$. Only one arm of the 180° transform, shown at the left and labelled "transform equilateral hyperbola," is traced out by a plot of two positive intercepts of the basis curves. The x-axis is coincident with the major (and transverse) axis, and the y-axis is coincident with the common LR; the coincident traditional foci are at the origin.

To illustrate the manner in which the transform curve is generated as the x and y radius vectors rotate about the coincident left traditional foci, the angles that the x radius vectors make with the x-axis are marked off along the transform at positions corresponding to lengths of the intercepts along the axes. As examples of correspondences between positions of the radius vectors and points on the transform curve, for $\theta = 0^{\circ}$, the x intercept for the ellipse is a(e+1)--the distance from the left focus to the right vertex--and the y intercept for the ellipse is a(1-e)--the distance from the left focus to the left vertex. The point on the transform curve corresponding to these lengths of the intercepts is at [a(e+1),a(1-e)], at the top of the figure to the left of the major-axis vertex, marked 0°, etc.. For the basis hyperbola, the point on the transform curve corresponding to 0° is at $[a(e-1), \infty]$. At $\theta = 90^{\circ}$, all radius vectors come into coincidence with the common LR of the basis curves (which are coincident with the y-axis), yielding a common point on the transform curve at $(b^2/a, b^2/a)$.

Note the differences in the spacing of corresponding positions along the transform curve for equal increments of θ for the x radius vectors of the two basis curves. These differences provide the basis for the *ordinal rank equations*, $\mathring{s}^2 = (ds/d\theta)^2 = f(x,y)$, and *subordinal rank equations* $\mathring{s}^2 = F(x)$ or $G(y)$, that serve for hierarchical circumpolar symmetry ranking at the next level of discrimination beyond the degree of the corresponding transform eq. (see Tables I-1 and V-1,2, and Appendix III). The differences in question are illustrated in another manner by plots of $ds/d\theta$ versus θ for various poles of the $e = 0.745$ ellipse in Figs. 7-16,17, and 18 (and for the parabola in Fig. 6-15). Corresponding plots for $ds/d\theta$ versus the x intercept are given in Fig. 7-19.

180°-Circumpolar Intercept Transform About The Traditional Foci of Central Conics

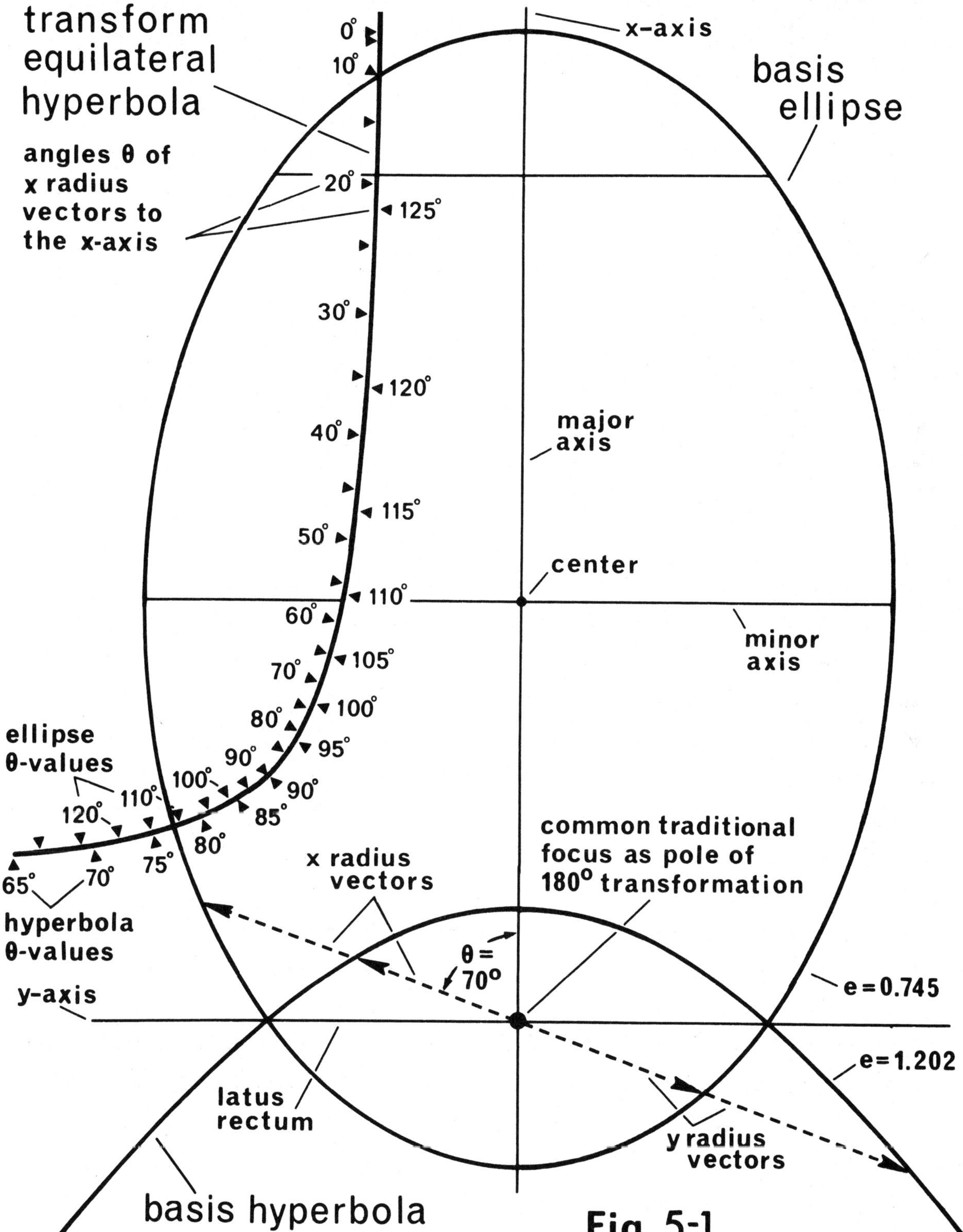

Fig. 5-1

CHAPTER VI

Fig. 6-1. 90° and 180°-circumpolar intercept transforms of the parallel line-pair for two positive intercepts. Inasmuch as the parallel line-pair has a true center (any point on the midline) and any point in the plane lies on a line of symmetry, all 90°-circumpolar intercept transforms are symmetrical about both coordinate axes, so that the transform in any given quadrant is congruent with those in the other 3 quadrants. Accordingly, only the 90° transforms for two positive intercepts are illustrated, and these only for poles lying between the two lines.

In the cases of the 180° transforms about poles lying between the lines, the transform for one positive and one negative intercept is the trivial transform of identity, $x=-y$. Accordingly, the trivial transform in the 2nd and 4th quadrants (which is a bisector of the coordinate axes) is not congruent with the transforms in the 1st and 3rd quadrants, which are for two positive and two negative intercepts. The algebraically complete 180° transform of

$$(x+y)^2[(d+k)x+(k-d)y]\cdot[(d+k)y+(k-d)x] = 0 \qquad (6\text{-}1\text{-}1)$$

algebraically complete 180° circumpolar intercept transform of the parallel line-pair

$$[(x^2+y^2)(k-d)^2-x^2y^2][(x^2+y^2)(k+d)^2-x^2y^2]\cdot[(x^2+y^2)(k^2+d^2)-x^2y^2-2kd(x^2-y^2)]\cdot[(x^2+y^2)(k^2+d^2)-x^2y^2-2kd(y^2-x^2)] = 0 \qquad (6\text{-}1\text{-}2)$$

algebraically complete 90° circumpolar intercept transform of the parallel line-pair

eq. 1 (see Chapter VI, *Negative Intercepts, Trivial Transforms, The Degree of Transforms, and The Duality of 0° and 180° Transforms*) represents a total of 3 lines--the trivial $x=-y$ and two lines in the 1st and 3rd quadrants, $(b+k)x=(b-k)y$ and $(b+k)y=(b-k)x$, of slopes $(b+k)/(b-k)$ and $(b-k)/(b+k)$. Four cases are represented in Fig. 1, transforms for $k=0$ (Fig a, 180°), $\frac{1}{4}d$, $\frac{1}{2}d$, and $3d/4$, where k is the displacement of the pole of transformation from the midline and d is the orthogonal distance from the midline to each line.

Taking the transform loci for $k=\frac{1}{2}d$ as an example, at $\theta=0^\circ$,

90° and 180°-Circumpolar Intercept Transforms of The Parallel Line-Pair

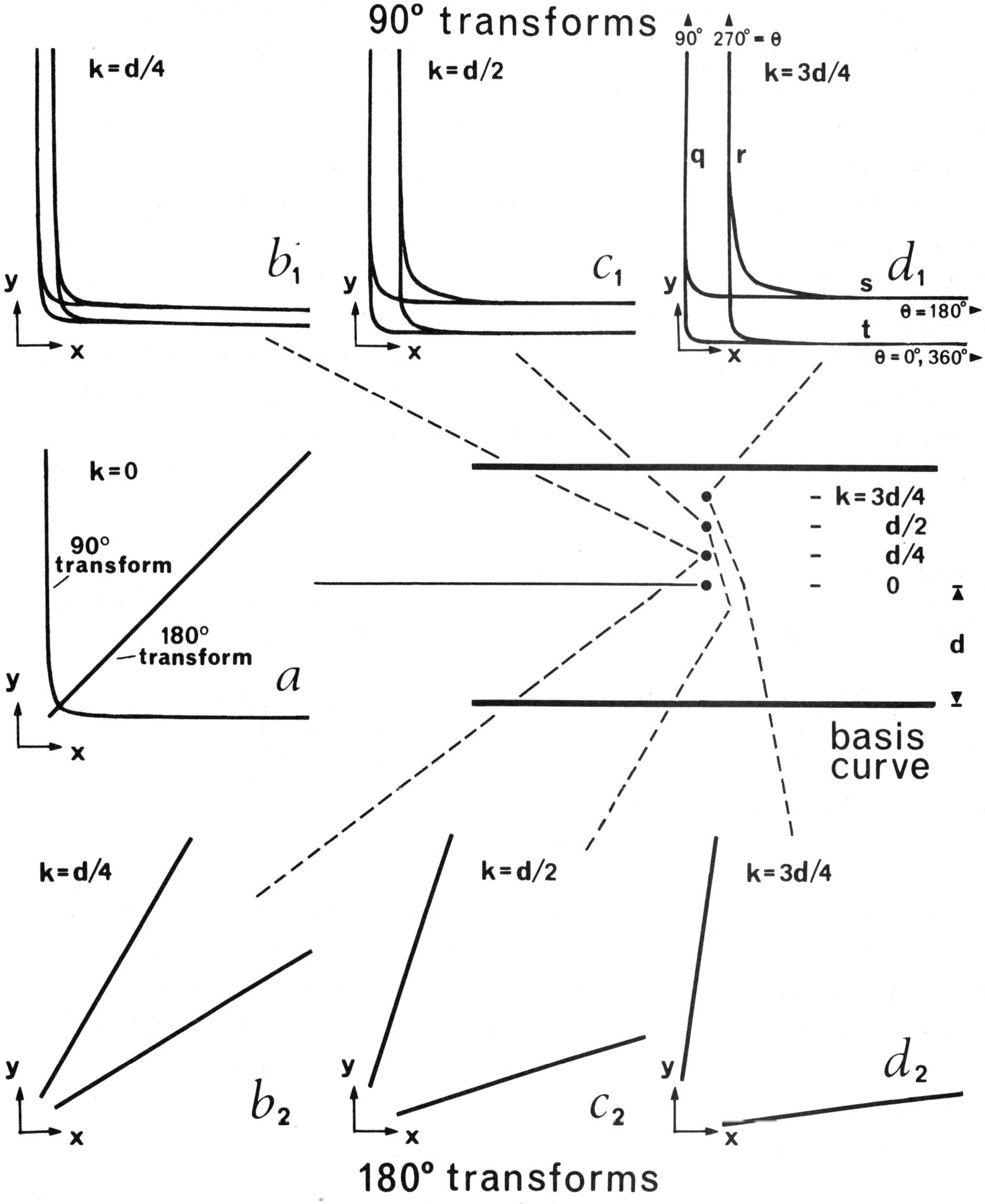

Fig. 6-1

$(x,y) = (\infty,\infty)$. As θ progresses to 90°, x and y decrease to $\frac{1}{2}d$ and $3d/2$, respectively, tracing the upper line segment of Fig. c_2 to its inner terminnus. As θ progresses from 90° to 180°, the upper line segment is retraced in the opposite direction to $(x,y) = (\infty,\infty)$.

In the cases of the 90° transforms, eq. 2 is the algebraically complete transform eq. Most commonly, each term of an algebraically complete transform eq. represents the transform for a particular combination of intercepts, including trivial combinations. This is not the case in the present instance (because of the simplicity of the simple intercept eq., $x\sin\theta = k \pm b$). Thus, the 1st and 2nd terms of eq. 2 represent the segments of the transform for two positive intercepts and two negative intercepts from 0° to 90° and 180° to 270°, but the 3rd and 4th terms represent the segments for two positive intercepts and two negative intercepts from 90° to 180° and 270° to 360°.

On the other hand, the 1st term represents segments for a positive x intercept and a negative y intercept from 90° to 180°, and a negative x intercept and a positive y intercept from 270° to 360°, etc. Accordingly, all 4 terms of the algebraically complete transform eq. contribute segments to the transform in each quadrant.

For example, one of each of the 4 asymptotic arms of the three 1st quadrant plots of Figs. b_1-d_1 is represented by each term. As θ progresses from 0° to 90°, the arm consisting of segments t and q (as labelled in Fig d_1) is traced; this is represented by the 2nd term. From 90° to 180°, the arm consisting of segments q and s is traced; this is represented by the 3rd term. From 180° to 270°, it is the arm consisting of segments s and r and represented by the 1st term that is traced, while from 270° to 360°, it is the arm consisting of segments r and t and represented by the 4th term that is traced. For the case $k = 0$ (Fig. a, 90°), the 4 arms are identical (congruent and coincident) and the 4 terms are identical.

90° transforms for the single line consist of arms such as the combination of segments q and t and the combination of segments r and s, to which, since $d = 0$, any of the 4 terms applies. For $d = 0$, the 180° transform becomes the trivial transform of identity $x = -y$.

Fig. 6-2. Circumpolar intercept transforms of the circle for two positive intercepts. This figure illustrates circumpolar intercept transforms of the circle about interior poles ranging from the center through points at ¼R, ½R, 3R/4, and 7R/8, to an incident point (as labelled in Figs. b and f), and at angles of 30°, 60°, 90°, 120°, 150°, and 180° (Figs. a-f, respectively). These transforms cover the full range of exponential degree and are essentially representative of the full range of form of sextic transforms of the circle (in the real plane) about points in the plane. Thus, the transforms at 210° are congruent with those at 150°, those at 240° are congruent with those at 120°, those at 270° are congruent with those at 90°, etc. In the cases of 0° transforms about exterior poles and 180° transforms about interior poles, all are segments of equilateral hyperbolas.

[In the cases of 4-quadrant transforms--i.e., transforms for all combinations of positive and negative intercepts--transforms for supplementary angles, θ and $180^\circ-\theta$, are congruent but rotated 90° relative to one another. In other words, the segments of the 30° transform in the 1st and 3rd quadrants also occur in the 2nd and 4th quadrants of the 150° transform, and vice versa, etc.]

Transforms about interior and exterior poles at other angles closely resemble one another for comparable ranges of intercept values, although a different subspecies is generated for each value of h (see eq. VI-15). For example, the 90° transforms for $h = 0.875$ and $h = 1.125$ are almost in coincidence. On the other hand, though all transforms about interior poles have only real solutions, those about exterior poles may have both real and complex solutions or only complex solutions.

Transforms about the center at all angles are of 0 exponential degree, consisting only of the point (R,R), as labelled in Fig. f. The plotting of this point transform serves to establish the scale, which is the same for each figure and is labelled in units of R along the abscissa and ordinate of Fig. c. [This point transform is not incident upon the $h = R$ transform of Fig. d nor the $h = \frac{1}{2}R$ transform of Fig. e.] As the location of the pole of the transform progresses from the center toward the perimeter, a series of gradually changing sextic loci are generated.

Incidence leads to reduction to quadratic loci, which are ellipses for transformation angles between 0° and 90° and between 90° and 180°, a circle of radius 2R for a transformation angle of 90°, and trivial transforms, $xy = 0$, for angles of 0° and 180°. The latter are equilateral hyperbolas

Circumpolar Intercept Transforms of The Circle

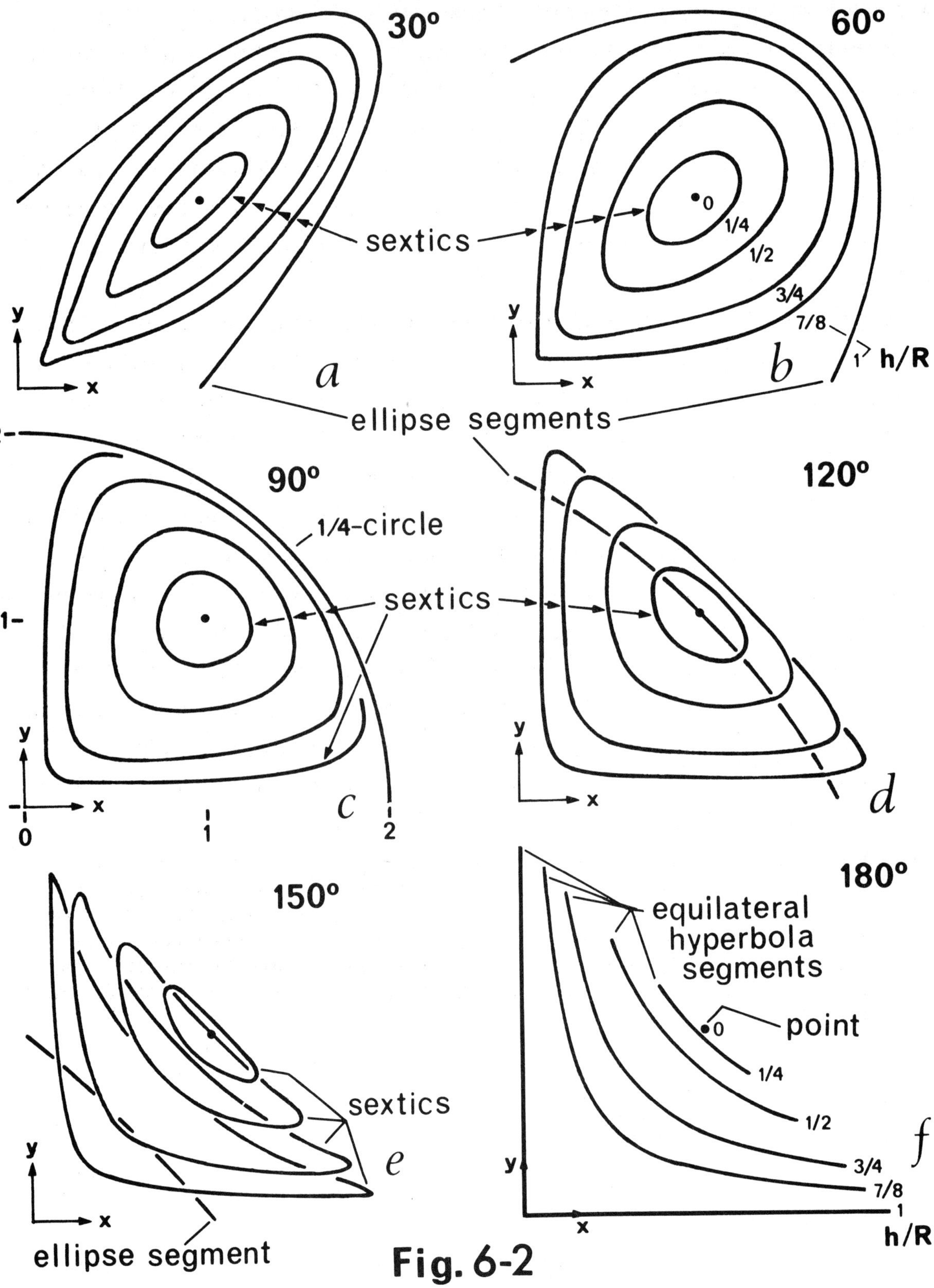

Fig. 6-2

with semi-axes of length 0, the locus of which is segments of lines that intersect at 90°, as illustrated in Fig. f (these are segments of the asymptotes of equilateral hyperbolas of non-zero semi-axes).

Whereas all the sextic transform ensembles about interior poles for given angles of transformation between 0° and 90° and 270° and 0° (including 90° and 270°) are non-intersecting and space-filling (i.e., they fill the space within the ellipses or circle), all those for angles between 90° and 270°, with the exception of the 180° transform, include intersecting subspecies and possess segments exterior to the elliptical transforms about incident poles.

Among point-in-the-plane circumpolar intercept transforms, those of lowest degree are the 1st-degree 0° and 180° transforms of the parallel line-pair and 90° transforms of the orthogonal line-pair, the 2nd-degree 0° and 180° transforms of the circle, and the 4th-degree 90° α-transforms of lines (i.e., transforms for an angle, α, which may take on any value). The sextic α-transforms of the circle possess the next-lowest degree, followed by the 8th-degree 0° and 180° transforms of the parabola and central conics (see Table VII-1).

90° point-in-the-plane transforms of the equilateral hyperbola are of 16th degree (giving lines, the circle, and the equilateral hyperbola the distinction of having the lowest-degree 90° point-in-the-plane transforms), while those of the parabola and central conics are of 20th degree. For comparison, the 0° and 180° point-in-the-plane transforms of QBI cubics and quartics are of 48th and 56th degree, respectively. The corresponding 90° transforms are of 68th and 72nd degree (see Table XII-2).

[The line, the circle, and the equilateral hyperbola have in common the properties of *polar-exchange symmetry* in *equidistant collinear tetrapolar coordinates* (pages IV-6,7 and *lxxii, lxxiii*) and of being the only known curves that are *simple curvilinear focal loci* (for a given non-trivial angle of transformation--90° for the equilateral hyperbola but any angle for the line and circle--no circumpolar point foci are incident upon the curve; see pages V-26,28).]

Fig. 6-3. Representative equilateral QBI curves. When an equilateral hyperbola (Fig. a) is inverted about the incident and axial poles and points in the plane indicated by the solid circles (taking the length of the semi-transverse axis, a, as the unit of linear dimension), one obtains the respective inverse loci depicted in Figs. b-o. The coordinates of the inversion poles (in units of a) are given below each inverse locus (all coordinates are represented as positive--as if all were for poles in the 1st quadrant--although, for convenience, some poles are depicted in other quadrants).

When the pole of inversion lies on both lines of symmetry of the basis curve, i.e., at the center (0,0) of the equilateral hyperbola, the inverse curve (the equilateral lemniscate of Fig. g) also possesses two lines of symmetry and, like the basis curve, does not self-invert about any pole in the finite plane. When the pole of inversion lies on only one line of symmetry of the basis curve, inversion leads to the loss of the orthogonal line of symmetry but to the gain of a 0° focus of self-inversion incident upon the remaining line of symmetry, as in the cases of Figs. b,h,i,l,m,n, and o.

When the pole of inversion is not incident upon a line of symmetry (Figs. c,d,e,f,j, and k), both lines of symmetry are lost; in their place, the inverse curves possess two 0° foci of self-inversion in the finite plane. These foci lie on lines concurrent at the DP and parallel to the lines of symmetry of the basis curve (see note, page XI-37). In all cases the reciprocal inversion pole, i.e., the pole about which the inverse curve inverts to the basis equilateral hyperbola, lies at the DP.

[These phenomena illustrate the exchange relationship between lines of symmetry and 0° foci of self-inversion in the finite plane (Maxim 8-20-2) or, what is equivalent, between (0°) foci of self-inversion on the ideal line at infinity and 0° foci of self-inversion in the finite plane (Maxim 8-20-2'). The loss of a line of symmetry is accompanied by the gain of a 0° focus of self-inversion in the finite plane, and vice versa. Lines of symmetry and 0° foci of self-inversion in the finite plane (or 0° foci of self-inversion on the line at infinity and 0° foci of self-inversion in the finite plane) thus are complementary aspects of symmetry with respect to inversion.

The basis for this exchange relationship can be elucidated by pursuing the consequences of inverting pairs of points reflected in a line of symmetry of a basis curve about an off-axis pole. It is a property of the inversion transformation that, though the connecting segments of all such point-pairs possess coincident midlines, the midlines of the inverse point-pairs are not coincident. Consequently, the inverse curve possesses no corresponding line of symmetry. By virtue of the same property of the inversion transformation, lines connecting each pair of inverse points have a common pole of concurrence.

Representative Equilateral QBI Curves

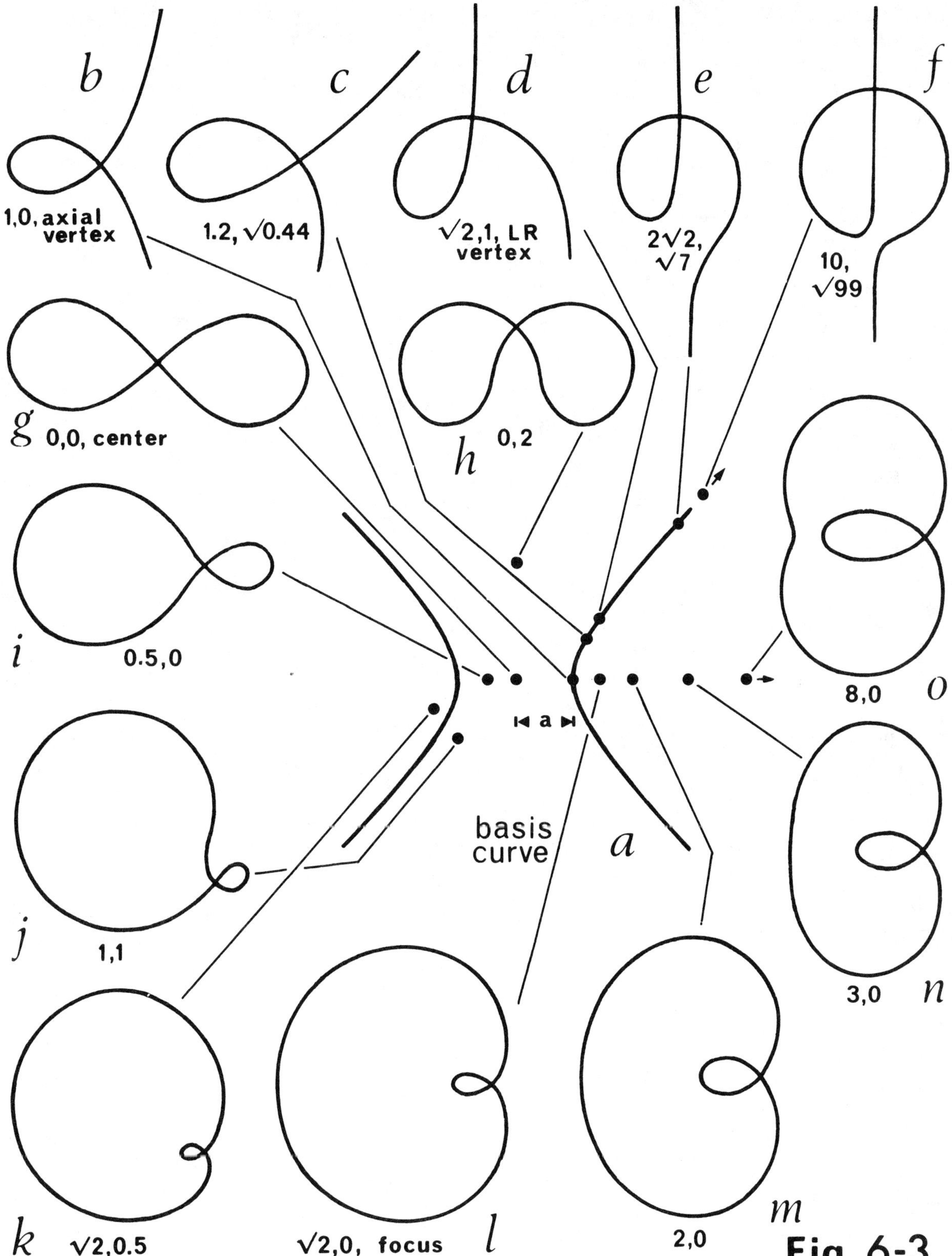

Fig. 6-3

In other words, inversion converts the property of coincidence of the midlines of pairs of image points of a basis curve to the property of possession of a common pole of concurrence of lines through the inverse point-pairs. Furthermore, the distances of the points of each inverse pair from the common pole of concurrence are related in the same manner for all pairs, namely all pairs self-invert about this pole (if x is the distance of the pole from the line of symmetry of the basis curve and j is the unit of linear dimension for the inversion of the basis curve, then the unit of linear dimension for self-inversion--the *modular value*--of the inverse curve about the pole in question is $\frac{1}{2}j^2/x$).]

The ensemble of inverse loci of the equilateral hyperbola depicted in the figure also is representative of the ensemble of inverse loci of every other curve depicted in the figure. In other words, poles exist about which every subspecies of equilateral QBI curve inverts to every other subspecies of equilateral QBI curve (see Maxim 6-4-7). With the sole exceptions of the equilateral hyperbola and the equilateral lemniscate, every equilateral QBI curve also self-inverts about at least one focus in the finite plane.

[For example, the equilateral strophoid of Fig. b inverts to the equilateral pseudo-exchange-limacon of Fig. m about an axial pole within the loop at 2/3rds of the distance from the DP to the loop vertex (2a/3); the reciprocal inversion pole lies at the small-loop vertex of the pseudo-exchange-limacon. The equilateral strophoid also inverts to the equilateral pseudo-exchange-limacon about an axial pole at a distance 2a from the DP on the asymptote side, but in this case it is the large-loop vertex that is the reciprocal inversion pole. Similarly, the equilateral strophoid of Fig. b inverts to the equilateral limacon (Fig. l) about a pole at a distance $2(1+\sqrt{2})a$ from the DP on the asymptote side and also about a pole at a distance $(1+\sqrt{2})a/(1\frac{1}{2}+\sqrt{2})$ from the DP on the loop vertex side. In the former case, the reciprocal inversion pole is the large-loop vertex, in the latter case, the small-loop vertex.]

Accordingly, every ensemble of inverse curves derived from any given basis subspecies of equilateral QBI curve through inversions about all points in the finite plane--together with the basis subspecies itself--possesses the same members as every other such ensemble, namely each ensemble possesses all of the equilateral subspecies. Similar relationships exist for QBI curves of all other eccentricities except 1 (i.e., inverse loci of the parabola; Figs. 8-34,35), for which none of the inverse loci about points on the line of symmetry self-inverts about a point in the finite plane.

A systematic illustration of these relationships is achieved by considering only incident inversions, as in the series of Figs. b-f, representative of inverse loci about poles from the vertex to a point at $(10,\sqrt{99})$. Of these curves, those of Figs b,d, and f are most notable:

(1) Fig. b is the equilateral strophoid, the only equilateral QBI cubic with a line of symmetry, and the curve obtained by axial-vertex inversions of all axial equilateral QBI curves. In fact, two incident poles exist about which all other equilateral QBI curves invert to the equilateral strophoid, and 4 incident poles exist about which all other equilateral QBI curves invert to every other equilateral QBI cubic. Moreover, the incident inversion ensembles derived from inversions about points on each of the two loops of all hyperbolic QBI curves possess the same members.

(2) Fig. d is the LR-vertex inversion cubic. The same subspecies of inversion cubic (see Fig. 6-5) is obtained at the LR vertices of the equilateral limacon (the poles of intersection with the curve of a line through the DP orthogonal to the line of symmetry). [The term "vertex" is used in a broader sense than "a point of relative extremal of the curvature;" see pages X-6 to 9.]

(3) The cubic of Fig. f --the *10a cubic*--is unexceptional except that its appearance is typical of that of many subspecies of equilateral QBI cubics, namely, it resembles a segment of a circle joined to segments of an extended diameter. The genesis of cubics having this appearance is understood readily in terms of incident inversions of the basis hyperbola about poles distant from the vertices. Since the arms of the basis curve are asymptotic, they increasingly approximate linearity with increasing distance from the vertices, so that inversions about distant incident poles possess a region that increasingly approximates linearity in the neighborhood of the reciprocal inversion pole (the arms of the inversion cubic always approach linearity at locations *distant* from the reciprocal inversion pole in any event, since they are asymptotic). The nearly circular region of the 10a cubic represents the inversion of the distant other regions of the two arms (which also increasingly approach linearity with increasing distance from the vertices). The approximately linear region of the 10a cubic assumes the position of a diameter because the distant regions of the two arms of the basis hyperbola approach an angle of 90° to one another. Visualization of the correspondences between regions of the basis hyperbola and the inverse 10a cubic are facilitated by recalling that an incident inversion of an orthogonal line-pair about any pole other than the point of intersection yields a circle, with the line upon which the pole of inversion is incident as an extended diameter.

Hyperbolic QBI curves of other eccentricities also invert to cubics that resemble a circle intersected by a line, but for them the "line" is "off center" (see Fig. 8-39a). Other looped cubics, not merely QBI cubics, possess incident inverse loci (see Figs. 8-40,41) that resemble a circle and an intersecting line. The basis for this resemblance can be understood most readily by analogy with the inverse loci of the corresponding hyperbola-like quartics to which non-QBI looped cubics invert about the DP (see Figs. 8-40g and 8-41h).

Hyperbola-like quartics differ notably from quadratic hyperbolas in that their vertex regions are incongruent and their asymptotes are not centered between the arms (in fact, the asymptotes may be internal to one arm and external to the other; see Fig. 8-40g). These deviations from quadratic hyperbolas are inconsequential, however, insofar as incident inverse loci about poles distant from the vertices are concerned. Accordingly such inverse loci also are cubics that resemble a circle intersected by a line. Since the incident inversion ensembles of all curves that invert to one another possess the same members (if an inverse locus about a point of intersection of the curve with the ideal line at infinity is included), the ensemble of incident inverse loci of the basis looped cubic from which the hyperbola-like quartic was derived also possesses curves resembling a circle intersected by a line (see also Figs. 6-4,5, 8-40,41).

Other notable curves in the point-in-the-plane inverse ensemble of the figure are the equilateral lemniscate of Fig. g (see also Figs. 2-5,8-17,19), the equilateral limacon of Fig. l (see also Figs. 6-4, 8-16,30), and the equilateral pseudo-exchange-limacon of Fig. m (see eqs. XI-15 and XIV-39). The latter curve is a member of the genus obtained by inverting central conics about an axial pole (a pole on the major or transverse axis) at $x = \pm 2a$.

Pseudo-exchange-limacons are distinguished equationally from other axial QBI quartics (quartics obtained by inverting conics about points incident upon the line of symmetry passing through the traditional foci) in that their rectangular eqs. cast about an axial vertex contain no cubic term (in the case of hyperbolic pseudo-exchange-limacons--Fig. 3m--it is the small-loop vertex that meets this condition). There also exists a group of minor-axis inversion quartics of the ellipse that possess this distinction, namely, the inverse loci about a pole at $y = \pm 2b$ cast about the $-j^2/3b$ vertex. Conjugate-axis inverse loci of the hyperbola about a pole $y = k$ possess no cubic term when cast about a pole at $K = -\frac{1}{2}j^2k/(b^2+k^2)$ but, of course, this is not a vertex. The cubic term does not vanish in the eq. cast about the vertex for either parabola axial-inversion quartics or parabola vertex-tangent inversion quartics.

In connection with the geometry of the equilateral pseudo-exchange-limacon of Fig. m, it was noted that the radius of curvature of the large-loop vertex is equal to the distance of this vertex from the DP. To investigate the possibility that this is a characteristic symmetry property of pseudo-exchange-limacons, the radii of curvature at the large-loop and small-loop vertices of *na post-vertex axial hyperbolic QBI quartics* (axial inversions of hyperbolas about poles at $h = na$, $n > 1$) were ascertained and compared with the corresponding distances to the DP (making use of the DP-asymptote distance of eq. IX-7a and Maxim 8-36-1). It is evident from eqs. 1a and 2a that for

parameters of *na post-vertex axial hyperbolic QBI quartics*

(a)	$j^2/(n-1)a$	distance from DP to large-loop vertex	(6-3-1)
(b)	$j^2/(n+1)a$	distance from DP to small-loop vertex	
(a)	$\rho = aj^2/(n-1)[2a^2-(n-1)b^2]$	radius of curvature of large-loop vertex	(6-3-2)
(b)	$\rho = aj^2/(n+1)[2a^2+(n+1)b^2]$	radius of curvature of small-loop vertex	

pseudo-exchange-limacons (i.e., $n = 2$) the noted property is not generic but is unique to the equilateral subspecies.

Note that in the equilateral subspecies the radius of curvature at the large-loop vertex of the *3a post-vertex axial hyperbolic QBI quartic* (Fig. n) (the *3a post-focal quartic* of Table VI-1) is infinite. In other words, the *equilateral 3a post-focal quartic* (Fig. n) demarcates between equilateral post-vertex inversion quartics in which the large-loop vertex is convex (Figs. l and m) and those in which it is concave (Fig. o).

Fig. 6-4. Incident inversions of the equilateral limacon. One is best introduced to equilateral QBI cubics by a systematic consideration of the incident inversions of the equilateral limacon, because of the ready visualization of the relationships between the incident inversions of this curve. Solid circles incident upon the small and large loops of the basis curve (upper left) identify the locations of the poles of inversion, while adjacent numbers identify the values of θ that are associated with these poles, where θ is the angle coordinate of the eq. $r = a(1-\sqrt{2}\cos\theta)$. The inverse loci (also identified by the value of θ asociated with them) are properly oriented relative to the basis curve, but the relative sizes of the curves have been altered to suit the purpose of illustration. The solid circles indicate the locations of the reciprocal inversion poles relative to each inverse curve. When a reciprocal inversion pole is distant from an inverse curve, a thin line is drawn from the reciprocal inversion pole to the neighborhood of the associated curve (for example, the 90° inversion).

Beginning at 0° (small-loop vertex) and 180° (large-loop vertex) the inverse curves are equilateral strophoids (see also Figs. 6-5a,b, 8-18), but these are of different size (the 0° inversion, for example, is 33.97 times larger than the 180° inversion) and orientation, and possess reciprocal inversion poles in different relative positions. As the poles of inversion progress along the loops from the vertices toward the DP, the inverse curves pass through the same sequence of shapes as they depart increasingly from the form of the equilateral strophoid and approach the form of the 44° cubic, which resembles a circle and an extended diameter. However, the size, orientation, and location of reciprocal inversion poles of corresponding subspecies differ in the two sequences.

Perhaps the most interesting question that arises in connection with these incident inversions concerns the comparative locations of *equivalent inversion poles* (poles of inversion to the same subspecies) along the two loops. In other words, what is the function $f(\theta_L,\theta_S) = 0$ that relates the coordinate angles θ_L on the large loop and θ_S on the small loop for inversion of the basis curve to the same subspecies of inversion cubics? Without knowledge of the mapping properties of the inversion transformation, the formidable process of equationally identifying the inversion cubics, themselves, would be necessary in order to derive this function. However, both the derivation of the function and the problem of visualizing the

Incident Inversions of The Equilateral Limaçon

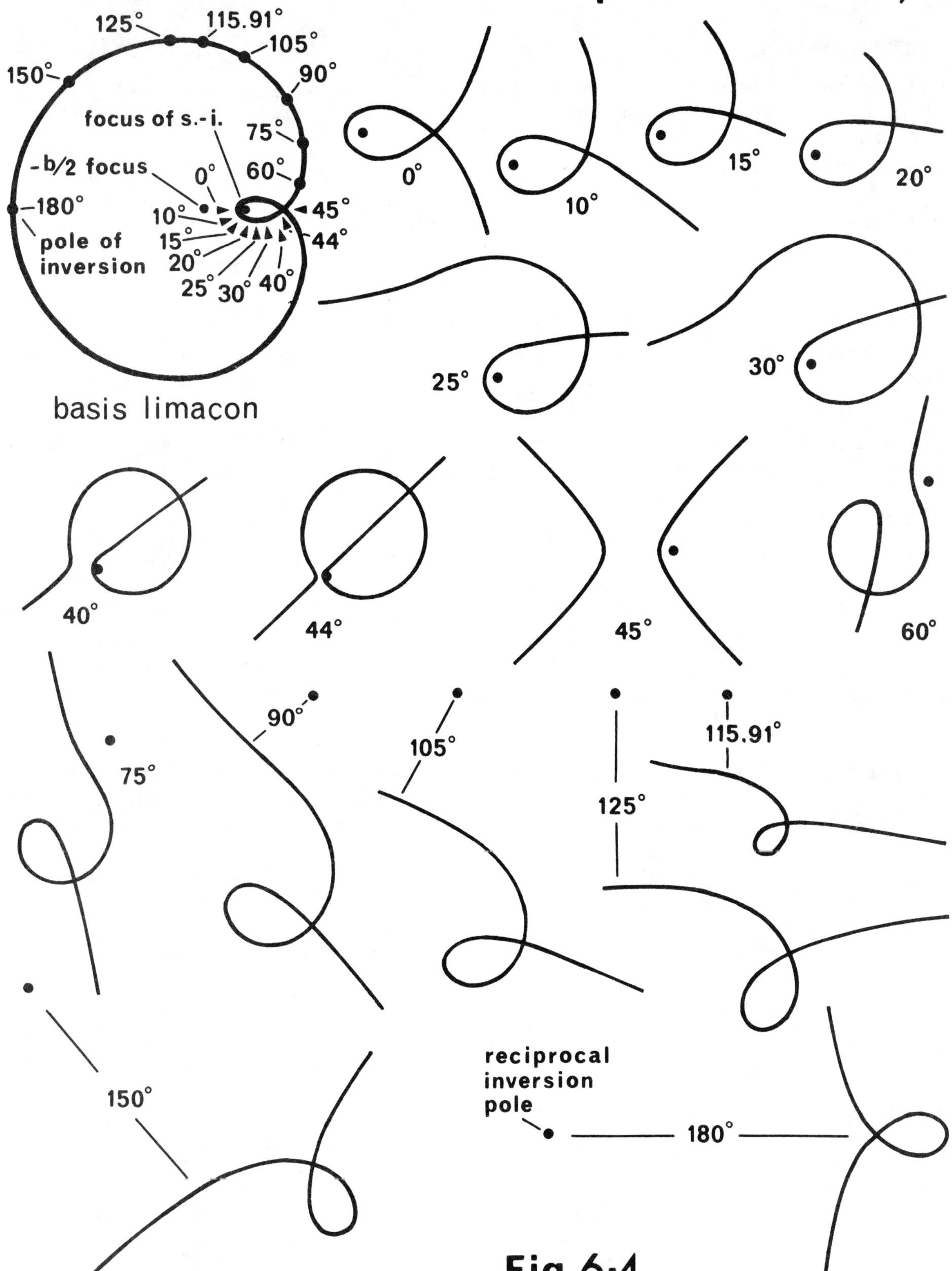

Fig. 6-4

locations of equivalent inversion poles are reduced to triviality by the following inversion-mapping Maxims and Maxims 6-5-1 and 6-7-5.

CIRCUMPOLAR MAXIMS 6-4-

1: *One obtains members of a single subspecies by inverting a basis curve about a given incident pole and by inverting any inverse curve about the pole inverse to the given incident pole.*

2: *One obtains two or 4 members of the same subspecies by inverting an inverse curve about poles inverse to equivalent* poles of a basis curve with one or two lines of symmetry, respectively.*

3: *One obtains members of only a single subspecies of curve from inversions of a basis curve about equivalent* poles and from inversions of any inverse curve about the poles inverse to the equivalent poles.*

4: *One obtains members of only a single subspecies of curve by inverting a basis curve (or any figure) about a given pole and by inverting any inverse curve or figure about the pole inverse to the given pole.*

5: *The equivalent inversion poles* for an ensemble of inverse curves obtained by inverting a basis curve B about all points on a line L comrise the circle L' inverse to L in the plane of any curve B' inverse to B about a pole not incident upon L.*

6: *The equivalent inversion poles* for an ensemble of inverse curves obtained by inverting a basis curve B about all points on a given locus L comprise the locus L' inverse to L in the plane of any curve B' inverse to B.*

7: *Inversion of a given basis curve B about all points in the plane generates an ensemble of inverse curves B', each member of which inverts to every other member but to no non-member. This is the property of* immediate closure *or* one-step closure.

**Equivalent poles* are poles that are mirror images in the single line of symmetry of an *axial* basis curve (2 poles), or in both lines of symmetry of a *central* basis curve (4 poles). *Equivalent inversion poles* (or *poles of equivalent inversion*) are poles for which the inverse loci belong to the same subspecies. All equivalent poles also are equivalent inversion poles but the reverse does not hold.

These Maxims and Maxims 6-5-1 and 6-7-5 are inversion-mapping Maxims of central importance. They make possible the identification of all the inverse curves (ensemble B') of any given basis curve B, and lay the foundations for many of the preceding (pages *xv-xix, Symmetry, An Analytical Treatment*) and following Maxims (Maxims 6-4-2 and 6-4-4 are proved in Appendix 3).

Among non-trivial point-in-the-plane transformations known to me, the properties encompassed by these Maxims are unique to the inversion transformation, and the property of *immediate closure* (Maxim 6-4-7) lies at the heart of the utility and pre-eminence of the inversion transformation for the classification of curves.

[Even though a basis curve does not self-invert about a point in the finite plane, it is a member of its own inversion ensemble since, for the appropriate (infinite) value of the unit of linear dimension, inversion of a basis curve about an infinitely distant pole yields a congruent curve (Maxim 6-7-5).]

The Locus Inverse To A Pole of Inversion

In the formulation of inversion Maxims, the locus, P', inverse to a pole P generally is referred to as the "inverse pole." However, when P is the pole of inversion itself, the inverse locus P' in the plane of the inverse curve is not a point-pole, nor is the reciprocal inversion pole in the plane of the inverse curve the inverse of a point-pole in the plane of the basis curve.

If one considers circles of ever decreasing radius centered on both P and the reciprocal inversion pole, it is evident that the former invert to circles P' of ever increasing radius, and that the latter are the inverse of circles of ever increasing radius. Points on these circles that are inverse to one another, and the directions in which they lie, are defined for all inverse pairs of circles of finite radii.

In the limit, however, one circle is of 0 radius and the other of infinite radius. All points in all directions of the circle of infinite radius in the plane of the inverse curve are inverse to the pole of inversion. Similarly, the reciprocal inversion pole is the inverse of all points in all directions of the circle of infinite radius in the plane of the basis curve. Points at infinity in specific directions, however, are specifiable as occupying the limiting inverse positions of points approaching the pole of inversion in corresponding directions.

This situation is dealt with in classical treatments by referring to the inverse locus P' of the pole of inversion P, i.e., the circle of infinite radius in the plane of the inverse curve, as the "line at infinity," or the "ideal line at infinity." In order to preserve a one-to-one relationship between points and inverse points for all points in the extended plane, some workers simply refer to P' as the "point at infinity."

In the present treatment, P' is referred to as the "ideal line at infinity," and points thereof are referred to as "points of the ideal line at infinity." The terminology, "the point at infinity," is eschewed both because the inverse of P is not a point and because the infinite number of points of which P' is composed are not equivalent. As we shall see, (legend to Fig. 6-8), these points not only can be distinguished from one another according to the directions in which they lie, but two distinct ideal points often do not possess equivalent circumpolar focal status. Whether foci of self-inversion are included among the points of the ideal line at infinity or not, however, it is the entire ensemble of points thereof that are inverse to the pole of inversion and the reciprocal inversion pole.

[It will be evident to some readers that these relationships, together with Maxim 6-4-4, lead to the conclusion that inversions of a basis curve about points of the ideal line at infinity yield a similar curve (see Maxim 6-7-5 and the legend to Fig. 6-7).]

Returning to the figure, the mapping function for the two loops in terms of the coordinate angle of the basis eq., $r = a(1-\sqrt{2}\cos\theta)$, is readily derived by reference to the basis equilateral hyperbola. The subspecies of inversion cubics for any conic of eccentricity e can be defined in terms of the abscissae of the incident poles along the lines of symmetry upon which the traditional foci lie. For example, the *equilateral 10a cubic* (Fig. 3f) is obtained by inverting an equilateral hyperbola (in standard position) about any of the 4 incident poles $(\pm 10a, \pm\sqrt{99}\,a)$.

To relate the angles for equivalent inversion poles on the two loops of the equilateral limacon, one need merely find the relationship between the angles of radii from one of the traditional foci of the basis hyperbola to the corresponding points on the two arms. Inversion of the basis arm--i.e., the arm for which the traditional focus is the pole of inversion--yields the large loop, while inversion of the contralateral arm yields the small loop. Since inversions about corresponding points on the two arms yield congruent curves, it follows from Maxim 6-4-1 that inversions about the poles on the limacon loops that are intersected by radii at the corresponding angles also yield the same subspecies.

The mapping function for equivalent inversion poles determined in this way is transcendental, and given by eq. 1, where θ_1 is the angle for the upper segment of the large loop and θ_2 is the angle for the lower segment of the small loop (see labelling of the basis limacon at the upper left). θ_1 pro-

$$8\tan^2\theta_2\tan^2\theta_1 - 6\tan\theta_2\tan\theta_1 - \tan^2\theta_2 - \tan^2\theta_1 = 0$$ mapping function between small and large loops (6-4-1)

$$\theta_2 = \arctan[(3\cos\theta_1 + 2\sqrt{2})/(9\sin\theta_1 - 1/\sin\theta_1)]$$ eq. 1 in another form (6-4-2)

gresses from 180° at the large-loop vertex to 45° at the DP, while θ_2 progresses from 0° at the small-loop vertex to 45° at the DP (see also Fig. 5c). Solving eq. 1 for θ_2 yields eq. 2 (since eq. 1 is symmetrical in θ_1 and θ_2, the same solution with θ_1 and θ_2 exchanged also holds for θ_1).

It is most interesting to note that, aside from the two pairs of boundary values, $(0^\circ, 180^\circ)$ and $(45^\circ, 45^\circ)$, only one other integral pair of angles within these limits satisfies eq. 1, namely $(15^\circ, 105^\circ)$. This finds its basis in the unusual circumstance that $\tan 15^\circ \tan 105^\circ = -1$, $\tan^2 15^\circ + \tan^2 105^\circ = 14$, and $\sin 15^\circ \cos 15^\circ = \frac{1}{4} = -\sin 105^\circ \cos 105^\circ$. The interesting feature associated with the angles 15° and 105° is that they define the locations of the peripheral vertices of the focus of self-inversion at $(a^2-b^2)/2b$, which is the only axial circumpolar focus of the equilateral limacon that has real vertices on both the large and small loops (see the related inversion-mapping Maxim 6-5-1).

Since a 90° radius from the traditional focus of the basis arm of the basis equilateral hyperbola meets the inverse curve at an LR vertex, inversion of the equilateral limacon about the 90° pole on the large loop yields an LR vertex cubic (the 90° curve and Fig. 6-3d). On the small loop, the same subspecies is produced by inversion about the pole at 19.4712206° (compare the 90° and 20° curves).

A peripheral vertex of the $-\frac{1}{2}b$ focus lies on the large loop at 115.9136462° and has circumpolar point-focal rank. The peripheral vertices of this focus on the small loop are imaginary in the equilateral subspecies, but one of the poles on the small loop that yields the same subspecies of inversion cubic (the 115.91° curve) lies at 12.25906143°. The latter pole lacks circumpolar point-focal rank. Analogous correspondences readily are derived for the loops or loop and lateral arms of the axial QBI curves that are obtained by inverting the equilateral hyperbola about other axial poles (and for other axial QBI curves obtained by inverting other conics about axial poles), an example of which is given in Fig. 6-5a for the equilateral strophoid.
[θ_1 also is the angle for the lower segment of the large loop and θ_2 the angle for the upper segment of the small loop, in which case θ_1 progresses from 180° to 315° and θ_2 from 360° to 315°. The integral angles corresponding to the peripheral vertices of the focus of self-inversion for these segments of the loops are 345° for the upper segment of the small loop and 255° for the lower segment of the large loop.]

As the DP is approached on both loops, the inversion cubics more nearly approximate circularity in their loop regions and their arms more nearly approximate linearity in the region of the loop, with the latter segments more nearly approximating the location of a diameter. The locus obtained at 44° is the equilateral 40.16587643a cubic. It is very close to being circular in the loop portion and linear in its arms (see the 44° curve). The locus for 46° (1° on the other side of the DP) is almost congruent to that at 44° (the same subspecies as the inversion cubic obtained at the 44° pole is obtained at the 46.01776281° pole) but has opposed orientation and the reciprocal inversion pole is in the other "vertex" region. This locus is not illustrated but the similar situation for 44° and 46° is illustrated in Fig. 8-40 for the folium of Descartes. At the DP (the 45° point) the inverse locus is the basis equilateral hyperbola.

Incident inversions of the equilateral strophoid are not illustrated (but see Fig. 6-5). They can be envisioned readily by reference to Fig. 8-40 for the folium of Descartes and also by reference to the present figure. Consider the following correspondences in the inversion of the equilateral limacon to an equilateral strophoid about the large-loop vertex: (a) the large loop of the limacon inverts to the asymptotic arms of the strophoid; (b) points near the large-loop vertex invert to points on the arms at locations distant from the DP; (c) points on the large loop near the DP invert to locations near the DP on the asymptotic arms; (d) the small loop of the limacon inverts to the loop of the strophoid (but with reversed orientation); and (e) the small-loop vertex and the DP of the limacon become the loop vertex and DP, respectively, of the strophoid.

To visualize the incident inverse loci of the equilateral strophoid, one need merely assign to its segments the same types of inverse curves and sequences of progressive changes in these curves that occur in the inverse loci about points on the corresponding segments of the equilateral limacon (or the equilateral hyperbola). By considering other such correspondences (see Maxim 6-4-1) one can deduce the nature of the incident inverse loci of all curves that invert to one another.

For example, points on the arms of hyperbolas distant from the vertices invert to regions near the DP of all of its inverse curves. Correspondingly, inversions in the neighborhood of the DP of its inverse curves produce the curves resembling intersecting lines and circles that are typical of inverse loci of hyperbolas about incident poles distant from the vertices. On the other hand, poles incident upon the arms of the equilateral strophoid but distant from the DP invert to poles neighboring the axial vertices of its axial inverse loci. Correspondingly, the inverse loci about poles in these regions resemble equilateral strophoids.

The focus of self-inversion of the equilateral limacon is located readily following the substance of Maxims 6-4-2 and 6-7-5. Thus, letting $j = a_c$ (the conic parameter, a), the contralateral focus of the basis hyperbola inverts to $\sqrt{2}a_c/4$ along a radius at 180°. Accordingly, this is the location of the focus of self-inversion of the inverse curve. Similarly, Maxims 6-4-2 and 6-7-5 reduce to triviality the derivation of the location of the 0° focus of self-inversion that replaces a line of symmetry of any curve inverted about a pole not incident upon that line.

It will be noted that no reciprocal inversion pole of incident inverse loci of the limacon is incident thereon, whereas *all* reciprocal inversion poles of incident inverse loci of QBI cubics are incident upon the inverse curves (see Fig. 6-6). This can be understood in terms of the following Maxims.

CIRCUMPOLAR MAXIMS 6-4-8 : *Inverse loci of QBI quartics about unexceptional incident poles are of 3rd degree (exceptional incident poles are the DP and DP homologue, about which all QBI quartics invert to quadratics).*

6-4-9 : *Inverse loci of QBI cubics about unexceptional incident poles also are cubics (inverse loci about exceptional incident poles are quadratics).*

6-4-10: *Inverse loci of cubics about non-incident poles in the finite plane are of higher degree than 3*.*

*In connection with these Maxims it is to be noted that the DP homologue of elliptical species of QBI cubics and quartics is an incident, though isolated, pole. For example, for the rectangular eqs. of these curves cast about the DP homologue, the point (0,0) is a double-root (see legend to Fig. 8-24).

It follows from a consideration of these Maxims that the inverse loci of QBI quartics about unexceptional incident poles are QBI cubics, that the reciprocal inversion poles cannot be incident, and that the reciprocal inversion poles of inverse loci of QBI cubics about incident poles cannot be non-incident.

Inversion-Mapping of Axial Segments Between The Equilateral Hyperbola And The Equilateral Limacon

Just as axial poles of the equilateral hyperbola occur in equivalent pairs, so too are there two poles on the axis of the equilateral limacon for inversion to any given axial hyperbolic QBI subspecies (transverse-axis inversions) except the hyperbola itself (for which the DP is the sole pole of inversion) and the equilateral lemniscate (for which the inter-loop modular pole is the only pole of inversion). For the focus of self-inversion, itself, the axial point at infinity is the equivalent inversion pole.

[In fact, the axial point at infinity is the only point of the ideal line at infinity for which the orientation of the inverse locus is reversed relative to the basis curve. All other subspecifically-identical inverse loci about equivalent inversion poles on the line of symmetry also have opposed orientation.]

The manner in which segments of the axis of the hyperbola map to segments of the axis of the limacon is represented in Table 6-4-1. The precise mapping of a given axial pole of the limacon into its equivalent inversion pole follows the rule that equivalent inversion poles occur in modularly-inverse pairs, both members of a given pair lying either to the right or to the left of the focus of self-inversion (with the two subspecifically-identical inverse loci having opposed orientation).

Table 6-4-1. Mapping of Axial Segments of The Equilateral Hyperbola Onto Axial Segments of The Equilateral Limacon

	segments of the transverse axis of the equilateral hyperbola	segments of the axis of the equilateral limacon
1.	from infinity at the right to the right traditional focus	from the DP to infinity at the right
2.	from infinity at the left to the left traditional focus	from the DP to the focus of self-inversion
3.	from the left traditional focus to the vertex of the left arm	from the focus of self-inversion to the vertex of the small loop
4.	from the vertex of the left arm to the center	from the vertex of the small loop to the inter-loop modular pole ($-\frac{1}{2}b$ focus)
5.	from the center to the vertex of the right arm	from the inter-loop modular pole to the vertex of the large loop
6.	from the right vertex to the right traditional focus	from the vertex of the large loop to infinity at the left

In summary:

(a) the entire left half of the axis of the hyperbola from the center to infinity maps onto the finite segment of poles from the DP to the inter-loop modular pole of the limacon

(b) poles on the finite segment from the center of the hyperbola to the right traditional focus map to poles on the limacon axis from the inter-loop modular pole to infinity at the left

(c) poles on the right half of the axis of the hyperbola from infinity to the traditional focus map onto poles on the limacon axis from the DP to infinity at the right.

CIRCUMPOLAR MAXIMS 6-4-

11: *A non-axial QBI curve of non-unit eccentricity inverts to axial QBI curves about points incident upon its two (orthogonal) circles of inversion (circles of modular radius centered upon its foci of self-inversion), excepting the two points of intersection of these circles. In these transformations, the circle upon which the pole of inversion is incident inverts to the line of symmetry of the inverse curve.*

12: *A non-axial QBI curve of non-unit eccentricity inverts to two subspecifically different central QBI curves about the two points of intersection of its two circles of inversion; one of these curves is a central conic, the other is a central quartic. In these transformations, the two circles of inversion invert to the two lines of symmetry of the central curves.*

13: *A non-axial QBI curve of non-unit eccentricity undergoes identical subspecific-inversion about 3 finite poles. Two of these poles are foci of self-inversion, the 3rd is a pole of congruent-inversion*. The latter lies on the line segment between the two foci of self-inversion.*

14: *The pole of congruent-inversion of a non-axial QBI curve of non-unit eccentricity lies midway between the two poles for inversion to central curves.*

15: *An axial QBI curve of non-unit eccentricity inverts to central QBI curves about the two points of intersection of its circle of inversion with its line of symmetry. In these transformations, the circle of inversion inverts to the orthogonal line of symmetry of the inverse central curve.*

16: *The two inverse central loci of an axial QBI curve of non-unit eccentricity consist of a central conic (the inverse locus about the DP or DP homologue) and a central quartic.*

17: *A non-axial QBI curve of unit eccentricity inverts to the entire ensemble of axial QBI curves of unit eccentricity about points on its circle of inversion.*

18: *About the pole on its circle of inversion diametrically opposite the pole for inversion to a parabola, a non-axial QBI curve of unit eccentricity inverts to the* $x=h$ *parabola axial-inversion quartic, where* h *is the abscissa of the reciprocal inversion pole in the plane of the basis parabola.*

19: *Equivalent inversion poles on the line of symmetry of axial inversions of central conics lie in modularly-inverse pairs, both members of a given pair lying either to the right or the left of the focus of self-inversion.*

*A *congruent-inversion* is an inversion in which, for the appropriate value of the unit of linear dimension (not necessarily finite), the inverse locus is of the same shape and size as the basis curve.

A Parametrization Procedure For Plotting Inverse Loci and Circumpolar Intercept Transforms

To obtain the loci depicted in this figure from a basis equilateral limacon by a straightforward incident inversion procedure requires finding the roots of eqs. of 3rd degree (eq. X-41a). However, for all cases for which the basis curve--no matter its rectangular degree--can be expressed in the form $r_b = f(\theta_b)$, quadratic eqs. for r_i and θ_i of the inverse curve can be cast in parametric form as follows. θ_p is the specific value of θ_b for the

$$r_i = j^2[f^2(\theta_b)+f^2(\theta_p)-2f(\theta_b)f(\theta_p)\cos(\theta_b-\theta_p)]^{-\frac{1}{2}}$$ polar coordinate r of inverse curve (6-4-3)

(a) $$\theta_i = \theta_b + \arcsin[j^{-2}r_i f(\theta_p)\sin(\theta_b-\theta_p)]$$ polar coordinate θ of inverse curve (6-4-4)

(b) or $$\theta_i = \theta_b+180^\circ-\arcsin[j^{-2}r_i f(\theta_p)\sin(\theta_b-\theta_p)]$$ alternate value of θ

position of the incident pole, p, and $r_p = f(\theta_p)$ is the specific value of r_b for the incident pole.

[The alternate eqs. 4a and 4b are required for use with computers that give the same values for the arcsin function for θ and $180^\circ-\theta$, in which case the correct value has to be ascertained by testing the alternate values. I am indebted to Christopher Kavanau for developing the parametrization technique leading to eqs. 6-3 to 6-8.]

For inverting about a non-incident point in the plane (r_p,θ_p), the eqs. become 5 and 6, where the correct alternate value among eqs. 6a-c must be ascertained by testing.

$$r_i = j^2[f^2(\theta_b)+r_p^2-2f(\theta_b)r_p\cos(\theta_b-\theta_p)]^{-\frac{1}{2}}$$ polar coordinate r of inverse curve (6-4-5)

(a) $$\theta_i = \theta_b+\arcsin[j^{-2}r_i r_p\sin(\theta_b-\theta_p)]$$ polar coordinate θ of inverse curve (6-4-6)

(b) or $$\theta_i = \theta_b+180^\circ-\arcsin[j^{-2}r_i r_p\sin(\theta_b-\theta_p)]$$ alternate value of θ

(c) or $$\theta_i = \theta_p+180^\circ-\arcsin[j^{-2}r_i r_p\sin(\theta_b-\theta_p)]$$ alternate value of θ

[If one makes use of the mapping properties of the inversion transformation, all the inverse loci in question also can be obtained directly as incident inverse loci of the equilateral hyperbola about poles of equivalent inversion (without recourse to the parametrization techique).]

The same approach in terms of parametric eqs. can be employed to obtain intercept transforms. In that event, one obtains the intercepts r_{bp} at the corresponding angles θ_{bp} of the basis curve about the pole, p, at (r_p,θ_p) by employing eqs. 7 and 8. The correct alternative among eqs. 8a-c, once again, must be ascertained by testing.

$r_{bp} = [f^2(\theta_b)+r_p^2-2f(\theta_b)r_p\cos(\theta_b-\theta_p)]^{\frac{1}{2}}$	polar coordinate r of intercept of basis curve about pole p	(6-4-7)
(a) $\theta_{bp} = \theta_b+\arcsin[r_{bp}^{-1}r_p\sin(\theta_b-\theta_p)]$	polar coordinate θ of intercept of basis curve about pole p	(6-4-8)
(b) or $\theta_{bp} = \theta_b+180^\circ-\arcsin[r_{bp}^{-1}r_p\sin(\theta_b-\theta_p)]$	alternate value of θ	
(c) or $\theta_{bp} = \theta_p+180^\circ-\arcsin[r_{bp}^{-1}r_p\sin(\theta_b-\theta_p)]$	alternate value of θ	

To derive intercept transforms for a given angle, α, between the radius vectors, it is necessary to pair values of intercepts r_{bp_1} and r_{bp_2} for which $\theta_{bp_1}-\theta_{bp_2} = \pm\alpha$. Since the values of θ_{bp} given by this procedure are dependent, rather than independent, one needs to obtain values of r_{bp} and θ_{bp} for small increments of θ_b, so that large numbers of these values are needed (values of θ_b have to be taken on a fine gridwork).

Fig. 6-5. Mapping of angles of incident poles for inversion to identical subspecies of equilateral cubics. Vertex inversions of the equilateral hyperbola to the equilateral strophoid (Fig. a) are unique in that the transcendental angle-mapping function (eq. 1) for incident *equivalent inversion poles* (poles

angle-mapping function for equivalent inversion poles
on the loop and arms of the equilateral strophoid

$\tan\theta_1\tan\theta_2 = -1$		(6-5-1)
$\theta_1 + 90^\circ = \theta_2$	solution to eq. 1 (θ_1 for upper half of loop, θ_2 for upper arm)	(6-5-2)
$r = j^2(2\cos^2\theta-1)/2a\cos\theta$	polar eq. of equilateral strophoid	(6-5-3)

Mapping of Angles of Incident Poles For Inversion To Identical Subspecies of Equilateral Cubics

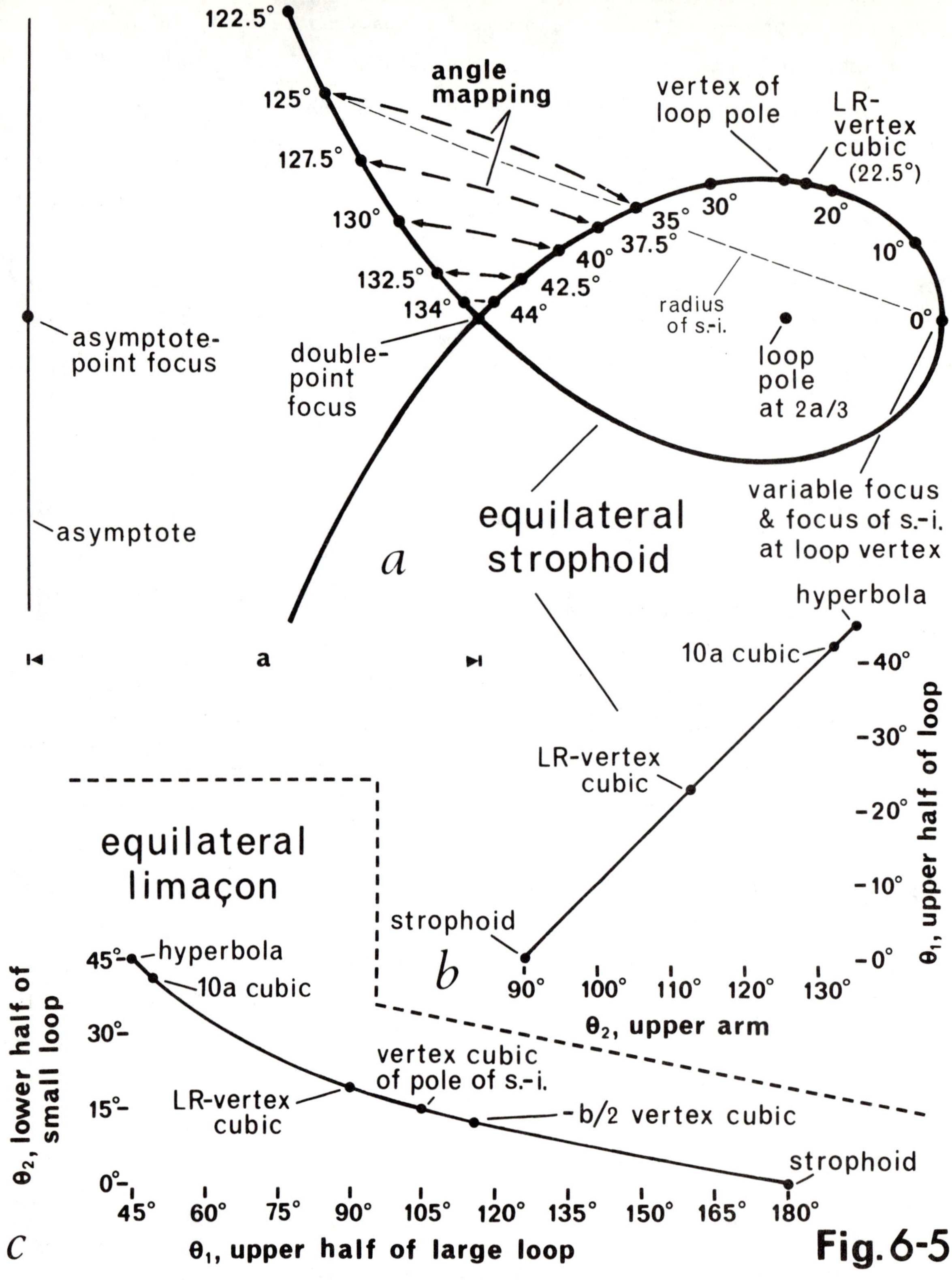

Fig. 6-5

general angle-mapping function for transverse-axis inversions of hyperbolas

$$(-b^2h \pm ab[(a^2-h^2)(1-1/\cos^2\theta_1)+b^2]^{\frac{1}{2}})(\sin\theta_1+\cos\theta_1\tan\theta_2) = 2h[a^2/\cos\theta_1-(a^2+b^2)\cos\theta_1]\tan\theta_2 \qquad (6\text{-}5\text{-}4)$$

about which inversions yield subspecifically identical curves) has a linear solution (eq. 2 and Fig. b). These eqs. hold for values of θ defined by the eq. of an equilateral strophoid (eq. 3) derived by inverting an equilateral hyperbola about the vertex of its left arm.

The entire strophoid is traced for θ ranging from 0° to 180°. The upper half of the loop from the loop vertex to the DP is traced for $\theta = 0^\circ$ to 45°, the lower arm is traced from the DP to infinity for $\theta = 45^\circ$ to 90°, the upper arm is traced from infinity to the DP for $\theta = 90^\circ$ to 135°, and the lower half of the loop is traced for $\theta = 135^\circ$ to 180°. Inversion poles at angles θ on the upper half of the loop map to poles at angles $(\theta+90^\circ)$ on the upper arm, as indicated by the arrows of correspondence (labelled "angle mapping") adjacent to the curve of Fig. a and the graph of Fig. b. Poles of inversion to specific equilateral curves are marked and labelled in Figs. a and b.

[Eq. 4 is the general angle-mapping function for transverse-axis inversions of hyperbolas (x+h translation). θ_1 is the angle that a radius from the origin to the right arm makes with the x-axis and θ_2 is the angle of a radius to the left arm (and the corresponding inverse segments). I am indebted to Christopher Kavanau for deriving this eq.]

The following question now is addressed. Do inversions about the peripheral vertices of the loop pole at $h = 2a/3$ (Fig. a), which is a point about which the equilateral strophoid has greater symmetry than about neighboring points (at the subfocal level; see page IX-12), yield LR-vertex cubics? Fig. a illustrates the fact that, though the peripheral vertices of this pole on the loop (at 24.09484255° on the upper half) are close to the points about which inversions yield LR-vertex cubics (at 22.5° on the upper half), they are not coincident with them. In other words, the peripheral vertices of the loop pole at $h = 2a/3$ yield unexceptional vertex cubics, and the poles on the loop about which inversions yield LR-vertex cubics lie opposite an otherwise unexceptional point on the line of symmetry.

Fig. c is a graph of the corresponding transcendental angle-mapping curve for the incident inversions of the equilateral limacon that are illustrated in Fig. 6-4. The poles that yield the principal inversion cubics are identified by labels adjoining the curve.

The relationships of Maxim 6-5-1, below, are evident from the geometry of the angle-mapping relationships of Fig. a for equivalent inversion poles (see the "radius of self-inversion" extending from the loop vertex through the 35° and 125° poles) and the fact that, for limacons, the angles 15° and 105° for inversion to identical subspecies define the locations of the peripheral vertices of the focus of self-inversion.

CIRCUMPOLAR MAXIM 6-5-1 : *The two inverse poles at which a given radius through a 0° focus of self-inversion (or given chord segments through a 180° focus of self-inversion) intersects a curve are equivalent inversion poles.*

Maxim 6-5-1 follows directly from Maxim 6-4-4 when corresponding inverse segments of a curve that self-inverts are regarded merely as inverse figures in the plane (the relationships of Maxim 6-4-4 are independent of continuity of curves--or figures--or of continuity of derivatives of any order). In that event, the poles of one segment of a curve that are inverse to the poles of the other segment are--by definition--corresponding points of intersection with the two segments of radii emanating from the focus of self-inversion. Following Maxim 6-4-4, the inverse curves (or segments of a curve) obtained by inversions about such pairs of poles belong to the same subspecies (are similar).

Fig. 6-6. Incident inversions of the equilateral 10a cubic. With the background of Maxim 6-4-1 and that provided by the discussions of Figs. 6-4 and 6-5, the incident inversions of the equilateral 10a cubic are readily understood. As in the case of Fig. 6-4, the orientations of the inverse loci relative to the basis cubic and the locations of the reciprocal inversion poles relative to the inverse curves are depicted in this figure.

Since the equilateral 10a cubic of Fig. 6-3f is the inversion of the equilateral hyperbola about an incident pole at $(10a, \sqrt{99}\,a)$, correlation of the inverse loci about poles on its segments with those about poles of corresponding segments of the equilateral hyperbola provides the most convenient framework for elucidating the properties of its incident inversions.

Incident Inversions of The Equilateral 10a Cubic

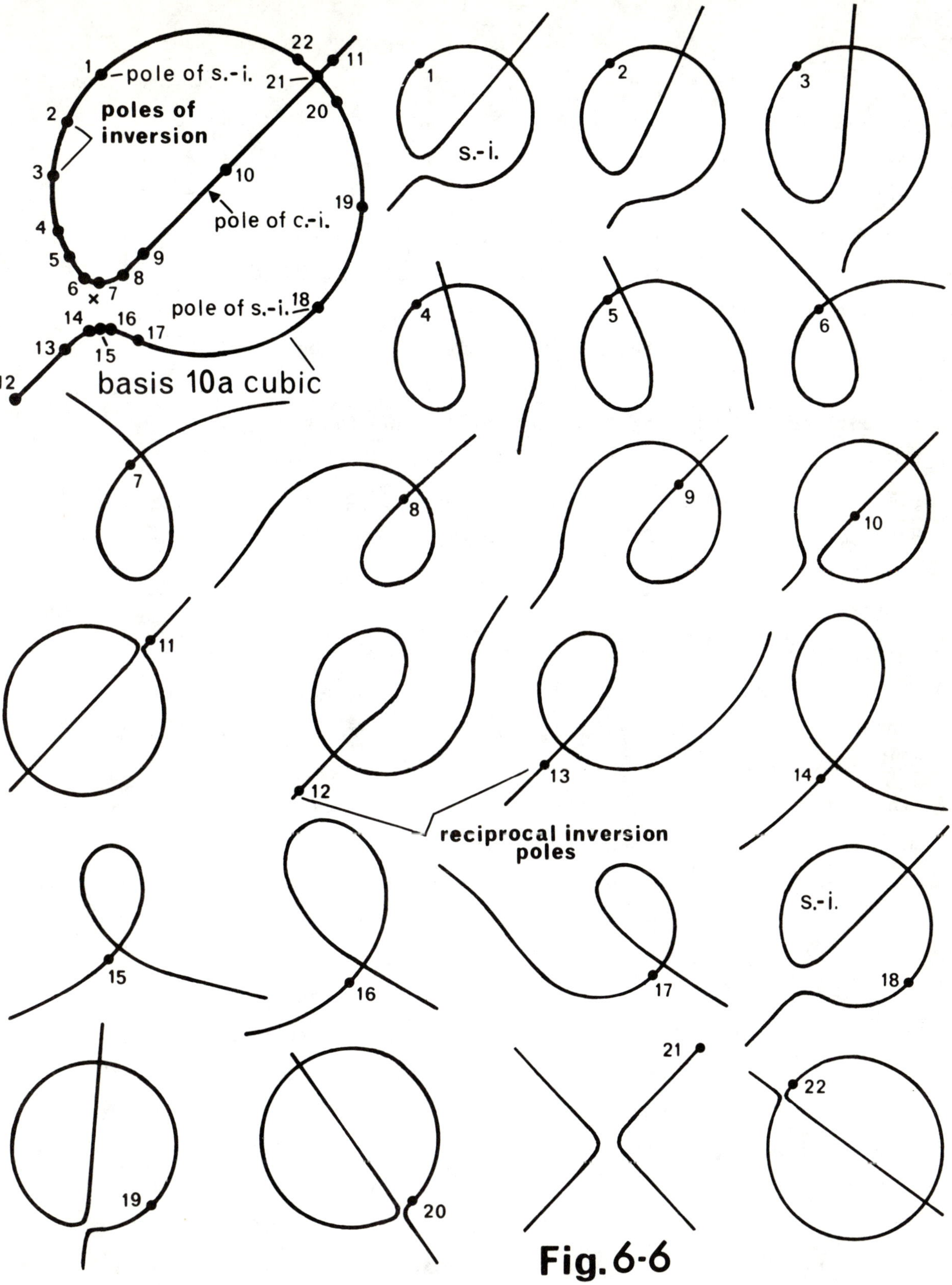

Fig. 6-6

The locations about which inversions to equilateral strophoids occur provide appropriate starting points. When the pole of inversion of the equilateral hyperbola at $(10a,\sqrt{99}a)$ is taken to be the origin, the eq. of the latter is $(x+10a)^2-(y+\sqrt{99}a)^2 = a^2$. Inversion gives the equilateral 10a cubic, $2(x^2+y^2)(\sqrt{99}y-10x) = a(x^2-y^2)$, where the unit of linear dimension, j, is taken to be equal to a, and the DP comes to lie at the origin. For this location, the incident inversion poles(numbered from 1 to 22 in the figure)lie in positive or negative directions along radii from the DP at angles to the x-axis of 0 (pole 1), 10, 20, 30, 35, 40, 42, 44, 44.5, 44.9, 45.01, (pole 11), 45.5 (pole 12), 46, 47, 48, 50, 55, 90, 110, 130, 135 (DP), and 140°, respectively.

Within this system, the two poles for inversion to equilateral strophoids (i.e., the inverse poles of the axial vertices of the equilateral hyperbola) lie on radii at 42.13041476° [$\arctan(\sqrt{99}/11)$] and 47.86958525° [$\arctan(\sqrt{99}/9)$], while the poles for inversion to LR-vertex cubics (i.e., the inverse poles of the LR vertices) lie on radii at 38.099618°, 43.81055494°, 46.18944506°, and 51.9000382° [$\arctan((\sqrt{99}\pm1)/(10\pm\sqrt{2}))$]. Accordingly, poles 7 and 15 lie within 0.13041476° of the inversion poles for equilateral strophoids, and poles 8 and 13 lie within 0.18944506° of the inversion poles for LR-vertex cubics (comparisons of the inverse curves for poles 7 and 15 and for poles 8 and 13 with the reference loci of Figs. 6-3b and 6-3d reveal that these pairs of curves deviate only slightly from equilateral strophoids and LR-vertex cubics).

As the poles of inversion approach the DP, the inverse curves increasingly resemble circles plus an extended diameter, inasmuch as poles approaching the DP correspond to increasingly more distant regions along the arms of the basis equilateral hyperbola. In other words, in the inversion of the basis equilateral hyperbola about the pole $(10a,\sqrt{99}a)$, the poles near the DP arise as inversions of the distant portions of the arms of the basis curve. It follows from Maxim 6-4-1 that inverse curves about poles near the DP are of the same subspecies as those about corresponding poles on distant portions of the arms of the basis curve; since the latter increasingly resemble circles and extended diameters as the distance from the vertices increases, the former increasingly resemble such curves as the DP is approached.

At the DP (pole 21), inversion, of course, yields the equilateral hyperbola, with the point at $(10a,\sqrt{99}a)$ as the reciprocal inversion pole. It follows from

Maxim 6-4-12 that the pole for inversion to an equilateral lemniscate lies at the non-incident position indicated by the small cross (between poles 7 and 14), this being the other point of intersection of the two circles of inversion of the basis 10a cubic (circles centered on the foci of self-inversion--poles 1 and 18--of radius equal to the corresponding unit of linear dimension for self-inversion).

Considerations of the nature of inversions about distant poles on the arms of the equilateral 10a cubic itself, in conjunction with the substance of Maxim 6-4-1, lead to Maxim 6-6-1. Additional considerations of symmetry lead to Maxims 6-6-2 and 6-6-3.

CIRCUMPOLAR MAXIMS 6-6-1: *For the appropriate value of the unit of linear dimension, the inverse curves obtained by inverting a non-axial asymptotic* looped cubic about incident poles increasingly distant from the loop approach congruence with the basis curve. In the limit, congruence is achieved (a congruent-inversion).*

6-6-2: *For the appropriate value of the unit of linear dimension, the inverse curves obtained by inverting an axial asymptotic* looped cubic about incident poles increasingly distant from the loop approach coincidence with the basis curve. In the limit, coincidence is achieved (a self-inversion).*

6-6-3: *For the appropriate value of the unit of linear dimension, the inverse of a basis asymptotic arm of an axial curve, obtained by inversion about a pole incident upon the arm at points increasingly distant from the nucleus,** approaches coincidence with the arm. In the limit, coincidence is achieved between the inverse arm and the basis arm.*

*Maxims 6-6-1 to 6-6-3 pertain to curves for which the asymptote is at a finite distance from the nucleus.

**The term *nucleus* refers to a loop, oval, or other closed segment of an open curve or to an exceptional point, such as a cusp. The term "oval" is not used in the strict sense of a "closed curve everywhere convex," as employed by differential geometers, but in the less restricted traditional sense of a closed curve or segment of a curve, not necessarily everywhere convex, as in the "ovals of Cassini."

The distant portions of the arms of the equilateral 10a cubic arise as inversions of the regions of the arms of the basis equilateral hyperbola in the neighborhood of the pole of inversion at (10a, $\sqrt{99}$a). The closer these regions are to the pole of inversion, the more distant the corresponding inverse regions of the arms of the 10a cubic. But it also is true that the closer a pole of inversion is to the pole for inversion to the 10a cubic, the more the inverse curves resemble the 10a cubic. Accordingly, it follows from Maxim 6-4-1 that for the appropriate value of the unit of linear dimension, inversions about poles at increasingly distant regions of the arms of the 10a cubic increasingly approach congruence with the 10a cubic. In other words, proceeding outward from poles 11 and 12 along the arms of the 10a cubic, the inverse curves change progressively from the form of the inverse loci about poles 11 and 12 toward congruence with each other and the 10a cubic itself.

Consider now the matter of self-inversion and, more generally, *identical subspecific-inversion* of central conic inversion cubics about poles in the finite plane. Since central conics possess two lines of symmetry, for any given incident pole--except the axial vertices--there exist 3 other symmetrically-located poles (the 4 poles forming a rectangle). Since the 4 poles in any set are symmetrically located about the true center, they are poles of equivalent inversion, i.e., the inverse curves about the 4 poles are subspecifically identical.

It follows from preceding Maxims that there exist 3 equivalent inversion poles in the finite plane of the cubic, B', obtained by inverting the basis central conic, B, about any one of the 4 symmetrically-located incident poles about which B' inverts to cubics similar to itself, i.e., to cubics that belong to the same subspecies as B'. For the appropriate value of the unit

[For example, if this result is derived from Maxims 6-4-2 and 6-6-1, one considers that the pole of inversion itself inverts to the ideal line at infinity, which includes the point at infinity on the asymptotic arms of the inverse cubic. Since the curves obtained by inverting about increasingly distant poles on these arms increasingly resemble the inverse cubic itself, the inverse curves about the 3 poles in the finite plane also must belong to the same subspecies as the inverse cubic.]

of linear dimension, two of these loci are identical with the inverse cubic, i.e., they are self-inversions (see Circumpolar Maxim 18); the 3rd subspecifically-identical locus is a *congruent-inversion*, falling short of being a self-inversion by virtue of its location and orientation.

In the case of the equilateral 10a cubic, the 3rd pole in question lies on a radius at an angle of 44.85604015° at the point on the basis curve between poles 9 and 10 (indicated by the labelled arrow in the figure), which is collinear with the two poles of self-inversion--poles 1 and 18 (see Maxims 6-4-11, 6-4-12, and those below.

The congruent locus obtained by inverting about this pole is a reflection of the 10a cubic in the line connecting the points of intersection of its two circles of inversion. This line passes through pole 21 and the cross (between poles 7 and 15) that marks the pole for inversion to an equilateral lemniscate.

CIRCUMPOLAR MAXIMS 6-6-4 : *A non-axial inversion cubic of a central conic or central quartic undergoes identical subspecific-inversion about 3 finite collinear poles. Two of the identical subspecific-inversions are self-inversions, the 3rd is a congruent-inversion (for the appropriate value of the unit of linear dimension). The pole for a congruent-inversion lies on the line segment connecting the foci of self-inversion.*

6-6-5: *A non-axial inverse curve of a central conic or central quartic possesses 3 finite collinear poles of identical subspecific-inversion. Two of these poles are foci of self-inversion. They lie on lines parallel to the lines of symmetry of the basis curve that pass through the pole of inversion. The 3rd is a pole of congruent-inversion (for the appropriate value of the unit of linear dimension). It lies at the point of concurrence of the line segment from the pole of inversion to the center of the basis curve, and the line segment connecting the foci of self-inversion.*

6-6-6: *The congruent inverse locus of a non-axial QBI basis curve of non-unit eccentricity (obtained by inverting about the pole of congruent-inversion) is the reflection of the basis curve in a line through the points of intersection of its circles of inversion. The unit of linear dimension for the congruent-inversion is one-half the length of the line segment connecting these points (the distance from the pole of congruent-inversion to either pole for inversion to a central curve).*

Fig. 6-7. Approach of inverse curves to subspecific identity with closed basis curves as inversion poles become increasingly distant. It is an inherent and conspicuous property of the inversion transformation that the relative distances from the pole of inversion to two points on a basis curve in any direction become reversed in the inverse curve. Accordingly, the more distant members of such pairs of points of a basis curve become the more proximate members of these pairs in the inverse curve. In essence, both a distortion and a reversal of the orientation of the basis curve take place.

Given that inversions of closed curves about distant poles increasingly resemble the basis curve with increasing distance, it follows from the fact of the relative distance reversals mentioned above, that the approach to subspecific identity characteristically is merely in the nature of an approach to similarity or congruence (for the appropriate value of the unit of linear dimension). Increasing resemblance of inversion to *self-inversion* for increasingly distant poles of inversion occurs only under certain conditions. For example, when a basis curve possesses two lines of symmetry, it occurs for poles of inversion that lie on either of these lines.

For increasingly distant poles on a line not orthogonal to a line of symmetry, the inverse curves become "rotated" (see Fig. 6-8) and it is similarity or congruence that is approached, not coincidence. Conditions for increasing approximation to self-inversion of inverse loci about *incident* poles increasingly distant from the nuclei of open curves already have been set forth in Maxims 6-6-2 and 6-6-3.

Some of the relationships characteristic of the inversion of closed curves about distant poles on a line of symmetry are illustrated in the present figure. In each case, the basis curve (Figs. a,b, and c) is inverted about a pole on the horizontal line of symmetry at distances of a, 2a, 5a, 10a, and 20a to the right of the nearest point on the curve (a vertex for Figs. a and c, the DP for Fig. b). For the basis ellipse, a is the length of the semi-major axis, for the basis $e = 2$ limacon it is the diameter of the small loop, while for the basis equilateral lemniscate it is the length of the semi-transverse axis (a loop diameter). Increasingly distant inverse loci are represented in sequence from top to bottom below the basis curves.

The sequence of Fig. a represents a basis ellipse (eccentricity, $\sqrt{3}/2$), in standard position, and 5 inverse loci thereof about increasingly distant poles on the extended major axis to the right of the vertex. Similarly the sequence of Fig. c represents a basis equilateral lemniscate and 5 inverse

Approach of Inverse Curves To Subspecific Identity With Closed Basis Curves As Inversion Poles Become Increasingly Distant

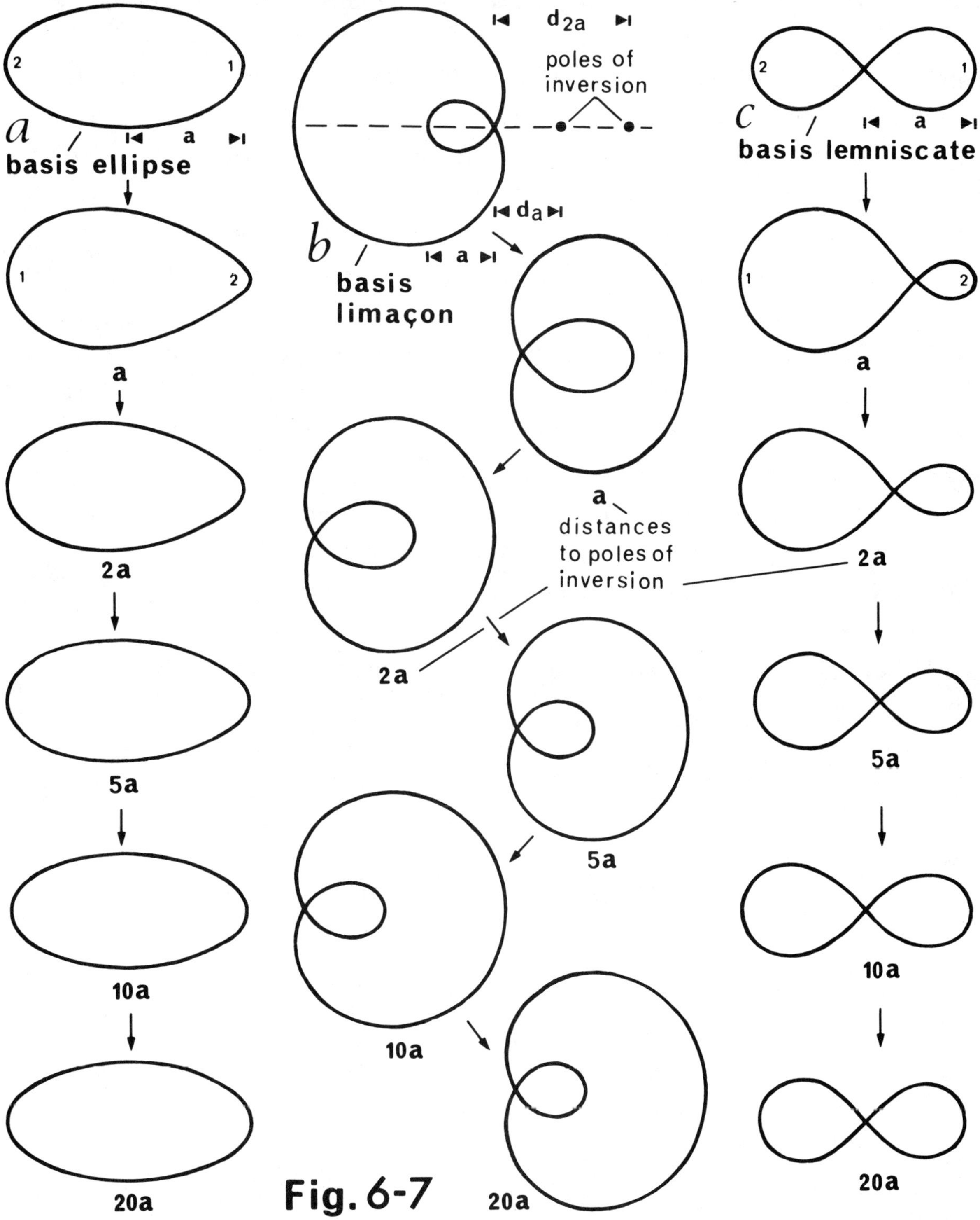

Fig. 6-7

loci thereof about increasingly distant poles incident upon the transverse axis to the right of the right axial vertex. Inasmuch as the basis curves of Figs. a and c possess two lines of symmetry, and the pole of inversion is incident upon one of these lines, the inverse curves approach *coincidence* with the basis curve with increasing distance of the pole of inversion from the curve (for the appropriate value of the unit of linear dimension).

Fig. b represents a basis $e = 2$ limacon [$r = a(1-2\cos\theta)$] and 5 inverse loci thereof about increasingly distant poles incident upon the line of symmetry to the right of the DP. Note that though subspecific identity is approached with increasing distance of the pole of inversion from the DP, the inverse curves have an opposed (mirror-image) orientation, ruling out an approach to coincidence for the inverse loci of axial curves about increasingly distant poles on the line of symmetry.

This example lends emphasis to the fact asserted above, that the approach of the inverse locus of a closed curve to subspecific identity with the basis curve, with increasing distance of the pole of inversion therefrom, characteristically is merely in the nature of an approach to similarity or congruence (for the appropriate value of the unit of linear dimension), not coincidence. In fact, as illustrated in Fig. 6-8, "rotations" occur for cases in which the poles of inversion recede along a line not orthogonal to a line of symmetry.

[In the examples of this figure, the lines of symmetry upon which the poles of inversion are incident are not lost, and the reversals of orientation are the equivalent of distorted reflections in lines orthogonal to these lines of symmetry (these reversals of orientation are indicated by the reversed labelling of the axial vertices for the first two figures in sequences a and c).]

CIRCUMPOLAR MAXIMS 6-7-

1: *For the appropriate value of the unit of linear dimension, inversions of a closed basis curve or finite segment of an open basis curve about distant poles yield inverse curves that increasingly approach congruence with the basis curve, the more distant the pole.*

2: *For the appropriate value of the unit of linear dimension, inversions of a basis curve about distant poles yield inverse curves that, in part or in their entirety, increasingly approach congruence with the basis curve (in part or in its entirety) the more distant the pole.*

3: *For the appropriate value of the unit of linear dimension, inversions of a closed central basis curve or finite segment of an open central basis curve about distant poles on either of the lines of symmetry yield inverse curves that increasingly approach coincidence with the basis curve (self-inversion), the more distant the pole.*

4: *For the appropriate value of the unit of linear dimension, inversions of an open central basis curve about distant poles on either of the lines of symmetry yield inverse curves, portions of which increasingly approach coincidence with portions of the basis curve, the more distant the pole.*

5: *For the appropriate (infinite) value of the unit of linear dimension, inversion of a basis curve or figure about an infinitely distant pole yields a congruent curve or figure.*

[For the basis for the qualifications in Maxims 6-7-3 and 4 regarding open basis curves, and in Maxim 6-7-2 for a basis curve, not specified as to being open or closed, see below and Figs. 8-34,35,39,40, and 41, particularly the legends for Figs. 8-34,35, and 39).]

Maxims 6-7-5 and 6-4-5 are fundamental inversion-mapping Maxims from which, together with elementary considerations of symmetry, follow most of the other mapping properties of the inversion transformation. They are not, however, independent; either can be derived from the other. Thus, if curve B' is obtained by inverting the basis curve B about the pole P, Maxim 6-4-5 asserts that inversion of curve B' about pole P', the pole inverse to P in the plane of B', also should yield curve B'. But since pole P' is the ideal line at infinity, it follows that B' inverts to subspecifically identical curves about infinitely distant poles.

Conversely, Maxim 6-7-5 asserts that a curve undergoes identical subspecific-inversion about a pole at infinity. Accordingly, given a basis curve B that inverts to curve B' about pole P; since pole P', inverse to P, is the ensemble of points of the ideal line at infinity in the plane of curve B', it follows from Maxim 6-7-5 that subspecifically identical curves are obtained by inverting the basis curve B about pole P and by inverting the curve inverse to B (curve B') about the pole inverse to P (points of the ideal line at infinity). In view of the equivalence of these Maxims, a proof of either is tantamount to a proof of the other.

The proof of Maxim 6-7-5 and, in consequence, Maxim 6-4-5 is almost trivial. A necessary and sufficient condition that the inverse of a curve or figure B about an infinitely distant pole, P, be a congruent figure B' (for the appropriate [infinite] value of the unit of linear dimension) is that the inverse element T' of a triangular element T (in the plane of B) about pole P be similar to T. It is a property of the inversion transformation that the inverse of T about a finite pole is a "triangle" consisting of circular arcs. It is a further property of the inversion transformation that as the pole of inversion recedes to infinity, these circular arcs approach linearity. For an infinitely distant pole P', and for the appropriate (infinite) value of the unit of linear dimension, the inverse locus is a triangle T' (which can be regarded as consisting of 3 circular arcs of infinite radius). But T' must be similar to T, because the inversion transformation is conformal. Hence, curve B' is congruent with the basis curve B.

[Christopher Kavanau first supplied a proof to this Maxim using vector algebra. An elegant geometrical proof, including the property that the congruent-inversion is a reflection in a line, was derived by my student Roozbeh Sadeghiyan.]

Thus, Maxim 6-7-5 is proved using only conformality of the inversion transformation, the property that lines invert to circles about non-incident poles, and the property that in the limit these circles are of infinite radius. The above arguments cannot be used in reverse, however, to prove conformality of the inversion transformation, in general, from Maxim 6-7-5, because the proof as set forth depends only on *conformality in the limit*.

Although Maxim 6-7-5 appears plausible from the examples of Figs. 6-6 through 6-8, other examples could have been selected that, upon the face of it would appear to be counterexamples. For example, the inverse loci of the parabola about distant incident (Fig. 8-34) and non-incident (Fig. 8-35) poles do not resemble parabolas. Nor do they appear to come increasingly to resemble parabolas as the pole of inversion becomes ever more distant.

These apparent counterexamples, however, are merely expressions of the relationships embodied in Maxims 6-7-2 to 6-7-4, which serve to emphasize the fact that congruence or coincidence is achieved only in the limit, for an infinite value of the unit of linear dimension. For finite values of the unit of linear dimension, only *portions* of most inverse loci of open curves about increasingly distant poles approach congruence with corresponding portions of the basis curves (see Figs. 8-34,35,39, and 41).

The situation may be likened to inverting a closed basis curve about a series of poles that approach and, finally, achieve coincidence with a point on the curve, for example, the sequence of curves o,m,n,l, and b of Fig. 6-3. The inverse loci, o,m,n, and l, about the corresponding poles are closed curves that do not resemble the inverse locus b, about the incident pole, except in the region of the small loop. The resemblance in this region becomes greater, the closer the approach of the pole of inversion to the incident pole, and the regions become virtually indistinguishable for poles very close to the incident pole. But only at the point of incidence does the large loop "open" to form the looped cubic. Similarly, for inversion of an open basis curve about increasingly distant poles, the inverse loci are closed. Only in the limit, for an infinitely distant pole and an appropriate (infinite) value of the unit of linear dimension, does the curve "open" to a locus congruent with the basis curve.

Fig. 6-8a,b. Non-axial inversions of closed curves about external poles, approach to subspecific identity, and angle-doubling inversion rotation. Fig. 6-7 illustrated cases in which increasingly distant poles of inversion of closed central and axial basis curves lie on a line of symmetry; it was emphasized that the reversal of orientation of an inverse locus, as compared to the basis curve, is an inherent property of the inversion transformation. We now consider cases of an axial (Fig. a) and central (Fig. b) basis curve inverted about poles increasingly distant on lines neither orthogonal to, parallel to, nor coincident with a line of symmetry.

This figure illustrates the progressive changes in the forms and orientations of the inverse loci of a basis $e = 2$ hyperbolic limacon and a basis $e = \sqrt{3}/2$ ellipse (each in standard position) with increasing distances of the poles of inversion along lines at 45° to the horizontal axes and passing through the DP (Fig. a) and the center (Fig. b), respectively. Distances from these reference points to the poles of inversion are a, 2a, 5a, 10a, and 20a in the sequences indicated by the arrows and labels. For the ellipse, a is the length of the semi-major axis, for the limacon, it is the diameter of the small loop. In the case of the basis ellipse, examples also are given for the inverse curves about poles lying at a distance of 20a along lines through the center at angles of $22\frac{1}{2}^\circ$ and $67\frac{1}{2}^\circ$ to the major axis.

In the case of the basis $e = 2$ limacon of Fig. a, inversion about a pole at $d = a$ leads to a curve resembling a distortion of the basis curve, with the appearance of having been rotated 90° CW. The transverse-axis vertices of the basis curve, and the poles inverse to them in the inverse curves, are marked by the solid circles to complement the DP as reference points.

The DP and the large-loop vertex of the basis curve are taken as the termini of a line segment, and the angles of a line to the midpoint of this segment from the pole of inversion are compared with the angles of a line connecting the inverse points of the inverse curves. The ratios of the latter angles to the former angles follow the progression, 3.141, 2.173, 2.093,

Non-Axial Inversions of Closed Curves About External Poles, Approach To Subspecific Identity And Angle-Doubling Inversion Rotation

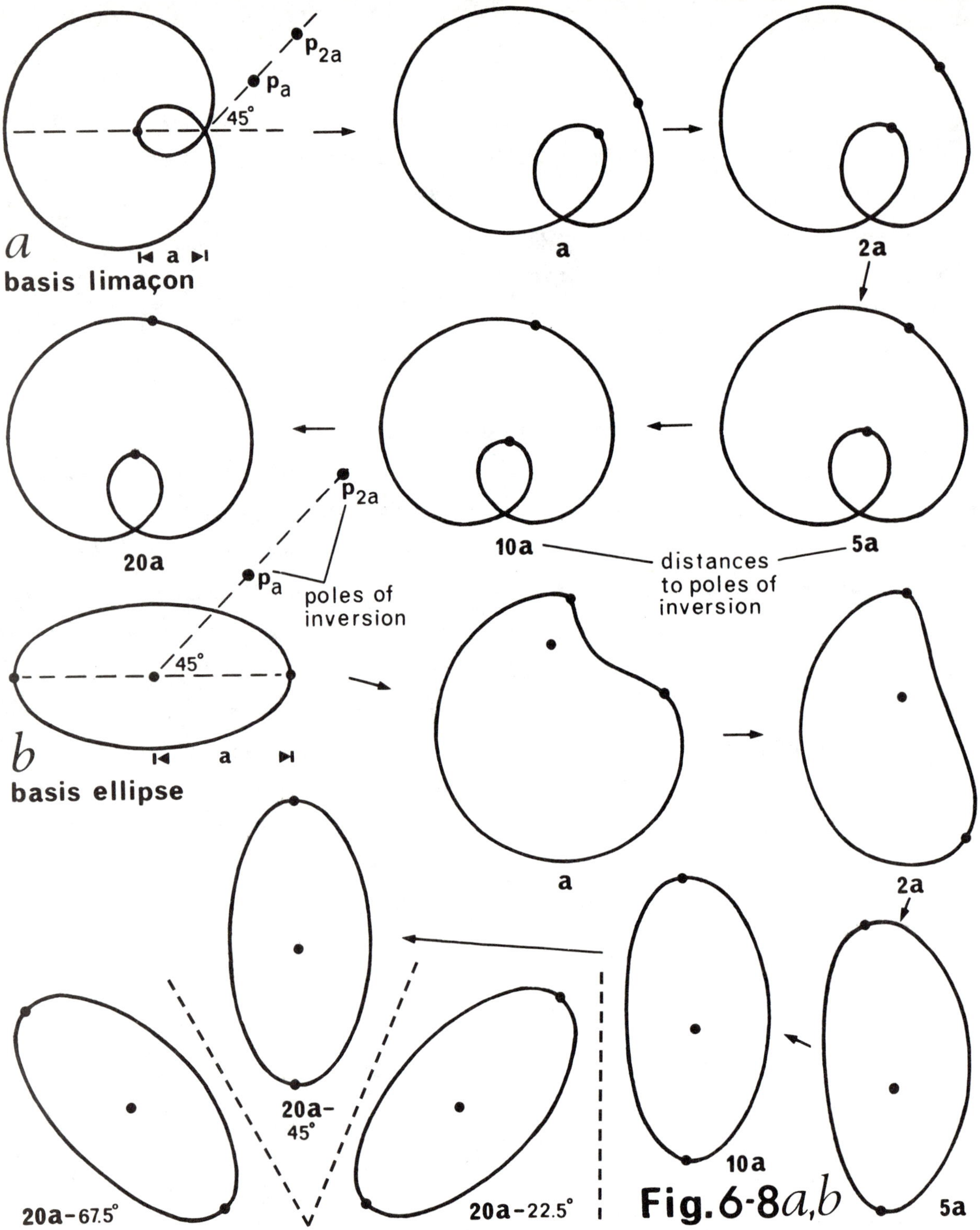

Fig.6-8*a,b*

2.026, and 2.007, approaching a limiting value of 2. Just as this ratio "rapidly" converges toward 2 for poles only a few units distant from the basis curve, so also does the form of the inverse curves rapidly converge toward congruence with the basis curve (as also was the case for the sequence of axial inversions of Fig. 7b).

Convergence of inverse curves--about distant poles of inversion at a given angle--toward congruence or subspecific identity with a closed basis curve, of course, implies convergence toward congruence with one another at different angles, and this is manifestly the case for angles of 0° and 45° (compare the sequences of Figs. 7b and 8a).

For the inversions of the ellipse, the reference poles that are marked by the solid circles are the center and the two major-axis vertices of the basis curve and their inversions in the plane of the inverse curves. As in the previous examples, convergence of the inverse curves toward elliptical form is "rapid" and the inverse curves for poles 20a distant from the center are very similar to the basis curve.

If one makes a comparison paralleling that made above, of the angles to the basis major axis of lines connecting the poles inverse to the major-axis vertices (135°, 104.04°, 92.29°, 90.57°, and 90.14°, for distances of a, 2a, 5a, 10a, and 20a, respectively) with the angle of the line on which the poles of inversion recede (45°), one obtains the angle ratios, 3, 2.312, 2.035, 2.013, and 2.003, for poles at distances of a, 2a, 5a, 10a, and 20a from the center, respectively. In the cases of the inverse curves about poles 20a distant on lines through the center at $22\frac{1}{2}^\circ$ and $67\frac{1}{2}^\circ$, the angles are 45.101° and 135.101°, yielding ratios of 2.0045 and 2.0015, respectively.

Accordingly, the following relationship emerges from these two examples. Given that increasingly distant poles of inversion lie on a line L at $22\frac{1}{2}^\circ$, 45°, and $67\frac{1}{2}^\circ$ to a line of symmetry, P_1P_2, of an axial or central curve, a line $P_1'P_2'$ connecting two points P_1' and P_2' inverse to points P_1 and P_2 on line P_1P_2 of the basis curves makes an angle with line P_1P_2 that increasingly approaches 45°, 90°, and 135°, respectively. Superficially, the curves appear to become rotated 45°, 90°, and 135° CW. These results strongly suggest the generality that, in the limit, the angle between the line $P_1'P_2'$ and the line P_1P_2 (i.e., the angle of "rotation") is twice the angle between line L and line P_1P_2 (see Fig. 6-8d).

[In the interests of simplicity, poles P_1 and P_2 are taken to be equidistant from the intersection of line L with line P_1P_2 but the results are independent of the locations of points P_1 and P_2 on line P_1P_2.]

Fig. 6-8c-e. "Angle-doubling rotation" by inversion about an infinitely distant pole is equivalent to reflection of the basis curve in a line--a segment of the circle of inversion (of infinite radius). It now is shown that the conclusion suggested above by the findings for inversion of axial and central curves at angles of $22\frac{1}{2}^{\circ}$, 45°, and $67\frac{1}{2}^{\circ}$ is valid in general.

Consider the example of Fig. 6-8d, in which a line L intersects a line segment P_1P_2 of length 2a at the midpoint between points P_1 and P_2, with the angle of intersection being θ. Let the pole of inversion of the two points P_1 and P_2 lie at a distance d from the point of intersection. The angle, β, at which the line $P_1'P_2'$, passing through the inverse poles P_1' and P_2', intersects the line P_1P_2 is given by the complicated expression of eq. 1, which simplifies to eq. 2. For $d \gg a$, i.e., for distant poles of inversion, eq. 2 simplifies still further to eq. 3, proving the "angle-doubling" property in the limit. The same results are obtained for any locations of the points P_1 and P_2 on line P_1P_2, for example, for Fig. e, in which line L passes through pole P_1.

angle of intersection of lines P_1P_2 and $P_1'P_2'$

$$\beta = \arctan \quad (6\text{-}8\text{-}1)$$

$$\left[\frac{\{[1-a^2/(a^2+d^2+2ad\cos\theta)] - [1-a^2/(a^2+d^2-2ad\cos\theta)]\}(\pm d\sin\theta)}{[1-a^2/(a^2+d^2+2ad\cos\theta)](d\cos\theta+a) - [1-a^2/(a^2+d^2-2ad\cos\theta)](d\cos\theta-a) -2a}\right]$$

$$\beta = \arctan[\pm d^2\sin 2\theta/(d^2\cos 2\theta-a^2)] \qquad \text{simplification of eq. 1} \qquad (6\text{-}8\text{-}2)$$

$$\beta = \arctan(\pm\sin 2\theta/\cos 2\theta) = \arctan(\tan 2\theta) = 2\theta \qquad \text{eq. 2 for } d \gg a \qquad (6\text{-}8\text{-}3)$$

CIRCUMPOLAR
MAXIM 6-8-1: *When two points P_1 and P_2 on the x-axis are inverted about increasingly distant points on a line L in the 1st quadrant that emanates from the origin at an angle θ to the x-axis, the line $P_1'P_2'$, connecting the inverse points P_1' and P_2', forms an angle with the x-axis whose limiting value is 2θ.*

The finding that "inversion rotation" about poles infinitely distant from the basis curve leads to angle-doubling (in the sense described) has a two-fold significance. First, it emphasizes the fact that the direction in which a pole recedes to infinity relative to a basis curve (or the direction in which it lies at infinity) is crucial for the orientation of an inverse

"Angle-Doubling Rotation" By Inversion About An Infinitely Distant Pole Is Equivalent To Reflection of The Basis Curve In A Line--A Segment of The Circle of Inversion (of Infinite Radius)

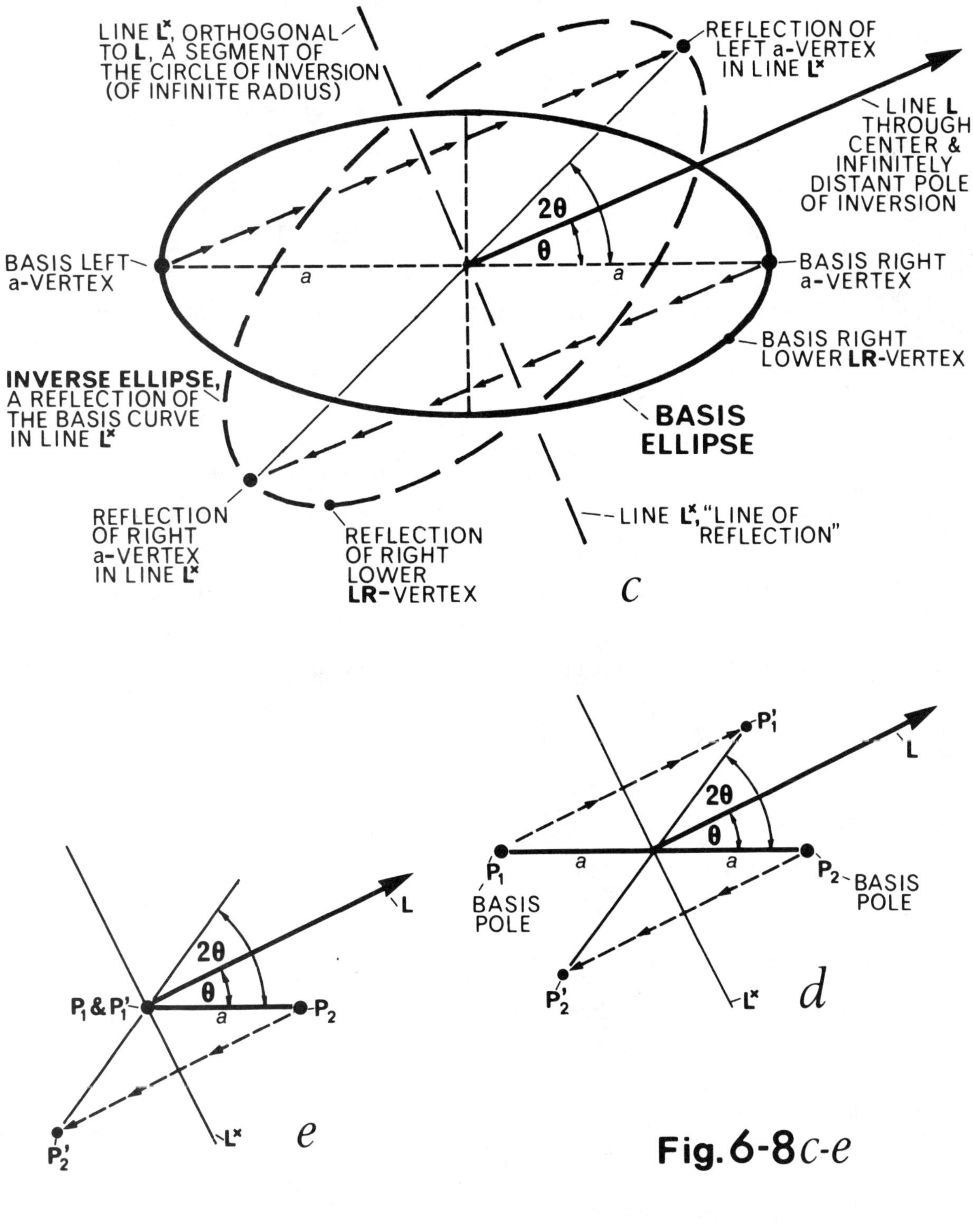

Fig. 6-8*c*-*e*

locus about it. Viewed in this context, it is not satisfactory to characterize all points of the ideal line at infinity simply as "the point at infinity."

Second, it reveals a fundamental property of inversion about infinitely distant poles. It is evident from Figs. d and e that angle-doubling is equivalent to a reflection of poles P_1 and P_2 in a line L*, orthogonal to the line L upon which the poles of inversion recede. [To double an angle, one simply reflects a point on one limb through the other limb, and draws a line from the apex of the angle to the point of reflection.]

In other words, for the appropriate (infinite) value of the unit of linear dimension, the inverse of a curve about an infinitely distant pole is a reflection of the curve in a line. Since the line in question is a segment of the circle of inversion, which is of infinite radius, this line is orthogonal to the line along which the pole recedes.

These relationships are illustrated in detail in Fig. c for a basis ellipse in standard position. Although reflection of the basis ellipse in line L* superficially resembles a rotation of the curve itself through an angle 2θ, it is evident that this is not the case. Thus, the basis right lower LR vertex, for example, comes to lie at a position in the inverse curve that is inconsistent with a rotation alone; the position can be achieved either by a reflection alone or by the combination of a rotation and a reflection.

The following Maxims incorporate these findings, those of previous figures, and aspects of the inversion of open curves about remote poles that are illustrated in Figs. 8-34,35,39,40, and 41. In the interests of simplicity, these Maxims are formulated with respect to closed basis curves, finite segments of open basis curves, and open basis curves that possess a nucleus (such as a loop or oval) or center.

CIRCUMPOLAR MAXIMS 6-8-

2: *For the appropriate (infinite) value of the unit of linear dimension, the inverse locus of a basis curve about an infinitely distant pole is a reflection of the curve in a line (specifically, the circle of inversion of infinite radius).*

3: *For the appropriate value of the unit of linear dimension, the inverse locus B' of an axial closed basis curve, or of finite segments of an axial open basis curve B about increasingly distant poles on a line orthogonal to the line of symmetry, increasingly resemble the reflection of B, or the reflection of finite segments of B, in the line of symmetry. In the limit, B' is a coincident reflection of B, whereupon B self-inverts.*

4: *For the appropriate value of the unit of linear dimension, the inverse locus B' of an axial closed basis curve, or of finite segments of an axial open basis curve B about increasingly distant poles on a line L not orthogonal to the line of symmetry, increasingly resemble a reflection of B, or a reflection of finite segments of B, in a line L* orthogonal to L. In the limit, B' is a congruent reflection of B in L* (a congruent-inversion).*

5: *For the appropriate value of the unit of linear dimension, the inverse locus B' of a central closed basis curve, or of finite segments of a central open basis curve B about increasingly distant poles on a line parallel to one of the lines of symmetry, increasingly resemble a reflection of B, or a reflection of finite segments of B, in the other line of symmetry. In the limit, B' is a coincident reflection of B, whereupon B self-inverts.*

6: *For the appropriate value of the unit of linear dimension, the inverse locus B' of a central closed basis curve, or of finite segments of a central open basis curve B about increasingly distant poles on a line L not orthogonal to either line of symmetry, increasingly resemble a reflection of B, or a reflection of finite segments of B, in a line L* orthogonal to L. In the limit, B' is a congruent reflection of B in L*.*

7: *For the appropriate value of the unit of linear dimension, the inverse locus B' of a non-axial closed basis curve, or of finite segments of a non-axial open basis curve B about increasingly distant poles on a line L, increasingly resemble a reflection of B, or a reflection of finite segments of B, in a line L* orthogonal to L. In the limit, B' is a congruent reflection of B in L*.*

It is evident from considerations of the above Maxims and Maxims 6-6-1 to 6-6-3 that the possession of a line of symmetry by a curve is equivalent to the possession of a 0° focus of self-inversion on the ideal line at infinity in the direction orthogonal to the line of symmetry; in fact, the line of symmetry is merely the circle of inversion (of infinite radius) of this 0° focus. Accordingly, a central curve with two lines of symmetry possesses two 0° foci of self-inversion on the ideal line at infinity, one in the direction of each line of symmetry.

In this view, the loss (or gain) of a line of symmetry (i.e., a circle of inversion of infinite radius) in an inversion (Maxim 18), is equivalent to the transfer of the corresponding 0° focus of self-inversion from the ideal line at infinity to the finite plane (or from the finite plane to the ideal line at infinity), with reduction (or increase) of the modular value for self-inversion to a finite (or infinite) value.

It is implicit in the present treatments relating to self-inversion that they apply only in circumstances in which self-inversion is defined. For example, the concept of self-inversion does not apply to any case in which a radius vector from a non-incident pole intersects a curve at an odd number of points, nor can a closed curve self-invert about an incident pole.

On the other hand, self-inversion is a phenomenon of extremely broad occurrence in inverse loci. Confining attention only to plane curves and figures, a non-axial inverse locus of any basis curve with a line of symmetry, regardless of the latter's degree or the number of segments that it possesses (ovals, loops, asymptotic arms, etc.), has at least one focus of self-inversion. This is a consequence of the fact that all point-pairs of the plane that are reflected in a line become self-inverse in a common point when inverted about a pole not incident upon the line. Furthermore, an inverse locus gains a focus of self-inversion for each line of symmetry of the basis curve that is lost, and this applies without limit (see Maxim 8-30-13; inverse loci of curves and figures with 3 and 4 lines of symmetry are treated in Fig. 8-30).

CIRCUMPOLAR MAXIMS 6-8-

8 : *The possession of a line of symmetry by a curve is equivalent to the possession of a 0° focus of self-inversion on the ideal line at infinity in the direction orthogonal to the line of symmetry.*

9 : *Central curves with two lines of symmetry possess two 0° foci of self-inversion on the ideal line at infinity, one in the direction of each line of symmetry.*

10: *The gain (or loss) of a line of symmetry in an inversion is accompanied by the transfer of a 0° focus of self-inversion to (or from) the ideal line at infinity from (or to) the finite plane* (see also Maxim 8-15-4a).

11: *In the finite plane, curves of odd degree self-invert only about incident foci; contrariwise, curves of even degree self-invert only about non-incident foci.*

Fig. 6-9. 90°-circumpolar axial intercept transforms of the parabola for two positive intercepts. All interior 90°-circumpolar axial intercept transforms of the parabola for positive intercepts possess two asymptotic segments (see pages VI-16 and 17). The *inner segment* (of which 10 examples are illustrated) is traced for angles of the x radius vector from 0° to 270°. The *outer segment* (of which two examples are illustrated) is traced for angles from 270° to 360°. The sequence in which the segments occur has been retained, whereas the locations have been shifted for the purpose of illustration (in its proper location, the $-\frac{1}{2}a$ outer segment intersects the loop of the 5a inner segment just inside the DP). For any given specific pole of transformation, the inner and outer segments have a common asymptote. 90° transforms with real solutions for exterior axial poles exist only to a distance of $-a$ from the vertex.

The basis curve (in standard position) and the locations of 10 axial poles of transformation are illustrated at the upper right. The manner of obtaining 90° transforms is illustrated for an axial pole at $h = 5a$, for which the positive x and y intercepts are shown for an angle $\theta = 135^\circ$. The rectangular plot of the positive x intercept against the positive y intercept yields the point on the 5a transform opposite the arrowhead labelled "135°." Points also are marked on the 5a transform corresponding to angles of the x intercept of 65°,90°,135°,180°, and 205°. In addition, for reference, the 90° and 180° points--which are extremals for transforms about interior poles--are marked by arrowheads for all of the inner segments for transforms about interior poles.

As the pole of the transformation approaches a distance of 4a from the vertex (4a is the axial distance from the vertex at which the abscissa equals the ordinates), the 90° and 180° extremal points approach one another along the arms of the inner segments. They become coincident at the DP when the pole of transformation is at 4a; for poles beyond 4a they recede from one another along the loop.

For interior poles, the line $x = y$ intersects all inner segments proximally (nearest the origin) at the 135° point. For exterior poles, for which only the segments of the $-\frac{1}{2}a$ transform are illustrated, it intersects both segments at the 315° point. The angles for the extremal points on the inner segments for exterior axial poles correspond to the tangent angles of the

90°-Circumpolar Axial Intercept Transforms of The Parabola

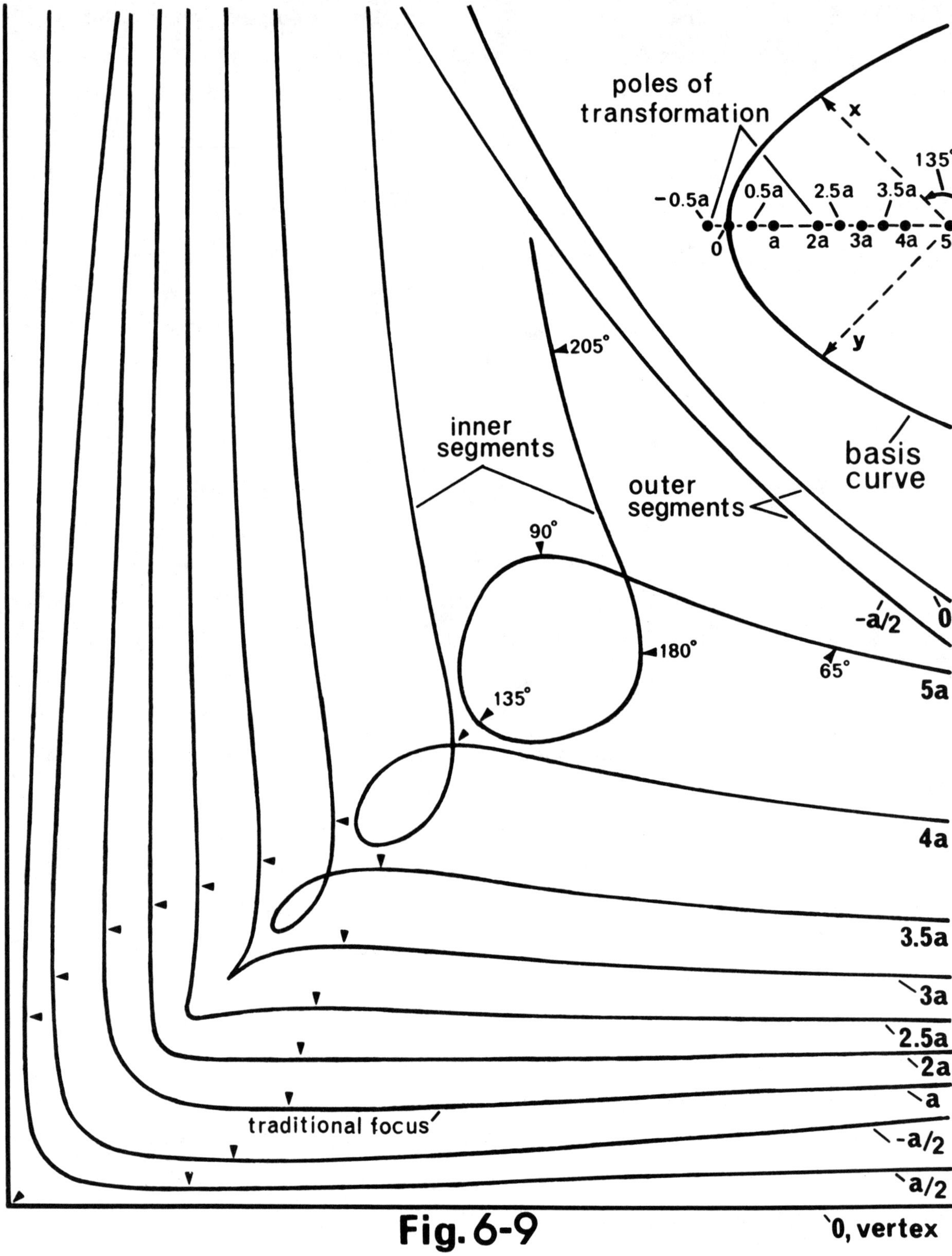

Fig. 6-9

x and y intercepts. For example, for the $-\frac{1}{2}a$ transforms these points are for angles of the x intercepts of approximately 305.3° and 324.7°. It is noteworthy that, whereas only one positive x intercept exists for interior poles for any given angle, two positive x intercepts exist for each angle for exterior poles (at all angles for which the intercepts are real).

In the case where the pole of transformation is at the vertex, the outer segment is traced for non-zero values of both intercepts, whereas the inner segment--consisting of the positive coordinate axes (the 90° and 180° points lie at the origin)--is degenerate, in the sense that one portion of it (from 0° to 90°) is traced for null values of the y intercept and the other (from 180° to 270°) for null values of the x intercept.

[The existence of this degenerate inner segment is an artifactual consequence of limiting the transform traces to those for only two positive (or two negative) intercepts. When all positive and negative intercepts are taken into account, no degenerate segment exists (see, for example, Figs. 6-13c,e).]

The form of the inner segments changes progressively as the pole of transformation progresses along the line of symmetry. In the neighborhood of the 3a pole, the elongating vertex region becomes a cusp which, thereafter, opens into a loop that enlarges increasingly for axial poles progressively more distant from the vertex; the segment for the 5a pole is the most distant of these segments that is illustrated. Although the relative locations of the outer segments for both interior and exterior poles proceed in sequence with the locations of the poles along the axis, this is not true for the inner segments. These proceed in sequence for the exterior poles alone and for the interior poles alone, but the two sequences interdigitate with one another.

Being for positive intercepts, the figure represents only the segments of the full 4-quadrant transform that lie in the 1st quadrant. When both positive and negative intercepts are taken into account, segments exist in all 4 quadrants, as depicted in Figs. 6-12 and 6-13. Other segments of 90° transforms about poles on the line of symmetry can be visualized as reflections through the coordinate axes and the origin.

The rectangular eqs. of 90°-circumpolar intercept transforms of the parabola about axial poles generally are of 16th degree, compared to 12th degree for those about incident poles and 20th degree for those about non-incident, non-axial poles (see Table VII-1, page VII-19, and eqs. VI-34,36 and 37b,c). Reductions of axial transforms occur only for two poles of transformation, and these reductions define the axial foci. For the traditional focus, the transform is of only 4th degree (eqs. VI-37b,c) and, accordingly, this pole is the highest-ranking focus of the curve; for the vertex the transform is of 8th degree.

The axial-vertex focus shares 2nd-highest focal rank with the LR vertices, which also have 8th-degree 90° transforms, but the symmetry of the curve is greater about the former pole than about the latter ones because the transform eqs. about the axial vertex are simpler (this is the criterion for establishing the relative *subfocal* symmetry of a curve about poles for which the transforms are of the same degree; see page VI-18).

It is a general but not invariable occurrence that the higher the degree of an intercept transform eq., the more the number of exceptional points (cusps, double points, triple points, inflection points, etc.) possessed by the curve. The situation for transforms about axial poles of the parabola is represented in Table 6-9-1, based upon eqs. 6-9-1a,b for the slopes of the axial transforms. None of the outer segments of the axial transforms is known to possess

slopes of 90°-circumpolar intercept transforms of the parabola about axial poles for positive intercepts

(a) $$\frac{dy}{dx} = \frac{-[2a^2/A+2A-2a\sin\theta-2a/\sin\theta]\tan^4\theta}{[2a^2/B+2B+2a\cos\theta+2a/\cos\theta]} \qquad (6\text{-}9\text{-}1)$$

$$A = (a^2\sin^2\theta+ah\cos^2\theta)^{\frac{1}{2}} \qquad B = (a^2\cos^2\theta+ah\sin^2\theta)^{\frac{1}{2}}$$

(b) $$dy/dx = [(1+\sin^2\theta)/(1+\cos^2\theta)]\tan^3\theta$$

eq. 1a for the axial vertex (rederived from the axial intercept eq. VI-23 with $k=0$)

an exceptional point. In the cases of the inner segments, the 4th-degree transform for the traditional focus possesses two inflection points (one in the range 0-135° and the other in the range 135°-270°).

Since the 8th-degree axial-vertex transform possesses only an outer segment, (or 4 outer segments, if all combinations of positive and negative intercepts are taken into account), it has no exceptional points. The 16th-degree transforms about other axial poles possess from 2 to 5 exceptional points. Those about $-\frac{1}{2}a$, $\frac{1}{2}a$, and 2a possess two inflection points, while that about the pole at $2\frac{1}{2}a$ possesses 4. For the pole at 3a, the point for $\theta=135^{\circ}$ is a cusp point

Table 6-9-1. Exceptional Points of 90°- Circumpolar Intercept Transforms of The Parabola About Axial Poles (positive intercepts only)

		θ for exceptional points in range 0-135°			
locations of axial poles	degrees of transforms	inflection points		cusps	double points
-½a	16		307.5		
0	8	none	none		
½a	16		64.2		
a	4		65.2		
2a	16	49.1			
2½a	16	38.4	76.9		
3a	16	32.5	76.1	135	
3½a	16	28.7	76.2		100
4a	16	25.9	76.4		90
5a	16	22.2	77.1		79.8

and the total number of exceptional points is 5. Transforms about more distant poles possess a looped region, the DP of which, plus 4 points of inflection, give all these transforms 5 exceptional points.

Fig. 6-10. Circumpolar symmetry of the parabola about the traditional focus (inner segments) for two positive intercepts. This figure illustrates the inner segments in the 1st quadrant for circumpolar transforms of the parabola about the traditional focus at 12 angles of transformation from 0° to 180°, as labelled (transforms for angles of 180° to 360° merely duplicate those from 0° to 180°). All the transforms are to the same scale and in proper sequence, though their locations have been shifted for the purpose of illustration (some segments have been spaced closer together, others farther apart).

No transform about the focus at any angle (nor about any pole more proximal to the vertex) possesses a cusp or looped portion. The linear 0° transform is simply the trivial transform of identity, $x = y$, for two identical positive intercepts. The 180° transform is a segment of an arm of the equilateral hyperbola and expresses the harmonic mean relationship, $a(x+y) = xy$, or $1/x + 1/y = 1/a$ (see pages xvi, xvii, VI-14 and VII-21 in *Symmetry, An Analytical Treatment*). The transforms for all other angles are of 4th degree (eqs. VI-37b,c).

Circumpolar Symmetry of The Parabola About The Traditional Focus (Inner Segments)

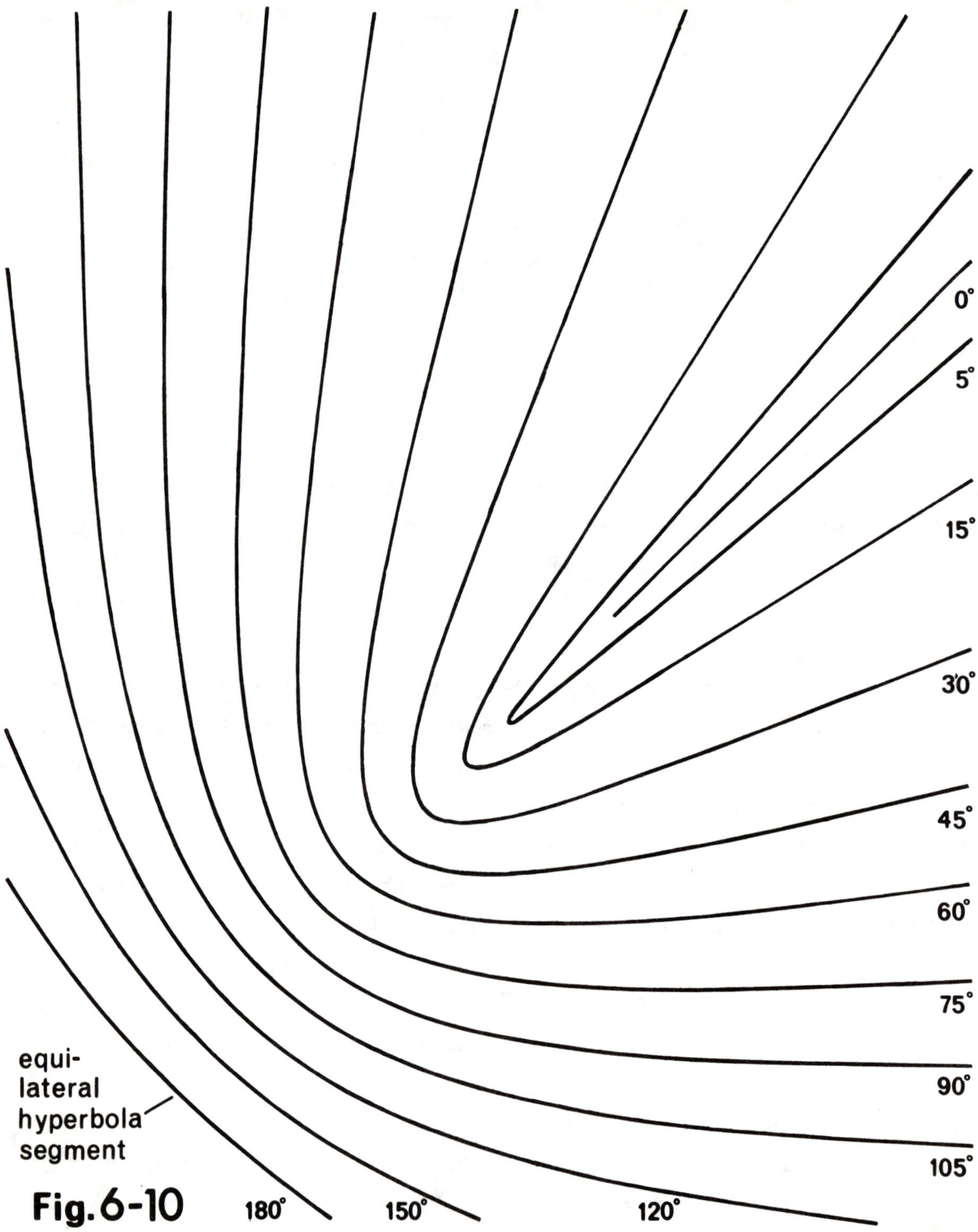

Fig. 6-10

The intercept transform of a curve about a given pole at a given angle gives a gross characterization of the symmetry of the curve about the pole at that angle. A more specific characterization of the symmetry of the curve about the pole is achieved by supplementation with the *intercept ordinal and subordinal rank eqs.* (see legends of Figs. 6-15 and 7-19). A complete implicit characterization is given by the parametric eqs. of a transform, i.e., $x = f(\theta,a,b,\ldots)$ and $y = g(\theta,a,b,\ldots)$, since all the information about the curve and its symmetry about the pole in question, as well as all other poles, is implicit in these eqs. (and in the α-transform).

On the other hand, two intercept transforms of a curve about a pole at two different angles (excluding trivial transforms) give more information about the symmetry of the curve about the pole than does but one transform, while an ensemble of transforms such as those of the present figure can be regarded as an essentially complete characterization of the symmetry, whence the title of the figure. This figure may be compared most aptly with the comparable figures (Figs. 7-5 and 7-6) for the circumpolar symmetry of the $e = 0.745$ ellipse and of the equilateral hyperbola about their traditional foci. It differs from Fig. 7-5 for the $e = 0.745$ ellipse in that the transforms are not complete, inasmuch as the outer segments are not illustrated.

Fig. 6-11. Circumpolar symmetry of the parabola about the axial pole at $2\frac{1}{2}a$ (inner segments) for two positive intercepts, as characterized by the transforms for 8 angles of transformation from 30° to 150°. All the transforms are to the same scale and in proper sequence, though their locations have been shifted for the purpose of illustration. Only relatively small segments of the 120° and 150° transforms are represented.

As well as characterizing the circumpolar symmetry of the parabola about the axial pole at $2\frac{1}{2}a$, this figure illustrates the fact that transforms of the parabola about a common axial pole at different angles may be of quite dissimilar appearance, including looped, cusped, and unlooped segments--though this is not the case for transforms about the traditional focus, nor for those about poles that are more proximal than the latter to the axial vertex.

All of the illustrated transforms have rectangular eqs. of 16th degree, inasmuch as degree reduction for non-trivial transforms about axial poles other than the traditional focus and axial vertex occurs only for transform-

Circumpolar Symmetry of The Parabola About The Axial Pole At 2.5a (Inner Segments)

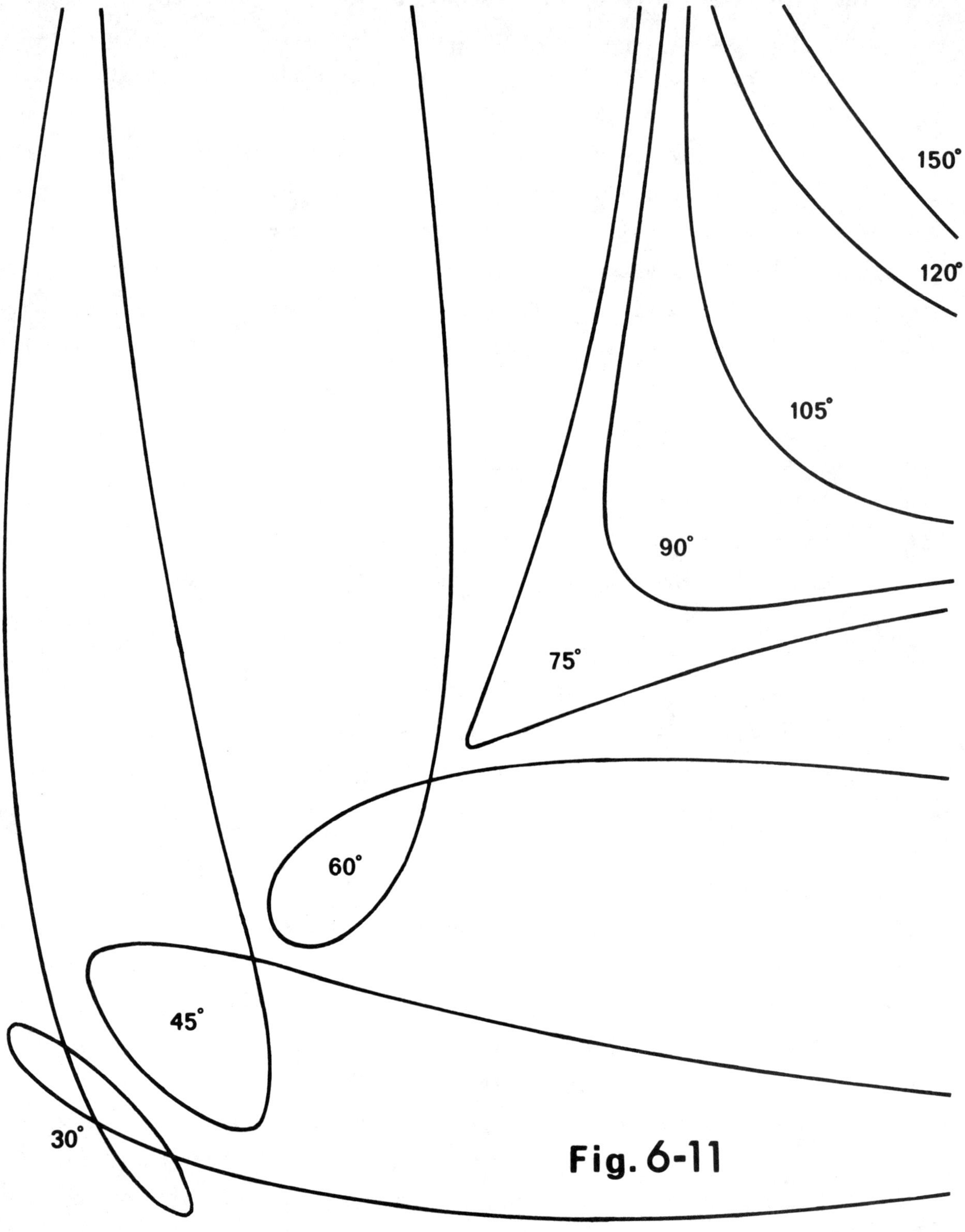

Fig. 6-11

ation angles of 180° and 0° (in the latter instance, for one positive and one negative intercept). This figure may be compared most appropriately with Fig. 7-10 for the circumpolar symmetry of the equilateral hyperbola about an axial pole at 3a. Although similarities are evident, distinct differences that have their origin in characteristic differences in the symmetry of the basis curves also are readily apparent.

Fig. 6-12. 90°-circumpolar intercept transforms of the parabola about the traditional focus and the axial pole at 4a, for both positive and negative intercepts and including both inner and outer segments. All of the illustrated segments are to the same scale and in correct relative positions. The transform segments for the traditional focus are represented by the heavy tracings, those for the pole at 3a by the light ones.

In the preceding figures of circumpolar intercept transforms of the parabola, only the segments for positive intercepts, i.e., segments in the 1st quadrant, have been illustrated, and these consisted almost exclusively of the inner segments. The present figure illustrates the complete 90° transforms, including all intercepts and segments in all quadrants.

For the previously illustrated transforms of the parabola in the 1st quadrant (Figs. 6-9 to 6-11), all of which are symmetric functions (they are symmetrical about the line $x = y$) and possess asymptotic segments, the 4-quadrant plots are visualized readily by imagining a reflection of the segments in the 1st quadrant through the coordinate axes and the origin. Accordingly, aside from depicting the transforms in proper relative positions and including all segments, the present figure is merely supplemental to Fig. 6-9. On the other hand, for 1st-quadrant transforms that are not symmetric functions, the full plot is less readily visualized (see Fig. 6-13).

Fig. 6-13. Non-axial incident and non-incident 90°-circumpolar intercept transforms of the parabola, including all segments and all intercepts. All of the segments in each of Figs. b-d are to the same scale, and within each transform the individual segments are in their proper relative positions. The locations of the poles for the 4 transforms relative to the basis curve are indicated by the solid circles in Fig. a. Locations of the coordinate axes are indicated by the labelled arrows, while the small solid circle in each of the Figs. b-d marks the position of the origin of the coordinate axes.

90°-Circumpolar Intercept Transforms of The Parabola About The Traditional Focus And The Axial Pole At 4a (All Segments, All Intercepts)

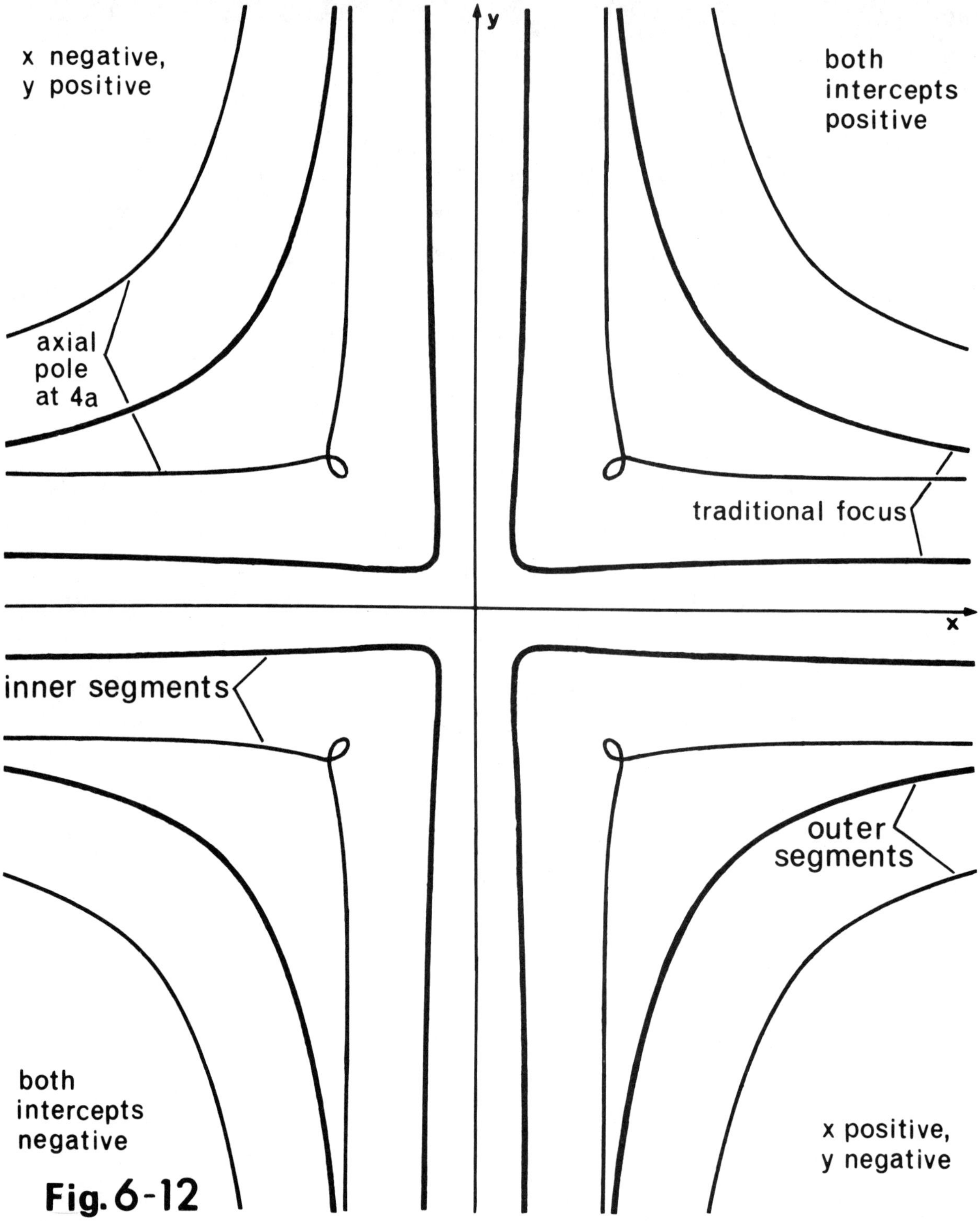

Fig. 6-12

The preceding plots of 90°-circumpolar intercept transforms of the parabola have been for transforms about axial poles. Axial transforms are symmetric functions. When represented for both positive and negative intercepts, these transforms exhibit symmetry about both coordinate axes and the origin. The present figure illustrates 4-quadrant plots of 90° transforms about non-axial poles. These are not symmetric functions and, accordingly, have symmetry only through the origin.

In cases of transforms about incident poles, that of Fig. c for the LR vertex--which has point-focal rank--is of 8th degree (eq. VI-42), whereas that of Fig. e, about a point opposite the axial pole at 5a--which does not have point-focal rank--is of 12th degree (page VI-15). [In these cases, degenerate portions of the transforms (for null values of one or the other intercept) do not exist, since positive and negative intercepts exist at all angles.] The transforms of Figs. b and d, about two interior poles that are both non-axial and non-incident, are both of 20th degree, which is the case for all 90° point-in-the-plane transforms of the parabola about unexceptional poles (see Table VII-1).

Fig. 6-14. 0° and 180°-circumpolar intercept transforms of the parabola about poles on the axis, the LR, and the directrix, and about miscellaneous external poles (for two positive intercepts). All of the illustrated transform curves are to the same scale. In the case of the 180° axial transforms of Fig. a, they also are plotted in their proper relative positions. The transforms of Fig. b have been shifted and placed in the sequence of their corresponding poles for the purpose of illustration. As in the case of the 90° transforms (Figs. 6-9, 1a, and 13), the 0° transforms about external poles possess both inner and outer segments. The segments of any given pair are plotted in correct positions relative to one another in Figs. c and d, but the positions of the 6 sets of segments have been shifted for the purpose of illustration. The locations of the poles of transformation relative to the basis curve are shown at the upper right of Fig. d and the transforms are identified by the accompanying coordinates of the poles of transformation (in units of the parameter, a).

Although the 180° axial transforms appear to consist of only a single segment, these curves are recurrent each 180°, so that each actually represents two congruent superimposed segments. In the cases of all the other

Non-Axial Incident And Non-Incident 90°-Circumpolar Intercept Transforms of The Parabola (All Segments, All Intercepts)

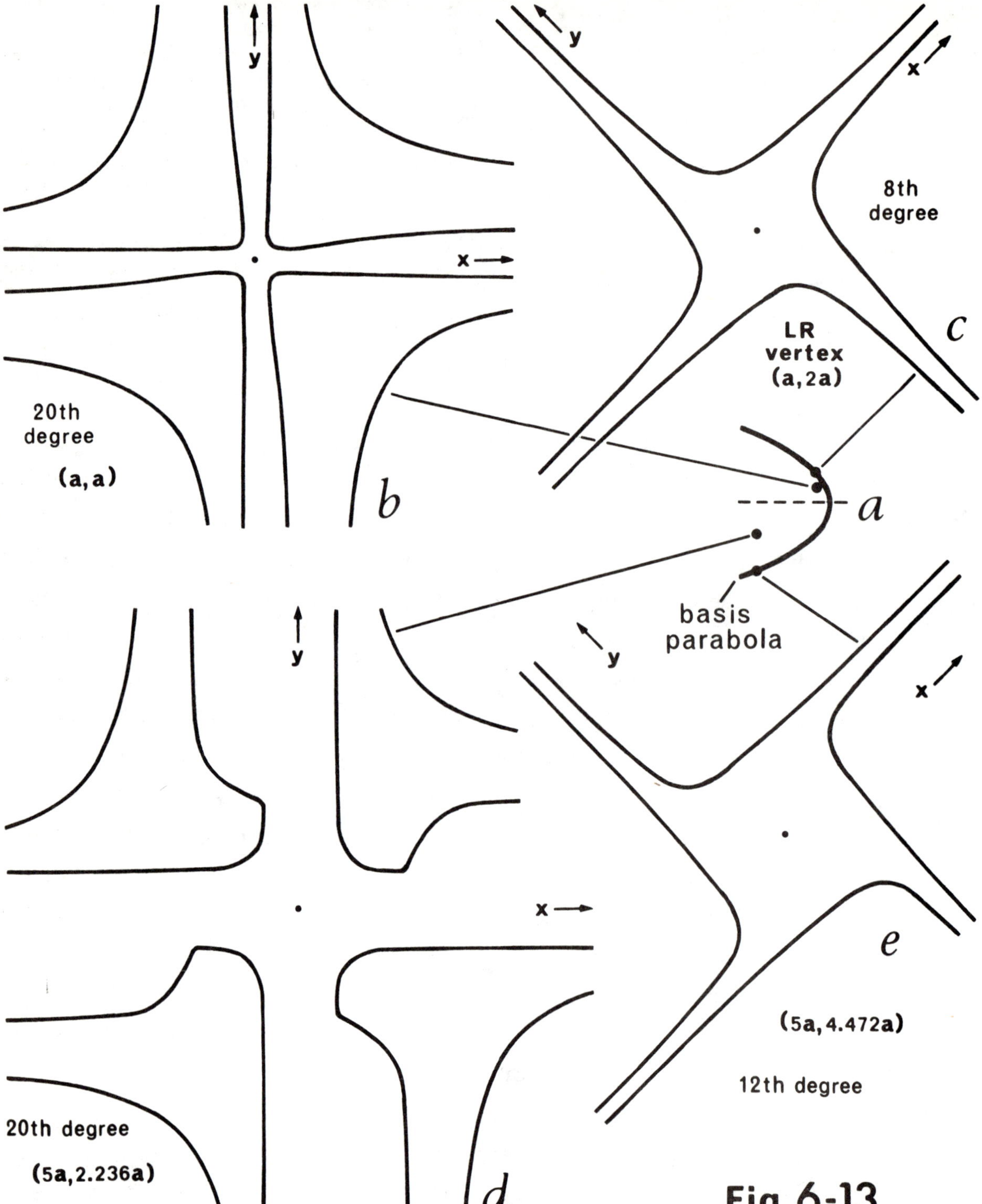

Fig. 6-13

transforms the poles of transformation do not lie on the line of symmetry, with the consequence that the two segments are incongruent. Interior non-axial 180°-intercept transforms typically consist of intersecting segments, as in Fig. b, whereas the segments of exterior, non-axial 0°-intercept transforms do not intersect (Figs. c and d).

[In cases in which alternatives exist for the labelling or pairing of intercepts, the practice is followed uniformly of plotting transforms that take into account both alternatives. This practice is consistent with equational findings, since 0° and 180° transform eqs. are symmetrical functions that represent all alternatives (see page V-8). In the present instances, which concern the 0° transforms of Figs. c and d, the alternative has to do with the labelling of the intercepts. The symmetry of the plots about the line $x = y$ (dashed line in Fig. c) derives from the fact that the two intercepts (coincident radii) with the parabola are indistinguishable. The portions of the inner and outer segments to the right of (below) the line $x = y$ are for the greater of the two intercepts taken to be the y intercepts. If an arbitrary fixed labelling were to be employed, only the portions of the segments above or below the line $x = y$ would exist, though the transform eqs. represent both segments.]

Fig. 6-15. $ds/d\theta$ versus θ for 0° to 180°-circumpolar intercept transforms of the parabola about the traditional focus (defined by the transcendental subordinal rank eqs.). For the purpose of illustration, most of the figures have been reduced greatly (to bring the outer segments into the plot). With the 0° curve taken as the reference standard, the reduction factors for the 30°, 45°, 60°, 90°, 135°, and 180° curves are 100, 50, 25, 10, 2, and 1, respectively.
As a result of these reductions it is not evident from the figures for the 30°, 45°, 60°, and 90° curves that (a) in all cases the vertices of the inner segments are regions of a concavity (although a cusp is attained only for the 0° curve), or (b) except for the 0° curve, the vertices are not at the origin. In all cases the vertices of the outer segments lie on a line through the origin at an angle of $-\frac{1}{2}$ the angle of the transformation (i.e., $-\frac{1}{2}\alpha$), corresponding to a (CW) rotation of the line of symmetry of the curves through that angle. The arrows labelled x and y mark the locations of rectangular coordinate axes; x also is the 0° polar reference axis [$x = (ds/d\theta)\cos\theta$ and $y = (ds/d\theta)\sin\theta$].

The eq. of the intercept transform of a basis curve about a given pole at a given angle is an index of the type of symmetry possessed by the basis curve about that pole at that angle, and is the most precise specification of the corresponding *specific symmetry class* of the curve (see page V-5). The degree

0° & 180°-Circumpolar Intercept Transforms of The Parabola

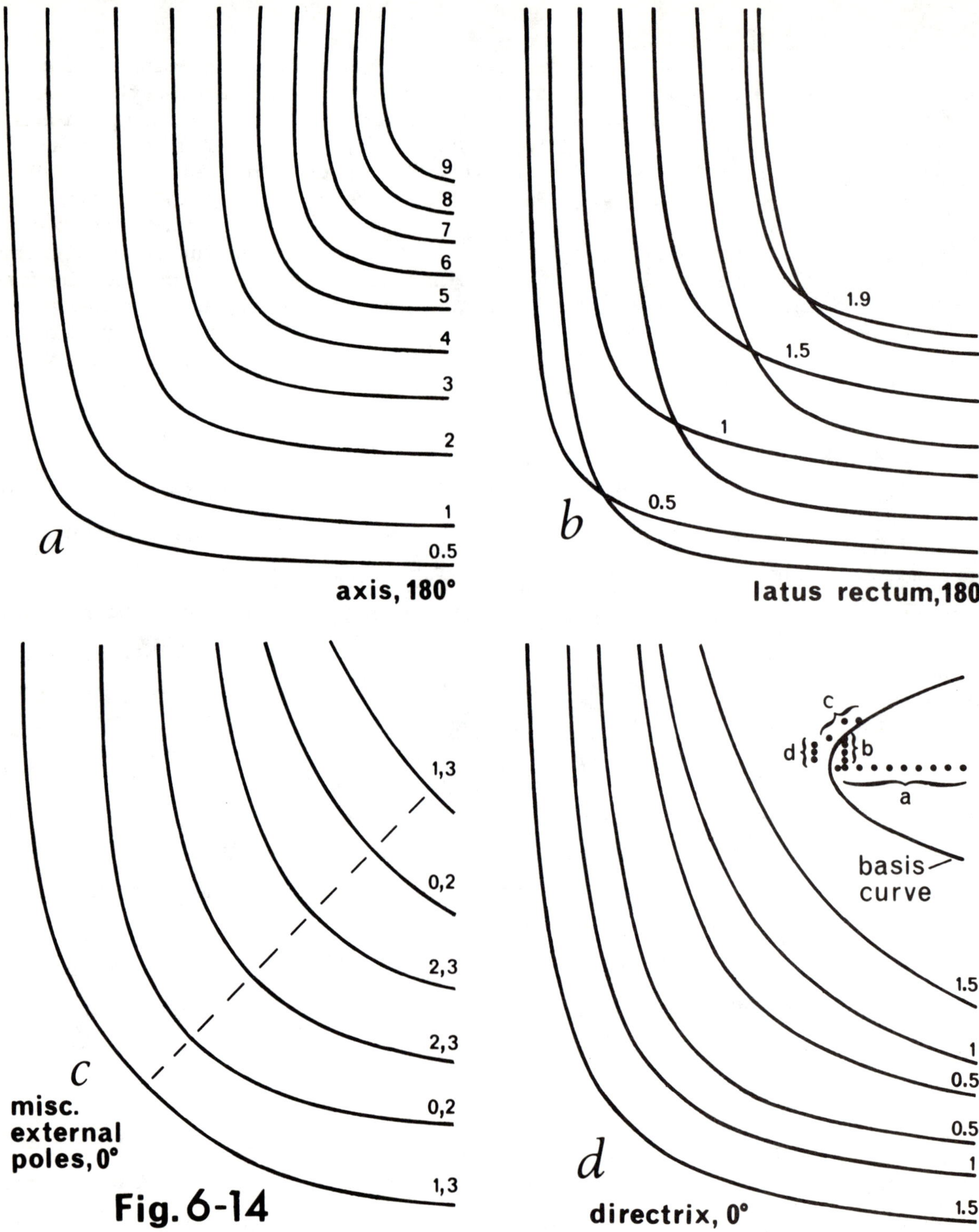

Fig. 6-14

of the eq. of the intercept transform provides a *focal* index for the type of symmetry.

However, the eq. of the intercept transform is not the most precise specification of the corresponding symmetry of the basis curve. Any given intercept transform eq. for a given angle describes the specific symmetry class of an infinite number of basis curves. A more precise specification of the circumpolar symmetry of a basis curve also employs a *subfocal* ranking. This ranking involves the use of $ds/d\theta$ of the transform curve, i.e., the instantaneous rate at which the transform curve is traced with respect to the angle θ of the x intercept of the basis curve. This rate differs for each basis curve. [These differences are illustrated for ellipse and hyperbola basis curves in Fig. 5-1, in which the values of θ for the x intercepts are given directly adjacent to the common circumpolar intercept transform of the basis central conics about their traditional foci.]

This subfocal ranking employs two indices--the *ordinal rank* and the *subordinal rank*. The *ordinal rank* for intercept transforms of a basis curve about a given pole at all angles is defined as the degree in x and y of the *intercept ordinal rank equation*, $(ds/d\theta)^2 = f(x,y)$, where x and y are the intercepts. The *subordinal rank* for a specific intercept transform, i.e., for a given angle of transformation, is defined as the degree in x (or y) of the *intercept ordinal rank equation*, $(ds/d\theta)^2 = F(x)$ or $G(y)$, where one of the intercepts is eliminated by substitution, using the intercept transform eq. itself, $g(x,y) = 0$, for that angle (see pages I-11 and V-12 to 20).

$(ds/d\theta)^2$ also can be expressed in transcendental form, $(ds/d\theta)^2 = h(\theta,\alpha)$, where α is the angle of the transformation. For transforms about the traditional focus of the parabola, the *intercept* and *transcendental ordinal rank equations* take the form of eqs. 1a and 1b.

(a) $a\mathring{s}^2 = (x^3+y^3)-a(x^2+y^2)$ *intercept ordinal rank equation* for transforms of the parabola about the traditional focus   (6-15-1)

(b) $\mathring{s}^2 = \dfrac{4a^2\sin^2\theta}{(\cos\theta-1)^4} + \dfrac{4a^2\sin^2(\theta+\alpha)}{[\cos(\theta+\alpha)-1]^4}$ *transcendental ordinal rank equation* (eq. 1a as a function of θ and α)

The present figure depicts the relationship between $ds/d\theta$ of the transform curve and θ of the x intercept of the basis curve for various angles of transformation, α, about the traditional focus. These plots serve to

ds/dθ Versus θ For 0° To 180°-Intercept Transforms of The Parabola About The Traditional Focus (Defined By Transcendental Subordinal Rank Equations)

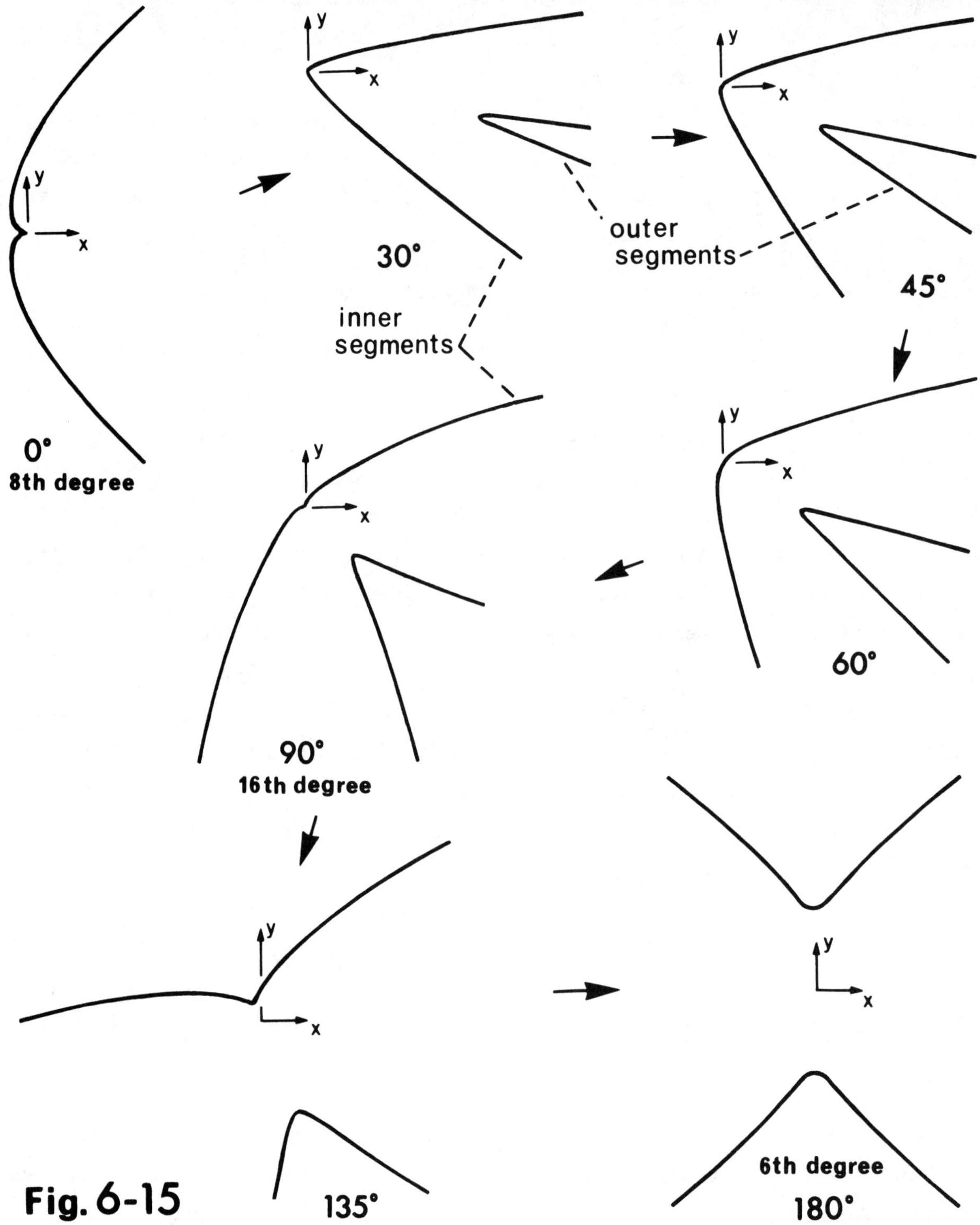

Fig. 6-15

illustrate the basis for a finer means of discrimination of the circumpolar symmetry of a curve about a given pole at a given angle than is provided by the mere eq. of the corresponding intercept transform (for plots of several *intercept subordinal rank equations*, i.e., plots of $ds/d\theta$ against the lengths of the x intercepts, see Fig. 7-19 for the $e = 0.745$ ellipse).

The most readily visualized case is that of the trivial linear 0° transform $x = y$ (the 0° curve). For this case eq. 1b becomes $(ds/d\theta)^2 = 8a^2\sin^2\theta/(\cos\theta-1)^4$. Since $(ds/d\theta)^2 = (dx/d\theta)^2 + (dy/d\theta)^2$, and $x = y$, $ds/d\theta$ is simply $\sqrt{2}\,dx/d\theta$. In other words, the 0° curve expresses changes in the length of the x intercept alone with changes in its angle, θ, to the x-axis. Except for being magnified by a factor of $\sqrt{2}$, then, the 0° curve is simply a polar plot of $dr/d\theta$ versus θ for the polar coordinate eq. of the parabola, $r = 2a/(1-\cos\theta)$, cast about the traditional focus. In keeping with this correspondence, $ds/d\theta$ of the 0° curve is 0 at 180°, since this is the angle of the vertex (the derivative, $dr/d\theta$, of the radius vector from the focus to the curve with respect to its angle with the x-axis is 0 at the vertex).

Similarly, $ds/d\theta$ ($\sqrt{2}\,dr/d\theta$) attains it greatest value in the neighborhood of $\theta = 0^\circ$, since r becomes infinitely great at 0°. Relative changes in the lengths of the radius vectors from the focus for angles between 0° and 180° follow the course shown in the figure, from which it is clear that exceptional changes do not occur at the LR vertices (the 90° and 270° points). As the axial vertex is approached, the magnitude of the change in length decreases along the arms of a cusp, attaining 0 at the vertex.

The 90° curve (as well as the other curves) can be constructed from the 0° curve by summing the contributions of the changes in two intercepts that are at 90° to one another. Thus, if the value of $ds/d\theta$ of the 0° plot is divided by $\sqrt{2}$ it becomes a plot of $dr/d\theta$ versus θ. $(ds/d\theta)^2$ for a given value of θ in all other plots now becomes merely the sum of $(dr/d\theta)^2$ for a radius vector at the angle θ of the 0° plot and $(dr/d\theta)^2$ for a radius vector at the angle $(\theta+\alpha)$ of the 0° plot.

For example, $(ds/d\theta)^2$ at 0° for the inner segment of the 90° curve ($\alpha = 90^\circ$) is infinite because it represents the sum of $(ds/d\theta)^2/2$ at 0° (∞) and $(ds/d\theta)^2/2$ at 90° ($4a^2$) of the 0° curve, the latter of which is small but the former of which is infinite. At $\theta = 90^\circ$ of the 90° curve, $(ds/d\theta)^2$ is the sum of $(ds/d\theta)^2/2$ at 90° ($4a^2$) and $(ds/d\theta)^2/2$ at 180° (0) of the 0° curve, the former of which is small and the latter of which is 0. Accordingly, $(ds/d\theta)^2$ at 90° for the 90° curve is small ($4a^2$).

At $\theta = 180^\circ$ of the 90° curve, $(ds/d\theta)^2$ is the sum of $(ds/d\theta)^2/2$ at 180° (0) plus $(ds/d\theta)^2/2$ at 270° ($4a^2$) of the 0° curve. This sum ($4a^2$) also is small because it is equal to the previous sum for $(ds/d\theta)^2/2$ at 90° and $(ds/d\theta)^2/2$ at 180° [since $(ds/d\theta)^2/2$ at 90° equals $(ds/d\theta)^2/2$ at 180°]. At $\theta = 270^\circ$ of the 90° curve, $(ds/d\theta)^2$ is the sum of $(ds/d\theta)^2/2$ at 270° ($4a^2$) and $(ds/d\theta)^2/2$ at 360° (∞) of the 0° curve, which, again, is infinite, since $(ds/d\theta)^2/2$ at 360° of the 0° curve is infinite. This completes the plot for the inner segment of the 90° curve. The plots of the *transcendental subordinal rank equations* of Figs. 7-16, 17, and 18 can be derived in similar fashion from the 0° curves.

The outer segment of the 90° curve is traced for a progression of θ for the x intercept from 270° to 360° of the 0° curve, and for the y intercept from 360° to 450° (90°) of the 0° curve. At 275° of the 90° curve, $(ds/d\theta)^2$ is the sum of $(ds/d\theta)^2/2$ at 275° ($5.717a^2$) and $(ds/d\theta)^2/2$ at 365° ($14.491\text{x}10^7a^2$) of the 0° curve. The former is small but the latter is very large, so that the sum is very large.

The vertex of the outer segment is attained for $\theta = 315^\circ$, at which point $(ds/d\theta)^2$ of the 90° curve is the sum of $(ds/d\theta)^2/2$ at 315° ($271.76a^2$) and $(ds/d\theta)^2/2$ at 45° ($271.76a^2$) of the 0° curve, both of the latter of which values are large but equal. Accordingly, the value for $(ds/d\theta)^2$ at the vertex of the 90° curve ($543.53a^2$) is large, though it is the smallest of the values for the outer segment [bear in mind that, though values of $(ds/d\theta)^2$ are referred to in the above descriptions, it is the values of $ds/d\theta$ that are plotted in the illustrations]. Values for the curves for other values of α are derived in similar fashion.

Although the *transcendental ordinal* and *subordinal rank equations* of intercept transforms of the parabola do not provide as high a degree of discrimination of its symmetry as the *intercept ordinal* and *subordinal rank equations*, i.e., the eqs. in $(ds/d\theta)^2 = f(x,y)$ and $(ds/d\theta)^2 = F(x)$ or $G(y)$, they are more easily plotted, the relationships that they convey are more readily visualized, and the types of reflective symmetry that they possess give information about the location of the pole of transformation that is not explicit in the intercept subordinal rank curves.

For example, if the pole of transformation lies at a true center, all of the transcendental subordinal rank curves have two lines if symmetry (Fig. 7-17); if it lies on a line of symmetry, all of them have only one line of symmetry (this figure and Fig. 7-16); while if it does not lie on a line of symmetry, none of them has a line of symmetry (Fig. 7-18)--though the 180° transform may be symmetrical about the origin (Fig. 7-18, the 180° curve). In the cases of intercept subordinal rank eqs., this information is not explicit in the symmetry of the corresponding curves but is implicit in their recurrence characteristics (see Fig. 7-19). [For example, if the pole of the transformation lies on two lines of symmetry, all of the intercept subordinal rank curves are recurrent and the 0°, 180°, and 90° curves are traced 4 times.]

On the other hand, the intercept ordinal and subordinal rank eqs. provide a greater degree of discrimination of the symmetry of the basis curves at the subfocal level. For example, in the cases of the parabola and central conics (excluding the circle and equilateral hyperbola), the eqs. of transforms about the traditional foci have the same degree for any given angle of transformation. The symmetry of the parabola about the traditional focus, however, is found to be greater than that of central conics about a traditional focus when the analysis is pursued to the subfocal level, namely, to the rectangular degrees of the intercept ordinal rank eqs. and the 0°, 180°, and 90° intercept subordinal rank eqs.

The former eqs. have a rectangular degree of 3 for the parabola but a degree of 4 for central conics, while the latter eqs. have degrees of 3, 6, and 16 for the parabola but 4, 8, and ≤ 24 for central conics (see Table V-1 and Appendix III, and Fig. 7-19). In all 4 cases, the symmetry of the parabola about the traditional focus is greater than that of central conics (since the rectangular degrees of the parabola's intercept ordinal and subordinal rank eqs. are lower).

Contrariwise, as assessed by the rectangular degree of the transcendental ordinal and subordinal rank eqs. (when transformed to rectangular coordinates), only the 180° (equilateral hyperbolic) symmetry of the parabola about the traditional focus is greater than the corresponding symmetry of central conics (because only for 180° is the rectangular degree of the transcendental subordinal rank eq. of the parabola less than the degree for central conics). Thus, the degrees for the rectangular equivalents of the transcendental ordinal rank eqs. are 16 for both the parabola and central conics, while for the subordinal rank eqs. they are 8, 6, and 16 for the parabola and 8, 8, and 16 for central conics (for transformation angles of 0°, 180°, and 90°, respectively; see Fig. 7-16).

Referring to the plots, the degree of the 0° curve (eq. 2) is 8, and the degree of the 180° curve (eq. 3) is 6, compared to a degree of 16 for the 90° curve (see also the more extensive treatment in the legend of Fig. 7-16).

transcendental subordinal rank eqs. for intercept transforms of the parabola about a traditional focus expressed in rectangular coordinates

$[x-(x^2+y^2)^{\frac{1}{2}}]^4 = 8a^2y^2$ eq. for the 0° transform (when the radical is eliminated, the degree is 8) (6-15-2)

$y^6 = 16a^2(2x^2+y^2)^2-8a^2y^4$ eq. for the 180° transform (6-15-3)

CHAPTER VII

Fig. 7-1. 180°-circumpolar intercept transforms of the $e = 0.745$ ellipse about interior poles on the lines of symmetry for two positive or two negative intercepts (transforms for two negative intercepts are congruent with those illustrated for two positive intercepts but reflected in the origin). All plots of this figure are to the same scale. Within each of the two sets of transforms (for poles on the major and minor axes), the individual plots are in their proper relative positions. However, the locations of the composite sets of transforms have been shifted for the purpose of illustration (the lines of symmetry of these curves should be coincident with the $x = y$ bisector). Corresponding pole locations (in units of the length of the semi-major axis, a) appear adjacent to the transforms.

Each individual plot represents the corresponding complete intercept transform curve in the 1st and 3rd quadrants (not merely a portion of it) although, of course, the plots cover only a portion of the loci represented by the corresponding intercept transform eqs. The locations of the poles of transformation on the lines of symmetry of the basis curve are shown in the illustration at the upper right. The first and last poles along each axis are labelled. The remaining poles can be matched to the transform curves by virtue of the fact that both the poles and curves are in sequence.

Transforms about poles on the minor axis have a quite different appearance from those for poles on the major axis; whereas the latter nestle, many of the former intersect. The transform about the pole at $k = 7a/12$ is an example of a transform for which there is no intersection with any of the transforms for $k \leq \frac{1}{2}a$. Within each set of transforms, the termini are collinear along a line normal to the 45° bisector, and all of the transforms are symmetric functions. [This is true of all 180° transforms and all transforms about poles on lines of symmetry (see page V-8).]

The degrees of 180° transforms about unexceptional interior poles on lines of symmetry is 4 (eqs. VII-7a and VII-9a). Reduction occurs only at the center and the traditional foci (see Table VII-1, page VII-19). No non-degenerate, non-trivial 180° transform exists for the a and b vertices or any other incident pole. However, these vertices have point-focal rank for a 90° angle of transformation.

180°-Circumpolar Intercept Transforms of The e=0.745 Ellipse About Interior Poles On The Lines of Symmetry

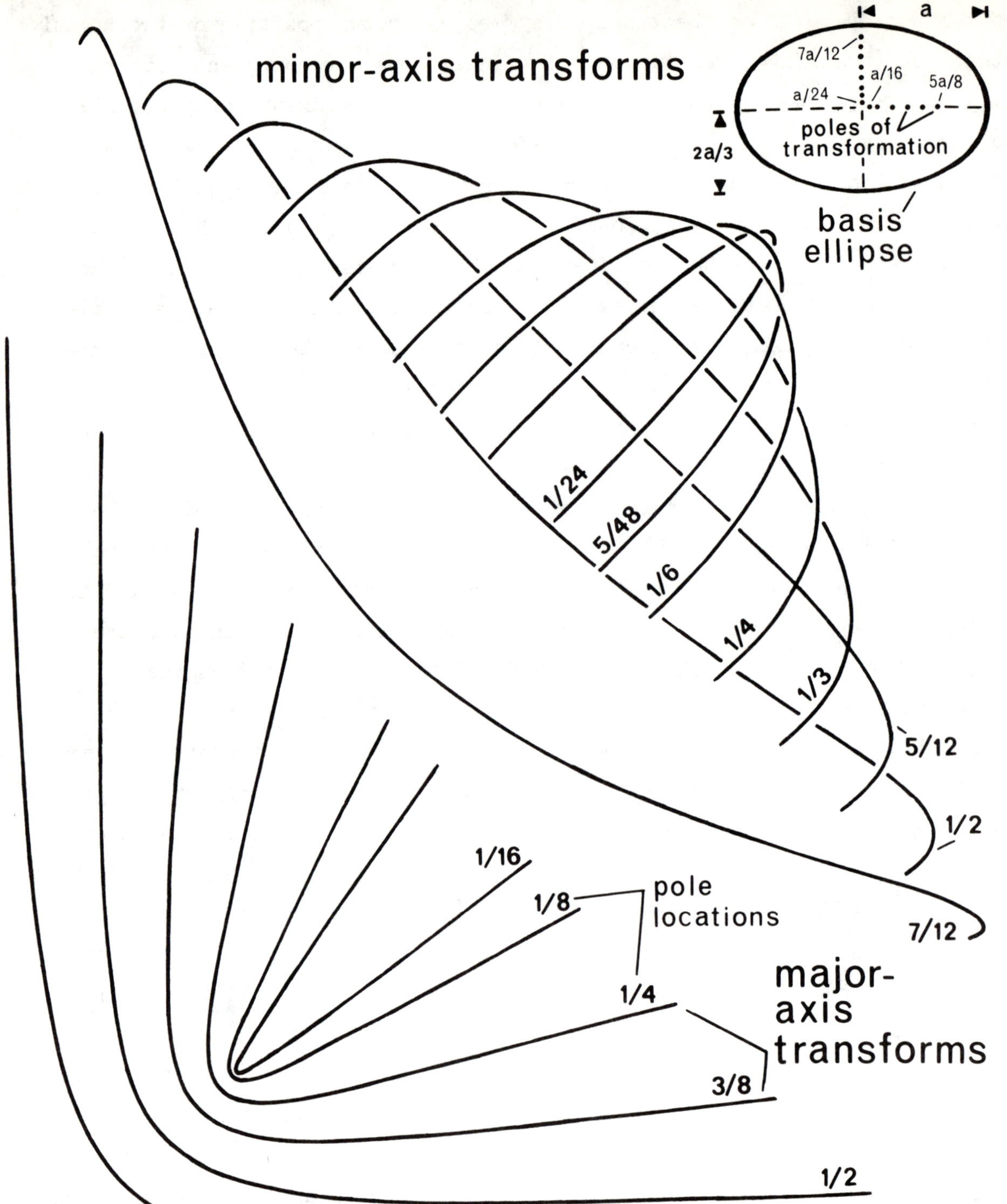

Fig. 7-1

Fig. 7-2. 180°-circumpolar intercept transforms of the $e = 0.745$ ellipse about interior poles for two positive or two negative intercepts. This figure complements Fig. 7-1 by presenting transforms for two interior non-axial linear sequences of pole progression; one sequence progresses from the traditional focus (transform at bottom left) along the LR to a pole at $k = 0.4a$ (5th transform from the bottom) and thence toward the minor axis along a chord at $k = 0.4a$, terminating at a pole on the minor axis (transform at the upper right). The pole locations relative to the basis ellipse are illustrated at the upper left of the figure. Corresponding pole locations (in units of the length of the semi-major axis, a) are labelled adjacent to the transform curves. All transform curves are to the same scale but their orientations and locations have been altered to suit the purpose of illustration.

All interior 180° transforms about poles not lying on a line of symmetry of the basis ellipse are closed curves that possess a DP. The closer a pole of transformation to a line of symmetry, the more flattened the transform curve becomes, until, when a pole becomes incident upon a line of symmetry, the transform curve consists of an open recurrent arc that is traced twice for a 360° sweep of the two radius vectors. The degree of 180° transforms of ellipses (eq. VII-6a) about unexceptional interior poles is 8 (see Table VII-1, page VII-19). Reduction occurs only for poles incident upon a line of symmetry.

Fig. 7-3. 0°-circumpolar intercept transforms of the $e = 0.745$ ellipse about exterior poles for two positive or two negative intercepts. This figure complements Figs. 7-1 and 7-2 in representing the corresponding transforms for exterior poles, for which it is the 0° rather than the 180° transformation that is of interest. [If all combinations of intercepts are taken into account, both 0° and 180° transforms exist about all non-incident poles; see Chapter VII, *Negative Intercepts, Trivial Transforms, The Degree of Transforms, and The Duality of 0° and 180° Transforms.*] The size, location, and orientation of the transform curves have been altered to suit the purpose of illustration. Corresponding pole locations (in units of the length of the semi-major axis, a) are labelled adjacent to the transform curves. The locations of the poles of transformation relative to the basis ellipse are shown in the illustration at the center.

180°-Circumpolar Intercept Transform of The e=0.745 Ellipse About Interior Poles

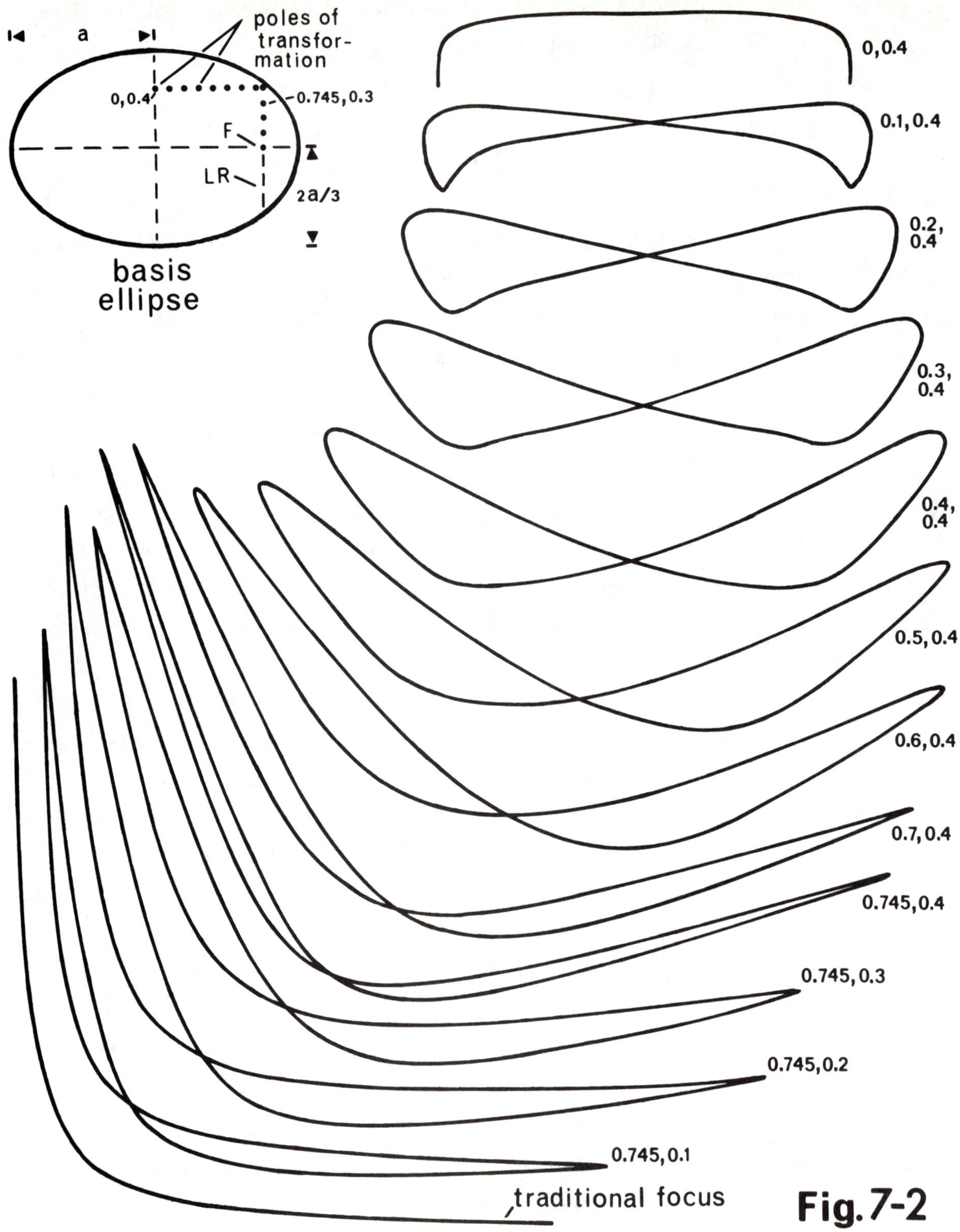

Fig. 7-2

Transforms about exterior poles that are incident upon 3 intersecting lines are illustrated. One of these lines is parallel to the minor axis (transforms at upper left), one is parallel to the major axis (transforms at upper right), and the 3rd is inclined at 45° to the other two (transforms at bottom). The lines intersect at the pole (4a/3,a).

For transforms about each set of collinear poles, the outermost curve is the transform about the common pole of intersection of the lines [at (4a/3,a)] and the inner curves are transforms about poles progressing in sequence along the lines referred to above. As is the case for 180° transforms about unexceptional interior poles, the degree of 0° transforms about unexceptional exterior poles (eq. VII-6a) is 8 (see Table VII-1, page VII-19). Reduction to 4 occurs only for exterior poles incident upon a line of symmetry.

Fig. 7-4. 0° and 180°-circumpolar intercept transforms of the equilateral hyperbola (inter- and intra-arm, two positive or two negative intercepts, all segments). The illustrations of this figure are roughly the equivalent for the equilateral hyperbola of those of Figs. 7-1,2, and 3 for the $e = 0.745$ ellipse, inasmuch as they comprise both the 0° and 180° transforms about poles interior to one arm and between the two arms, including points on the lines of symmetry--in collinear sets. The curves within each set are to the same scale and in correct relative orientation, but the locations of the curves in the LR and conjugate-axis sets have been altered and the sequence for the LR set has been reversed for the purpose of illustration.

The locations of the poles of transformation relative to the basis hyperbola are shown in the illustration above and to the right of the transverse-axis transforms. Corresponding pole locations (in units of the length of the semi-transverse axis, a) are labelled adjacent to the transform curves. One of the 4 collinear pole sequences is incident upon the transverse axis (within an arm), another on the conjugate axis, a 3rd on the LR, and the 4th on an asymptote. Transform curves that terminate in arrowheads should extend to infinity but have been terminated because their extensions too closely overlap other transforms. On the other hand, the proximal terminations of the 0° inter-arm segments of transforms about poles on the transverse axis are true terminations of the transform curves (the loci represented by the eqs. of the transforms, of course, do not terminate).

0°-Circumpolar Intercept Transforms of The e=0.745 Ellipse About Exterior Poles

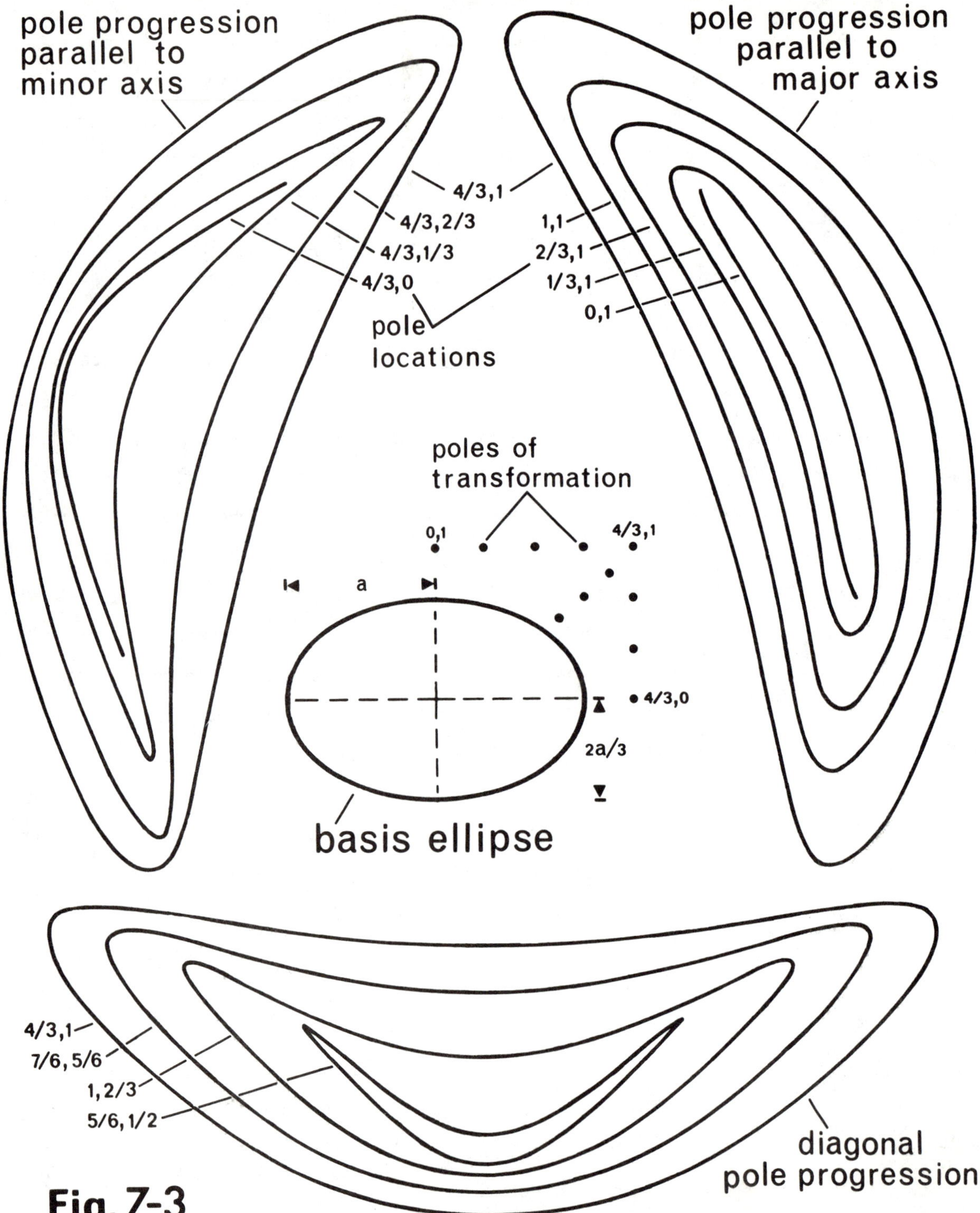

Fig. 7-3

Both inter and intra-arm transform curves exist for all of the illustrated pole locations, but whether one or the other type of curve represents the 0° transform or the 180° transform depends upon the specific pole location (inasmuch as the transform curves are only for either two positive or two negative intercepts). The principal basis for differences in the nature of hyperbola and ellipse transforms is that hyperbolas consist of two asymptotic arms whereas ellipses are closed curves without real asymptotes.

The most notable consequences of this difference are that: (a) circumpolar intercept transforms of ellipses generally are closed curves, whereas those of hyperbolas are open; and (b) both 0° and 180° transforms of hyperbolas generally exist about non-incident poles, whereas for the ellipse only 0° transforms exist about exterior poles and only 180° transforms about interior poles. With the explicit recognition of these basic differences at this juncture, systematic comparisons between the transforms of the two species of central conics are not carried out in the legends of the following figures.

The 180° transforms about poles on the LR (upper left), like those about non-axial interior poles of an ellipse (Fig. 7-2), possess a DP. The corresponding 0° inter-arm transforms consist of two hyperbola-like arms that have no line of symmetry individually, and that become increasingly narrow with approach of the pole of transformation to the transverse axis. At the transverse axis (the position of the traditional focus) the segments "close" to recurrent traces that are very similar to those for the axial pole at $h = 3a/2$ of the transverse-axis plots (lower left). Since the LR is not a focal locus, the degree of the 0° and 180° transforms (eq. VII-45) about unexceptional poles incident upon it is 8, the same as for other unexceptional points in the plane (Table VII-1, page VII-19).

The 0° intercept transforms about poles on the asymptotes (non-axial) are open curves that superficially resemble an arm of a hyperbola. The 180° inter-arm transforms consist of two hyperbola-like arms that have no line of symmetry individually. As the pole of transformation approaches the center, these arms approach one another and become more narrow. At the center they become coincident to form a recurrent linear transform, $x = y$, extending from the point (a,a) to infinity. The degree of 0° and 180° transforms about unexceptional poles incident upon the asymptotes (eq. VII-10) is 6, the asymptotes being focal loci (Table VII-1).

0°&180°-Circumpolar Intercept Transforms of The Equilateral Hyperbola (Inter & Intra-Arm, Positive Intercepts, All Segments)

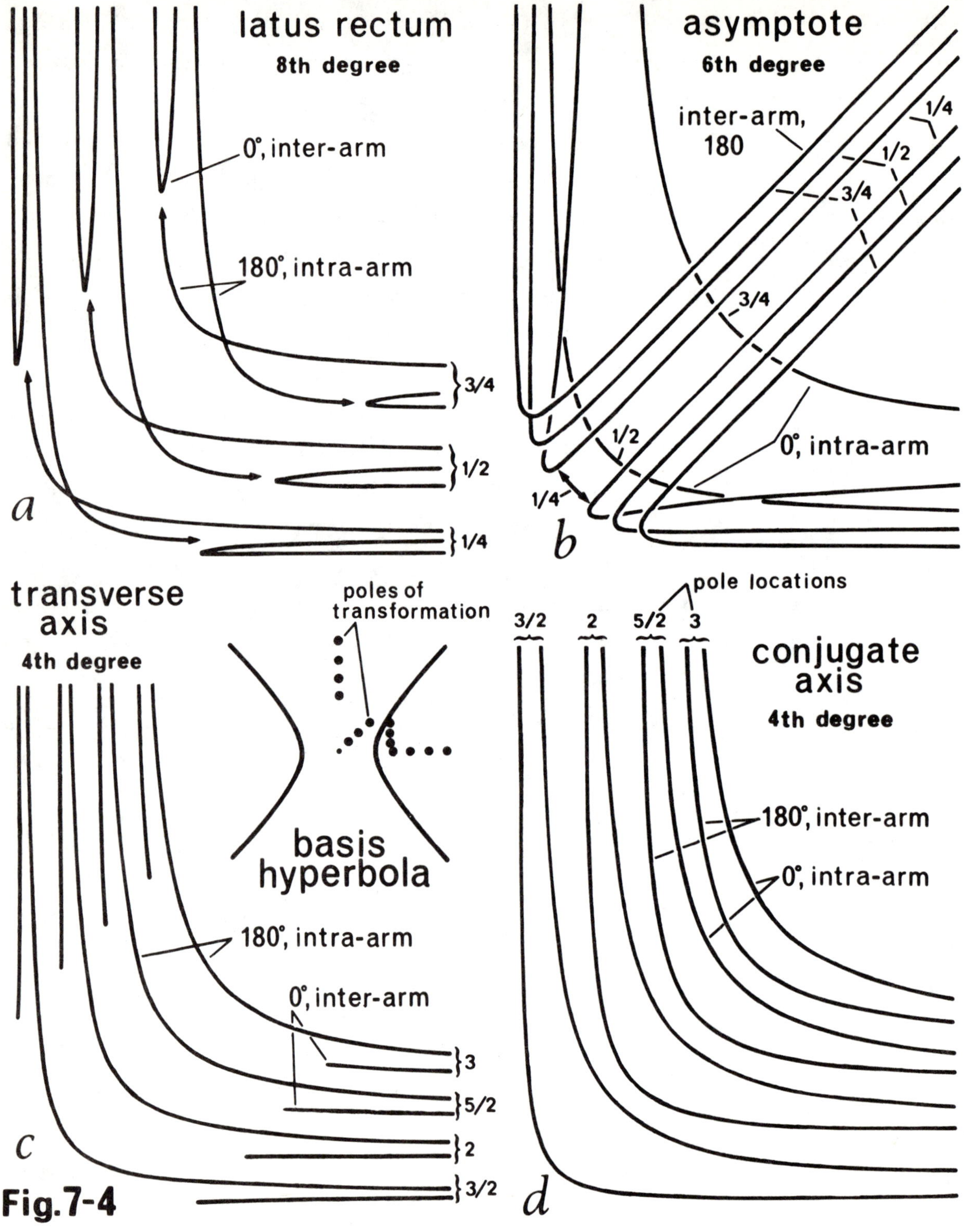

Fig.7-4

180° transforms about poles on the transverse axis (lower left) represent intercepts on the same arm--the basis arm--and superficially resemble an arm of a hyperbola. At the traditional focus, the transform is an arm of the equilateral hyperbola (the harmonic-mean relationship; page VII-4). The 0° transforms consist of terminated segments that represent intercepts on both arms. [The intercepts at the termini are for $\theta = 0^\circ$ for the left focus and $\theta = 180^\circ$ for the right focus.] The degree of 0° and 180° transforms (eq. VII-7b) about unexceptional poles on the transverse axis is 4, compared to 8 for unexceptional points in the plane (Table VII-1).

All 0° and 180° transforms about unexceptional poles on the conjugate axis (lower right) superficially resemble an arm of a hyperbola. The 0° transforms are for intercepts on a single arm, while the 180° ones are for intercepts on both arms. Except for the transforms about a traditional focus (Fig. 7-6), the degree of all transforms about points on the conjugate axis (eq. VII-9b) is 4. Hyperbolas have 5 curvilinear focal loci--the two lines of symmetry, the two asymptotes, and the curve itself (Tables VII-1 and X-2).

Fig. 7-5. Circumpolar symmetry of the $e = 0.745$ ellipse about a traditional focus for two positive or two negative intercepts. The transformation angles for the 10 transforms of this figure range from 0° to 180°. The transforms have been shifted, reduced, and reoriented for the purpose of illustration. When all curves are to scale and at their proper locations relative to the coordinate axes, they form an extensively intersecting and overlapping network. The trivial 0° transform consists of a line segment, $x = y$, while the 180° transform (eq. VII-8a') consists of a segment of an arm of the equilateral hyperbola. At an angle of 90° (eq. VII-43a'), and other intermediate angles of transformation, the curves are of 4th degree (Table VII-1).

This figure can be compared most aptly with Figs. 6-10 and 7-6 for the circumpolar symmetry of the parabola and the equilateral hyperbola about their traditional foci. The basis for regarding an ensemble of intercept transforms about a given pole for a wide range of transformation angles as a characterization of the *circumpolar symmetry* of a curve about that pole is discussed in the legend of Fig. 6-10.

Circumpolar Symmetry of The e=0.745 Ellipse About A Traditional Focus (Positive Intercepts)

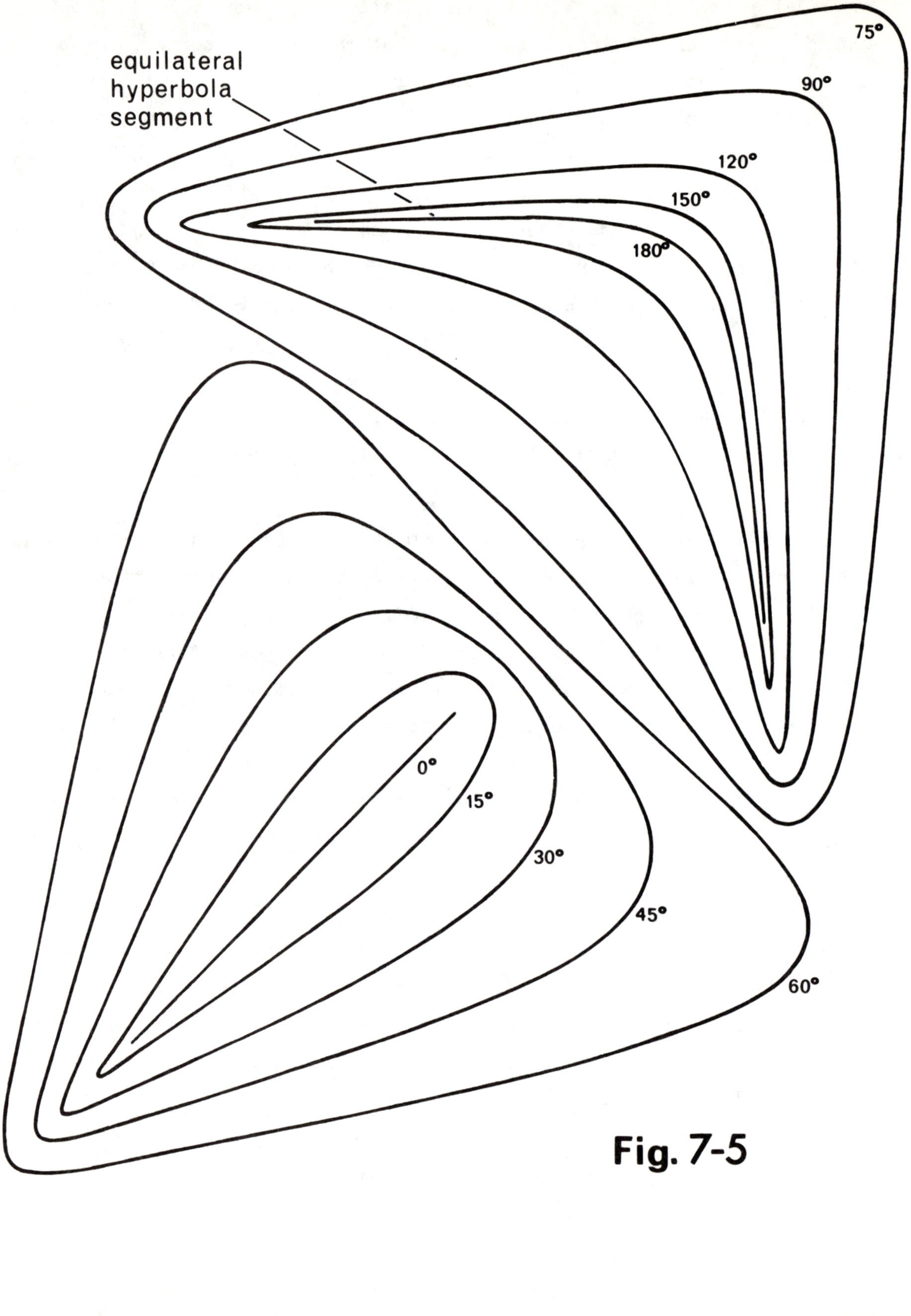

Fig. 7-5

Fig. 7-6. Circumpolar symmetry of the equilateral hyperbola about a traditional focus for two positive or two negative intercepts (intra-arm and inner segments). Twelve intercept transforms of the equilateral hyperbola for transformation angles of 0° to 180° are illustrated. All of the curves are to the same scale and in proper orientation but their positions have been shifted for the purpose of illustration (because of extensive overlapping and intersection). The trivial 0° transform consists of a segment of the line, $x = y$, extending to infinity, while the 180° transform consists of an arm of the equilateral hyperbola (the harmonic-mean relationship). For a transformation angle of 90° (eqs. VII-26b and 43c') the degree of the transform is 3 (an example of *eccentricity-correlated rank*; see pages VII-8 to 12 and Figs. 7-8,9,10, and 12). For all other angles the transforms are of 4th degree (eq. VII-23 with $a = b$).

This figure may be compared most aptly with the corresponding Figs. 6-10 for the parabola and 7-5 for the $e = 0.745$ ellipse. It differs from Fig. 7-5 for the ellipse in that the transforms are not all complete, inasmuch as the outer segments--that exist for transformation angles between 90° and 180°--are not illustrated (in the case of the 180° transform the curve is recurrent, being traced twice). The basis for regarding an ensemble of intercept transforms of a curve about a given pole for a wide range of transformation angles as a characterization of the *circumpolar symmetry* of the curve about the pole is discussed in the legend of Fig. 6-10.

Fig. 7-7. 90°-circumpolar intercept transforms of the $e = 0.745$ ellipse about interior and incident poles on the lines of symmetry for two positive or two negative intercepts. The 11 90° transforms illustrated in this figure have been shifted and reduced for the purpose of illustration. The positions of the poles of transformation (in units of the length of the semi-major axis, a) on the lines of symmetry of the basis ellipse are shown in the illustration at the lower left. The ordinates of the poles on the minor axis (Fig. a) and abscissae of the poles on the major axis (Fig. b) are given adjacent to the corresponding transforms.

The transform about the center is recurrent and traced 4 times. For a very small departure of the pole of transformation from the center the traces become slightly different, thereby departing from coincidence. An example for

Circumpolar Symmetry of The Equilateral Hyperbola About A Traditional Focus (Intra-Arm, Inner Segments)

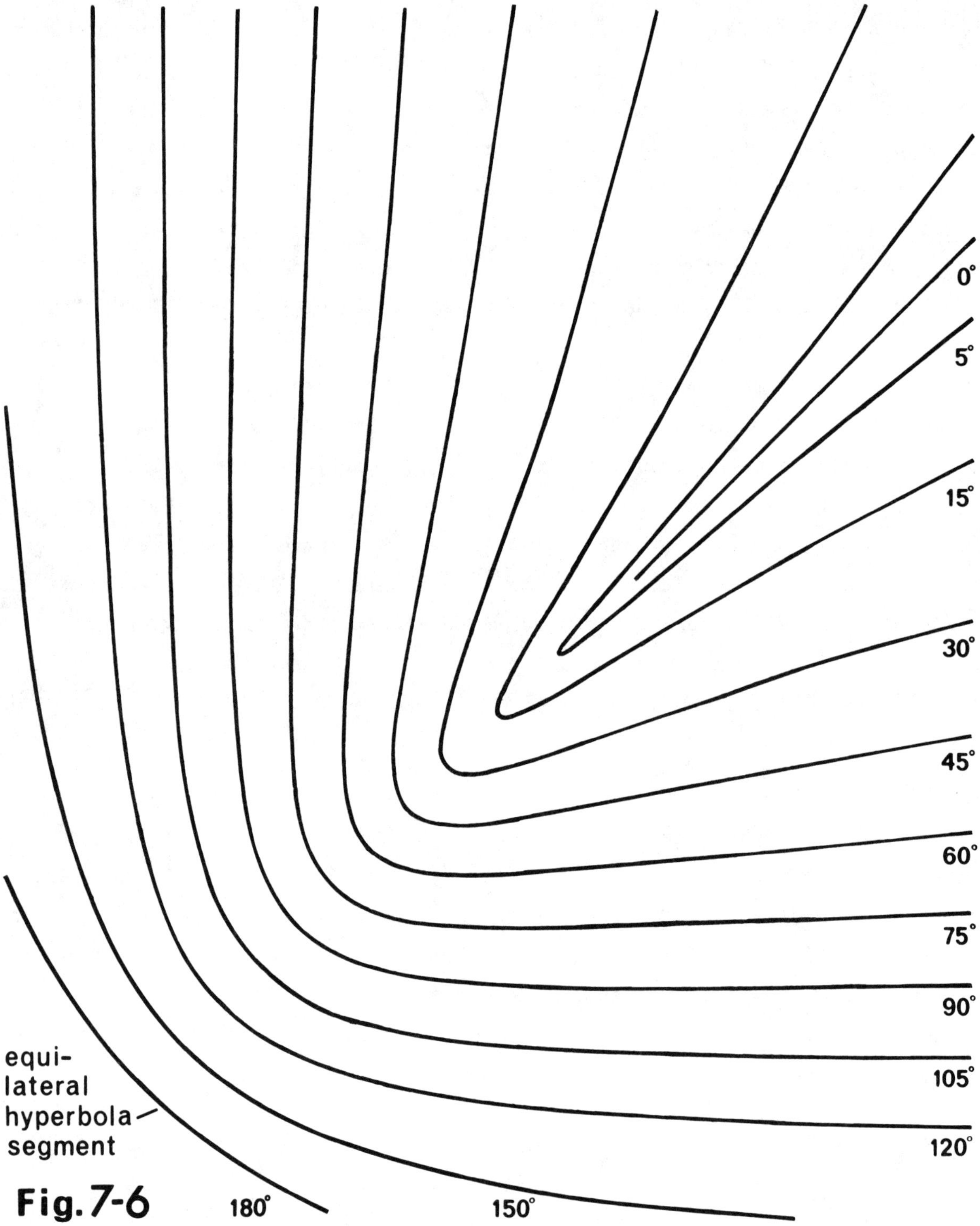

Fig. 7-6

a pole at (0,a/24) is illustrated in Fig. a. As the axial pole of transformation departs increasingly from the center, the segment of the curve on one side of the DP progressively decreases in size while the segment on the other side progressively increases in size. With increasing departure from the center the smaller looped segment is lost, leaving first a cusp, then an "exceptional" region of characteristic appearance (Figs. 7-7a-½a,7-7b-½a, 7-9-1.375a, and 7-10-75°) possessing extremals and inflection points, and then a region of regular convex (outward) appearance (Figs. 7-7a-2a/3 and 7-7b-3a/4).

Transforms about the a and b vertices are illustrated with degenerate axial segments, representing angles, θ, for which only one non-zero intercept exists (except at the origin, which represents a range of values for which both intercepts are 0). In a full 4-quadrant plot (for all combinations of positive and negative intercepts), the degenerate axial segments are lost, the transforms take on the form of Figs. 1-2g (a vertex) and 1-2i (b vertex) (see *Symmetry, An Analytical Treatment*), and there are only 4 values of θ for which one of the intercepts is 0 (namely, the values for the 4 points at which the transform curve crosses the coordinate axes). The degree of 90°-intercept transforms of the ellipse about unexceptional axial poles is 16 (eq. VII-27, $\alpha = 90^{\circ}$). Reduction occurs only at the center, the traditional foci, and the vertices.

Fig. 7-8. 90°-circumpolar axial intercept transforms of the equilateral hyperbola for two positive or two negative intercepts (intra-arm). The 12 90° transforms illustrated in this figure are to scale and properly oriented but their positions have been shifted for the purpose of illustration. These transforms are homologous to those of the ellipse of Fig. 7-7. But the homology is not complete because: (a) the equilateral hyperbola is a highly exceptional subspecies; (b) the angle (90°) of transformation is the same as the angle between the asymptotes, as a consequence of which no outer segments of the transform curves exist and all axial transforms have *eccentricity-correlated* rank (see pages VII-8 to 12 and Figs. 7-9,10, and 12) [outer segments exist only when the degree of the transform (for $0^{\circ} < \alpha < 180^{\circ}$) is greater than the angle between the asymptotes]; and (c) only the intra-arm segments are represented.

It is evident from the figure that none of the 11 90°-interior axial intra-

90°-Circumpolar Intercept Transforms of The e = 0.745 Ellipse About Interior And Incident Poles On The Lines of Symmetry (Pos. Intercepts)

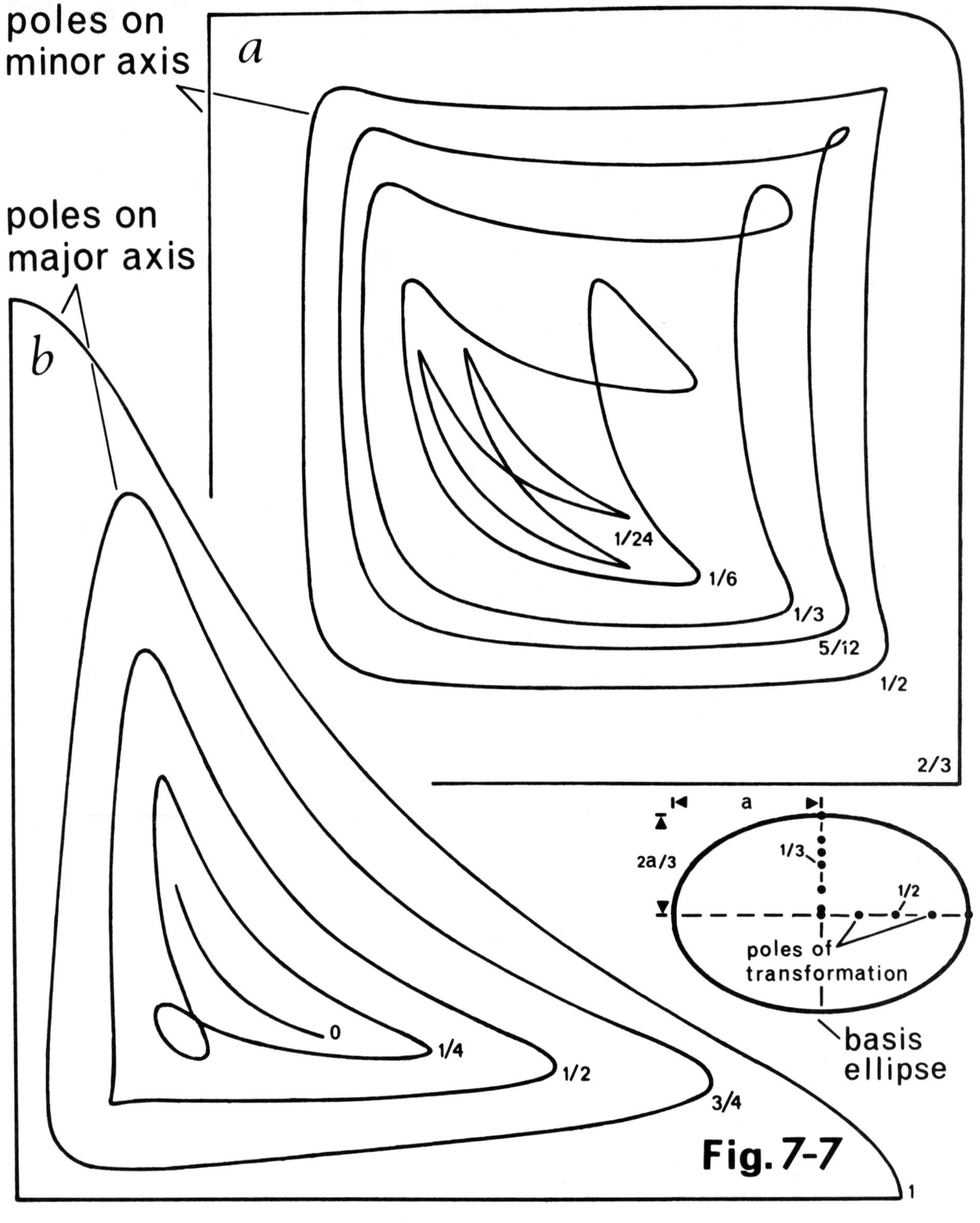

Fig. 7-7

arm transforms of the equilateral hyperbola is looped. It already has been noted that no intra-arm transform about a traditional focus of the equilateral hyperbola is looped at any angle (Fig. 7-6). However, this is not a characteristic of all axial poles of the equilateral hyperbola, as shown in Fig. 7-10. The degree of 90° transforms of the equilateral hyperbola about unexceptional axial poles is 12 (for $\alpha = 90^\circ$, $a = b$, and upper alternate signs, eq. VII-27 reduces to 12th degree). Reduction occurs at the vertices, the traditional foci, and the center (Table VII-1) but the 2nd-degree transform about the latter pole (eq. VII-44c') has only imaginary solutions.

Fig. 7-9. 90°-circumpolar axial intercept transforms of the $e = 1.118$ hyperbola for two positive or two negative intercepts (intra-arm, inner segments). This figure is the counterpart of Fig. 7-8 but for a hyperbola of lesser eccentricity, with asymptotes at an angle of 53.13°. As with Fig. 7-8, the 8 intercept transforms of this figure are to scale and properly oriented but their positions have been shifted for the purpose of illustration. The axial locations of the poles of transformation relative to an arm of the basis hyperbola are illustrated at the upper right.

In the cases of ellipses, the DP of the transform curve is lost as the pole of transformation recedes from the center. In the present case, the DP is lost with increasing approximation to the center. On the other hand, the characteristics are the same for the transforms of both curves if the axial vertices are taken as reference poles. In both cases the DP is lost with increasing approximation to the vertices.

The 90° transforms of this curve differ from those of the equilateral hyperbola in that some of them are looped (possess a DP), whereas none is looped for the latter curve. In fact, it appears that no axial 90° transform is looped for hyperbolas for which the angle of the asymptotes is 90° or greater. However, the question of whether axial transforms are looped or not is not dependent solely upon whether the angle of the asymptotes is less than the angle of the transform. It also depends upon the location of the pole.

For example, no transform about a traditional focus of the equilateral hyperbola is looped for any angle of transformation (Fig. 7-6), whereas for the pole at 3a the 30°, 45°, and 60° intercept transforms are looped (Fig. 7-10). The loop (DP) is acquired for some angle less than 75°; for 60° the loop

90°-Circumpolar Axial Intercept Transforms of The Equilateral Hyperbola (Intra-Arm)

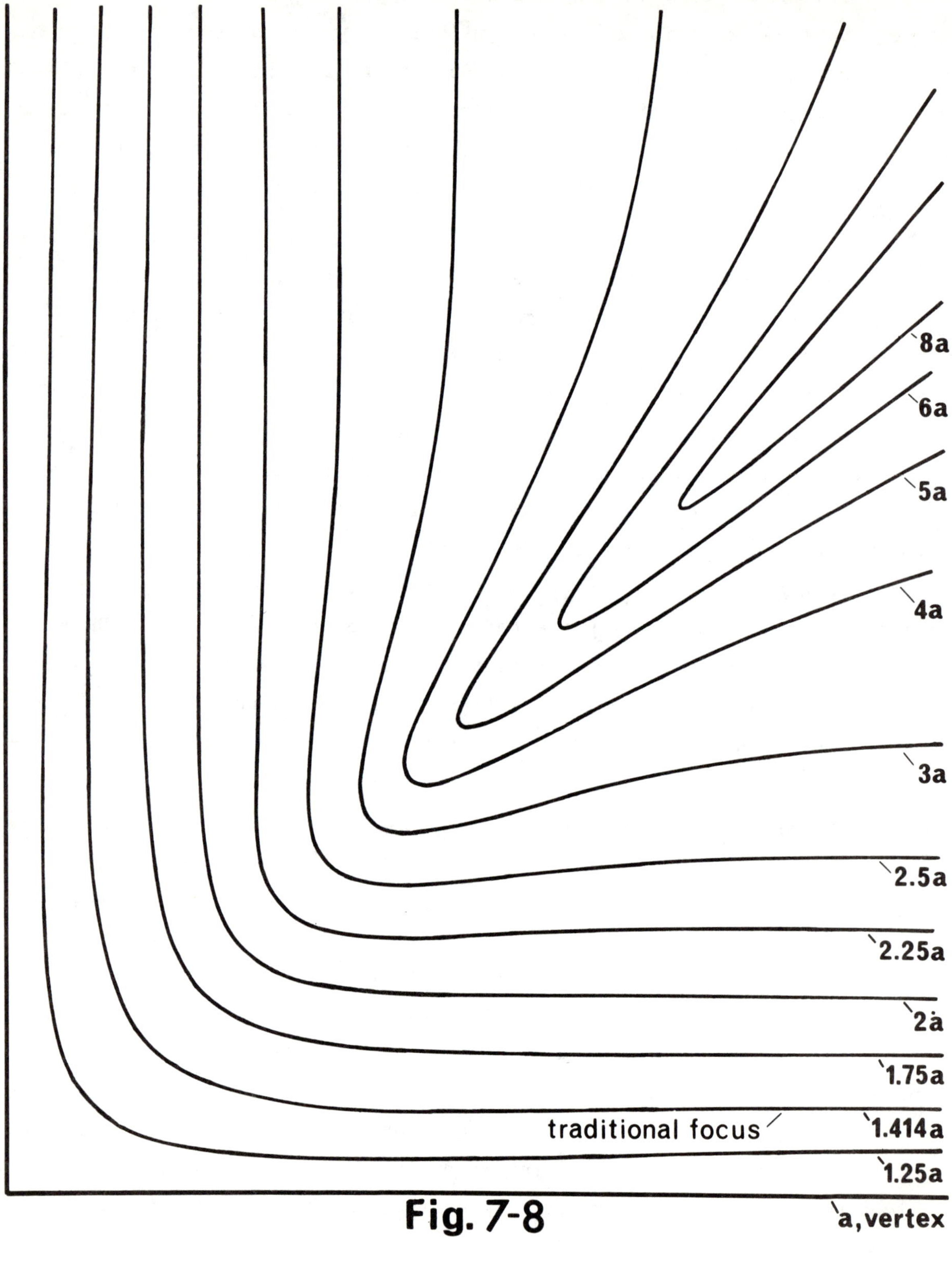

Fig. 7-8

already is quite large. No systematic investigation of the conditions under which looping occurs has been undertaken.

The degree of 90°-intercept transforms about unexceptional poles on the transverse axis of all hyperbolas except the equilateral subspecies is 16 (eq. VII-27 for $\alpha = 90^\circ$). Reduction occurs at the traditional foci, the vertices, and the center (Table VII-1). On the other hand, excluding 0° and 180° transforms, the 90° transform is not the lowest degree axial transform of the $e = 1.118$ hyperbola about unexceptional poles. In general, angles exist at which the degrees of transforms of hyperbolas about points on the transverse axis reduce below the value for the 90° transform.

These angles, which are specified by eq. 1 (see pages VII-8 to 12), are the angles between the asymptotes [i.e., $2\arctan(b/a)$ and $180^\circ - 2\arctan(b/a)$]. Accordingly, aside from the corresponding 0° and 180° transforms, circumpolar intercept transforms of hyperbolas about points on the transverse axis, for angles equal to either of the angles between the asymptotes, generally are the transforms of lowest degree.

$$\cos\alpha = \pm(2-e^2)/e^2$$ eq. defining the angles of reduction for variable-angle transforms of hyperbolas (7-9-1)

Fig. 7-10. Circumpolar symmetry of the equilateral hyperbola about the axial pole at 3a for two positive or two negative intercepts (intra-arm, inner segments). This figure, for the non-focal (i.e., non-point-focal) axial pole at 3a is the counterpart of Fig. 7-6 for a traditional focus. The transforms are in correct relative orientation but their locations have been altered and the 90°-150° transforms have been reduced to half-size for the purpose of illustration.

A direct comparison of the transforms of this figure with those of Fig. 7-6 for a traditional focus illustrates the much higher focal rank of the latter pole, i.e., the much greater circumpolar symmetry of the equilateral hyperbola about a traditional focus than about an unexceptional pole on the transverse axis. Thus, the curves of Fig. 7-6 are of much lower degree than those of the present figure. The degree of the α-transform for unexceptional poles on the transverse axis of the equilateral hyperbola (eq. VII-27) is 16, compared to 4 for the α-transform about a traditional focus (eq. VII-23). Reduction of

90°-Circumpolar Axial Intercept Transforms of The e=1.118 Hyperbola (Intra-Arm, Inner Segments)

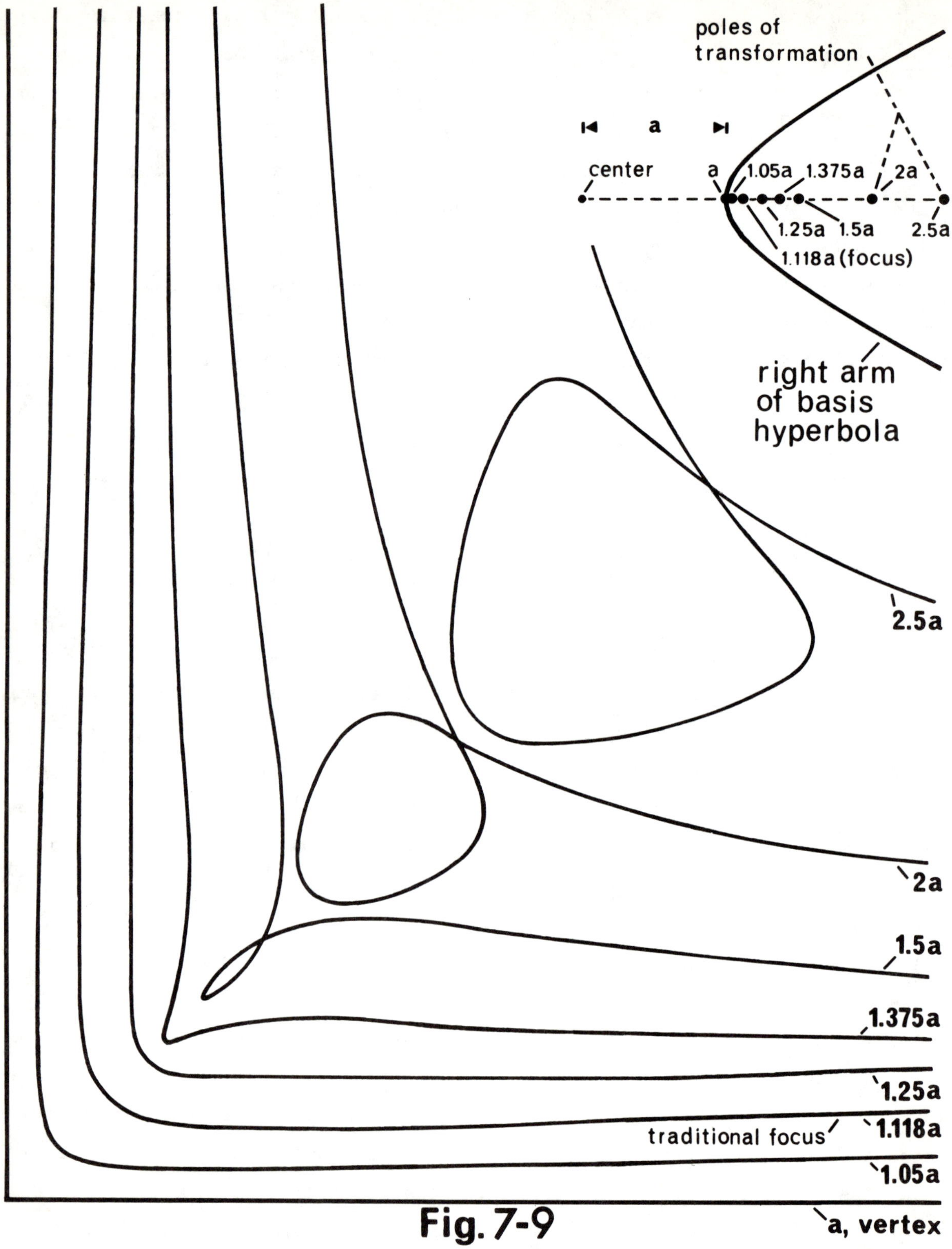

Fig. 7-9

the former occurs only at the center, vertices, and traditional foci. In addition to the reductions for angles of 0° and 180°, both α-transforms reduce for an angle of 90°--the angle between the asymptotes--to values of 3 and 12, respectively (see legend to Fig. 7-9 and the discussion of eccentricity-correlated rank on pages VII-8 to 12).

Fig. 7-11. Non-axial non-incident 90°-circumpolar intercept transforms of the $e = 0.745$ ellipse for two positive (or two negative) intercepts. The orientation, size, and location of these transforms have been altered to suit the purpose of illustration, and the transforms have been assembled in groups for the efficient utilization of space. The locations of the poles of transformation relative to the basis ellipse are illustrated at the lower right. Horizontal and vertical spacings of the interior poles are in units of ¼a, while adjacent collinear sets of poles are offset from one another by amounts a/8. The abscissa and ordinate of the pole of transformation for each transform (in units of the semi-major axis, a) are given adjacent to the corresponding transform.

It already has been noted that, as the poles for 90°-circumpolar intercept transformations of the $e = 0.745$ ellipse recede along the axes from the center, the DP of the corresponding transforms is lost (Fig. 7-7). This also is seen to be true in the case of recession of non-axial poles. Only the transforms for poles at (a/8,a/8) and (a/8,3a/8) have a DP. Although axial transforms about poles at (0,¼a) and (¼a,0) have double points (Fig. 7-7), the non-axial transform for a pole at (¼a,¼a) does not (center transform at lower left).

It also will be noted that, whether in possession of a DP or not, the transforms for poles nearest the center are the most "angular" and asymmetrical, and that the 3 illustrated transforms about exterior poles are of rather different and more elongated shape than those about interior poles.

The degree of 90°-intercept transforms of the ellipse about unexceptional points in the plane (the eliminant between eqs. VII-31 and VII-33) is 20. Reduction occurs only for points incident upon the curve and the axes (Table VII-1).

Circumpolar Symmetry of The Equilateral Hyperbola About The Axial Pole At 3a (Intra-Arm, Inner Segments)

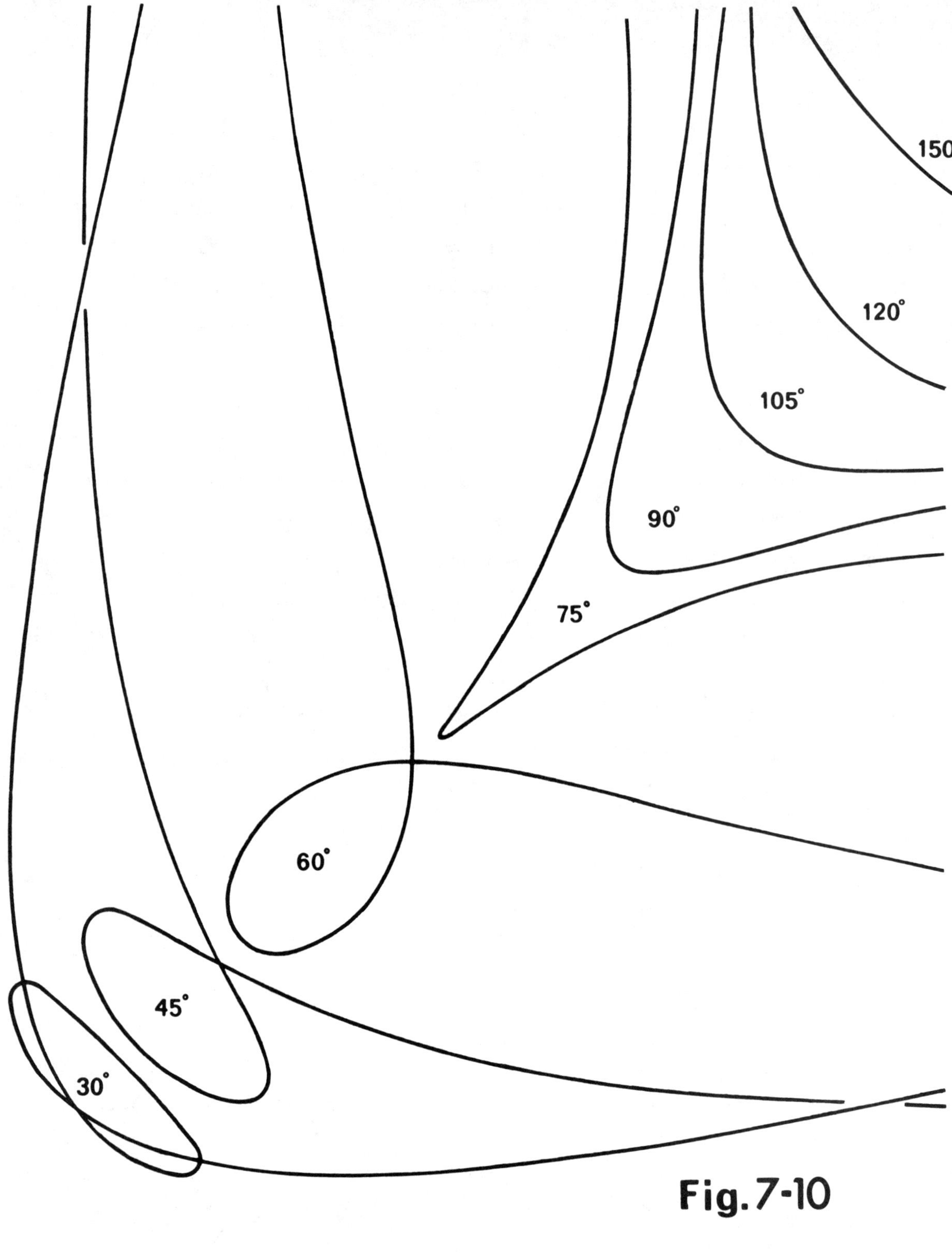

Fig. 7-10

Fig. 7-12. Incident 90°-circumpolar intercept transforms of hyperbolas and eclipsing of vertex foci (all intercepts and all segments). This figure illustrates full 4-quadrant plots of 90°-circumpolar intercept transforms about: (1) an axial vertex (Figs. a and d); (2) an intermediate incident pole with the abscissa of the LR vertices (Figs. b and e); and (3) an LR vertex (Figs. c and f). The transforms are for both the equilateral hyperbola (Figs. a-c) and the $e = 1.26$ (1.260472412) hyperbola (Figs. d-f). The small solid circles mark the origin of coordinates for the transform plots, and the coordinate axes extend horizontally and vertically therefrom. The transforms of Figs. a-c are to the same scale, whereas those of Figs e and f have been reduced relative to that of Fig. d.

The most conspicuous difference between the incident transforms of these two subspecies lies in the much greater simplicity of the transforms of the equilateral hyperbola than of the $e = 1.26$ hyperbola. Also conspicuous is the fact that only the axial vertex transforms are symmetric functions about the coordinate axis bisectors ($x = y$ and $x = -y$). Also noteworthy is the fact that all 3 transforms of the equilateral hyperbola are of 4th degree, whereas for the $e = 1.26$ hyperbola the transforms about the LR and axial vertices are of 8th degree, and the transform about the intermediate incident pole is of 12th degree (Fig. e).

Comparisons Between The Transforms of The Equilateral And $e = 1.26$ Hyperbolas

The comparative complexity of the curves can be explained by the fact that variable-angle focal conditions exist for hyperbolas (see pages VII-8,9 and Fig. 7-9), whereupon the degree of the α-transform about a specific pole and the degrees of transforms for specific angles depend not only upon the specific value of α but upon the eccentricity of the basis hyperbola. In essence, when the angle of transformation is equal to the angle between the asymptotes (or to the supplementary angle), the transforms simplify by the loss of segments.

For example, one readily understands that for 90° transforms about any interior point on the transverse axis of an arm of the equilateral hyperbola there can be no transform curve for two positive intercepts for values of θ between 270° and 360°. Thus, when the x radius vector is at 270°, it has a positive intercept with the curve, but the y radius vector, which lies along

Non-Axial Non-Incident 90°-Circumpolar Intercept Transforms of The e=0.745 Ellipse (Positive Intercepts)

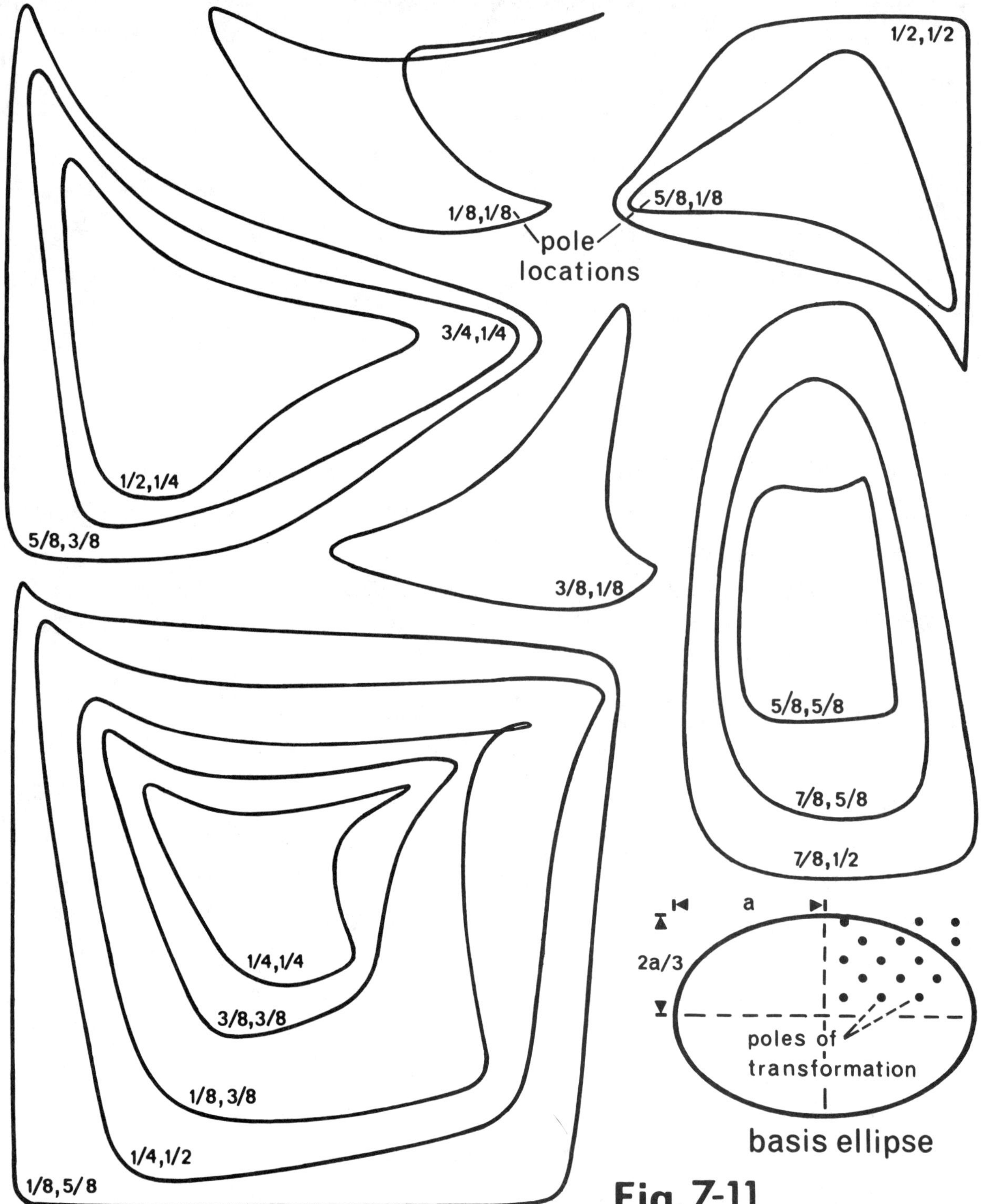

Fig. 7-11

the positive x-axis does not. The same situation holds as θ progresses from 270° to 315°. When the x intercept is at 315°, the y intercept is at 45° (405°) and neither one has a real intercept with the curve--since they are parallel to the asymptotes. As the y radius vector progresses from 45° to 90°, it has intercepts with the curve but the x radius vector does not. As a consequence, the transform curve in the 1st quadrant possesses no intra-arm outer segments. This also is true of the 90° transform about other points in the plane of the equilateral hyperbola.

Furthermore, what is true for two positive intercepts also is true for two negative intercepts, so there also are no intra-arm outer segments in the 3rd quadrant. It is unique to the equilateral hyperbola, however, that both the angle between the asymptotes (in the sectors in which the arms lie) and the supplementary angle are 90°. Accordingly, what is true of the outer segments on the intra-arm transforms for two positive and two negative intercepts also is true of them for one positive and one negative intercept.

It follows that no interior axial 90°-circumpolar intercept transform of the equilateral hyperbola has intra-arm outer segments; in fact, this is true for all of its 90° transforms. Moreover, there are no intra-arm segments of any kind for 90° transforms about incident poles, because when one radius vector intersects the basis arm, the other one intersects the contralateral arm.

The above treatment explains why all three 90° 4-quadrant transforms of the equilateral hyperbola about incident poles (Figs. a-c) have only inter-arm segments. The eqs. of these transforms (eqs. VII-36,40,42c) are repeated here for convenience as eqs. 1-3. All 3 eqs. are of 4th degree, giving all 3 inci-

90° transforms of the equilateral hyperbola about incident poles

$(x^2-y^2)^2 = 4a^2(x^2+y^2)$	axial vertices	(7-12-1)
$[x^2-y^2+4\sqrt{2}\,xy]^2 = 108a^2(x^2+y^2)$	LR vertices	(7-12-2)
$[a^2(x^2-y^2)+4h(h^2-a^2)^{\frac{1}{2}}xy]^2 = 4(2h^2-a^2)^3(x^2+y^2)$	intermediate pole at (1.207a,0.676a) [h = 1.207a]	(7-12-3)

dent poles the 90°-focal rank of ¼. Since this also is the focal rank of a transform about any pole incident upon the curve (with abscissa, h; eq. 3),

Incident 90°-Circumpolar Intercept Transforms of Hyperbolas And Eclipsing of Vertex Foci (All Intercepts & Segments)

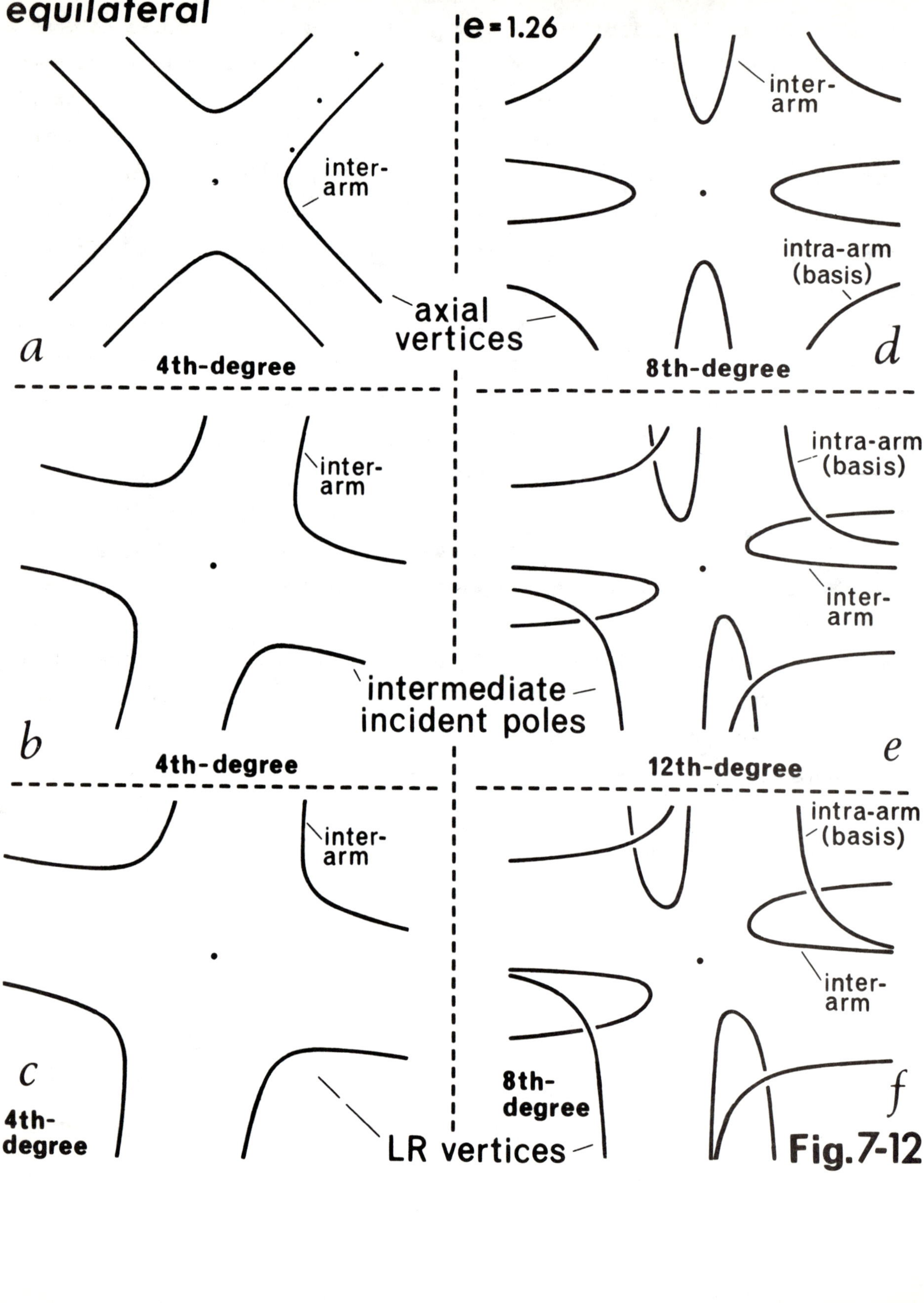

Fig.7-12

the axial and LR vertices, which have point-focal rank for a 90° angle of transformation of other hyperbolas, do not qualify as point foci in the equilateral subspecies. Point-foci and focal loci that lose their focal rank in a subspecies are referred to as *eclipsable foci*.

The 90° axial-vertex transform of Fig. d is the counterpart of the transform of Fig. a for the $e = 1.26$ hyperbola, the angle between the asymptotes of which is 75° (with a supplementary angle of 105°). Since the 90° angle of transformation does not correspond to either of the angles between the asymptotes, the intra-arm segments of the basis arm are not lost. Accordingly, the 4-quadrant transform consists of 8 segments, rather than 4, and is of 8th degree rather than 4th degree; like the transform of Fig. a--and all 90° transforms about poles on a line of symmetry--the axial-vertex transform of Fig. d is a symmetric function (about the bisectors of the coordinate axes). Similarly, the 90° transform about an LR vertex of the $e = 1.26$ hyperbola possesses 4 intra-arm segments and 4 inter-arm segments and is of 8th degree. Like the transform of Fig. c, that of Fig. f is not a symmetric function about the coordinate-axes bisectors but has symmetry about the origin.

The 90° transform of Fig. e is for an incident non-focal intermediate pole with abscissa midway between the vertex and the corresponding traditional focus. This transform greatly resembles that of Fig. f for an LR vertex but is of 12th rather than 8th degree. Since it is the transform for a non-point-focal incident pole, its degree establishes the degree for all other unexceptional incident points as 12 and, at the same time, establishes that the axial and LR vertices are point foci for 90° transformations (since the degree of transforms about them is only 8).

The difference in degree between the transform of Fig. e and those of Figs. d and f is not based upon differences in the number of segments--since all possess 8--but upon differences in other properties of the transforms. The fact that the 90° transform about the LR vertices is only symmetrical about the origin, whereas that about an axial vertex is symmetrical about the coordinate-axes bisectors--and both are of 8th degree--parallels the situation for 90° transforms of the equilateral hyperbola (Figs. a and c), though in the latter cases both transforms have only 4 segments and are of only 4th degree. [In fact, the 90° transforms of the equilateral hyperbola about all incident poles also are symmetric functions when referred to appropriately rotated axes (see below).]

4-Quadrant Circumpolar Intercept Transforms of The $e = 1.26$ Hyperbola For The Asymptote Angle

Having considered transforms of the equilateral hyperbola for an angle of transformation equal to the angle of the asymptotes, and transforms of the $e = 1.26$ hyperbola for an angle not equal to the asymptote angle (or its supplement), the complementary situations now are considered, namely, 4-quadrant circumpolar intercept transforms of the equilateral hyperbola for other angles than 90° (Fig. 7-13) and 4-quadrant circumpolar intercept transforms of the $e = 1.26$ hyperbola for the asymptote angle. In the latter case, various illustrative transforms merely are described; in the former case they are illustrated in Fig. 7-13.

Consider first a 75°-axial-vertex transform of the $e = 1.26$ hyperbola, which is the counterpart of Fig. d but for the asymptote angle rather than for an angle of 90°. When the radius vectors are at an angle of 75° to one another, there can be no case of two positive nor two negative non-zero intra-arm intercepts (for reasons discussed above). Accordingly, the intra-arm segments in the 1st and 3rd quadrants are lost, leaving a total of 6 segments rather than 8. Of these 6 segments, the 4 inter-arm ones are congruent and the two intra-arm ones are congruent but they differ in shape from those of Fig. d. Each segment, however, retains a line of symmetry and the transform retains symmetry about both coordinate axes. On the other hand, the transform also differs from the transform of Fig. d in that each intra-arm segment intersects an inter-arm one. Since the degree of the 8-segment α-transform for an a vertex of hyperbolas is 16, and only 6 segments are present in the transform for the angle of the asymptotes, the degree of the latter may be 12. [The degree of the α-transform for points on the transverse axis is a maximum of 14 (see page VII-12) but neither the degree for the α-transform nor the asymptote-angle transform of the a vertices has been established.]

A similar situation prevails for transforms about the LR vertex at the asymptote angle (and for the supplement of this angle) in that intra-arm segments in the 1st and 3rd quadrants are lost and the transform remains symmetrical about the origin. The situation differs in that all 4 of the inter-arm segments no longer are congruent with one another but are congruent in pairs (reflected through the origin), and in that two of the inter-arm segments no longer are intersected by intra-arm segments (since the intra-arm segments that would have been intersecting are lost). The degree of the α-transform about the LR vertices is a maximum of 32; a reduction of the degree of the asymptote-angle transform to 3/4ths of the value for the α-transform seems likely.

Mere reduction of the angle of the transformation about a given pole to an angle less than the asymptote angle does not lead to a reduction in the number of intra-arm segments for two positive and two negative intercepts. Though it leads to the loss of the outer segments for the basis arm, these segments are replaced by the corresponding intra-arm segments for the contralateral arm, which do not exist for transforms at the angle of the asymptotes or for greater angles. A reduction in the number of segments for two positive and two negative intercepts occurs only for transforms at the asmptote angle.

[In the case of transforms at the supplement of the asymptote angle, the curves are rotated 90° (see Fig. 7-13) so that it is the intra-arm segments for one positive intercept and one negative intercept that are lost, rather than those for two positive and two negative intercepts.]

Transforms about an unexceptional point in the plane now are treated briefly, using the $e = 1.26$ hyperbola and a pole of transformation on the LR at $k = 0.2a$. The degree of the α-transform is a maximum of 32 and the transforms for unexceptional angles, for example 60°, consist of 16 segments; 4 of these are in each quadrant--two intra-arm segments and two inter-arm segments (see Fig. 7-15 for an example of a 16-segment transform of a hyperbola). The 60° transform is symmetrical about the origin but the segments in the 1st and 3rd quadrants are incongruent with those in the 2nd and 4th quadrants, i.e., the transform consists of 8 pairs of congruent segments.

The 75° transform is the transform for the angle of the asymptotes. It very much resembles the 60° transform, with the notable exception that the intra-arm outer segments in the 1st and 3rd quadrants are lost. Although the degree of this transform has not been ascertained, 28 is a reasonable expectation (based upon a value of 7/8ths of the probable degree of the α-transform). The transform of next-lowest degree is the 20th-degree 90° transform. This locus resembles the 60° transform in being symmetrical about the origin and in not being symmetrical about either bisector or either coordinate axis. Its symmetry is greater than that of the 60° transform, however, in that it possesses 4 quartets of congruent segments (rather than 8 pairs)--one member of each quartet in each quadrant.

The 4-quadrant transforms of the $e = 1.26$ hyperbola of lowest degree for a pole on the LR at 0.2a are the 0° and 180° transforms (see Fig. 7-4 for comparable transforms of the equilateral hyperbola for the 1st and 3rd quadrants). These curves are of 8th degree and possess 8 segments, but the latter consist of two congruent quartets; in other words, only two differently-shaped loci are present. These transforms not only are symmetrical about the origin, they are symmetrical about both bisectors as well. The two segments of the 0° transform in each of the 1st and 3rd quadrants are interarm segments. One segment of each pair is for one labelling of the intercepts, the other segment for the opposite labelling.

[For example, in one labelling, the intercept with the basis arm is labelled x, and the intercept with the contralateral arm is labelled y; in the opposite labelling the labels are reversed.]

In the 2nd and 4th quadrants the curves for the 0° transforms are intraarm segments, one for one labelling and the other for the opposite labelling. The 180° transform is congruent but rotated 90° CW or CCW (all 4-quadrant circumpolar intercept transforms for supplementary angles are congruent but rotated 90°; see Fig. 7-13a,d).

The Incident 90° Transforms of The Equilateral Hyperbola, Another Homology With The Circumpolar Symmetry of The Circle

Before proceeding to Fig. 7-13, which treats incident transforms of the equilateral hyperbola for unexceptional angles (i.e., angles for which the degree of the α-transform does not reduce), a comparison of the incident 90° transforms of the equilateral hyperbola (Figs. a-c and eqs. 1-3) is made. It has been noted that, since all the incident 90° transforms of the equilateral hyperbola are of 4th degree, the 90° axial and LR-vertex foci that exist in other subspecies are eclipsed in the equilateral subspecies. The question then arises as to whether the 90° incident transforms might be differentiated from one another in such a manner as to delineate the 90° symmetry of the basis curve about incident poles at the *subfocal* level.

To investigate this question, the general incident 90° transform (eq. 2) is rotated CCW through an angle θ (eq. 4) to eliminate the xy-crossproduct term (which brings the vertices of the 4 segments onto the coordinate axes). This leads to the general eq. 5a (which is a symmetric function). When this eq. is converted to polar coordinates (eq. 5b) it is seen to be that of a single subspecies, since changes in the value of h for a given value of a (i.e., for a given size of equilateral hyperbola) merely lead to a different-sized curve (Fig. 5c). In essence, as the pole of transformation moves outward

$\cos\theta = h/(2h^2-a^2)^{\frac{1}{2}}$ angle of rotation to bring vertices of segments onto coordinate axes (7-12-4)

(a) $(x^2-y^2)^2 = 4(2h^2-a^2)(x^2+y^2)$ general eq. of 90° incident transforms with same orientation (7-12-5)

(b) $r^2 = 4(2h^2-a^2)/(\cos^2\theta-\sin^2\theta)^2$ polar eq. of 5a

(c) $r \approx (2h^2-a^2)^{\frac{1}{2}}$

along an arm from its axial vertex, the transform curve rotates and progressively enlarges but does not change in form, i.e., it remains a member of a single subspecies.

This remarkable result is the 6th planar homology between the circle and the equilateral hyperbola that has been revealed by symmetry analyses (see page V-8). Thus, the circle and the equilateral hyperbola are the only known curves for which the incident 90° transforms belong to only a single subspecies. The 90° transform of the orthogonal line-pair falls short of possessing this property; the incident 90° transforms are lines for all poles except the point of intersection, at which the transform is a point (see pages V-30 and VI-3,4).

[A previously known planar homology between the circle and the equilateral hyperbola is that $dA/d\theta$ about the center is constant for both, where A is the area of a sector swept out by a central radius vector (for the equilateral hyperbola, θ is the hyperbolic angle).]

Since all the incident 90° transforms of the equilateral hyperbola are members of the same subspecies, the question of 90°-circumpolar symmetry about incident poles at the subfocal level reduces solely to a question of "for which poles does the general eq. of the curves (eq. 2) achieve its greatest simplicity?" The answers are clear. The curve has by far the greatest subfocal 90°-circumpolar symmetry about the axial vertex because only for $h = \pm a$ does the xy term of eq. 2 vanish (eq. 1).

The subspecies has the 2nd-greatest subfocal 90°-circumpolar symmetry about the LR vertices because, aside from the values for $h = \pm a$ for the vertex, $h = \pm\sqrt{2}\,a$ leads to the simplest expression for the coefficient of the xy term --$4\sqrt{2}$--which involves a radical with the smallest achievable integral radicand, together with the smallest achievable integral value of the coefficient of the (x^2+y^2) term, namely 108.

Inverse 4-Leaf Roses

Lastly, we consider the other affinities of the 90°-transform locus of eq. 1 and Figs. a-c. In the case of the circle, the 90° incident transforms are circles of twice the radius of the basis circle. In the present case, the 4 arms of the 90° incident transforms resemble arms of hyperbolas and also have a semi-transverse axis that is twice the length of that of the basis curve.

If the values, r, of the polar eqs. of the two curves are compared (where the equilateral hyperbola is taken to have a semi-transverse axis length of 2a), one finds that they differ only in the exponent of the denominator; thus, $r^2 = 4a^2/(\cos^2\theta - \sin^2\theta)$ for the hyperbola and $r^2 = 4a^2/(\cos^2\theta - \sin^2\theta)^2$ for the 90° transforms. Since $|(\cos^2\theta - \sin^2\theta)|$ never exceeds unity, this means that for a given angle, $|r|$ for the hyperbola never exceeds $|r|$ for the 90° transform.

In other words, except at the vertices, the arms of the 90° transforms lie within (i.e., distal to) the arms of equilateral hyperbolas that have the same semi-transverse axis length and coincident vertices (the asymptote angle for both is 90°). This is indicated in Fig. a for the upper portion of the right arm of the 90° transform, where the small solid circles lie on the arm of an equilateral hyperbola with the same center and vertex.

Since the incident 90°-circumpolar intercept transforms of the equilateral hyperbola are the homologues of the corresponding transforms of circles, which also are circles, the identity of the former transforms is of great interest, and it can be anticipated that they are curves of very high circumpolar symmetry about the center. They are readily identified by a central inversion, which produces the well-known 4-leaf rose. In other words, they are *inverse 4-leaf roses*.

If one refers to Table A-III (page *xii* in *Symmetry, An Analytical Treatment*), the 4-leaf rose is seen to have the 2nd-highest known linear circumpolar symmetry about a true center, the same as for the sextic symmetry mimic of tangent circles and other members of the $\cos^2\theta$-group. All of these curves have an intercept ordinal and subordinal rank of 2. For 90° and 180° angles of transformation, the only known curve with greater circumpolar symmetry about a true center is the circle, for which the intercept ordinal and subordinal rank is 0 for all angles of transformation (not listed in Table A-III).

Since the 4-leaf rose has such a high degree of linear circumpolar symmetry about a true center, one expects the same to be true of the inverse locus about the true center (Circumpolar Maxim 3, page VI-9). This expectation is fulfilled by the results of a circumpolar symmetry analysis. Not only does the inverse 4-leaf rose also have linear circumpolar symmetry about the true center at 90° and 180°, but the intercept ordinal and subordinal rank for these angles of transformation is only 4 (eq. 6b).

transcendental ordinal rank eq. of the inverse 4-leaf rose about the true center

$$\text{(a)}\quad \overset{\circ}{s}{}^2 = 64a^2\left[\frac{\sin^2\theta\cos^2\theta}{(2\cos^2\theta-1)^4} + \frac{\sin^2(\theta+\alpha)\cos^2(\theta+\alpha)}{[2\cos^2(\theta+\alpha)-1]^4}\right] \qquad (7\text{-}12\text{-}6)$$

$$\text{(b)}\quad \overset{\circ}{s}{}^2 = \frac{(x^4+y^4)-4a^2(x^2+y^2)}{a^2}$$

intercept ordinal rank eq. of the inverse 4-leaf rose about the true center

Thus, the inverse 4-leaf rose outranks (has greater linear circumpolar symmetry about a true center) 14 of the 19 entries in Table A-III for curves with a true center, including central conics and the parallel line-pair. The latter curve and the inverse-$\cos^2\theta$ group also have intercept ordinal and subordinal ranks of 4 for a 180° angle of transformation, but their 90° transforms are not linear.

Fig. 7-13. Incident 60° and 120°-circumpolar intercept transforms of the equilateral hyperbola (all intercepts and segments). All 4 figures are to the same scale. The same poles of transformation are represented as in Figs. 7-12a-c; only the degrees of the transforms differ.

This figure illustrates aspects of the relationships represented in Table VII-1 (page VII-19), namely, that for an unexceptional angle (in this case any angle but 0°, 180°, and 90°) the circumpolar symmetry of the equilateral hyperbola about incident poles is not exceptional, as compared to the comparable symmetry of other hyperbolas for unexceptional angles. Thus, insofar as the properties of the 60° transform are concerned, the a vertex curve of Fig. a differs in no significant manner from the corresponding 60° transform of the $e = 1.26$ hyperbola (for which subspecies 60° is an unexceptional angle). Both consist of 8 segments, both have symmetry about the origin, both are symmetric functions, and neither has symmetry about the coordinate axes (and the same

Incident 60°& 120°-Circumpolar Intercept Transforms of The Equilateral Hyperbola (All Intercepts & Segments)

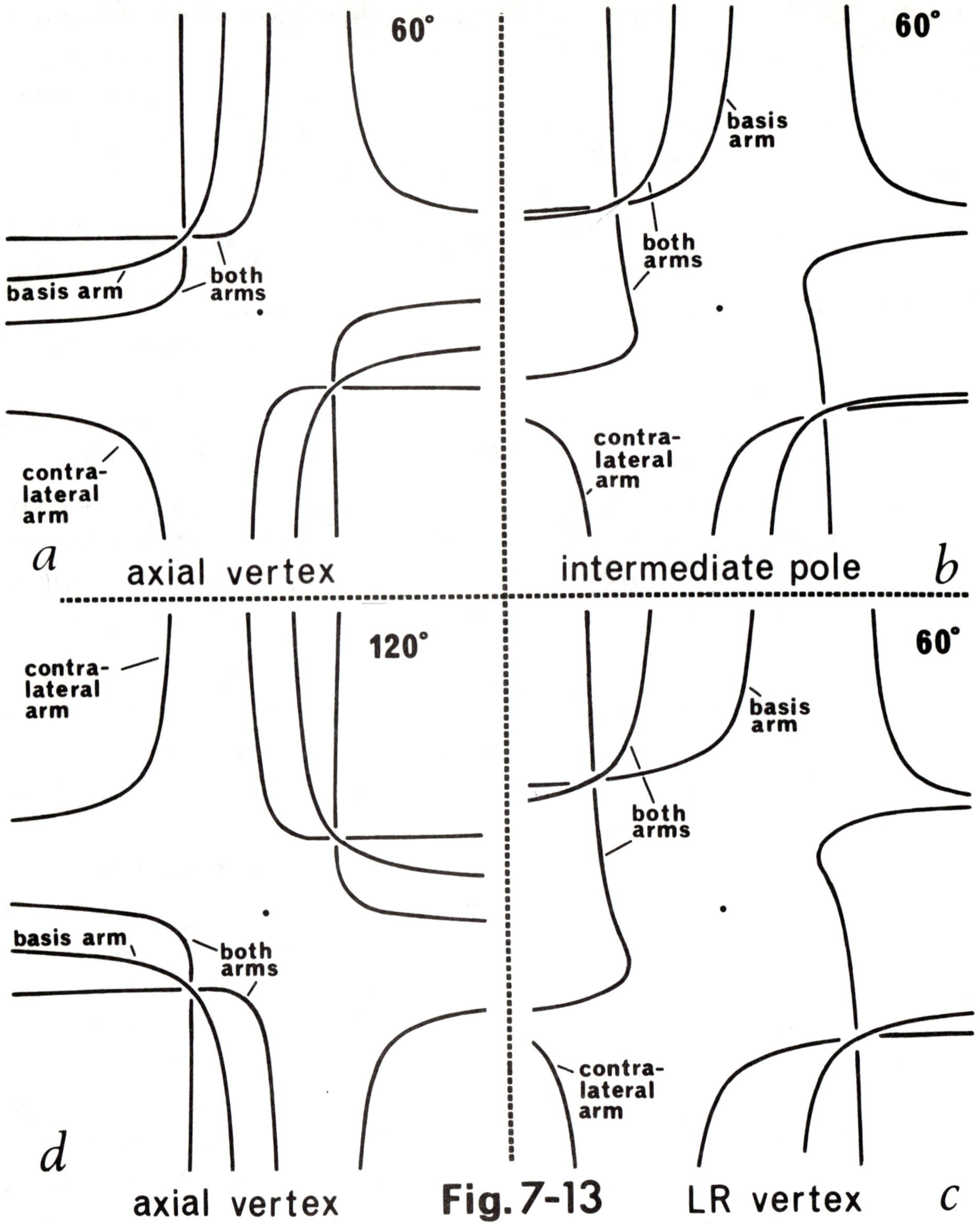

Fig. 7-13

is true of both hyperbolas at other unexceptional angles).

On the other hand, the 60° transforms about both the LR vertex and the intermediate pole (Figs. b and c), though they have the same numbers of segments and retain symmetry about the origin, do not possess symmetry about the coordinate axis bisectors. Lastly, the 120° transform about an axial vertex (Fig. d) is seen to be congruent with the 60° transform about the same pole, as is true of all 4-quadrant transforms about any given pole for angles that are supplementary (this generally is not the case, of course, for transforms that represent the relationships between two positive intercepts).

Fig. 7-14. 90°-circumpolar intercept transform of the equilateral hyperbola about a traditional focus (all segments and intercepts). Although the 4-quadrant 90° transforms of other hyperbolas about a traditional focus possess 16 segments, the transform of the equilateral hyperbola possesses only 12. This is a consequence of the fact that the 90° angle of transformation is the same as the angle of the asymptotes (a case of eccentricity-correlated rank; see pages VII-8 to 12 and Fig. 7-9). This transform is used to illustrate the manner of equational representation of segments of 4-quadrant transforms.

The 12 segments of the transform can be represented by 4 different eqs. (eqs. 1-4), each eq. representing 3 segments, as indicated by the labels adjacent to

eqs. of segments	quadrants of represented segments		
	intra-arm	inter-arm	
$a(x^2+y^2) = 2xy(x+y)$	1st	2nd and 4th	(7-14-1)
$a(x^2+y^2) = -2xy(x-y)$	2nd	1st and 3rd	(7-14-2)
$a(x^2+y^2) = -2xy(x+y)$	3rd	2nd and 4th	(7-14-3)
$a(x^2+y^2) = 2xy(x-y)$	4th	1st and 3rd	(7-14-4)

algebraically complete 90° transform of the equilateral hyperbola about a traditional focus (12 segments)

$$[a(x^2+y^2)-2xy(x+y)][a(x^2+y^2)+2xy(x-y)]\cdot \qquad (7\text{-}14\text{-}5)$$

$$[a(x^2+y^2)+2xy(x+y)][a(x^2+y^2)-2xy(x-y)] = 0$$

the segments. The complete 4-quadrant transform is represented by the degenerate eq. 5, the *algebraically complete transform equation.* Note that the loci represented by the individual eqs. are not symmetrical about either coordinate axis or the origin but are symmetrical about one or the other of the coordinate-axes bisectors (eqs. 1 and 3 are symmetrical about the line $x = y$, and eqs. 2 and 4 are symmetrical about the line $x = -y$). On the other hand, the complete 4-quadrant transform (eq. 5) is symmetrical about the origin, both coordinate axes, and both coordinate-axes bisectors.

Fig. 7-15. 90°-circumpolar intercept transform of the $e = 1.118$ hyperbola about a traditional focus (all segments and all intercepts). The present figure is the counterpart of Fig. 7-14 but for a hyperbola of eccentricity 1.118, rather than for the equilateral hyperbola (eccentricity, $\sqrt{2}$). Since the 90° angle of transformation does not correspond to the asymptote angle or its supplement (53.13° or 126.87°), the 4-quadrant transform consists of 16 segments, the 4 additional segments representing the intra-arm outer segments. As in the case of the transform of Fig. 14, the loci of this transform are represented by a degenerate eq. consisting of the product of 4 factors equated to 0 (eq. VII-57c). In the present instance, however, each factor is of 4th degree and, when equated to 0, represents 4 segments, rather than being of 3rd degree and representing 3 segments. The symmetry properties of the loci represented by the individual factors equated to 0 are the same as for the loci represented by eqs. 7-14-1 to 4, i.e., the loci are not symmetrical about the origin or either coordinate axis but they are symmetrical about one of the coordinate-axes bisectors.

Fig. 7-16. $ds/d\theta$ versus θ for 0° to 180°-intercept transforms of the $e = 0.745$ ellipse about a traditional focus for two positive and two negative intercepts, i.e., for the portions of the transforms in the 1st and 3rd quadrants. The transformation angle of the transform to which each transcendental subordinal rank plot applies is given adjacent to the plots at the left. The heavy arrows mark the sequence of transformation angles for the corresponding transforms progressing from 0° to 180°. The small arrows labeled x mark the locations of the polar reference axes, while those labelled y mark the 90° polar axes.

90°-Circumpolar Intercept Transform of The Equilateral Hyperbola About A Traditional Focus (All Segments, All Intercepts)

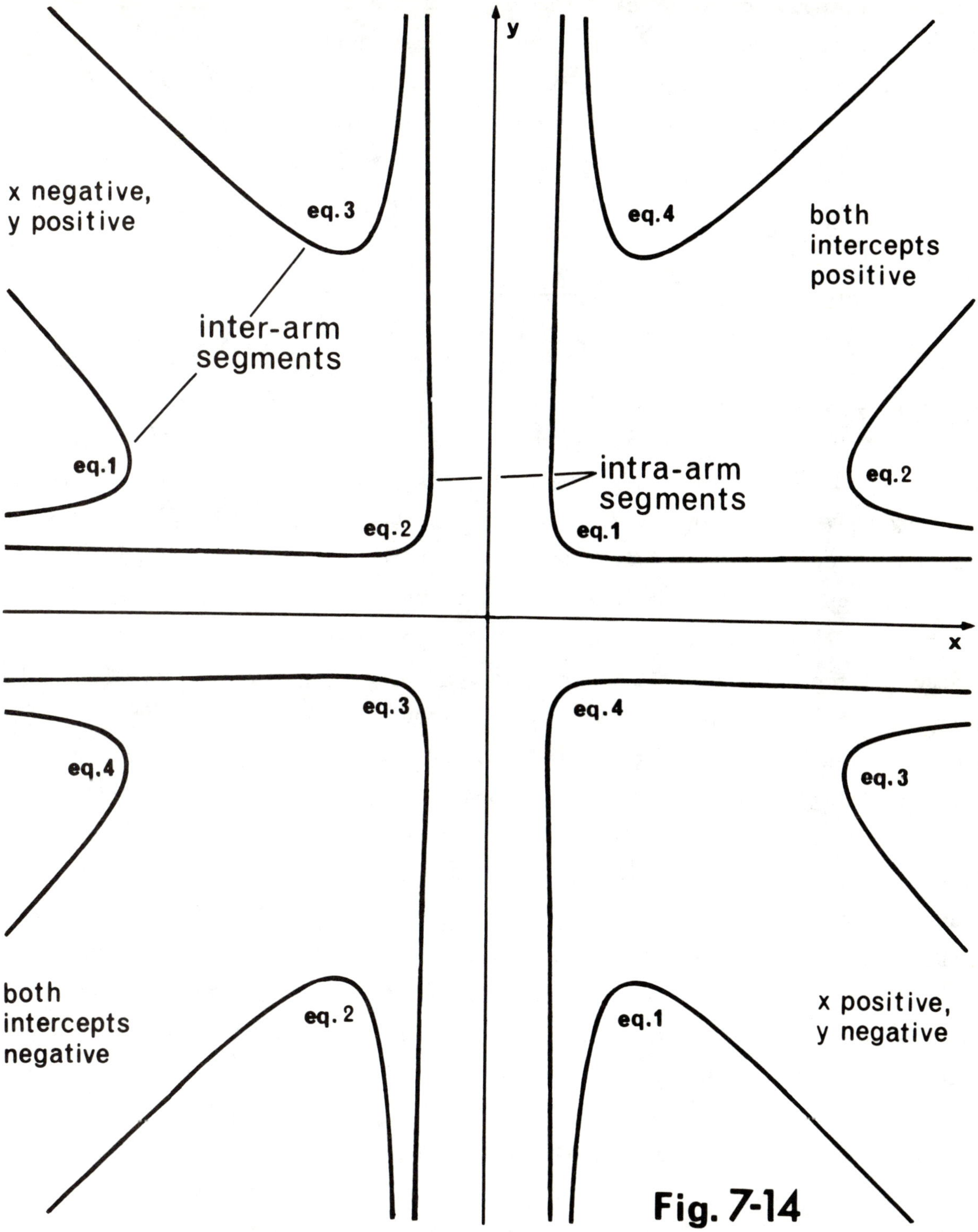

Fig. 7-14

When an intersection of the arrows is shown, the point of intersection is the position of the origin. [In a rectangular plot, $x = (ds/d\theta)\cos\theta$ and $y = (ds/d\theta)\sin\theta$]. As in the case of the corresponding curves for the traditional focus of the parabola, the line of symmetry is rotated CW relative to the x-axis by an amount equal to half the angle of the transformation, a property that appears to be characteristic of transcendental subordinal rank curves.

As pointed out in the legend of Fig. 6-15, although the *transcendental ordinal* and *subordinal rank equations* (eq. 1) of intercept transforms do not

transcendental ordinal rank eq. for transforms
of an ellipse about a traditional focus

$$(ds/d\theta)^2 = \overset{o}{s}{}^2 = \frac{b^4(a^2-b^2)\sin^2\theta}{[(a^2-b^2)^{\frac{1}{2}}\cos\theta-a]^4} + \frac{b^4(a^2-b^2)\sin^2(\theta+\alpha)}{[(a^2-b^2)^{\frac{1}{2}}\cos(\theta+\alpha)-a]^4} \qquad (7\text{-}16\text{-}1)$$

provide as high a degree of discrimination of the subfocal symmetry of a curve about the pole of the transformation as do the *intercept ordinal* and *subordinal rank equations*, they are more readily plotted, the relationships that they convey are more readily visualized, and the types of reflective symmetry that they possess give information about the location of the pole of transformation (and information about the symmetry of the basis curve) that is not explicit in the intercept subordinal rank curves.

These relationships are illustrated in Figs. 7-16 to 7-19 for transforms of the $e = 0.745$ ellipse about a traditional focus, the center, and an interior pole, $(-a/8,-a/8)$. Although the intercept and transcendental subordinal rank curves differ greatly in appearance, overt correspondences do exist (based upon the fact that one variable, $ds/d\theta$, is the same for both).

The ordinal rank eq. 1 is made specific for an angle of transformation, α, by substituting the value of the angle (where $a = 1$ and $b = 2/3$). The plots of the present figure represent the curves of eq. 1 for the indicated values of α. The fact that all of these plots possess a line of symmetry reveals that the pole of the transformation lies on a single line of symmetry of the basis curve. Inasmuch as all 180° transcendental subordinal rank curves possess symmetry about the origin (see Fig. 7-8-180°), the combination of this property and the possession of a line of symmetry by all transcendental subordinal rank curves for poles of transformation incident upon a line of symmetry leads to a 180° plot that possesses two lines of symmetry.

90°-Circumpolar Intercept Transform of The e=1.118 Hyperbola About A Traditional Focus (All Segments, All Intercepts)

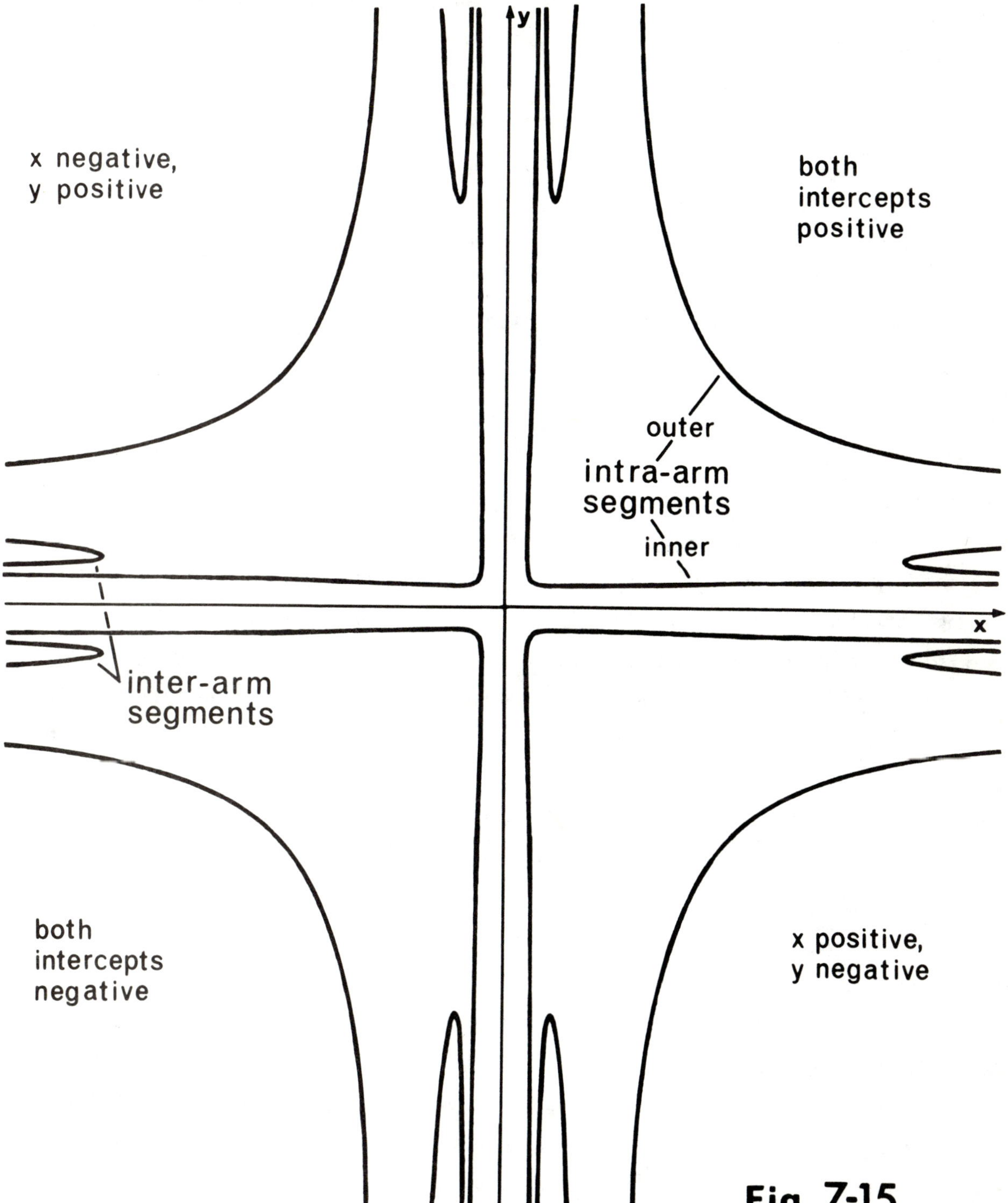

Fig. 7-15

All of the plots of this figure are of 16th degree except those for 0° and 180°, which are of 8th degree. The rectangular exponential degree of these curves provides some measure of discrimination of the relative symmetry of basis curves that possess the same intercept transforms about a given pole at a given angle (for example, for which all of the transforms are equilateral hyperbolas). However, it is the degree of the corresponding intercept subordinal rank eqs. (Fig. 7-19, upper plots for a traditional focus) that provide the highest degree of discrimination for such cases. For example, the 180° transforms of the parabola and central conics about the traditional foci are eqs. 2a and 3a.

180° equilateral hyperbolic circumpolar intercept transforms
of conic sections about the traditional foci

(a) $a(x+y) = xy$ (b) $(x-\sqrt{2}a)^2-y^2 = 2a^2$ parabola (7-16-2)

(a) $b^2(x+y) = 2axy$ (b) $[x-b^2\sqrt{2}/2a]^2-y^2 = b^4/2a^2$ central conics (7-16-3)

Rotation of the coordinate axes CCW through 45° yields eqs. 2b and 3b, from which it can be seen that both of the transforms of eqs. 2a and 3a are cast about the left vertices, with the transverse axes rotated 45° CW. Accordingly, the circumpolar intercept transform eqs. provide no basis for discriminating between the relative circumpolar symmetry of the 3 species of conic sections about the traditional foci.

If one appeals to the ordinal and subordinal rank eqs., however, discriminations between these transforms are possible. For example, if one converts the *transcendental ordinal* and *subordinal rank equations* (eqs. 6-15-1b and eq. 1 above for the ordinal rank, and the same eqs. with the angle, α, specified for the subordinal rank) to the corresponding rectangular coordinate eqs., the degree of these eqs. provides a subfocal index. For $\alpha = 180^\circ$, the angle of the transformation, the rectangular equivalents of the transcendental subordinal rank eqs. of the parabola are of 6th degree (Fig. 6-15-180°), whereas the corresponding eqs. for both species of central conics are of 8th degree (the 180° curve).

Accordingly, the 180° transcendental subordinal rank eqs. provide a means of differentiating between the 180° circumpolar symmetry of conic sections about the traditional foci at the subfocal level. On the other hand, for the

ds/dθ Versus θ For 0° To 180°-Intercept Transforms of The e=0.745 Ellipse About A Traditional Focus

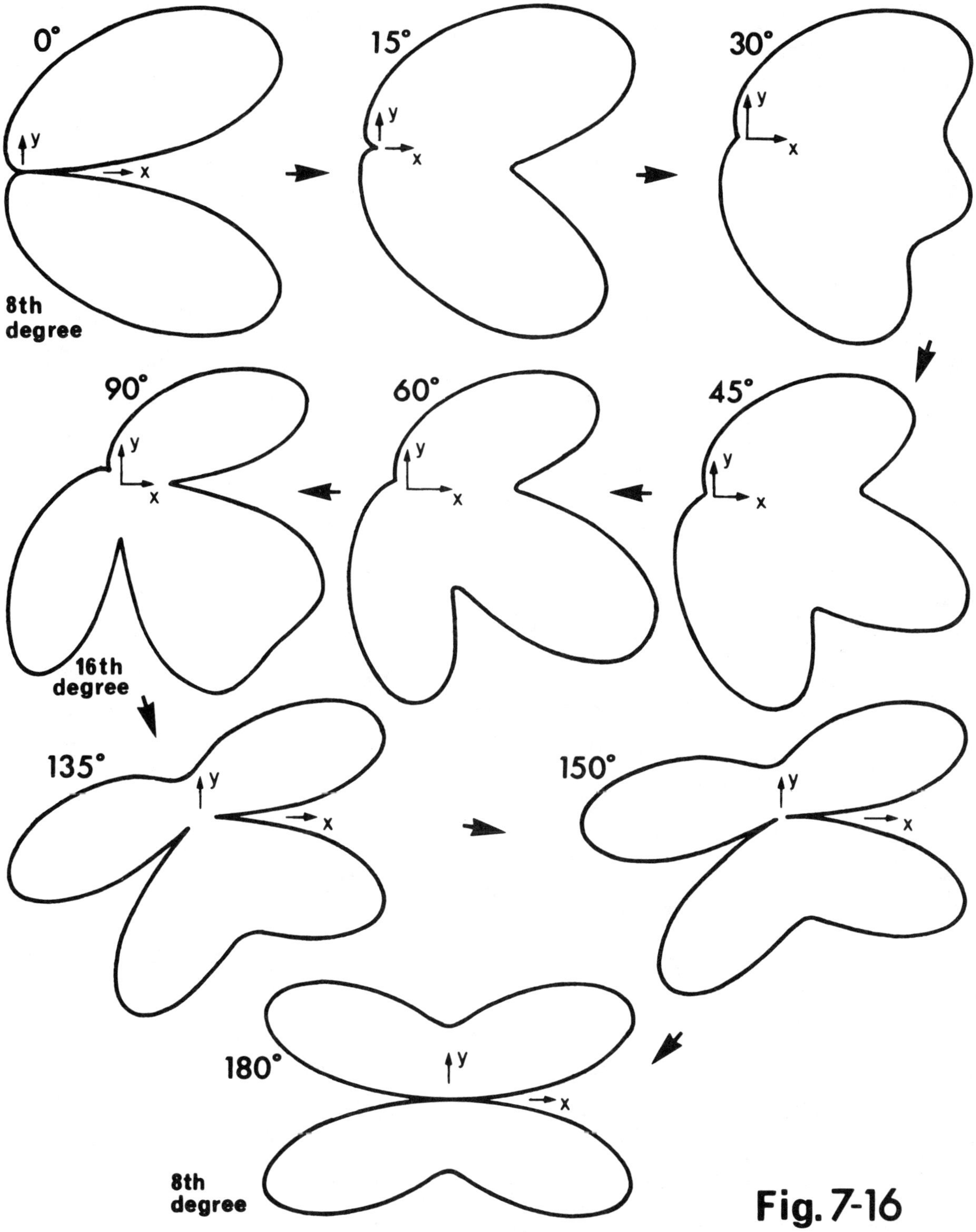

Fig. 7-16

linear 0° transforms and quartic 90° transforms about the same pole, the transcendental subordinal rank eqs. provide no measure of discrimination, since the degrees of their rectangular coordinate equivalents are the same (both are 8 for 0° and 16 for 90°). Similarly, the equivalent rectangular degree of the transcendental ordinal rank eqs. is 16 for both the parabola and central conics.

If, now, the ordinal rank eqs. are expressed in terms of the x and y intercepts rather than θ and α--the *intercept subordinal rank equations* (eqs. 6-15-1 and VII-46)--the degree in x and y provides an *intercept subordinal rank* index of symmetry at the subfocal level. This degree is 3 for the parabola but 4 for central conics. Accordingly, by this measure intercept transforms of the parabola about the traditional focus are expected to have greater subfocal symmetry than those of central conics at all angles of transformation.

This expectation is supported by a derivation of the degrees of the *intercept subordinal rank equations*. This is accomplished by eliminating y (or x) from the ordinal eqs. by substituting $y = g(x)$ from the transform eq. Upon this substitution, the degrees of the intercept subordinal rank eqs. for 0°, 180°, and 90° angles of transformation become 3, 6, and 16, respectively, for the parabola, each of which is less than the corresponding values, 4, 8, and ≤ 24, respectively, for central conics.

Since it already has been noted that 4-quadrant transforms for supplementary angles are congruent but rotated 90°, and the present figures are for two positive and for two negative intercepts, it is clear that curves for supplementary angles such as 0° and 180°, 30° and 150°, and 45° and 135°, taken together represent the full 4-quadrant transform for either of the two supplementary angles. For example, the 0° curve characterizes $ds/d\theta$ for a segment of the line $x = y$ extending from (b,b) to (a,a) in the 1st quadrant and (-b,-b) to (-a,-a) in the 3rd quadrant. The 180° curve not only characterizes the 180° transform for two positive and two negative intercepts but also represents $ds/d\theta$ for the 0° transform for one positive and one negative intercept. The latter consists of segments of an equilateral hyperbola in the 2nd and 4th quadrants.

The correlation of $ds/d\theta$ with θ for the 180° curve of this figure is elucidated by reference to Fig. 5-1 for the 180° transform about the traditional foci of central conics, in which equal increments of θ are marked off adjacent to the transform equilateral hyperbola. Inspection of the latter curve confirms that, as indicated by the present 180° curve, $ds/d\theta$ is least for $\theta = 0^\circ$ and 180° (namely, at the termini of the arm segments), it increases to a maximum in the region of the segment for θ between 25° and 30° (and between 150° and 155°), and attains a relative minimum at 90°. Similar considerations apply to the transcendental subordinal rank curves represented in other figures for supplementary angles (for example, Figs. 6-15, 7-17, and 7-18).

Fig. 7-17. $ds/d\theta$ versus θ for 0° to 180°-intercept transforms of the $e = 0.745$ ellipse about the center. The format of this figure is the same as for Figs. 6-15 and 7-16 for other transcendental subordinal rank curves. The α-transform of central conics about the center is of 8th degree. Reduction occurs for transformation angles of 0°, 180°, 90°, and the angle of the asymptotes of hyperbolas (and the supplementary angle).

Inasmuch as the pole of the transformation is a true center, i.e., it lies at the point of concurrence of two lines of symmetry,* all transcendental subordinal rank curves also have two lines of symmetry. The degrees of the equivalent eqs. in rectangular coordinates provide a basis for comparisons between the symmetry of curves at the subfocal level for transforms about true centers, for example for comparisons of the linear 180° symmetry of curves about their true centers.

However, such comparisons do not always provide the same measure of discrimination as the intercept subordinal rank eqs. This is seen by examining the general transcendental subordinal rank eq. of ellipses (eq. 1) for central

general transcendental subordinal rank equation for
central transforms of ellipses at the angle α

$$\frac{(b^2x^2+a^2y^2)^3[(a^2-b^2)(x\cos\alpha-y\sin\alpha)^2-a^2(x^2+y^2)]^3}{a^2b^2(a^2-b^2)^2} = \tag{7-17-1}$$

$$x^2y^2[(a^2-b^2)(x\cos\alpha-y\sin\alpha)^2-a^2(x^2+y^2)]^3 -$$

$$(b^2x^2+a^2y^2)^3(x\cos\alpha-y\sin\alpha)^2(x\cos\alpha-y\sin\alpha)^2$$

*[i.e., two orthogonal lines of symmetry]

ds/dθ Versus θ For 0° To 180°-Intercept Transforms of The e=0.745 Ellipse About The Center

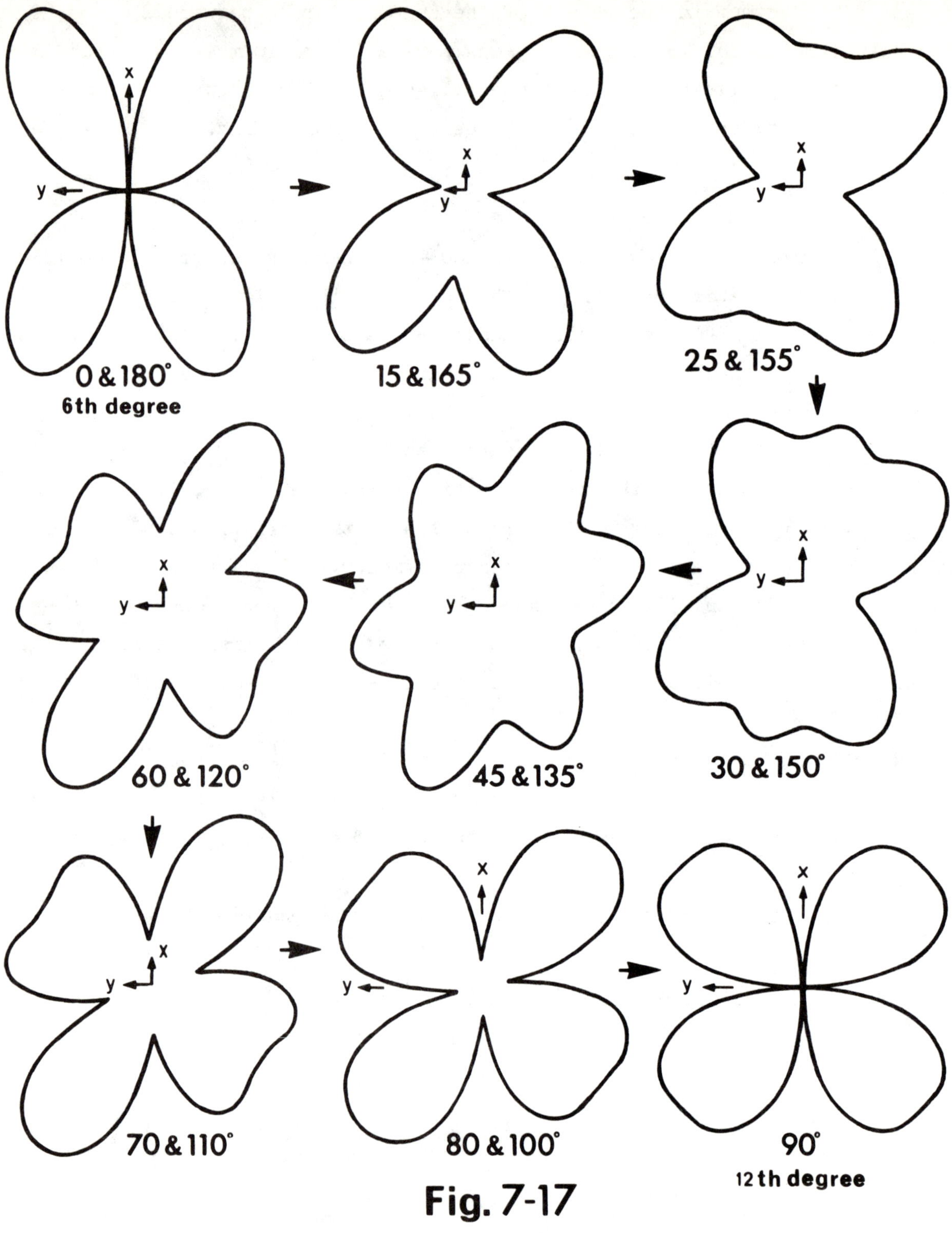

Fig. 7-17

transforms at an angle, α, expressed in rectangular form [where $(ds/d\theta)^2 = (x^2+y^2)$ and $\theta = \arctan y/x$]. It readily is ascertained that reduction of the degree of this eq. from the general value of 12 can occur only by factoring a term in $(b^2x^2+a^2y^2)^3$, and this can occur only for $\alpha = 0^\circ$ and 180°, whereupon the degree reduces to 6. On the other hand, the intercept subordinal rank eqs. provide more exponential-degree categories. Though the degrees of the 0°, 180°, and 90° eqs. are 6, 6, and 12, respectively (Fig. 7-19), the degree is higher for transforms at other angles (since the α-transform about the center is of 8th degree, compared to only 4th degree for the 90° transform).

Fig. 7-18. $ds/d\theta$ versus θ for 0° to 180°-intercept transforms of the $e = 0.745$ ellipse about the pole at $(-a/8,-a/8)$. The format of this figure is the same as that of previous figures for transcendental subordinal rank plots. All of the curves are to the same scale, but those for angles of 135° to 180° have been rotated for efficient utilization of space. The curves of this figure, as well as those of Fig. 7-19, were obtained from the intercept eq. about a point in the plane and derivatives ($dx/d\theta$ and $dy/d\theta$) obtained therefrom [where x is the intercept for the angle θ and y is the intercept for the angle $(\theta+\alpha)$].

Since the pole of the present transformation does not lie on a line of symmetry, none of the transcendental subordinal rank curves possesses a line of symmetry. The only apparent symmetry to be seen among these curves is that of the locus for a 180° angle of transformation, which possesses symmetry about the origin. The degrees of the curves of this figure have not been ascertained.

Fig. 7-19. Subordinal rank-determining curves, $(ds/d\theta)^2 = f(x)$, for various intercept transforms of the $e = 0.745$ ellipse about focal and non-focal poles. The curves of this figure are not to scale, having been reduced where necesary for the purpose of illustration. Values of $ds/d\theta$ are plotted as ordinates and values of the x intercepts as abscissae. Corresponding values of θ are indicated in small numerals at various points adjacent to the curves. The value of the angle of transformation, α, to which each subordinal rank curve applies is given in large numerals at the upper left of each curve. The exponential degrees of the rectangular eqs. of the curves, where known, are indicated beneath the plots. In cases in which the curve is recurrent, the

ds/dθ Versus θ For 0° To 180°-Intercept Transforms of The e=0.745 Ellipse About The Pole At (-1/8,-1/8)

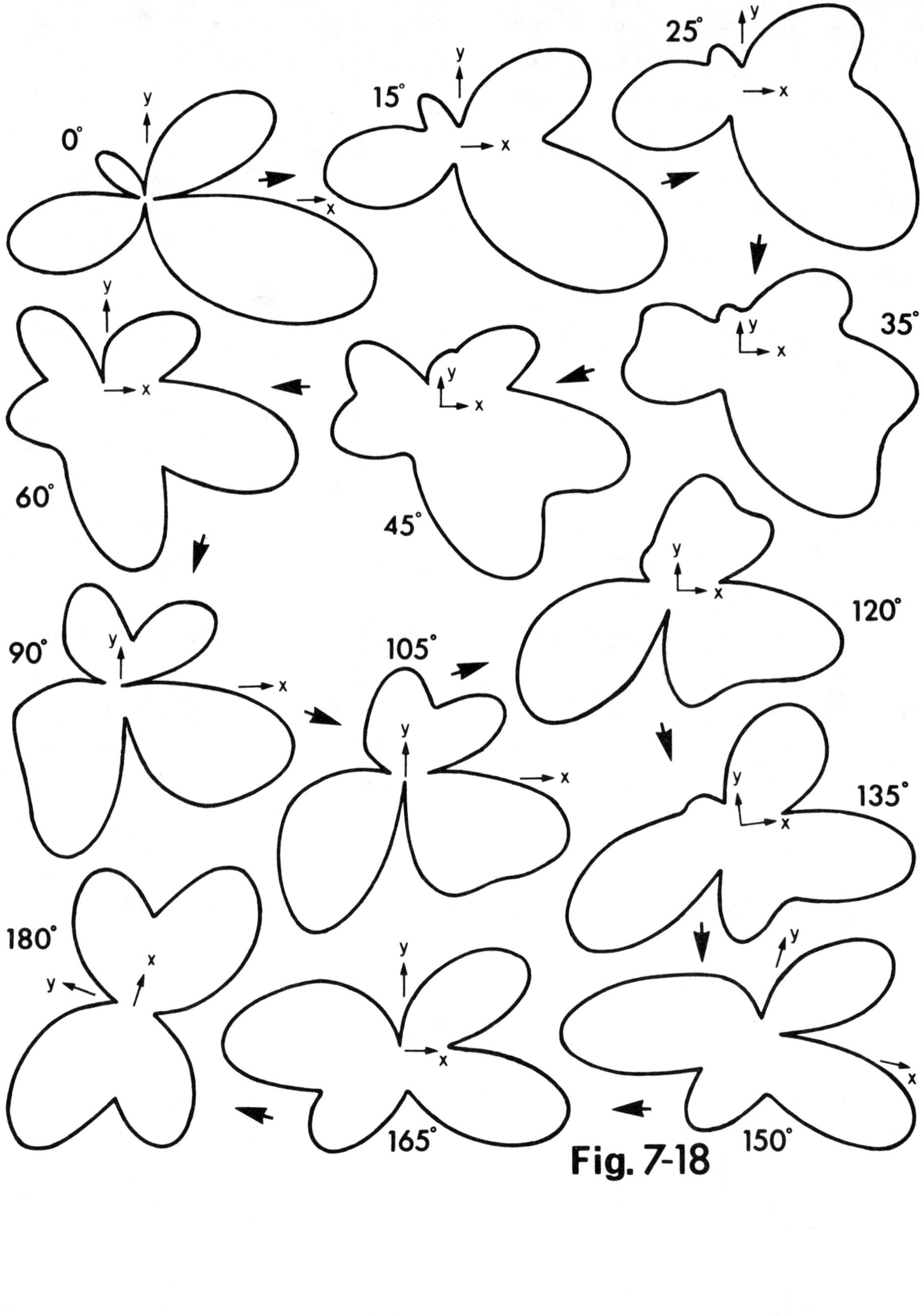

Fig. 7-18

number of traces is indicated adjacent to the curve.

The exponential degrees of the intercept subordinal rank eqs. of this figure define the subordinal rank of transforms about the corresponding poles at the corresponding angles and, as already discussed in the legends of Figs. 6-15 and 7-16, give a greater degree of discrimination of the symmetry of the curve at the subfocal level about the given pole at the given angle than do the transcendental subordinal rank eqs.

Certain aspects of the symmetry of the basis curve about the pole of transformation are evident from mere inspection of *transcendental* subordinal rank curves. If these curves have two lines of symmetry, the pole lies at a true center; possession of only one line of symmetry indicates that the pole lies on a single line of symmetry, while the absence of a line of symmetry indicates that the pole does not lie on a line of symmetry. Curves for a transformation angle of 180° (or, strictly speaking, for intercepts along oppositely-directed radius vectors) characteristically are symmetrical about the origin. Accordingly, the 180° curves for transforms about a pole incident upon a line of symmetry possess two lines of symmetry.

The same information is implicit in the present curves and depends upon the possession of information about their recurrent properties. Thus, for poles that lie on a single line of symmetry (upper set of curves) the 0° and 180° curves characteristically are traced twice, while the 90° curve is traced but once. For poles coincident with a true center, the 0°, 180°, and 90° curves characteristically are traced 4 times, while curves for all other angles are traced twice. For poles not lying on a line of symmetry, the curves characteristically are traced but once.

Non-recurrent intercept subordinal rank curves for 180° transforms about poles not lying on a line of symmetry, such as the 180° curve for the pole (-a/8,-a/8) (bottom center) have an unusual property not shared by any of the recurrent 180° curves. This property is readily related to the fact that the corresponding transcendental subordinal rank curves are symmetrical about the origin but do not have a line of symmetry. The property is that all lines parallel to the x-axis intersect the curve an even number of times, and the intersections are for values of θ that are 180° apart. If there are maxima or minima, they must occur in pairs and have the same ordinates. For example, refer to the maxima near the 160° and 340° points and 30° and 210° points, and

Subordinal Rank-Determining Curves, $(ds/d\theta)^2 = f(x)$, For Various Intercept Transforms of The e = 0.745 Ellipse About Focal And Non-Focal Poles

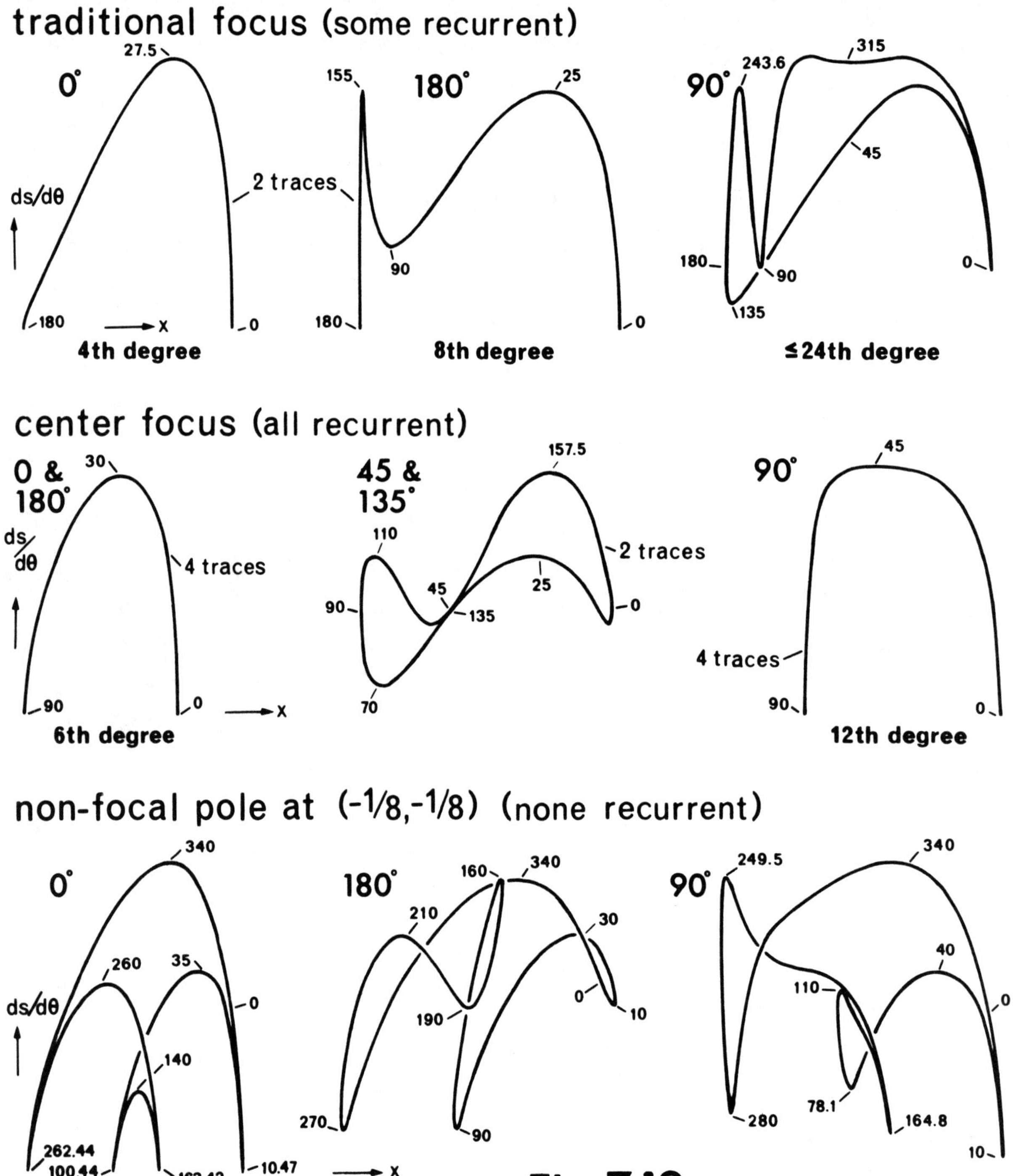

Fig. 7-19

the minima near the 90° and 270° points and 10° and 190° points of the 180° curve for the pole, (-a/8,-a/8).

Although the transcendental and intercept subordinal rank curves for transforms about a given pole at a given angle are quite different in appearance, they also possess similarities that are based upon the fact that one coordinate variable is the same for both. Thus, there is a correspondence between the number and size of the lobes of the transcendental curves and the number and size of the segments of the intercept curves. In addition, the transitions between lobes and segments occur for the same values of $ds/d\theta$.

For example, if one considers the 0° transcendental and intercept subordinal rank curves for the pole, (-a/8,-a/8), the large segment of the latter curve (the curve at the bottom for 0°) corresponds to the large loop of the former curve (Fig. 7-18-0°) and both portions are delineated by minima of $ds/d\theta$ for 10.47° and 262.44°; the small segment of the latter corresponds to the small loop of the former and both portions are delineated by minima of $ds/d\theta$ at 100.44° and 163.42° and a maximum at 140°, etc.

In the cases of 180° intercept subordinal rank curves that are traced twice, all points on the curves represent identical values of $ds/d\theta$ for angles 180° apart, while for quadruply traced curves all points represent identical values of $ds/d\theta$ 90°, 180°, and 270° apart.

θ Self-Inversion Loci of The e=0.745 Ellipse For $j = b^2/a = 4a/9$

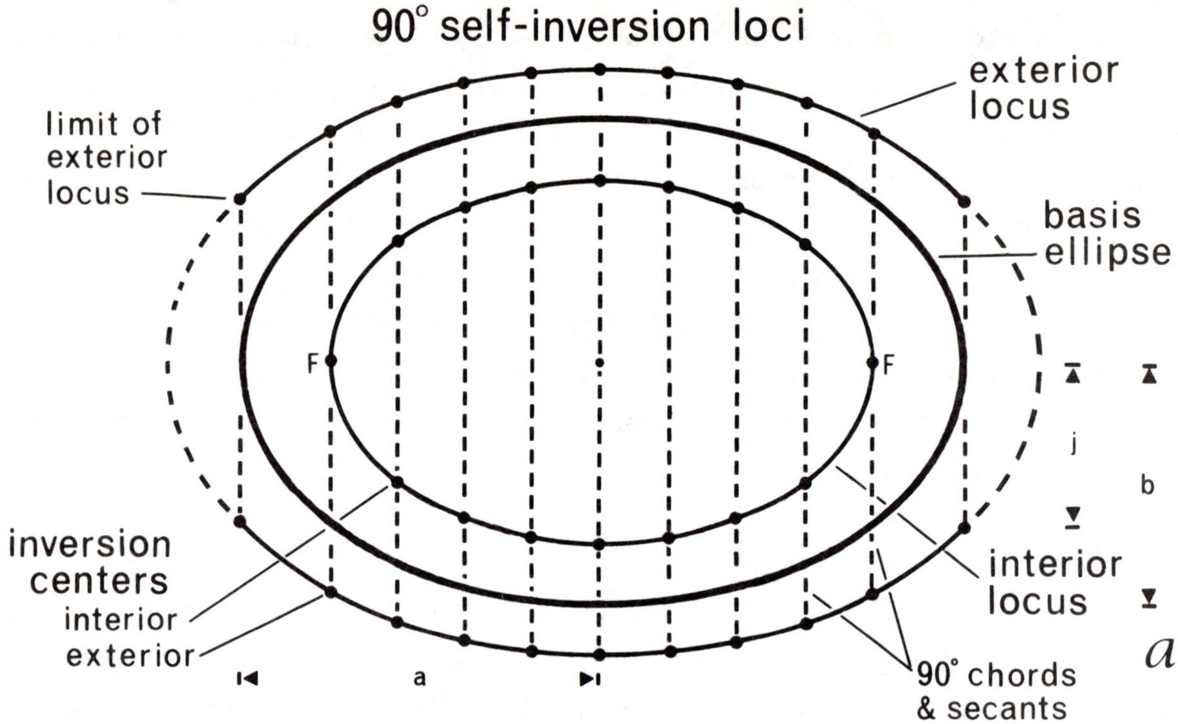

a

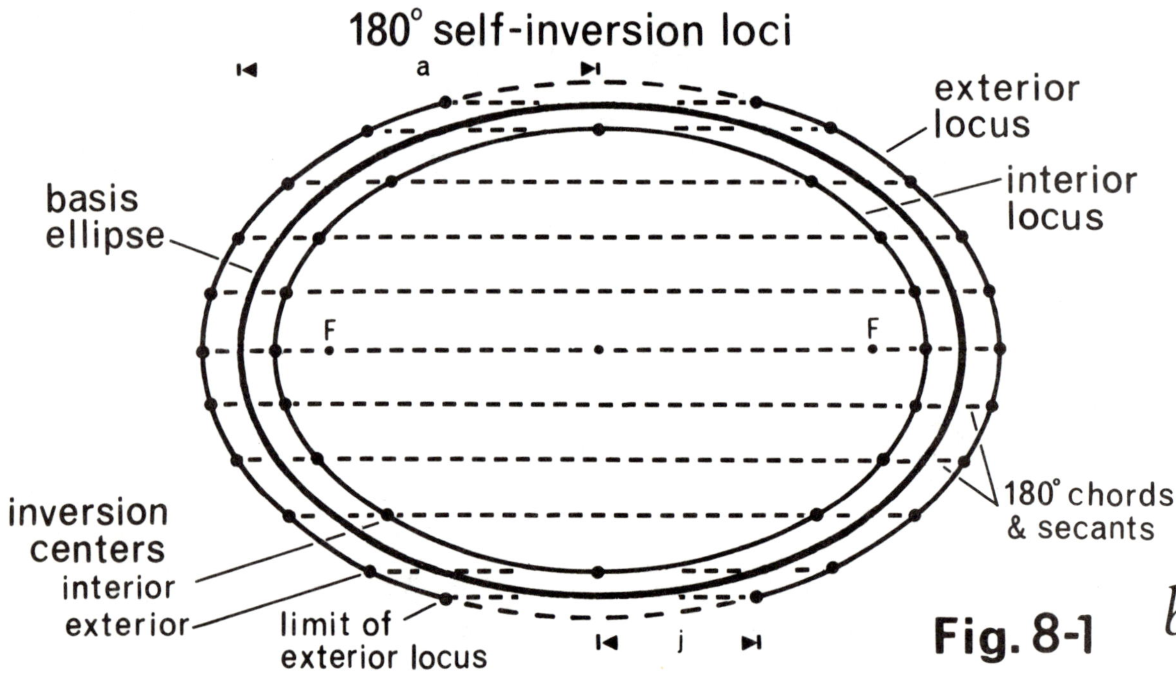

Fig. 8-1 *b*

CHAPTER VIII

Fig. 8-1. θ self-inversion loci of the $e = 0.745$ ellipse for $j = b^2/a = 4a/9$. If parallel secants are drawn through an ellipse at any angle θ, to the x-axis there generally will exist (for a given value of the unit of linear dimension less than a) two interior and two exterior points on each secant (the *interior* and *exterior inversion centers*) about which the points of intersection with the basis curve are inverse to one another (see pages VIII-1 to 4). The interior inversion centers define the *interior self-inversion locus* and the exterior inversion centers define the *exterior self-inversion locus* for the angle θ, in question, i.e., the θ *self-inversion loci*. For conic sections these loci are curves or segments of curves that belong to the same subspecies (have the same eccentricity) as the basis curve.

The present figure illustrates the 90° and 180° self-inversion loci of the $e = 0.745$ ellipse for a value of the unit of linear dimension, j, equal to the length of the hemi-LR, in this case 4a/9. The locations of the interior and exterior inversion centers are marked on each secant by the large solid circles, which define the θ self-inversion loci. The locations of the traditional foci and center of the basis curve are marked by the small solid circles.

Both of the exterior self-inversion loci consist only of segments of $e = 0.745$ ellipses, inasmuch as they terminate at the a and b vertices of the basis curve. The distances of the termini from the corresponding vertices are equal to the length of the hemi-LR, since this is the unit of linear dimension. In contrast to the exterior loci, the interior loci are complete ellipses. The limits of these interior loci, i.e., their a and b vertices, are determined by the length of the LR, since the interior loci cannot extend beyond the level of chords having this length (i.e., a *2j chord*).

The interior and exterior inversion centers that define the θ self-inversion loci also are of significance in connection with the *non-focal self-inversion loci* (see pages VIII-6 to 17) treated in Figs. 8-2 to 15. The latter loci pass through the *interior* and *exterior inversion centers* on any axis (or secant) that passes through the reference pole for the non-focal self-inversion. In fact, any given θ self-inversion locus is itself a non-focal self-inversion locus relative to an infinitely distant reference pole along a line at the angle θ.

Fig. 8-2. Interior non-focal self-inversion loci of the parallel line-pair. The classical self-inversion of a curve along concurrent lines with respect to the point of concurrence is a special case of self-inversion along concurrent lines with respect to points on the lines that do not necessarily coincide with the point of concurrence. In other words, when the points on the lines concurrent at the reference pole about which the curve self-inverts coincide with the reference pole, one has the classical, or *focal*, type of self-inversion; when they do not, one has *non-focal self-inversion.*

For example, the parallel line-pair of Fig. d "self-inverts" along lines concurrent at the reference pole (open circle) with respect to the points of intersection of the lines with the hour-glass-like curve (the *non-focal self-inversion locus*), for a unit of linear dimension, $j = d$ (where d is the distance from the midline to each line). Thus, the chord-segment length, a, times the chord-segment length, b, equals d^2, and similarly for all chord segments cut off by the interior non-focal self-inversion locus on lines concurrent at the reference pole.

Three examples for each of two cases are given in the figure. In Figs. a and b, the reference pole is at $h = \frac{1}{2}d$ and the interior non-focal self-inversion loci are given for values of $j = 3d/4$, d, and $3d/2$ illustrating, respectively, a locus with a cusp, a loop and DP, and two intersecting arms. In all cases, the lateral arms of the loci are asymptotic to the lines of the basis curve. The eq. of these curves--given at the center of the figure --is the same as eq. VIII-20, but for a locus rotated 90°. Since the reference pole is located asymmetrically with respect to the parallel line-pair, the interior non-focal self-inversion loci do not have reflective symmetry about any line parallel to the lines of the basis curve.

Figs. c and d give 3 examples of interior non-focal self-inversion loci for the case in which the reference pole is on the midline, for values of $j = \frac{1}{2}d$, d, and $3d/2$. Since the reference pole is on the midline, these loci are symmetrical with respect to this line. For $j = \frac{1}{2}d$, there are two branches that lie close to and are asymptotic to the lines of the basis curve. For $j = d$, there are two cusped asymptotic arms that meet at the reference pole (or two branches with their vertices tangent at the reference pole), while for $j = 3d/2$, there are two intersecting asymptotic branches.

The construction of *exterior* non-focal self-inversion loci of the parallel

Non-Focal Self-Inversion Loci of The Parallel Line-Pair (Interior, Various j)

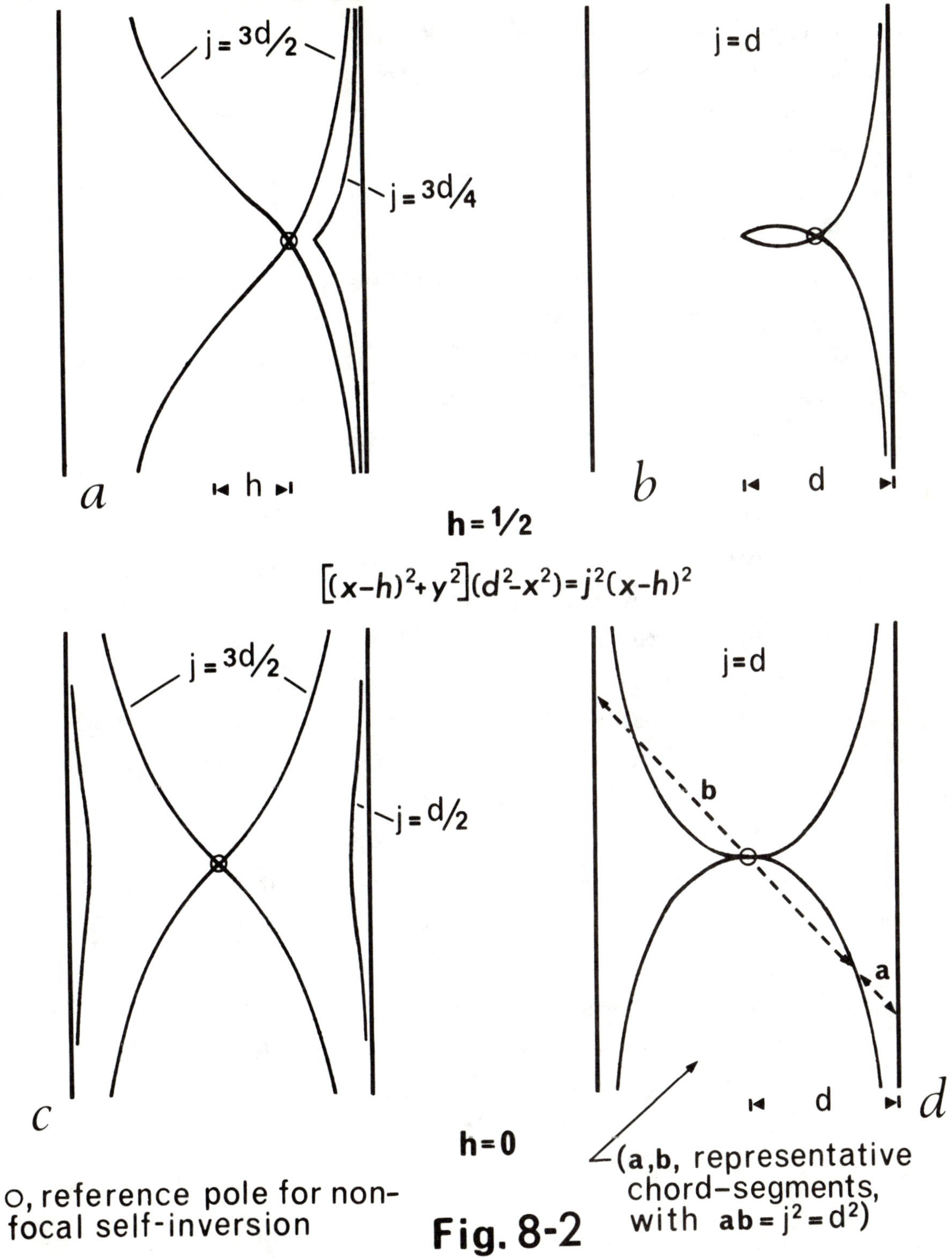

Fig. 8-2

line-pair about an interior pole is carried out in a similar manner. In that case, points are located on the exterior extensions of lines concurrent at the reference pole (one such point on each line on each side of the parallel line-pair) such that coincident radii a and b, to the line from each point, satisfy the relationship $ab = j^2$. The constructions proceed in an analogous fashion for exterior reference poles ($h > d$ in the eq. given at the center). As for interior poles, the points on the interior loci satisfy the relation $ab = j^2$ for chord segments, and the points on the exterior loci do so for coincident radii.

Interior non-focal self-inversion loci for other curves are plotted by selecting a reference pole and a value for j and passing chords through the pole at all angles. These chords then are divided into two segments, a and b, such that $ab = j^2$. The points at which these segments meet are connected, giving the interior non-focal self-inversion loci. If the basis curve self-inverts classically (i.e., undergoes *focal* self-inversion) about the point-pole, the non-focal self-inversion loci "collapse" to a point coincident with this point-pole when the reference pole comes into coincidence with it (illustrated in Fig. 8-9).

Interior non-focal self-inversion loci of the parallel line-pair are a type of polar-bilinear construction, $uv = Cd^2$, in which: (a) the two reference lines are parallel; (b) the distances u and v are measured off along chords concurrent at the reference pole from the points of intersection of these chords with the line-poles; and (c) the sum of u and v for any chord equals the chord length. The most symmetrical curve for these construction rules, $u = v$, is the midline.

A related classical construction is the cissoid construction. In this construction, lines are passed through a reference pole at all angles. Along each line, a distance is laid off from the reference pole equal to the distance between the points of intersection of that line with the two basis curves. The cissoid of the parallel line-pair, for example, is a parallel line, while that of a line and a circle (for a reference pole at the center of the circle) is the conchoid of Nicomedes (Fig. 3-7a).

Fig. 8-3. Non-focal self-inversion loci of the parabola for $j = 2a$ (interior, 1st quadrant, axial reference poles). Eight loci are plotted to scale in the composite figure running diagonally from lower left to upper right. Nine individual loci (omitting the 0.9a locus but including loci for 1.002a and 1.05a) are plotted in sequence from 0 to 2a, beginning at the upper left and terminating at the lower right. The locations of the axial reference poles are marked by the solid circles (except for those at 1.002a and 1.05a). The complete loci include the reflections of the illustrated segments in the axis of the basis curve, since the basis curve and the non-focal self-inversion loci have a common line of symmetry.

Inasmuch as the parabola is an open curve, the axial inversion centers lie at the vertex and infinity, and all axial non-focal self-inversion loci (eq. VIII-19) pass through these two poles. The loci also pass through the reference pole, provided that there exists a chord through this pole for which an inversion center coincides with the pole (because of the symmetry of the basis curve about the axis, if one such chord exists, a second one does also). Such chords exist for all reference poles from the vertex to $h = a$, at which point the two chords become orthogonal to the axis, coming into coincidence with the LR. This is a consequence of the choice $j = 2a$; in other words, axial non-focal self-inversion loci of the parabola pass through their axial reference poles for all poles that precede and include the point of concurrence of the 2j chord with the axis (i.e., a chord of length 2j orthogonal to the axis).

For all chords through reference poles beyond the traditional focus (or beyond the position of intersection of the 2j chord, for any other selection of j), one inversion center lies on one side of the axis and the other lies on the other side. The positions of the nearest of such centers to the axis "rapidly" recede from the axis as the reference pole recedes from the point of intersection of the 2j chord. In the present instance, it will be noted that for a reference pole only 0.002a beyond the focus, the distance of the points on the locus nearest the reference pole is already as great as 0.063a, while for a pole 0.05a beyond the focus this distance is 0.314a.

All the points at which the axial non-focal self-inversion loci of the parabola intersect or otherwise make contact with its axis are exceptional points of the loci. In the case of the locus for the vertex as reference

Non-Focal Self-Inversion Loci of The Parabola For j=2a (Interior, 1st Quadrant, Axial Reference Poles)

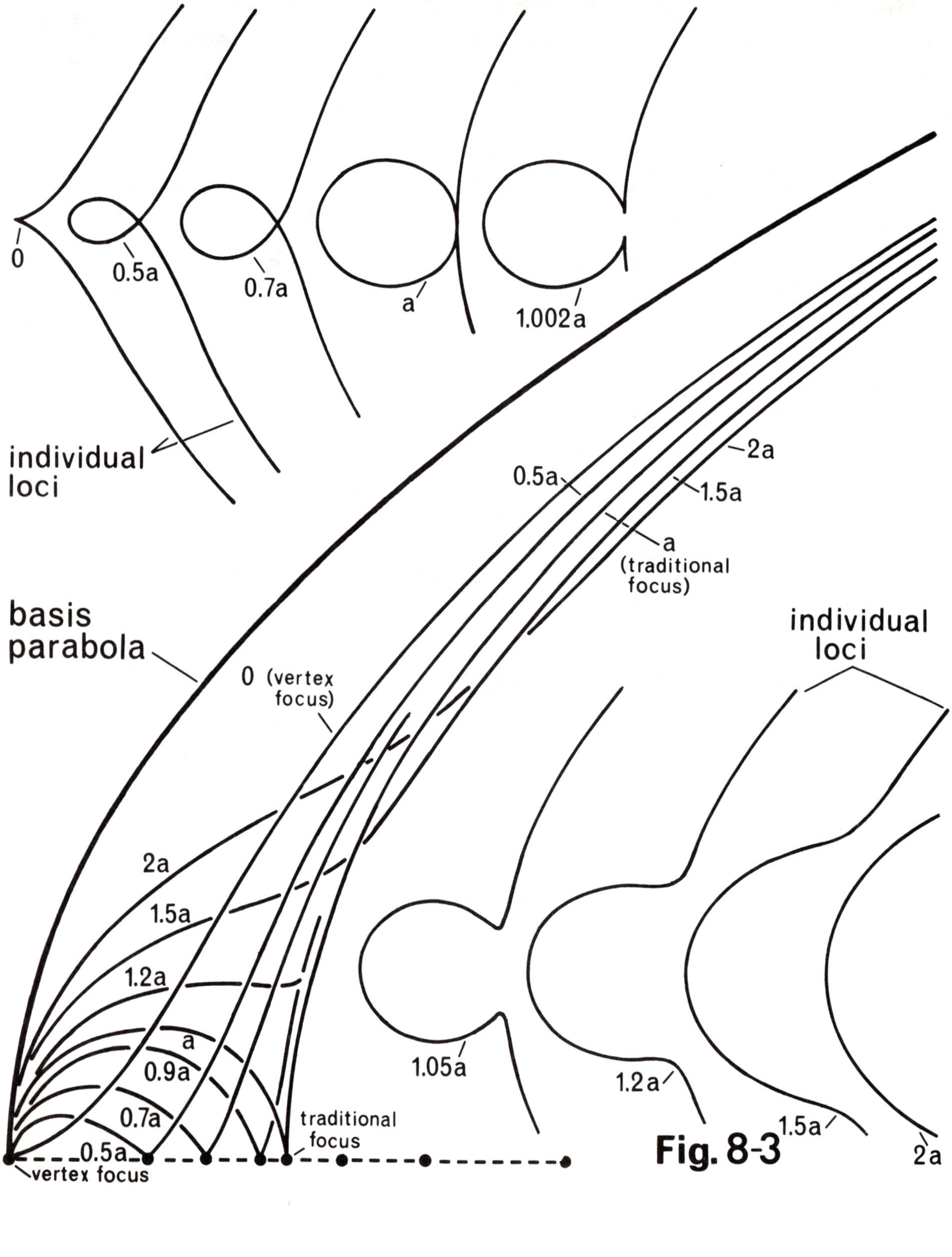

Fig. 8-3

pole, the curve is tangent to the axis at the vertex, forming a cusp. For all reference poles between the vertex and the 2j chord, a DP is formed. The latter is the point of intersection of a closed loop and an "open loop." Because of the symmetry of the basis curve about the axis, the DP-tangents at any given point make the same (acute) angle with the axis (if one DP-tangent intersects the axis at an angle θ, the other does so at an angle $-\theta$), this being the angle of the two chords that possess an inversion center incident upon the axis at the given point. At the position of the 2j chord, however, the two loops of the locus are tangent to the chord and one another at the point of intersection with the axis (which also may be regarded as a point of contact of two opposed cusps).

The loop "opens rapidly" for axial reference poles beyond the 2j chord (in this case, beyond the LR), and for a pole at 2a, little trace of the previous loop region remains. The exceptional points of these loci for poles beyond the 2j chord consist solely of inflection points, and these loci make no contact with the axis of the basis curve other than at their point of intersection at the vertex.

Fig. 8-4. Non-focal self-inversion loci (eq. VIII-18) of the $e = 0.707$ ellipse for $j = b^2/a = a/2$ (interior and exterior single-quadrant segments, axial reference poles). The axial reference poles are marked by the large solid circles, while the small solid circles mark the locations of pertinent points that are not reference poles. All segments, inversion centers, and other poles or points are identified by associated labels.

In the sole case of the upper traditional focus, the figure illustrates segments of the interior non-focal self-inversion locus in two quadrants (left upper and lower portions of the figure; see Figs. 8-6 and 8-7 for complete 4-quadrant interior and exterior loci). These, of course, characterize the entire locus, since the other two segments are reflections of these segments in the major axis. In all other cases, the figure illustrates only a quarter of the complete locus. In each case the illustrated segment is for the same half of the ellipse that contains the reference pole. For example, for the 5 poles incident upon the right half of the minor axis ($k = 0.25a$ to $k = 0.707a$), the ¼-segments are illustrated at the upper right (portions of all 5 interior loci and of two exterior loci).

In cases in which extensive overlap of the loci would occur, only portions of some of the segments are illustrated. The manner in which the loci are constructed is illustrated for a point (labelled p) on the locus for the center at the upper left. The location of this point is such that the product of the lengths of the chord segment x, from point p to the near point of intersection with the basis curve, and the chord segment y, from point p to the far point of intersection, equals j^2. As in the case of the parabola of Fig. 8-3, the unit of linear dimension is taken to be the length of the hemi-LR.

The non-focal self-inversion loci for the center are the only such loci that have two lines of symmetry (since the center is the only reference pole that lies on both lines of symmetry of the basis curve). Accordingly, the interior and exterior segments of the locus for the center are representative of all 4 interior and exterior center segments; the complete locus can be visualized by reflecting the upper right segments in both axes (see also Fig. 8-6a). In all other cases, illustrated ¼-segments are representative of only half of the complete loci, since other loci have only one line of symmetry.

The interior and exterior axial inversion centers are established as described previously: the interior centers on both axes demarcate chord segments to the basis curve, the product of whose lengths is j^2; the corresponding exterior centers demarcate coincident radii to the basis curve, the product of whose lengths is j^2. All interior (or exterior) loci for reference poles on a given axis intersect the axis at both corresponding interior (or exterior) inversion centers.

In addition, certain interior non-focal self-inversion loci also pass through their corresponding reference poles and consist of two loops. These are the loci for reference poles for which an intersecting chord exists upon which an inversion center coincides with the reference pole. This is the case for all axial reference poles lying between the position of the 2j chord and the position of the corresponding interior axial inversion center (and including the pole at the position of the 2j chord). Accordingly, loci for all reference poles on the major axis that satisfy the relation $0.707a \leq h < 0.866a$ also intersect the major axis at the reference pole, and loci for all reference poles on the minor axis that satisfy the relation $0.5a < k \leq 0.612a$ also intersect the minor axis at the reference pole.

Non-Focal Self-Inversion Loci of The e=0.707 Ellipse For $j=b^2/a=a/2$ (Interior & Exterior Single-Quadrant Segments, Axial Reference Poles)

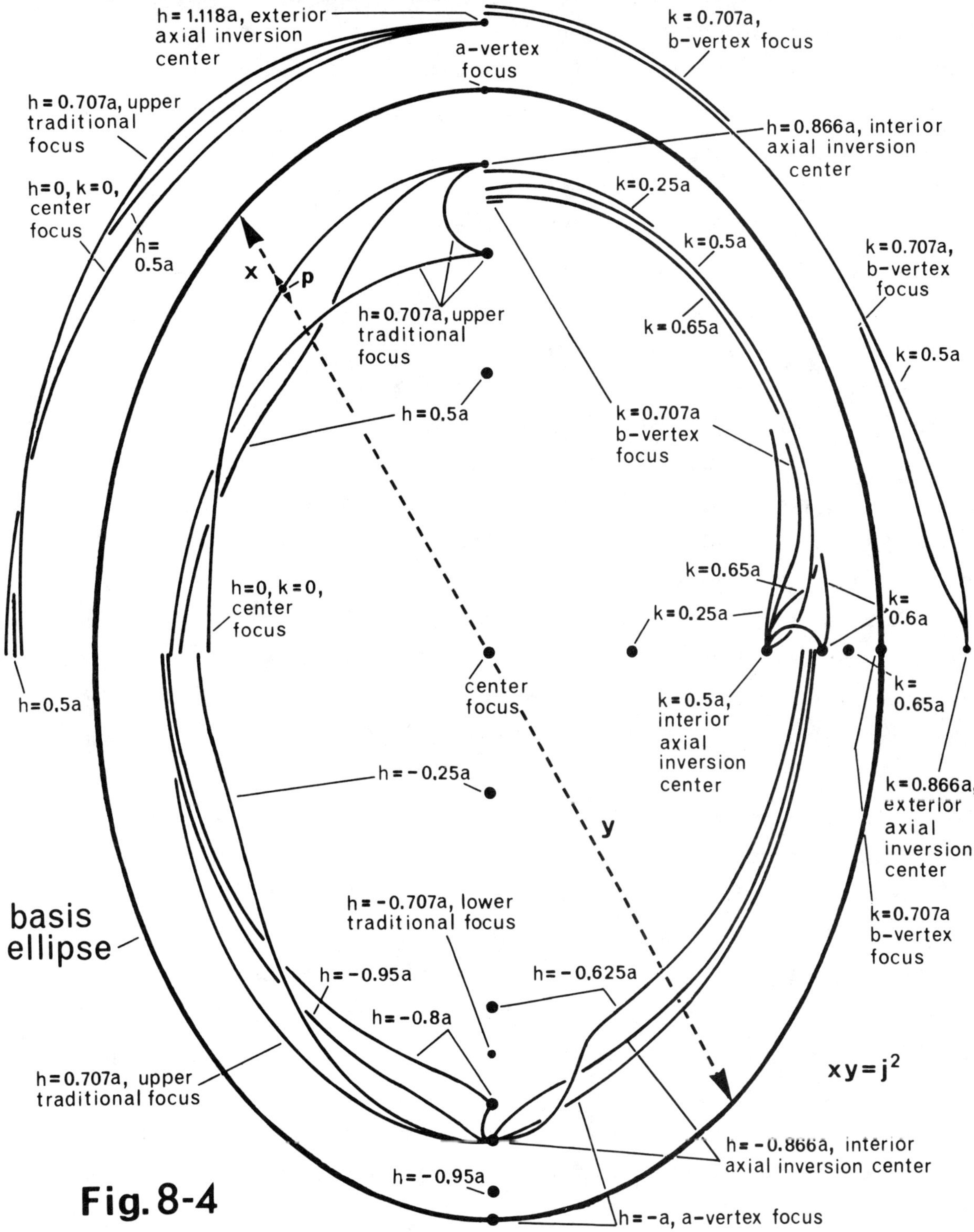

Fig. 8-4

One or the other of the above conditions is satisfied in the figure by the axial reference poles at $h = 0.707a$ and $-0.8a$, and $k = 0.6a$. Accordingly, the loci for these reference poles possess two loops. In cases in which the reference pole lies at the position of the $2j$ chord, the two loops of the locus are tangent to one another and the $2j$ chord at the reference pole; otherwise the loci meet the axis at a DP, with the DP-tangents being at the angles of those two chords for which an inversion center is coincident with the reference pole.

Segments of only 5 exterior non-focal self-inversion loci are illustrated, two segments of loci for poles in the minor axis, two for poles on the major axis, and one for a pole on both axes (the center). Exterior loci also may possess two loops (see Fig. 8-7f) but this is not the case for any of the exterior loci of the present figure.

Fig. 8-5. Non-focal self-inversion loci for the $e = 0.707$ ellipse for various values of j (1st quadrant, reference poles on the major axis). The present figure for the non-focal self-inversion loci of 3 major-axis point foci and an unexceptional pole at $h = a/2$--for values of j^2 ranging from $1/16$ to $7/8$--illustrates the influence of the value of the unit of linear dimension on the form and size of the loci. The value of j^2 for each locus is obtainable from the numbers just to the right of the axial inversion center for each curve (the termini of the loci at the right of each ensemble), which represent $16j^2/a^2$. Thus, for the number 1, $16j^2/a^2 = 1$, whence $j^2 = a^2/16$, etc.

Positions along the major axis are indicated (in units of a) by the numbers running vertically at the center, while positions along the minor axis (also in units of a) are indicated by the numbers at the bottom. The large solid circles mark the locations of the axial reference poles (not shown for the vertex), while the small solid circles in the lower right corners of each set of curves mark the center. A segment of the basis ellipse (major-axis vertex at a and minor-axis vertex at $\sqrt{2}\,a/2$) is illustrated in Fig. c.

Reference to Fig. a for the interior non-focal self-inversion loci with the center as reference pole reveals that the curves generally consist of two loops for values of $2j$ equal to or greater than the length of the orthogo-

Non-Focal Self-Inversion Loci of The e=0.707 Ellipse For Various j (1st Quadrant, Reference Poles On Major Axis)

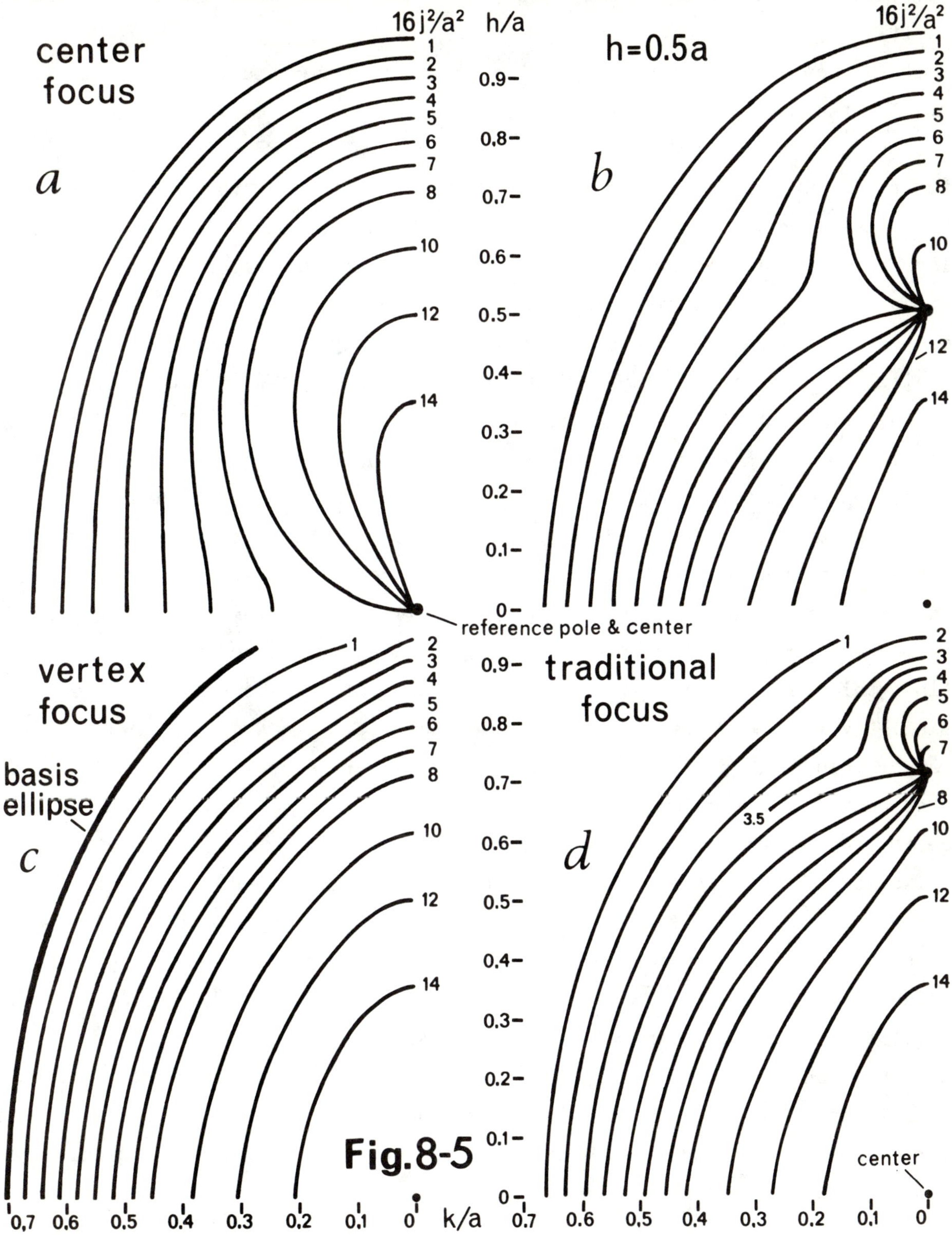

Fig.8-5

nal chord at the reference pole (the minor axis), but of a single oval for values of $2j$ less than the length of this chord. The largest value of j for which an interior locus exists is a, the length of the semi-major axis, for which the locus is the reference pole itself, but only the two a vertices "self-invert."

It is clear from this consideration that the larger the value of j, the smaller the segment of the basis curve that undergoes non-focal self-inversion with respect to points on the corresponding locus. In fact, the segments of the basis curve that undergo self-inversion with respect to the non-focal self-inversion locus are demarcated by the DP-tangents. This means that the largest value of j for which the entire basis curve self-inverts with respect to the corresponding non-focal self-inversion locus is $j = b = 0.707a$, one-half the length of the orthogonal chord through the reference pole.

Similar considerations apply for the loci of Figs. b-d. The portions of the basis curve that self-invert with respect to the plotted non-focal self-inversion loci are demarcated by the intersections with the basis curve of lines that pass through the reference pole and are tangent to these loci.

For the unexceptional pole at $h = \frac{1}{2}a$ (Fig. b), the loci consist of single ovals for values of j less than half the length of the orthogonal chord at the reference pole. They consist of two loops for values of j varying from equality with this hemi-length $[(3/8)^{\frac{1}{2}}a]$ up to but not including $\sqrt{2}$ times this hemi-length ($\sqrt{3}a/4$). For the latter value the curve becomes a single oval with a cusp at the reference pole (the curve of Fig. b-12). For larger values of j the cusp is lost and the curves consist of single ovals that contract in size with increasing j. For $j = a$, the locus becomes the center and only the a vertices "self-invert."

The loci for the reference pole at the traditional focus, as well as those for a pole at any other axial point except the center and vertices, differ in no significant manner from those at $h = \frac{1}{2}a$. The absolute values of j at which the loci break up into two ovals and return to single ovals are different, of course, but their relative values are the same, namely, the hemi-length of the orthogonal chord through the reference pole, and $\sqrt{2}$ times this length. Additionally, the largest value of j for which the entire basis curve undergos non-focal self-inversion is the length of the orthogonal hemi-chord through the reference pole.

The situation for an a-vertex focus as reference pole (Fig. d) differs from that for non-incident locations. All the non-focal self-inversion loci now consist of single ovals, none of which possesses a cusp, and the entire basis curve does not undergo non-focal self-inversion for any finite value of j (although the portion not undergoing such self-inversion may be made as small as desired by taking j sufficiently small). In addition, for small values of j, "irregular" inflections occur in those regions of the non-focal self-inversion loci (the curves of Figs. c-1 to c-5) that are proximal to the reference pole for the inversion (the a vertex). These are indicated in Fig. c but are not evidenced well at the scale employed.

The changes of size and form that some of the non-focal self-inversion loci undergo with changes in j are somewhat reminiscent of the changes undergone by, for example, the 4th-degree ovals of Cassini (eqs. XII-11 and 8-31-5), all of which possess a true center. For $C<\frac{1}{4}$ the curves consist of two congruent ovals. For $C=\frac{1}{4}$ the ovals come into contact at a DP, forming an equilateral lemniscate--a *curve of demarcation* (see pages VIII-19 to 21 and Figs. 8-16 to 8-21,8-24,8-25,8-28, and 8-42). For $C>\frac{1}{4}$ the curves consist of a single oval (Figs. 8-17 and 8-19). In the present instance, only the loci for the center-focus (eq. 1b) possess a true center. For them, the pro-

major-axis non-focal self-inversion loci of
the $e=0.707$ ellipse

(a) $(a^2-x^2-2y^2)[y^2+(x-h)^2]-j^2[(x-h)^2+2y^2] = 0$ about pole at $x=h$ (8-5-1)

(b) $a^2(x^2+y^2)-(x^2+y^2+j^2)(x^2+2y^2) = 0$ eq. 1a for center focus

(c) $a^2y^2+(x^2+y^2)(x^2+2y^2) = 0$ eq. 1b for $j=a$

(d) $a^2y^2+[(x-h)^2+y^2](x^2+2y^2) = 0$ eq. 1a for $j=a$

gression with increasing j is from single ovals to two tangent ovals (for $j=\sqrt{2}a/2$) to two opposed lemniscate-like loops with a DP, to a point locus, $x=y=0$, for $j=a$ (eq. 1c).

For points on the major axis other than a center or a vertex, the curves (eq. 1a) have only a single line of symmetry and the progression of loci is more complex. The loci progress from single ovals for small j, to two tangent ovals for $j^2=(a^2-h^2)/2$, to two opposed lemniscate-like loops with a DP, to a single oval with a cusp at $j^2 = (a^2-h^2)$, to single ovals and, finally, to a point locus, $x = y = 0$, for $j = a$ (eq. 1d).

Fig. 8-6. Non-focal self-inversion loci of the $e = 0.707$ ellipse ($j = \frac{1}{2}a$, complete interior and exterior loci, reference poles incident upon the line $x = y$). Each set of curves represents the complete interior and exterior non-focal self-inversion loci of the $e = 0.707$ ellipse about the indicated reference pole (coordinates at lower left or right of each set). In Fig. a the basis ellipse also is shown (dashed tracing). The locations of the reference poles in the plane of the basis ellipse are shown at mid-upper center.

The general eq. VIII-18 for the non-focal self-inversion loci of ellipses is repeated here for convenience as eq. 1a, where the upper alternate sign is for

non-focal self-inversion loci of ellipses
about the pole (h,k)

$$\text{(a)}\quad \mp[(x-h)^2+(y-k)^2](a^2b^2-b^2x^2-a^2y^2) = j^2[b^2(x-h)^2+a^2(y-k)^2] \qquad (8\text{-}6\text{-}1)$$

$$\text{(b)}\quad \mp(x^2+y^2)[a^2b^2-b^2(x+h)^2-a^2(y+k)^2] = j^2(b^2x^2+a^2y^2) \qquad \text{eq. 1a recast about pole at } (h,k)$$

$$\text{(c)}\quad \mp[a^2b^2-b^2(r\cos\theta+h)^2-a^2(r\sin\theta+k)^2] = j^2(b^2\cos^2\theta+a^2\sin^2\theta) \qquad \text{eq. 1b in polar form}$$

the exterior loci. This eq. is a quartic with cubic, quadratic, and linear terms in the variables, and a constant term. Accordingly, plotting of loci from this eq. would require the use of a zeros of functions computer program. However, $(x+h)$ and $(y+k)$ translations of the loci, which recast the eqs. about the reference poles (as opposed to the center of the basis ellipse) lead to eq. 1b. In polar form this becomes eq. 1c, which is quadratic in r, and leads readily to the plots of this figure and of Fig. 8-7.

Reduction of the degree of the polar eq. in r from 4 to two by recasting eq. 1a about the point (h,k) identifies the latter pole as a high-ranking circumpolar focus of the curve--in fact, the highest-ranking. This typically is the case for the poles of point transformations of conic sections that yield curves of low degree. Thus, since the transformation (i.e., the construction rule for the transform curves) is specified with respect to the highest-ranking focal pole, it is simpler than the construction rules with respect to all other (lower-ranking) poles.

Non-Focal Self-Inversion Loci of The e = 0.707 Ellipse (j = a/2, Complete Interior & Exterior Loci, Reference Poles Incident Upon Line x = y)

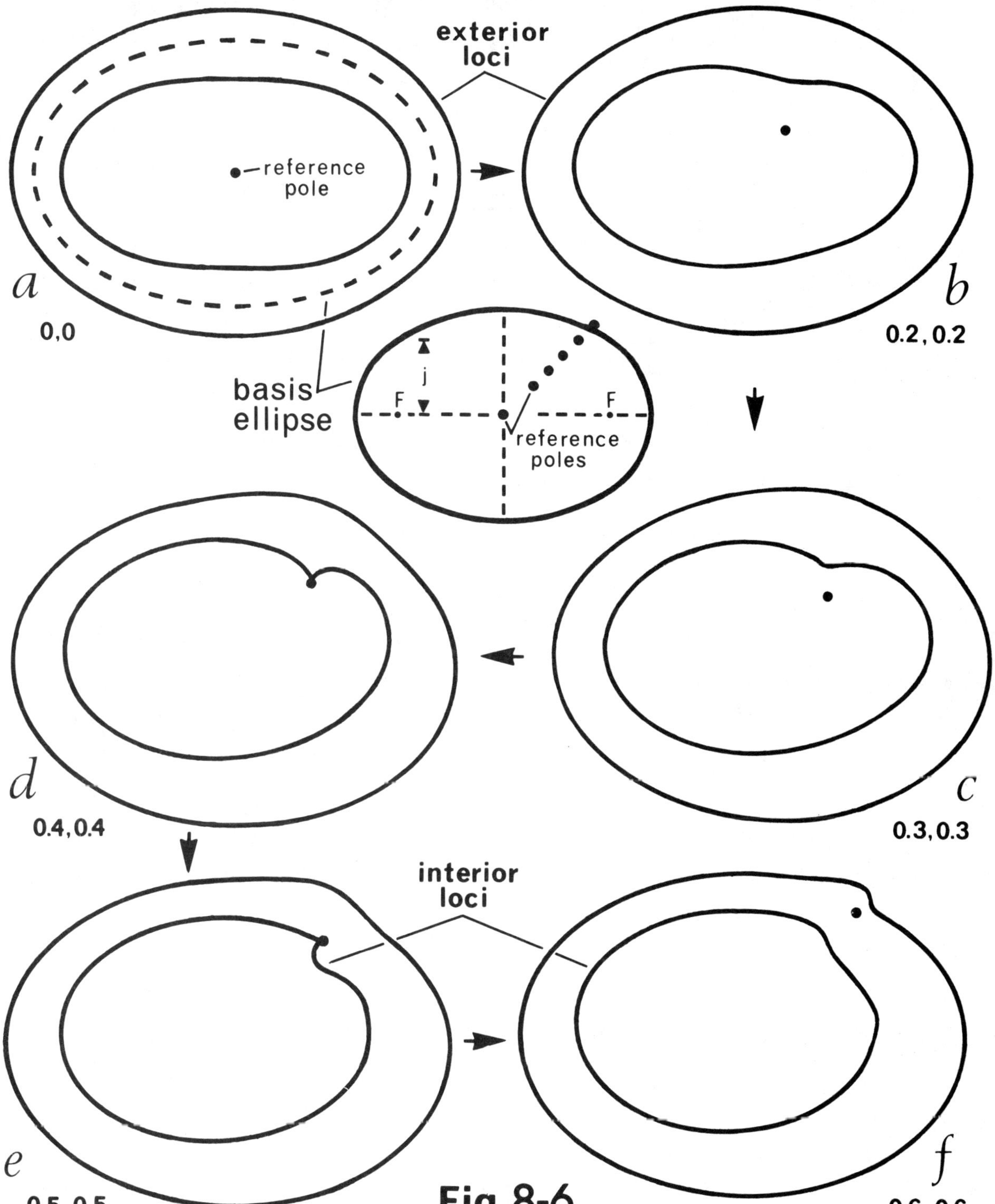

Fig. 8-6

The individual illustrations of this figure are essentially self-explanatory. Only the non-focal self-inversion loci about the center of an ellipse have two lines of symmetry, the interior locus being greatly flattened in the region of the "minor-axis" vertices (see also the larger scale illustration in Fig. 8-4, upper left). Displacement of the reference pole from the center along the line $x = y$ leads to asymmetry of both the interior and exterior loci. The markedly asymmetrical locations of these loci with respect to the reference poles emphasize the essentially different nature of the point transformation of non-focal self-inversion from other commonly dealt with point transformations (such as traditional inversions and pedals).

This different nature has its basis in the fact that the location of the reference pole has very little overall influence on the location of the transform curves (the more distant the pole, the lesser the influence), though the location of the pole considerably influences the nature of a transform curve in its near vicinity. The reference pole is proximal to the interior loci in Figs. d and e, but is not incident thereupon, nor does either interior locus possess a cusp in the neighboring region.

Fig. 8-7. Non-focal self-inversion loci of the $e = 0.707$ ellipse ($j = \frac{1}{2}a$, complete interior and exterior loci, reference poles incident upon the right LR). The present figure complements Fig. 8-6 and follows the same format. It differs notably in presenting the non-focal self-inversion loci about reference poles incident upon the LR, including the right traditional focus and a reference pole that also is incident upon the basis curve. The factors that influence the nature of the loci (such as the presence of two tangent ovals for $j = \frac{1}{2}a$, when the traditional focus is the reference pole) are treated in Figs. 8-4 and 8-5, as well as in the text (pages VIII-11 to 13).

It is instructive to follow the relationships between non-focal self-inversion loci and the LR for poles incident thereon. Since all poles lie on the LR, and the length of the hemi-LR is the unit of linear dimension, all the interior loci must be tangent to the LR at the traditional focus. Accordingly, a tangent from the reference pole to the curve at the point of tangency of the curve and the LR must be of the same length as the ordinate of the reference pole (shown by the dashed lines from the reference pole to the curve along the position of the LR). For the traditional focus this

Non-Focal Self-Inversion Loci of The e=0.707 Ellipse (j=a/2, Complete Interior & Exterior Loci, Reference Poles Incident Upon Latus Rectum)

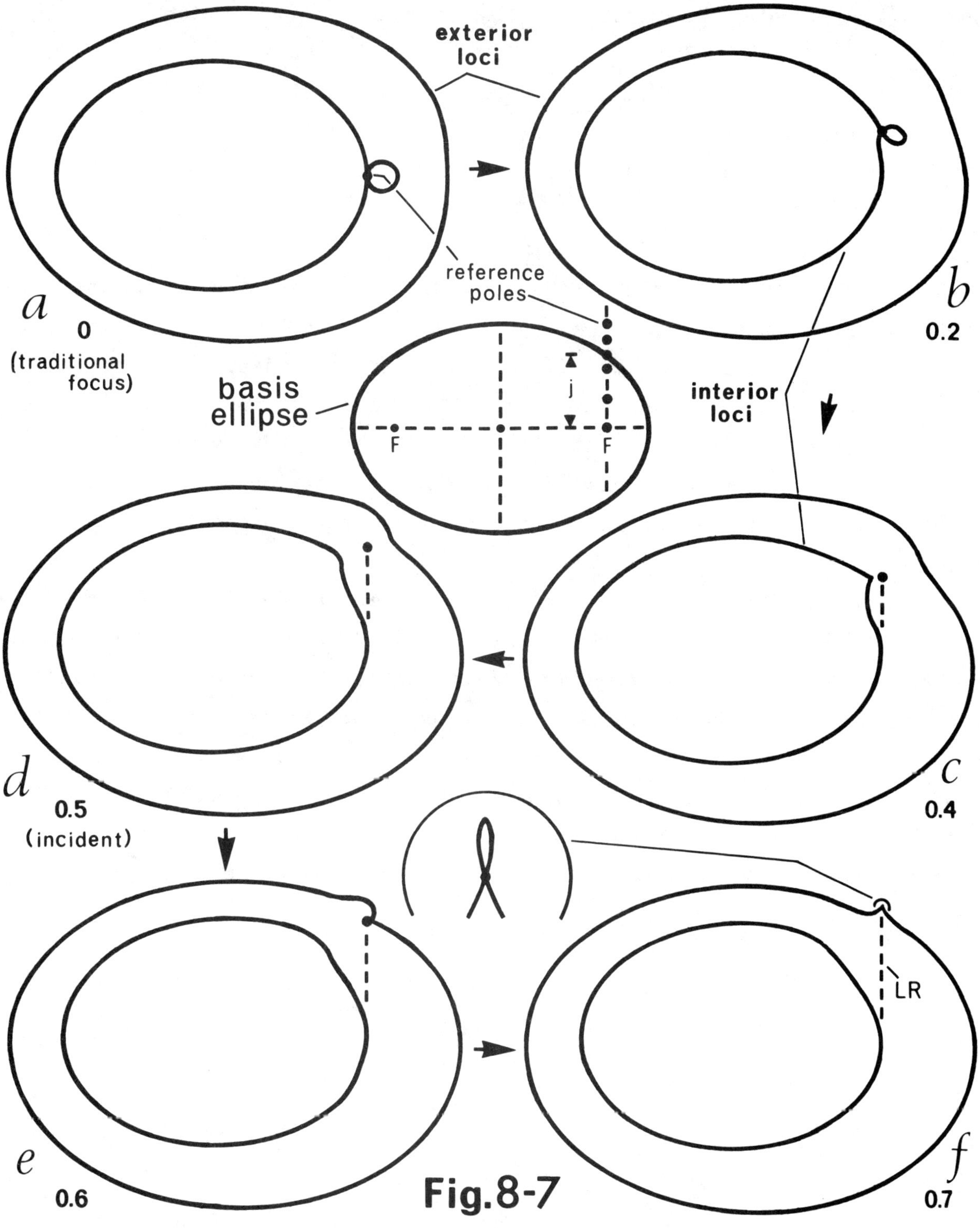

Fig.8-7

length is 0, for the pole at $k = 0.2a$ its length is 0.2a (not shown), for the pole at 0.4a its length is 0.4a, etc.

On the other hand, intersections of the exterior loci with the extended LR always must be at a distance, $L = 0.207a$, from the upper right LR vertex, where $L(L+a) = a^2/4$. Accordingly, this intersection is at a distance $(0.7071a - k)$ from the reference pole. For $k = 0$ in Fig. a, this distance is 0.7071a, for $k = 0.4a$ in Fig. c, it is 0.3071a, while for $k = 0.6a$ in Fig. e, it is 0.1071a.

An interesting situation arises in Fig. f, for which the distance of the reference pole from the locus is only 0.0071a. The corresponding region of Fig. f (exterior locus) has been enlarged greatly in the insert to show this situation. The diameter of the tiny loop--from the reference pole (at its DP) to its vertex--is 0.0071a. The DP-tangents pass through the reference pole at the angles for which the product of the lengths of coincident radii from the reference pole to the near and far points on the basis ellipse is $a^2/4$. In the case of the tangent through the loop DP from lower left center to upper right, the exterior radius vector shortens and the interior one lengthens; for the other DP-tangent, the lengths of the radius vectors change in the opposite sense.

Whether or not the exterior non-focal self-inversion locus for a given exterior reference pole passes through the pole itself depends entirely upon whether an angle exists for which the product of the lengths of the radius vectors from the pole to their near and far points of intersection with the curve is $j = a^2/4$. For $k = 0.7a$, two such angles exist, with the consequence that a loop is formed. The interior locus passes through the reference pole for all locations up to, and slightly beyond, $k = 0.3a$; beyond the limiting ordinate no chord segments through the pole exist for which the product of the lengths is $a^2/4$.

Fig. 8-8. Non-focal self-inversion loci of the equilateral hyperbola ($j = a$, 0° and 180° loci, transverse-axis reference poles, and detail for the traditional foci). Since the non-focal self-inversion loci of hyperbolas are the most complicated of those for any conic species, these loci are shown in several aspects in the present illustration. At the lower right (Fig. a) is shown a composite of single-quadrant plots of loci for 5 axial reference poles; $h = 0$ (center), $h = a$ (vertex), $h = 1.1a$, $h = \sqrt{2}\,a$ (traditional focus), and $h = 2a$.

A segment of the centered basis curve is shown by the heavy tracing, an asymptote by the heavy dashed line emanating from the origin at 45° to the positive x-axis, the positive x-axis by the horizontal light dashed line emanating from the origin, the upper right hemi-LR by the light dashed line emanating from the right traditional focus at 90° to the x-axis (and terminating at the upper right LR vertex), and segments of various 0° and 180° loci by the light tracings. The labels adjacent to the latter segments identify the corresponding reference poles.

An overall view of the nature of complete intra-arm and inter-arm loci referred to a single pole, as well as the relationships of these loci to one another and the basis curve, is shown at the upper left (Fig. b) in a full 4-quadrant plot referred to the two traditional foci. The 5 axial foci and latera recta also are shown (the key to the labels is at the lower right).

The loci can be regarded as consisting of 3 segments: (a) a central lemniscate-like portion with its loop-vertex foci coincident with the conic traditional foci, with its DP coincident with the center, and with its loop foci coincident with the hyperbola vertex foci; and (b) two lateral-arm segments, each consisting of an arm terminating in a cusp at the traditional focus, as well as a hyperbola-like arm with its vertex at the traditional focus.

These loci are dissected into their component segments (dotted segments in Figs. c-g)based upon their non-focal self-inversion properties with respect to the basis curve. For example, consider the segment of the lemniscate-like locus (Fig. c) in the 1st quadrant that extends from the center to its point of intersection (the "45° point") with the right upper basis arm. This segment is the 180° inter-arm non-focal self-inversion locus of the entire upper left arm of the basis curve (from its vertex to infinity) and a segment of the upper right arm (from the vertex to the 45° point).

The two chord segments labelled x and y show how this portion of the 180° inter-arm locus is related to the arm segments in question. Fig. d illustrates the portion of the right 0° inter-arm locus (from the traditional focus to the same 45° point of intersection with the basis curve) that is formed by virtue of coincident radii emanating from the same segments of the basis curve as those illustrated in Fig. c.

The portion of the 180° intra-arm locus shown in Fig. e is a continuation of the portion of the 180° inter-arm locus of Fig. c. In this case the locus is formed by virtue of chord segments emanating from points thereon (along lines through the right traditional focus) to two segments of the right arm of the basis curve--one segment extends from the lower LR vertex to infinity and the other extends from the upper 45° point to infinity.

The 0° intra-arm dotted segment of Fig. f is a continuation of the dotted 180° intra-arm segment of Fig. d, while the dotted 180° intra-arm segment of Fig. g is the complete right upper segment of the hyperbola-like portion of the locus. The formation of these portions of the locus in relation to corresponding segments of the basis curve is evident from the illustrations. These 5 dotted segments complete the non-focal self-inversion loci of the 1st quadrant relative to the right traditional focus as reference pole.

The manner of the origin of the other loci illustrated in Fig. a is elucidated readily by analogy with the illustrated relations of Figs. c-g. If the right upper arm were to be shown extending to a greater distance, it would be seen that the inter-arm and intra-arm loci for the 5 reference poles become closely spaced and "collect" into narrow parallel bands running along parallel to and just above and below the basis arm. Their near-to-far sequence from the basis curve eventually assumes the pole sequence itself (0, a, 1.1a, $\sqrt{2}$a, and 2a).

Non-Focal Self-Inversion Loci of The Equilateral Hyperbola (j=a, 0° & 180° Loci, Transverse-Axis Reference Poles & Detail For Traditional Foci)

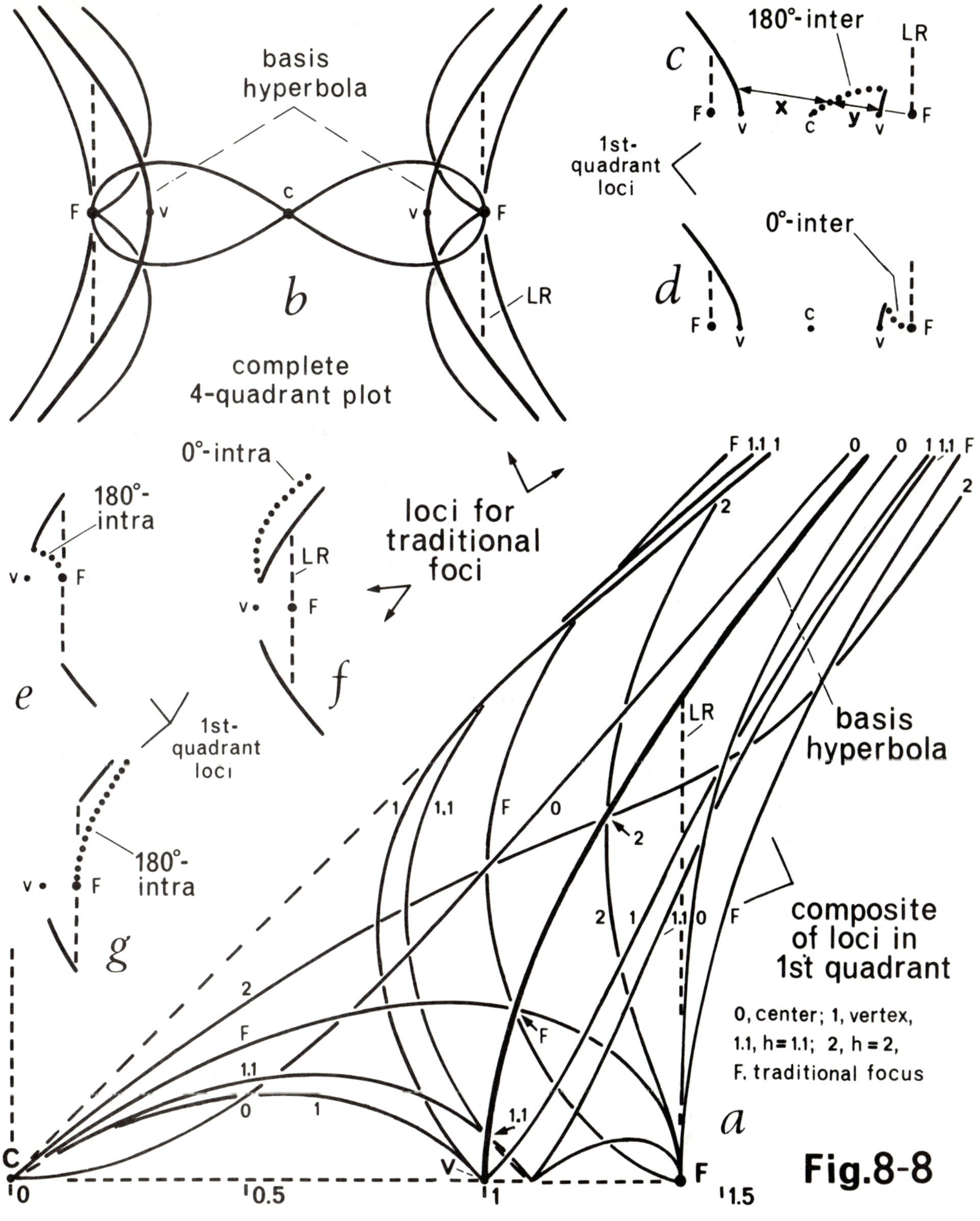

Fig.8-8

Fig. 8-9. Non-focal self-inversion loci of the $e=3$ and $e=3/4$ limacons for various values of the unit of linear dimension. Both interior and exterior loci are illustrated. These are referred to 3 axial reference poles that include the $-\frac{1}{2}b$ focus, the focus of self-inversion at $(a^2-b^2)/2b$, and a pole lying midway between them, as illustrated in Figs. a and b. The relative dimensions of the curves in the sequences of loci c,f,i, d,g,j, and e,h,k for the 3 values of j^2, have been retained.

It was noted above (legend to Fig. 8-2) and on page VIII-1 that focal self-inversions can be regarded as limiting cases of non-focal self-inversions, namely, those cases in which the non-focal loci have "collapsed" to point loci coincident with the reference poles. "Collapse" and coincidence only can occur, however, if two conditions are met: (1) a focus of self-inversion must exist in the region of the curve in question (interior, exterior, inter-oval, etc.); (2) the unit of linear dimension employed in the derivation of the non-focal self-inversion loci must have the same value as for self-inversion (the unit of linear dimension for a self-inversion is referred to hereafter as *the unit for self-inversion* or *the modular value*). If either of these conditions is not satisfied, a focal locus of self-inversion, i.e., a point locus, is not obtained.

The present figure illustrates all the alternatives, namely: neither condition obtains; condition (1) but not condition (2) obtains; condition (2) but not condition (1) obtains; and both conditions obtain. The most interesting cases have to do with the loci of non-focal self-inversion, other than the focus of self-inversion, that are obtained when the focus of self-inversion itself is taken as reference pole (for modular j); in other words, the loci that exist in other regions of a curve for the same conditions that lead to a point-pole as the non-focal self-inversion locus in a given region (these are designated as *modular* non-focal self-inversion loci).

The nature and affinities of the non-focal self-inversion loci of curves other than conics have been studied for only a few genera, in fact, only for several genera of Cartesian self-inverters (see pages X-1 to 3, eqs. X-1a-c, and Figs. 8-10 to 15), and mostly for a focus of self-inversion as reference pole. All non-focal self-inversion loci that are referred to a focus of self-inversion also are found to be Cartesian self-inverters but, with few exceptions, these loci do not belong to the same species or genera as the basis curves.

Non-Focal Self-Inversion Loci of The e = 3 & e = 3/4 Limaçons (Various j, Interior & Exterior Loci, Axial Reference Poles, Including Foci of Self-Inversion)

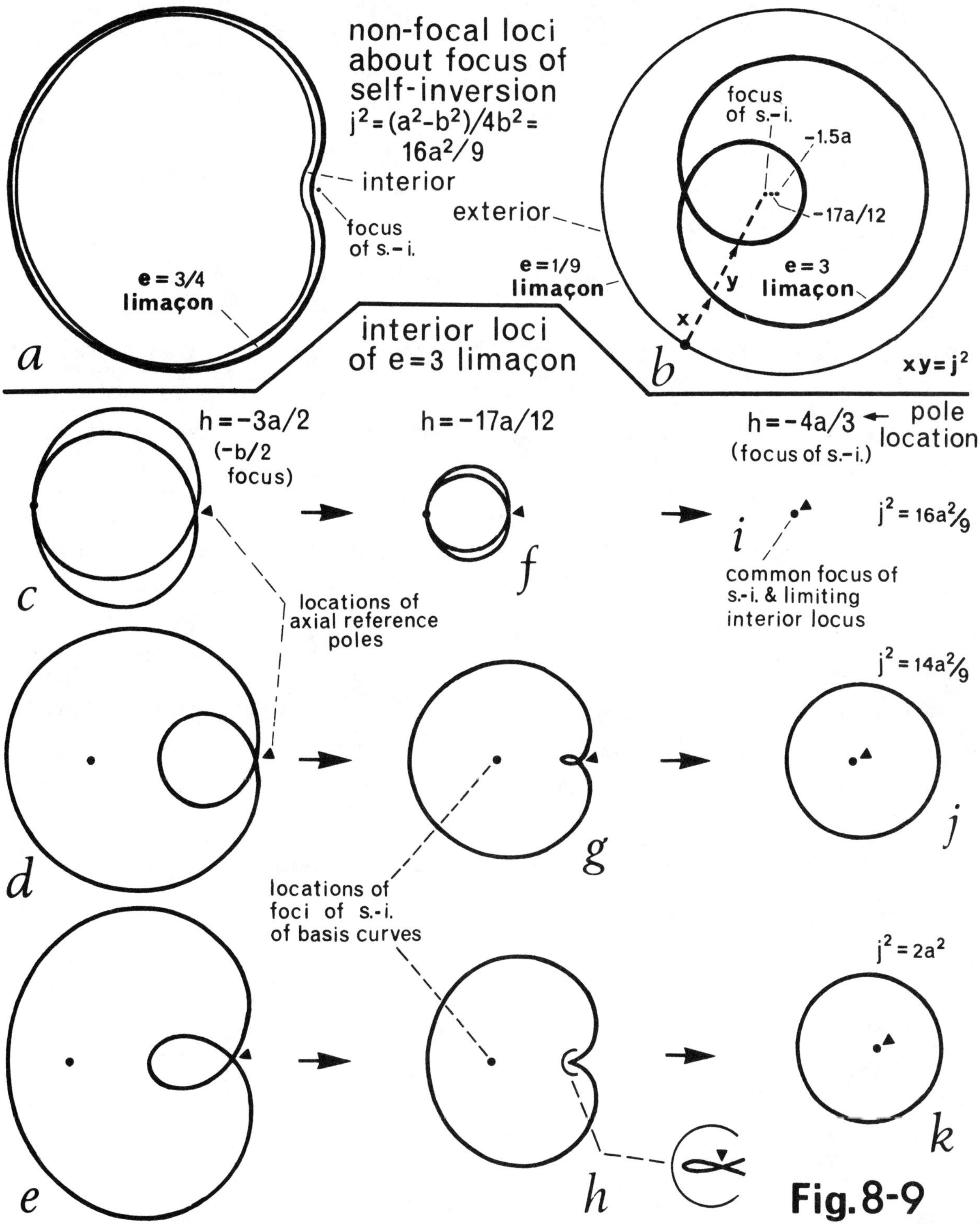

Fig. 8-9

Derivation of The Non-Focal Self-Inversion Loci of Limacons Referred to Axial Poles

A purely geometrical derivation of plots--as opposed to derivation of the corresponding eqs.--of the non-focal self-inversion loci of curves other than conics is a straightforward procedure. One merely draws lines through the reference pole at all angles, finds their points of intersection with the basis curve, locates the exterior, interior, inter-oval, or other points of self-inversion along these lines (for a given value of the unit of linear dimension), and draws a continuous curve through these points.

Semi-analytical procedures (illustrated below) for deriving these plots--as opposed to the eqs.--frequently require calculations of zeros of functions. Wholly analytical (algebraic) derivations of the eqs. of non-focal self-inversion loci can become quite formidable, involving, as they frequently do, derivations of the analytical solutions of eqs. of greater than 2nd degree. The present illustrations (Figs. 8-9 to 8-15) are limited to the non-focal self-inversion loci of Cartesian self-inverters about axial poles.

The non-focal self-inversion loci of the $e = 3$ limacon referred to axial poles at $h = -3a/2$ (the $-\frac{1}{2}b$ focus; Figs. c-e) and $h = -17a/12$ (Figs. f-h) were obtained by employing the semi-analytical procedure. First, the simple intercept eq. for a point on the line of symmetry (eq. X-12) was solved using a zeros of functions computer program for the given locations of the reference poles ($h = -3a/2$ and $-17a/12$). This led to values for the roots for various values of θ, i.e., values for the intercepts with the curve of radius vectors passing through the reference poles at various angles to the x-axis.

With the values of the intercepts for given angles at hand, the corresponding points on the non-focal self-inversion loci were found by employing the appropriate root eq. (eqs. 1) for basis curves consisting of a single oval or two loops or annular ovals (for examples of cases represented by eqs. 1, see Fig. 8-11c).

	root equation	locations of points of non-focal self-inversion loci on radius vectors from reference poles	
(a)	$(r-b)(r-a) = j^2$	beyond both intercepts in question	(8-9-1)
(b)	$(b-r)(r-a) = j^2$	between the intercepts in question	
(c)	$(a-r)(b-r) = j^2$	between the reference pole and the intercepts in question	
(d)	$(a+r)(b+r) = j^2$	on the opposite side of the reference pole from the intercepts in question	

$$r = \frac{(a+b)\pm[(a+b)^2-4(ab\mp j^2)]^{\frac{1}{2}}}{2}$$ solutions to cases 1a (upper alternate sign within radical) and 1b (8-9-2)

(a) $(a+b)^2 > 4ab\mp 4j^2$ (b) $(a-b)^2 > \mp 4j^2$ conditions for real roots of equation 2 (8-9-3)

Inasmuch as cases 1a and 1c lead to the same eq. and case 1d merely gives the negative roots of cases 1a and 1c (the non-focal self-inversion locus is reflected through the reference pole), all the loci are obtained by considering only cases 1a and 1b, for which the solutions are eq. 2. Examination of the conditions for real roots of the radical (eqs. 3a,b) reveals that two real roots always exist for case 1a.

For case 1b, however (for points on the locus lying between the intercepts), real roots exist only for $(a-b)^2 > 4j^2$, which simply defines the range of the unit of linear dimension for which real interior or inter-oval non-focal self-inversion loci exist. Accordingly, the substitution of sets of intercepts obtained from eq. X-12 into eq. 2 gives points on the interior loci--the loci plotted for poles at $h = -3a/2$ and $-17a/12$ (inter-oval and exterior loci also exist for appropriate values of j).

Some of the interior non-focal self-inversion loci derived as described above are seen to resemble axial QBI curves (compare Figs. d,e, and g with loci in Fig. 6-3). Neither the degree, nor other affinities of these curves, however, is known, though the degree cannot be lower than 6. It is noteworthy that when the unit of linear dimension is *modular* (i.e., equal to the unit for self-inversion of the basis curve), the non-focal self-inversion loci referred to the $-\frac{1}{2}b$ focus and the midpole consist of two tangent ovals (Figs. c and f). These are tangent at the focus of self-inversion of the basis curve but also come into contact at the reference pole for the non-focal self-inversion (the only QBI curves that have two loops that come into contact at more than one point are intersecting quartic circles; see pages XI-26 to 28). Using these two points for reference, the size of the loci compared to the basis curve is ascertained readily (since these points are marked by solid circles on the basis curve of Fig. b).

On the other hand, for *non-modular* values of j (i.e., j not the unit for self-inversion), though the DP of each locus coincides with the axial reference pole for the inversion, the focus of self-inversion no longer is incident upon the loci (Figs. d,e,g,h,j, and k). The fact that the axial reference pole for the inversion lies at the DP of the non-focal self-inversion loci is not unexpected, since the pole for a point transformation often becomes the highest-ranking circumpolar focus of the transform curve (for example, the inversions and pedals of conic sections).

Derivation of The Non-Focal Self-Inversion Loci of Limacons Referred To The Focus of Self-Inversion

When the reference pole for the non-focal self-inversion is taken to be the focus of self-inversion itself, a purely algebraic analysis of the non-focal self-inversion loci readily is achieved. Thus, the cosθ simple intercept

$$\cos\theta = \frac{4b^3x-4b^2x^2-(a^2-b^2)^2}{4a^2bx}$$ simple intercept format of limacons cast about the focus of self-inversion (8-9-4)

$$r = \frac{(b^2-a^2\cos\theta)\pm[(b^2-a^2\cos\theta)^2-(b^2-a^2)^2]^{\frac{1}{2}}}{2b}$$ polar form of eq. 4 in r (8-9-5)

$$(r-r_1)(r-r_2) = \pm j^2$$ root eq. for non-focal self-inversion loci of monovular, annular, and two-looped curves (8-9-6)

$$4b^2r^2 + 4br(a^2\cos\theta-b^2) + [(b^2-a^2)^2\pm 4b^2j^2] = 0$$ general eq. for non-focal self-inversion loci of limacons about the focus of self-inversion as reference pole (8-9-7)

(a) $$\cos\theta = \frac{4b^3x-4b^2x^2-[(b^2-a^2)^2\pm 4b^2j^2]}{4a^2bx}$$ cosθ simple intercept format of eq. 7 (8-9-8)

(b) $$\cos\theta = \frac{x^2+Ex+F^2}{Gx}, \quad E,F^2,G \neq 0$$ cosθ simple intercept format for Cartesian self-inverters

format of limacons about this pole is eq. 4 (x is the intercept at the angle θ), and the corresponding explicit solution for r is eq. 5. If eqs. 1a and 1b are combined to give eq. 6, the latter is the root eq. for non-focal self-inversion loci of monovular, biovular-annular, and two-looped curves.

Substituting for r_1 and r_2 from eq. 5 (one root being given by the positive and the other by the negative alternate sign) gives eq. 7 for the non-focal self-inversion loci. The corresponding simple intercept format is eq. 8a, which identifies these loci as Cartesian self-inverters (see page X-2), the general simple intercept format for which is eq. 8b (E,F^2, and G positive or negative; see treatments in legends of Figs. 8-10 to 8-15).

[In eqs. 7 and 8a the upper alternate sign is for inter-loop loci of hyperbolic limacons and interior loci of elliptical limacons. The lower alternate sign is for interior and exterior loci of hyperbolic limacons and exterior loci of elliptical limacons (see also last paragraph below).]

For the special case of the lower alternate sign in eqs. 7 and 8a, and $4b^2j^2 = (a^2-b^2)^2$, which corresponds to j being modular, eq. 7 becomes eq. 9a, which has the two solutions, eqs. 9b and 9c. Eq. 9b represents the limit-

(a)	$4b^2r^2+4br(a^2\cos\theta-b^2) = 0$	eq. 7 for $j^2 = (b^2-a^2)^2/4b^2$ and the lower alternate sign	(8-9-9)
(b)	$r = 0$	point non-focal self-inversion locus of elliptical and hyperbolic limacons, which is coincident with the focus of self-inversion as reference pole, for modular j	
(c)	$r = b-(a^2/b)\cos\theta$	exterior non-focal self-inversion locus of basis hyperbolic limacon, with focus of self-inversion as reference pole andmodular j--the *modular locus*	

ing cases referred to above (Fig. i for hyperbolic limacons, Fig. a for elliptical limacons) in which the non-focal self-inversion locus has "collapsed" into coincidence with the reference pole, which is the focus of self-inversion.

The second solution, 9c, of eq. 9a, which represents the *exterior* non-focal self-inversion locus of hyperbolic limacons about the focus of self-inversion as reference pole (for modular j), is seen to be an elliptical limacon of eccentricity a^2/b^2 (Fig. b, outer curve). Thus, the *modular* exterior non-focal self-inversion locus of a hyperbolic limacon is another limacon, with eccentricity equal to the reciprocal square of the eccentricity of the basis curve. This means that the *modular locus* always is an elliptical limacon.

This case, for modular j, is the only instance in which the non-focal self-inversion locus of a limacon about the focus of self-inversion is another limacon. In all other cases (except point solutions) the non-focal self-inversion loci are apparently previously unstudied Cartesian self-inverters (see also Figs. 8-10,11a,14, and 15) or segments thereof, with the general eqs. 7 and 8a.

If modular j is employed with the positive sign, eqs. 7 and 8a become eqs. 10a,b. It will be noted that eq. 10b differs from the corresponding eq. for limacons (eq. 4) only in that the $(a^2-b^2)^2$ term is replaced by a term of

(a) $4b^2r^2+4br(a^2\cos\theta-b^2)+2(a^2-b^2)^2 = 0$ eq. 7 for plus sign and modular j (8-9-10)

(b) $\cos\theta = [4b^3x-4b^2x^2-2(a^2-b^2)^2]/4a^2bx$ eq. 8a for plus sign and modular j

twice the magnitude. In the case of the $e=3/4$ elliptical limacon, the corresponding curve is illustrated in Fig. a (see also Fig. 8-14). An inter-loop locus of the $e=3$ hyperbolic limacon does not exist for modular j (the maximum value of j that can be accommodated between the loops is a, whereas the value for modular j is $4a/3$).

The interior non-focal self-inversion loci of eqs. 7 and 8a with the negative sign and two other values of j are illustrated in Figs. j and k. Though both of these loci resemble elliptical limacons and are quartics, they are not members of the limacon genus.

The Bipolar Equation of Non-Focal Self-Inversion Loci of Limacons About The Focus of Self-Inversion As Reference Pole

As noted above, all the non-focal self-inversion loci of limacons with the focus of self-inversion as reference pole have two foci in common. This being

[For the modular curves, which are elliptical limacons (exterior non-focal self-inversion loci of hyperbolic limacons for modular j), the two foci are the DP and the $-\frac{1}{2}b$ focus (defined by the $\cos^2\theta$-condition), while for the non-modular curves they are the focus of self-inversion and the $-\frac{1}{2}a^2/b$ focus (defined by the $\cos^2\theta$-condition). It is pointed out below (Figs. 8-19 and 8-20) that when the $\cos^2\theta$-condition does not define 2 discrete foci, the single pole that is defined is a double-focus.]

the case, a bipolar eq. can be cast to reproduce all the loci in the super-genus represented by eqs. 7 and 8a. Expressing eq. 7 in rectangular coordinates gives eq. 11. Letting p_v be the origin and p_u the $-\frac{1}{2}a^2/b$ focus,

rectangular eq. of non-focal self-inversion loci of limacons about the focus of self-inversion as reference pole

$$[4b^2(x^2+y^2)+4a^2bx+(b^2-a^2)^2\pm4b^2j^2]^2 = 16b^6(x^2-y^2) \quad (8\text{-}9\text{-}11)$$

(a) $v^2=x^2+y^2$ (b) $x=(b/a^2)(u^2-v^2-a^4/4b^2)$ conversion eqs. between rectangular and bipolar coordinates (8-9-12)

$$[4b^2u^2-a^4+(b^2-a^2)^2\pm4b^2j^2]^2 = 16b^6v^2 \quad \text{eq. 11 in bipolar form} \quad (8\text{-}9\text{-}13)$$

(a) $(4b^2u^2-a^4)^2 = 16b^6v^2$ (b) $4b^2u^2-a^4 = 4b^3v$ eq. 13 for modular elliptical limacons (8-9-14)

(a) $(u^2-\frac{1}{4}b^2)^2 = a^2v^2$ general bipolar eq. of limacons cast about the DP or DP homologue and $-\frac{1}{2}b$ focus (8-9-15)

(b) $u^2-\frac{1}{4}b^2 = av$ bipolar eq. of elliptical limacons cast about the DP homologue and the $-\frac{1}{2}b$ focus

it is shown readily that the bipolar and rectangular coordinates are related as in eqs. 12a,b, whence it follows that the desired bipolar eq. is the 4th-degree eq. 13. Though both sides of this eq. are perfect squares, a 2nd-degree eq. does not, in general, represent all the loci, since the bracketed expression can be negative for some points on some loci.

For the exterior modular loci of hyperbolic limacons, $r = b-(a^2/b)\cos\theta$, eq. 13 becomes eq. 14a, where the parameters a and b are those of the basis limacon (eq. 4). For comparison, the general bipolar eq. of limacons cast about the DP (p_v) and the $-\frac{1}{2}b$ focus (p_u) is eq. 15a. It is shown readily, however, that the terms in parentheses on the left side of eqs. 14a and 15a cannot be negative for elliptical limacons, whereupon roots may be taken yielding eqs. 14b and 15b. Similar considerations apply to eq. 13, i.e., for some of the non-focal self-inversion loci (in addition to the modular elliptical limacons), a 2nd-degree bipolar eq. is representative, whereas for others, a 4th-degree eq. is required to represent the entire loci.

Non-Focal Self-Inversion Loci of Individual Ovals Or Individual Loops of Biovular And Two-Looped Curves

One also can consider the non-focal self-inversion loci of individual ovals or individual loops of biovular and two-looped curves (see also Fig. 8-12). The case of such loci referred to the focus of self-inversion of limacons is taken as an example. Examination of eq. 5 for limacons cast about the focus of self-inversion reveals that for hyperbolic limacons the roots(intercepts) are positive for all values of θ, and that the upper alternate sign is for intercepts with the large loop and the lower sign for intercepts with the small loop.

The general root eq. (analogous to eqs. 1a-d) is eq. 16, where r_1 is for the intercepts with either the large loop or the small loop. Substituting for r_1 from eq. 5 leads to the starting eq. 17, where the upper alternate sign root eq. for non-focal self-inversion loci of a single loop or oval of a two-looped or biovular curve

$$[r_1(\theta)-r][r+r_1(\theta+180°)] = j^2 \tag{8-9-16}$$

non-focal self-inversion loci of individual loops of hyperbolic limacons referred to the focus of self-inversion

$$\{(b^2-a^2\cos\theta)\pm[(b^2-a^2\cos\theta)^2-(b^2-a^2)^2]^{\frac{1}{2}}-2br\}\cdot \tag{8-9-17}$$
$$\{(b^2+a^2\cos\theta)\pm[(b^2+a^2\cos\theta)^2-(b^2-a^2)^2]^{\frac{1}{2}}+2br\} = 4b^2j^2$$

of the radical is for the large loop and the lower for the small loop. In elimination of the radical from the polar eq., the alternate signs are lost. Other affinities of the curves represented by eq. 17 are unknown, but the degree of the corresponding rectangular eq. is a maximum of 16.

Fig. 8-10. Partial circumpolar symmetry and inversion analyses of the interior non-focal self-inversion loci of the $e = 3/4$ limacon about the focus of self-inversion as reference pole, for $j^2 = (a^2-b^2)^2/2b^2$. The present figure serves as an introduction to the circumpolar symmetry and inversion properties of monovular Cartesian self-inverters (eqs. 8-9-10a,b). More detailed treatments of these topics appear in the legends for Figs. 8-13,15,17, and 19.

If one solves the general eq.(8-9-7) for non-focal self-inversion loci of limacons referred to the focus of self-inversion, one obtains eq. 1. The loci

$$r = \frac{(b^2-a^2\cos^2\theta)\pm[(b^2-a^2\cos^2\theta)^2-(b^2-a^2)^2\mp 4b^2j^2]^{\frac{1}{2}}}{2b} \quad \text{solution of eq. 8-9-10 for } r \tag{8-10-1}$$

$$r_1r_2 = (b^2-a^2)^2/4b^2 \pm j^2 \quad \text{product of the roots of eq. 1} \tag{8-10-2}$$

represented by eq. 1 are seen to self-invert, inasmuch as the product of the roots thereof is constant (eq. 2), at a value equal to the product of the roots for the basis curve, $(b^2-a^2)^2/4b^2$, augmented or diminished by j^2, the square of the unit of linear dimesnion for the non-focal self-inversion locus.

Since the loci of eq. 1 and its special case, eq. 8-9-10a, both self-invert and have a line of symmetry, it follows from Maxims 38 and 6-4-2 that there exist on the line of symmetry of each represented locus two poles about which the locus inverts to curves with true centers. These poles lie at modular distance on each side of the focus of self-inversion (one pole is interior and the other exterior to the locus). These relationships are embodied in Maxims 8-10-1 and 8-10-2. Maxim 8-10-3 is the equivalent of Maxim 8-10-1 for axial QBI curves, while Maxim 8-10-4 anticipates the findings for axial bicircular quartics. The terminology employed in Maxim 8-10-4 and hereafter is elucidated in Table 8-10-1.

CIRCUMPOLAR MAXIMS 8-10-1: *There are two axial poles about which a self-inverting monovular basis curve with a line of symmetry inverts to curves with true centers. These poles are reflections of one another in the focus of self-inversion at the modular distance.*

8-10-2: *For the appropriate value of the unit of linear dimension, the two curves with true centers, derived by inverting monovular self-inverting curves with a line of symmetry, are central inversions of one another. These curves may be subspecifically identical or dissimilar.*

8-10-3: *If a self-inverting QBI curve with a line of symmetry is inverted about the reflection of the DP or DP homologue in the focus of self-inversion, a central quartic is obtained.*

8-10-4: *There are two pairs of axial poles about which axial bicircular quartics invert to central curves. The members of one pair of poles are modular reflections of one another in one 0° focus of self-inversion, and the members of the other pair are modular reflections of one another in the other 0° focus of self-inversion.*

Curve B of Fig. a is the *central curve* (i.e., a curve with a true center) obtained by inverting the interior non-focal self-inversion locus about the interior pole at the modular distance from its focus of self-inversion $[-(a^2-b^2)/\sqrt{2}b]$, for an arbitrary value of j. A central curve (curve C of Fig. b) also is obtained by inverting the interior non-focal self-inversion locus about an exterior pole at the same distance from the focus of self-inversion. In the case of the QBI (quadratic-based-inversion) superfamily, the two member genera of central curves are the central conics and central quartics. The central conics are taken to be the basis curves for the superfamily because they are of only 2nd degree, compared to 4th degree for central quartics.

In the cases of inversion superfamilies that possess only quartic central species, such as all central Cartesian-based inversion superfamilies, all of these species are taken to be members of the basis genus. Since monovular

Table 8-10-1. Nomenclature And Properties of Members of Inversion Superfamilies Based Upon Central Cartesian Genera

$(x^2+y^2)^2+B^2x^2+C^2y^2+D^4=0$, $B^2, C^2, D^4 \neq 0$, real — eq. of central Cartesians in standard form

MONOVALS (consist of single oval, unspecified number of lines of symmetry, self-invert in 0, 1, or 2 modes)

1. *MONOVULAR CENTRAL CARTESIANS* (2 lines of symmetry, self-invert in 0 or 2 modes)
   - A. *MONOVULAR CENTRAL CARTESIANS THAT SELF-INVERT*
     - a. *CASSINIAN MONOVALS* (self-invert about center by 90° and 270° intra-oval modes)
     - b. other species unknown (individual subspecies known; Fig. 8-10b,h)
   - B. *MONOVULAR CENTRAL CARTESIANS THAT DO NOT SELF-INVERT*
2. *AXIAL MONOVALS* (1 line of symmetry, self-invert by 0° intra-oval mode about a single pole)
   - A. *MONOVULAR CARTESIAN SELF-INVERTERS* ($\cos^2\theta$-condition foci in coincidence as double-focus, simple intercept format is eq. 8-13-5)
   - B. all others
3. *NON-AXIAL MONOVALS* (no line of symmetry, self-invert by 0° intra-oval-mode about two different poles)

SERPENTINE CUBICS (consist of serpentine arm only, unspecified number of lines of symmetry, self-invert in 1 or 2 intra-segment modes)

1. *AXIAL SERPENTINE CUBICS* (1 line of symmetry, self-invert about axial vertex by the 0° intra-segment mode)
2. *NON-AXIAL SERPENTINE CUBICS* (no line of symmetry, self-invert about 2 incident poles by the 0° intra-segment mode)

SERPENTINE CIRCULAR CUBICS (consist of oval and serpentine arm, unspecified number of lines of symmetry, self-invert in 3 or 4 modes)

1. *AXIAL SERPENTINE CIRCULAR CUBICS* (1 line of symmetry, self-invert in 3 modes)
2. *NON-AXIAL SERPENTINE CIRCULAR CUBICS* (no line of symmetry, self-invert in 4 modes)

BICIRCULAR QUARTICS (consist of 2 ovals, unspecified number of lines of symmetry, self-invert in 2, 3, or 4 modes)

1. *BIOVULAR CENTRAL CARTESIANS* (*CENTRAL BICIRCULAR QUARTICS*) (2 lines of symmetry, self-invert in 2 modes)
2. *AXIAL BICIRCULAR QUARTICS* (1 line of symmetry, self-invert in 3 modes)
   - A. *BIOVULAR CARTESIAN SELF-INVERTERS* ($\cos^2\theta$-condition foci in coincidence as double-focus, simple intercept format is eq. 8-9-8b, all representatives annular)
   - B. all others
3. *NON-AXIAL BICIRCULAR QUARTICS* (no line of symmetry, self-invert in 4 modes)

and biovular species possess different properties in the real plane, separate treatments are in order for them. When only one monovular central basis species exists, of course, the two central curves obtained by inverting any non-central member of the superfamily about the appropriate two poles belong to the same subspecies. In general, however, these central curves are neither congruent nor similarly oriented.

Axial biovular species invert to central curves about 4 axial poles. The two poles yielding one species are reflections in one 0° focus of self-inversion, while the two poles yielding the other species are reflections in the other 0° focus of self-inversion (see treatments of Figs. 8-19 to 8-22). For any given axial biovular species the two subspecifically identical central member curves in each pair are incongruent.

To obtain the eq. of the centered central Cartesian, curve B, one first employs an $[x-(a^2-b^2)/\sqrt{2}b]$ shift of the basis curves of eq. 8-9-10a, yielding eq. 3. This is inverted, giving a central Cartesian basis curve, B, of Fig. a

eq. 8-9-10a cast about an interior axial pole at $-(a^2-b^2)/\sqrt{2}b$ (curve A, Fig. a, for $b=3a/4$)

$$\{b^2r^2+br[a^2-\sqrt{2}(a^2-b^2)]\cos\theta+(a^2-b^2)(a^2-b^2-a^2/\sqrt{2})\}^2 = b^6[r^2-2(a^2-b^2)(r\cos\theta)/\sqrt{2}b+(a^2-b^2)^2/2b^2] \qquad (8\text{-}10\text{-}3)$$

central Cartesian basis curves derived by inverting eq. 3 (curve B, Fig. a, for $b=3a/4$)

$$\{b^2j^4+bj^2[a^2-\sqrt{2}(a^2-b^2)]r\cos\theta+(a^2-b^2)(a^2-b^2-a^2/\sqrt{2})r^2\}^2 = b^6r^2[j^4-2(a^2-b^2)j^2r(\cos\theta)/\sqrt{2}b+(a^2-b^2)^2r^2/2b^2] \qquad (8\text{-}10\text{-}4)$$

curves of eq. 4 centered at origin (curve B, Fig. b, for $b=3a/4$)

(a) $$8(a^2-b^2)^4[(a^2-b^2-a^2/\sqrt{2})^2-b^4/2]r^4+4b^2j^4(a^2-b^2)^2(a^4-4a^2b^2+b^4)r^2 + 8b^6j^4(a^2-b^2)^2r^2\cos^2\theta-2b^4j^8(a^2-b^2-a^2/\sqrt{2})^2+b^8j^8=0 \qquad (8\text{-}10\text{-}5)$$

(b) $r^4+16.04170351a^2r^2-10.87345585a^2r^2\cos^2\theta-29.08384926a^4=0$ curve B, Fig. b

(c) $r^4-0.551567414j^4r^2/a^2+0.373865775j^4r^2(\cos^2\theta)/a^2 - 0.034383344j^8/a^4=0$ curve C, Fig. b (eq. 4 with $b = 3a/4$) for the corresponding superfamily (an assemblage of all the curves to which the loci of eqs. 8-9-10a, or eq. 3 or 4, above, invert about all points in the plane). An $[x+\sqrt{2}bj^2/2(a^2-b^2)]$ shift of the central curves brings the centers of the loci of eq. 4 to the origin, giving eq. 5a. The specific central curve B of Fig. b is obtained by letting $b = 3a/4$ (eq. 5b). Inverting the latter curve gives the central curve C of Fig. b.

The central curve B of Fig. b inverts to curve A about poles at $\pm\sqrt{2}bj^2/2(a^2-b^2)$. These poles are marked by the solid circles just beyond the axial vertices of curve C. The corresponding axial poles about which curve C inverts to curve A (not marked) lie outside both curves at $x = \pm 2(a^2-b^2)/\sqrt{2}b$.

[The unit of linear dimension for the inversion of curves B and C of Fig. b is not the same as for curves A and B of Fig. a. The value of j in each case is the distance from the pole of inversion to one of the points of intersection of the basis and inverse loci.]

The real axial "loop foci" of curves B and C lie on the major axes and are specified by the $\cos^2\theta$-condition and $\sin^2\theta$-condition, respectively. If eqs. 5b,c are written in the general form of eqs. 6b,c, the real axial "loop foci" of the loci of eq. 6b and curve B of Fig. b lie at $h = \pm N/2$. For the k-axis, the foci defined by the $\sin^2\theta$-condition are imaginary and lie at $k = \pm Ni/2$. For curve C of Fig. b the k-axis is the major axis, and the "loop foci" lie at $k = \pm Nj^2/2P^2$ (not shown), whereas the foci on the minor axis (defined by the $\cos^2\theta$-condition) are imaginary and lie at $h = \pm Nj^2 i/2P^2$.

(b) $r^4+M^2r^2-N^2r^2\cos^2\theta-P^4= 0$ eq. 5b in general form (8-10-6)

(c) $r^4-M^2j^4r^2/P^4+N^2j^4r^2(\cos^2\theta)/P^4-j^8/P^4 = 0$ eq. 5c in general form

The highest-ranking focus of each of curves B and C is the true center, for which the simple intercept format for curve B, derived from eq. 6b, is eq. 7a. This will be recognized as being of the same form as that of the format for mutually-inverting quartic circles (eqs. XII-57 and XIV-43e).

(a) $\cos^2\theta = (x^4+M^2x^2-P^4)/N^2x^2$ simple intercept format of curve B about the true center (8-10-7)

(b) $\cos\theta = \dfrac{x^4+(M^2+N^2/2)x^2+(M^2N^2/4-3N^4/16-P^4)}{N(N^2/2-M^2)x-2Nx^3}$ simple intercept format of curve B about the N/2 axial focus

Partial Circumpolar Symmetry & Inversion Analyses of The Interior Non-Focal Self-Inversion Locus of The e = 3/4 Limaçon About The Focus of Self-Inversion As Reference Pole

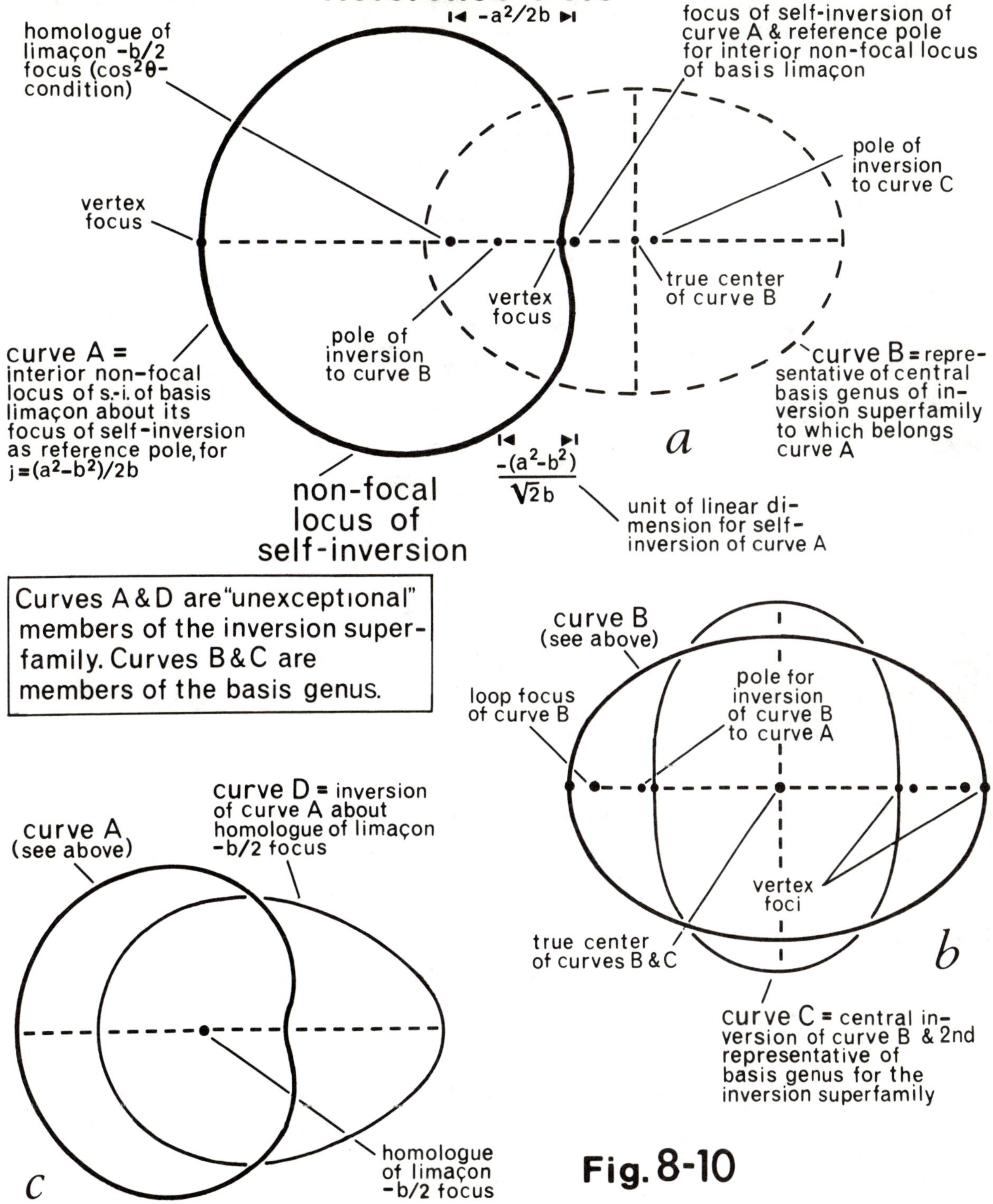

Fig. 8-10

Returning to the topic of the interior non-focal self-inversion locus of elliptical limacons (curve A) cast about the focus of self-inversion (eq. 8-9-10a), the homologue of the limacon $-\frac{1}{2}b$ focus is defined by the $\cos^2\theta$-condition and lies at $h = -a^2/2b$ (Fig. a, leftmost interior pole). It is, in fact, coincident with the $-\frac{1}{2}b$ focus of the basis $e = 3/4$ elliptical limacon. The simple intercept format of the curve about this focus is eq. 8. The

simple intercept format of curve A about the $-a^2/2b$ focus

$$\cos\theta = \frac{-16b^4x^4+8b^2(a^4-A^4+2b^4)x^2+4a^4b^4+2a^4A^4-(a^8-A^8)}{16a^2b^5x} = 0 \qquad (8\text{-}10\text{-}8)$$

$$A^4 = 2(b^2-a^2)^2$$

$-a^2/2b$ focus may coincide with a vertex (see Fig. 8-11a for a case of near coincidence), as does its $-\frac{1}{2}b$ homologue in the $e = 2$ limacon and in all true exchange-limacons (see Figs. 8-25,26), and this is common for foci defined by the $\cos^2\theta$-condition. The inversion of curve A about this focus produces the egg-shaped curve, D, of Fig. c.

Circumpolar Maxims

The formulation of certain results as Circumpolar Maxims is for summarization, organization, emphasis, and ready reference. It should be pointed out that some of the Maxims merely are working hypotheses--based upon knowledge to date--subject to modifications and amplifications as new information becomes available. For example, several Circumpolar Maxims based primarily on studies of QBI curves require modification and restriction when viewed in the light of results with non-QBI quartics (see legend to Fig. 8-20). Although the great majority of the findings incorporated into the Maxims are original, others were known previously; in a few cases the Maxims merely express more or less obscure properties of polar eqs. (for example, Maxims 23 and 37).

Some of the inversion properties of serpentine circular cubics and bicircular quartics incorporated into the Maxims have been known for many years. However, previous studies of inversion were carried out without knowledge of the mapping properties of the inversion transformation, without benefit of the perspective provided by Inversion Taxonomy, and without a realization of the fundamental significance of the inversion transformation for the classification of curves.

The main reasons for these lacunae are that previous workers dealt primarily with inversions of ensembles of lines and/or circles, and rarely concerned themselves with either the precise identity of specific inverse loci of other curves or with relationships between the inverse loci of other curves about specific poles. The most notable exception is considerations of the inverse loci of other curves about traditional foci and vertices, which provide the chief examples in the literature.

[Lack of awareness of principles of Inversion Taxonomy has allowed the perpetuation in the literature of the erroneous assertion that the trisectrix of Maclaurin inverts about its loop focus to Tschirnhausen's cubic. Since the trisectrix is the inversion of a quadratic, it cannot invert to a curve that is an inversion of Cayley's sextic. Nor can any cubic invert to another cubic about a non-incident pole (Maxim 6-8-11).

The richest single 20th-century textual source of information about the inversions of cubics and quartics (and the classical properties of plane algebraic curves) is Hilton's *Plane Algebraic Curves* (1920).]

Fig. 8-11. Compounded, common non-focal self-inversion loci with common foci of self-inversion. Fig. a illustrates 4 interior non-focal self-inversion loci of the $e = 3/4$ limacon. All of these curves, including the basis curve, also are interior or exterior non-focal self-inversion loci of one another. Fig. b illustrates 4 inter-oval and two exterior non-focal self-inversion loci of the $C = 4\frac{1}{4}$ parabolic Cartesian (reference pole = pole 1).

The inner exterior locus is the *modular* non-focal self-inversion locus of the basis curve and is unique among the non-trivial non-focal self-inversion loci of this subspecies in that (a) it is a limacon, $r = 4\frac{1}{4}d - 2d\cos\theta$, (b) the reference pole is not a focus of self-inversion (as it is for all of the other loci), and (c) it does not participate as a mutual non-focal self-inversion locus with all of the other curves. Each of the other ovals, including those of the basis curve, is--in one respect or another--a non-focal self-inversion locus of every other curve, i.e., the curves are *mutual non-focal self-inversion loci*.

The manner of derivation of non-focal self-inversion loci is illustrated in Figs. a and b by the dashed radius vectors, x and y, emanating at 180° to one another from a pole on one of the non-focal self-inversion loci to the corresponding basis curve (and, if extended, passing through the corresponding reference pole). The product of the lengths of the x and y vectors from all points on a non-focal self-inversion locus to the corresponding basis curve is constant.

Fig. c illustrates the basis curve of Fig. b (heavy solid ovals), one of the monovular inter-oval loci (light solid oval), and the annular locus (dashed ovals). These are enlarged to facilitate depicting the various modes of non-focal self-inversion of each of the 5 individual ovals to each of the other curves (namely, two annular loci and one monovular locus). The 3 foci of self-inversion of the basis curve are marked by the open circles.

If one compares the eqs. for limacons and their non-focal self-inversion loci referred to the focus of self-inversion (for example, eqs. 8-9-4 and 8-9-8), it will be noted that they differ only in the value of the constant term. [It is because of this difference that the non-focal self-inversion loci of limacons generally are other Cartesian self-inverters rather than other limacons.] Following this format, the 2nd-generation non-focal self-inversion loci of the 1st-generation non-focal self-inversion loci of limacons (eq. 8-9-7) are specified by eq. 1, where j_1 is the unit of linear dimension for

2nd-generation non-focal self-inversion loci of limacons referred to the focus of self-inversion

$$4b^2r^2+4br(a^2\cos\theta-b^2)+[(b^2-a^2)^2\pm4b^2j_1^2\pm4b^2j_2^2] = 0 \qquad (8\text{-}11\text{-}1)$$

the 1st generation and j_2 is the unit for the 2nd generation. Accordingly, the 2nd-generation loci also are Cartesian self-inverters.

The only exceptions to the representation of the 2nd-generation loci by eq. 1 occur when the 1st-generation curves are for the exterior non-focal self-inversion loci of hyperbolic limacons and j_1 is modular. In that case, the 1st-generation exterior non-focal self-inversion loci are *modular* elliptical limacons, $r = b-(a^2/b)\cos\theta$, while the interior non-focal self-inversion loci are points coincident with the focus of self-inversion (i.e., they are instances of the limiting cases in which non-focal self-inversion loci become *focal* self-inversion loci).

Compounded, Common Non-Focal Self-Inversion Loci With Common Foci of Self-Inversion

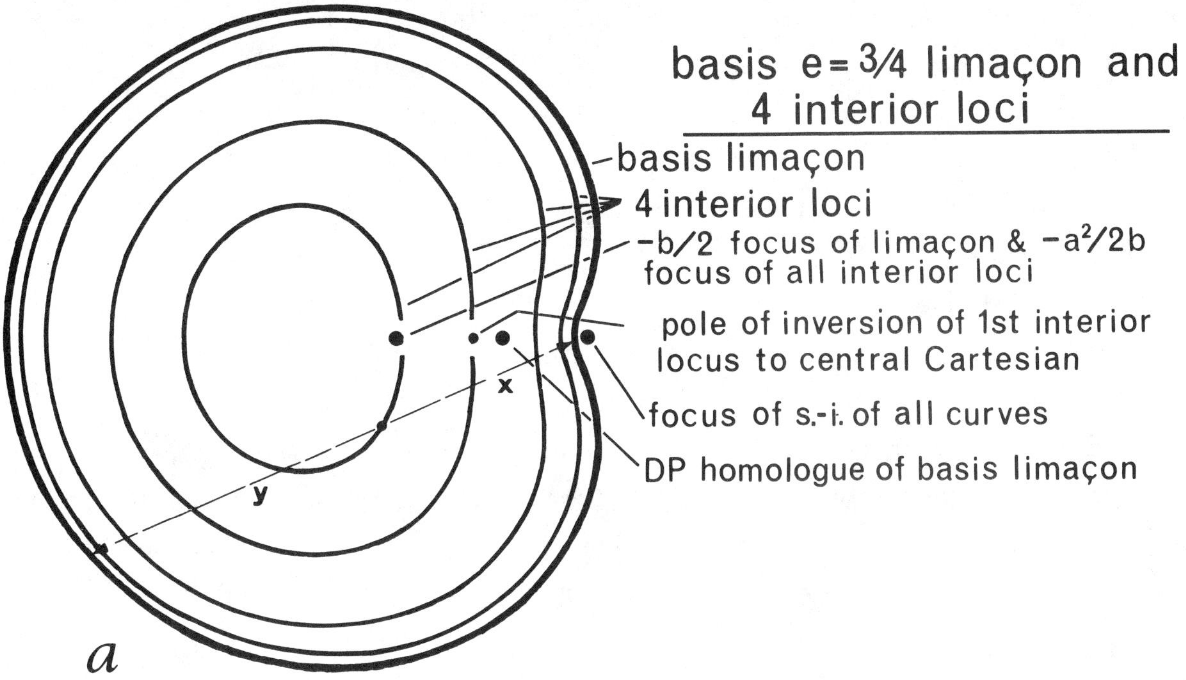

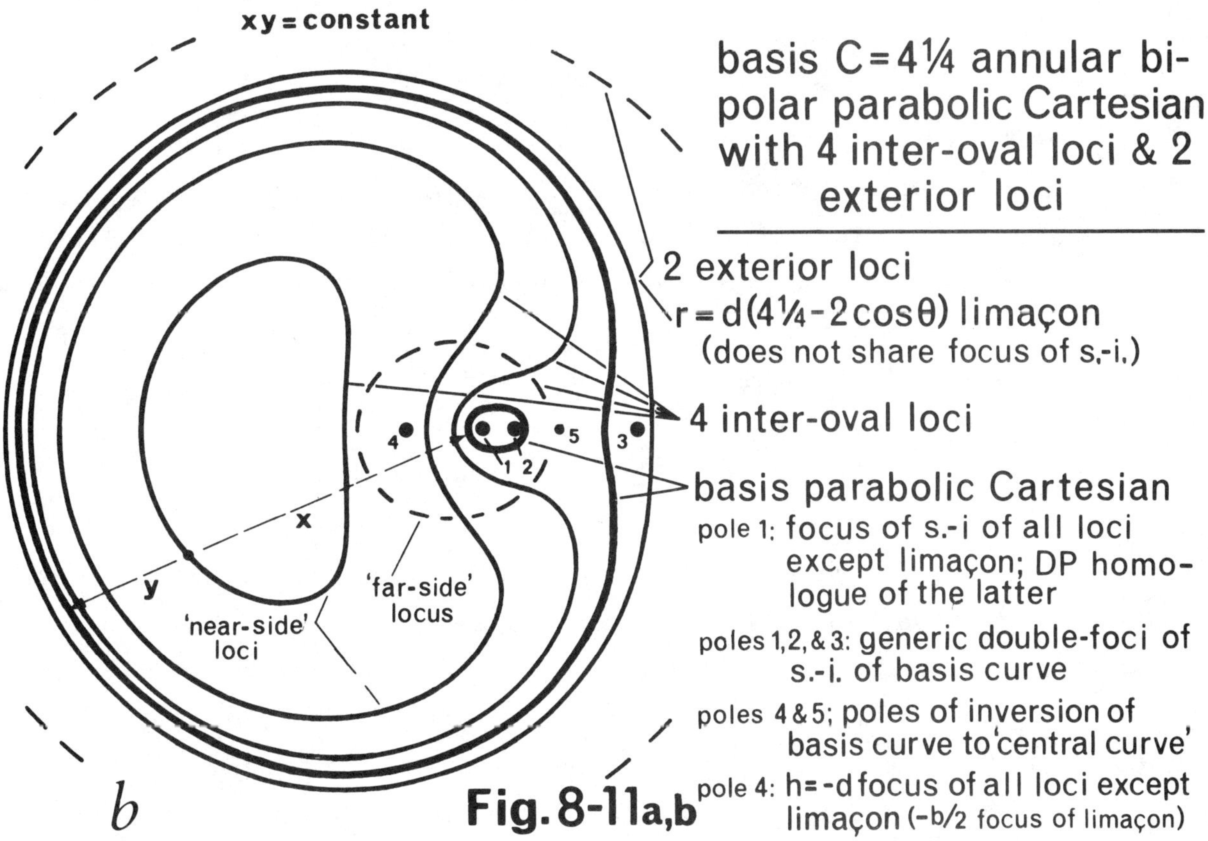

Fig. 8-11a,b

In the case of the basis limacon of Fig. a, j^2 for self-inversion about the reference pole is $(a^2-b^2)^2/4b^2$. For the 1st-generation interior locus closest to the basis curve, the square of the unit for self-inversion about the same pole is $2j^2$, for the next-smaller 1st-generation interior locus the value is $4j^2$, for the next smaller it is $8j^2$, while for the smallest interior locus it is $11j^2$. Considered as 1st-generation non-focal self-inversion loci of the basis curve, the corresponding xy products (the products of the intercepts for non-focal self-inversion) for these curves are j^2, $3j^2$, $7j^2$, and $10j^2$.

Thus, in all cases $j^2_{s.-i.} = j^2_{basis} + j^2_{xy}$, i.e., j^2 for focal self-inversion of the non-focal self-inversion locus is equal to the sum of j^2 for the focal self-inversion of the basis curve plus j^2 (the xy product) for non-focal self-inversion of the non-focal self-inversion locus. If, now, each successive 1st-generation smaller interior non-focal self-inversion locus of the basis curve is considered as the next-generation interior locus of the next-larger locus, then the xy products follow the sequence j^2, $2j^2$, $4j^2$, and $3j^2$. In each case, except the last, the value chosen for the xy product was twice the value for the previous generation (for the last case, no interior non-focal self-inversion locus exists for $xy = 8j^2$, so a value of $3j^2$ was employed).

Thus, if the practice were followed consistently of employing for each successive generation of interior non-focal self-inversion locus an xy product equal to j^2 for self-inversion of the previous locus, the xy products with respect to the previous generation would follow the sequence j^2, $2j^2$, $4j^2$, $8j^2$, $16j^2$, etc. The values of j^2 for self-inversion of the successive non-focal self-inversion loci would be $2j^2$, $4j^2$, $8j^2$, $16j^2$, etc., i.e., twice the value of the corresponding xy product, while the successive xy products relative to the basis curve would follow the sequence j^2, $3j^2$, $7j^2$, $15j^2$, $31j^2$, etc.

These results merely are an expression of the following relationship, which applies to any Cartesian self-inverter that is a non-focal self-inversion locus of a given basis curve about the latter's focus of self-inversion: *the square of the unit of linear dimension for self-inversion of the non-focal self-inversion locus is the sum of the corresponding modular value for the basis curve plus (or minus) the xy product used in the derivation of the non-focal self-inversion locus* (see eq. 8-10-2).

The position of the focus defined by the $\cos^2\theta$-condition for limacon non-focal self-inversion loci referred to the focus of self-inversion is independent of the constant term of eq. 8-9-7. Accordingly, all the non-trivial non-focal self-inversion loci of limacons that possess a common focus of self-inversion, including the basis curve, share a 2nd common focus, namely the pole at $h=-a^2/2b$ (interiormost pole of Fig. a). This only applies, however, when the non-focal self-inversion locus and the basis curve self-invert about a common pole. Accordingly, the modular $e=1/9$ limacon non-focal self-inversion locus of Fig. 8-9b is excluded. Such exceptions occur only for the *modular* loci.

The correspondences of the focus of self-inversion and the $-a^2/2b$ focus in the basis curve and all non-modular subspecies of non-focal self-inversion loci indicate the existence of special symmetry relationships between basis curves and non-focal self-inversion loci when the latter are referred to a focus of self-inversion.

[The groups of curves represented by eq. 8-9-7 and eq. 2, below, comprise supergenera (or subfamilies) that possess an infinite number of genera. The partitioning of the non-focal self-inversion loci into genera can be envisioned in two ways: (1) each genus would consist of all the non-focal self-inversion loci of a given subspecies of basis curve (i.e., taking all values of xy products and both alternate signs); (2) each genus would consist of one corresponding non-focal self-inversion locus of each subspecies of basis curve (i.e., taking one value of the xy product for one of the alternate signs).]

Non-focal self-inversion loci of parabolic Cartesians (eq. VII-56a-c and Figs. b and c) referred to the $h=0$ focus of self-inversion take the form of eq. 2a. The product of the roots is eq. 2b and the corresponding simple intercept format is eq. 3, where j^2 is the value of the xy product for the non-focal self-inversion and d^2 is the square of the modular value for the basis curve.

polar equation of non-focal self-inversion loci of parabolic Cartesians referred to the $h=0$ focus of self-inversion

(a) $r^2-d(C+2\cos\theta)r+(d^2\pm j^2) = 0$ (8-11-2)

(b) $r_1r_2 = (d^2\pm j^2)$ product of the roots of eq. 2a

$\cos\theta = \dfrac{x^2-Cdx+(d^2\pm j^2)}{2dx}$ simple intercept format of eq. 2a (8-11-3)

Eq. 3 identifies these non-focal self-inversion loci also as being Cartesian self-inverters (compare with eq. 8-9-8b). The upper alternate sign for the j^2 term corresponds to the locations of the non-focal self-inversion loci that hold for eq. 8-9-1b and the lower sign for the locations that hold for eqs. 8-9-1a,c,d. It follows from the above considerations of the non-focal self-inversion loci of Fig. a that all non-focal self-inversion loci of a parabolic Cartesian basis curve that have a common focus of self-inversion (pole 1), also are non-focal self-inversion loci of one another.

CIRCUMPOLAR
MAXIM 8-11-1: *All the non-focal self-inversion loci of a basis curve that self-invert about a focus of self-inversion of the latter, taken as reference pole, also are non-focal self-iversion loci of one another.*

The basis curve of the illustrations of Figs. b and c is the $C = 4\frac{1}{4}$ parabolic Cartesian, which consists of an annular locus (the heavily-traced ovals), with one oval very much larger than the other. These possess 3 foci of self-inversion, one each with the 0° inter-oval (pole 1) and intra-oval (pole 3) modes and one with the 180° inter-oval mode (pole 2).

Returning to Fig. b, 2 poles are marked in addition to the 3 foci of self-inversion of the basis curve (poles 1, 2, and 3), namely poles 4 and 5, both equidistant from the $h = 0$ 0° inter-oval-mode focus of self-inversion of the basis curve (pole 1). Pole 4 is the $h = -d$ focus of the basis curve and the homologue of the limacon $-\frac{1}{2}b$ double-focus. In fact, the $-\frac{1}{2}b$ double-focus of the $r = d(4\frac{1}{4} - 2\cos\theta)$ limacon, which is the *modular* non-focal self-inversion locus of the basis curve, is coincident with the $h = -d$ focus. Moreover, the $h = -d$ focus of the basis curve also is the $h = -d$ focus of all of its non-limacon non-focal self-inversion loci (all of which are Cartesian self-inverters). This follows because the constant term of eq. 3 does not enter into the relationship of the $\cos^2\theta$-condition, which is the condition that defines this focus (actually, it also is a double-focus; see Fig. 8-20).

Accordingly, since pole 1 of Fig. b is the DP homologue of the modular non-focal self-inversion locus, the present basis curve (like the limacon of Fig. a) shares two axial foci with all of its non-focal self-inversion loci. Pole 4 also is one of the two poles--pole 5 being the other--about which the basis curve inverts to central Cartesians. The possession of focal rank by one of these poles is characteristic of parabolic Cartesians (see legend to Fig.

8-22); in general, however, neither pole of Cartesian self-inverters possesses focal rank.

The Bipolar Equation of Non-Focal Self-Inversion Loci of Parabolic Cartesians Referred To The Focus of Self-Inversion

Inasmuch as all of these loci have two foci in common, a bipolar eq. can be derived for the entire supergenus represented by eq. 2 or 3. Casting eq. 2 or 3 in rectangular form gives eq. 4. Letting the focus of self-inversion at the origin be p_v and the $\cos^2\theta$-condition focus at $h = d$ be p_u, the relations of eqs. 5a,b readily follow. Substituting the relations of eqs. 5 in eq. 4 yields the 4th-degree eq. 6, which represents the entire supergenus.

[The illustrated curves are rotated 180° with respect to those represented by eq. 2, so that the $h = -d$ focus of Figs. a-c is at $h = +d$ for eq. 2. In this connection, the orientation of bipolar Cartesians as influenced by the signs of terms is illustrated in Fig. 8-13.]

$(x^2+y^2-2dx+d^2\pm j^2)^2 = C^2d^2(x^2+y^2)$	eq. 2 or 3 in rectangular form	(8-11-4)
(a) $v^2 = x^2+y^2$ (b) $x = (u^2-v^2-d^2)/-2d$	conversion eqs. between rectangular and bipolar coordinates	(8-11-5)
$(u^2\pm j^2)^2 = C^2d^2v^2$	eq. 4 in bipolar form	(8-11-6)
(a) $u^2\pm j^2 = Cdv$ (b) $u^2-d^2 = Cdv$		(8-11-7)

As in the case of eq. 8-9-13 for the non-focal self-inversion loci of limacons about the focus of self-inversion as reference pole, a 2nd-degree root eq. (7a) holds for some of the represented loci but not for others. For example, for the exterior modular elliptical-limacon loci, eq. 7b is valid. Note that eqs. 7 differ from the bipolar eq. of parabolic Cartesians only in the presence of the constant term. In the case of the lower alternate sign, eq. 7b differs from the bipolar eq. of *displaced parabolic Cartesians* (see Fig. 8-13) only in the magnitude of the constant term, while the loci of eq. 7a for the lower alternate sign and $j^2 = 2d^2$ are displaced parabolic Cartesians.

The Modular Non-Focal Self-Inversion Locus of The $C = 4\frac{1}{4}$ Parabolic Cartesian

As in the case of limacon non-focal self-inversion loci referred to the focus of self-inversion, when the xy product for non-focal self-inversion is taken to be equal to the square of the modular value for self-inversion of the basis curve about the reference pole (in this case, $j^2 = d^2$, and the $h = 0$ focus), and the negative alternate sign is employed (eq. 2 or 3), a *modular* limacon exterior non-focal self-inversion locus is obtained (the other non-focal self-inversion locus is the limiting focal case that consists of a point coincident with the focus of self-inversion).

The $r = d(4\frac{1}{4} - 2\cos\theta)$ limacon is the *modular* locus in question for the $C = 4\frac{1}{4}$ parabolic Cartesian. For other subspecies, the modular loci are given by the eq., $r = d(C - 2\cos\theta)$. Since the equilateral limacon is the curve of demar-

[The same species of limacon is the modular inter-oval locus for the $C = 4\frac{1}{4}$ *displaced parabolic Cartesian* (eqs. VIII-56b,c with the minus sign for the d^2 term; see also Fig. 8-13). In the case of the $C = 4\frac{1}{4}$ parabolic Cartesian, the non-focal xy products of self-inversion involve coincident radii to the nearest intercepts with each oval. In the case of the displaced parabolic Cartesian, the xy product involves a chord segment to the near point on the large oval but the far point on the small oval.]

cation of parabolic Cartesians (for $C = 4$; see pages VIII-19 to 21 and Fig. 8-16), and it is known from the treatment above (legend to Fig. 8-9) that the eccentricity of the the *modular* limacon non-focal self-inversion locus is the reciprocal square of that of the basis curve, the results of the two derivations can be checked against one another. The first derivation yields $e = (1/e_B)^2 = \frac{1}{2}$, while the second yields $e = 2/C = 2/4 = \frac{1}{2}$.

It is shown readily that when any one of the 3 foci of self-inversion of biovular Cartesian self-inverters (including bipolar linear Cartesians; eqs. XIV-90a,91a,92a) is employed as the reference pole, the *modular* non-focal self-inversion locus (eqs. corresponding to eqs. 3 and 4) is a limacon.

In the cases of the near-side inter-oval loci of Fig. b (eq. 3 or 4 with the positive alternate sign), taking $(xy)_1 = j^2$ of the basis curve yields the largest solid-line interior locus, which self-inverts about pole 1 for $j_1^2 = 2j^2$. Letting this 1st-generation locus be the basis curve for a 2nd-generation locus, with $(xy)_2 = 2j^2$ yields the intermediate solid-line near-side inter-oval locus, for which $j_2^2 = 4j^2$ for self-inversion. Lastly, letting the 2nd-generation locus be the basis curve for a 3rd-generation locus, with $(xy)_3 = 4j^2$ yields the smallest solid-line near-side inter-oval locus, for which $j_3^2 = 8j^2$ for self-inversion about pole 1. The xy products for these curves, when all are taken to be non-focal self-inversion loci of the basis curve, can be evaluated as in the examples of Fig. a (d^2, $3d^2$, and $7d^2$, respectively). As in the examples of Fig. a, these curves all are mutual non-focal self-inversion loci.

Other exterior non-focal self-inversion loci of the $C = 4\frac{3}{4}$ parabolic Cartesian are biovular Cartesian self-inverters. Fig. b illustrates the biovular locus obtained for $j^2 = 7d^2$. This consists of an inter-oval segment (dashed oval lying between the ovals of the basis curve) and an exterior segment (dashed exterior oval, partially shown in Fig. b but shown in its entirety in Fig. c).

[All biovular Cartesian self-inverters are of the annular type, i.e., one oval is within the other, and all of them self-invert in 3 different modes about 3 different foci (see legend to Fig. 8-13).]

Modes of Mutual Non-Focal Self-Inversion of Cartesian Self-Inverters

Fig. c illustrates the modes by which the annular and monovular non-focal self-inversion loci and the basis curve itself are non-focal self-inversion loci of one another, employing only one transform curve of each type from Fig. b (including the intermediate-sized monovular locus, for which j^2 for self-inversion$=4d^2$). The examples, shown by radius vectors that emanate from poles on the various curves, include one instance of each mode by which each of the 5 discrete ovals is a non-focal self-inversion locus of the other 3 curves. Since no annular oval is a non-focal self-inversion locus of its conjugate oval, nor of a single oval of another annular locus, there are only two examples for each of the 5 ovals, namely one coupling with each of the other two Cartesian self-inverters (see below).

These examples illustrate the mutual nature of non-focal self-inversion of Cartesian self-inverters. They also illustrate the fact that both ovals of an annular locus may comprise the non-focal self-inversion locus of a monovular basis curve (for the same value of the xy product). But only certain segments of the ovals of each annular locus are generated by a given monovular locus. The bases for this situation can be elucidated in two ways--in terms of: (a) the relative positions of the reference pole for the inversions, the intercepts, and the points on the non-focal self-inversion loci; and (b) *tangents of demarcation* from the reference pole to the monovular locus.

Following the first alternative, and referring to the loci as *basis*, *annular*, and *monovular*, and to points as *poles* (i.e., points on a curve considered as the non-focal self-inversion locus of another curve; numbers in Fig. c) and *intercepts* (i.e., points on a curve considered as the basis curve for a non-focal self-inversion; numbers with letter suffixes in Fig. c).

1. Poles (7) on the interior annular oval comprise a non-focal self-inversion locus of the monovular locus only if they lie on the opposite side of the reference pole from the intercepts (7a,7b).

2. Poles (10) on the interior basis oval comprise a non-focal self-inversion locus of the monovular locus only if they lie between the reference pole and the intercepts (10a,10b).

3. Poles (2) on the exterior annular (and exterior basis) oval are non-focal self-inversion loci of the monovular locus only if the intercepts (2a,2b) lie between these poles and the reference pole.

4. Poles on the monovular locus are non-focal self-inversion loci of the annular and basis loci only for combinations of intercepts with the latter loci that include one intercept on the interior oval (5b,6b) and one intercept on the exterior oval (5a,6a), and only when these intercepts lie on the segments of the annular and basis loci that can be generated as non-focal self-inversion loci of the monovular loci (i.e., only on segments made up of intercepts that, when considered as poles, satisfy rules 1, 2, and 3).

5. The interior basis oval is a non-focal self-inversion locus of the annular locus provided that the intercept (9a) with the exterior annular locus is on the same side of the reference pole as the pole (9) for the inversion and that the intercept with the interior annular locus (9b) is on the other side of the reference pole (this applies to any two annular non-focal self-inversion loci of the basis curve when the smaller of the interior ovals is considered to be a non-focal self-inversion locus of the larger annular locus).

6. The interior annular oval is a non-focal self-inversion locus of the basis curve provided that both the intercepts with the large oval (8a) and the small oval (8b) of the basis curve are on the opposite side of the pole (8) from the reference pole.

7. Poles (1) on the exterior annular oval are non-focal self-inversion loci of the basis ovals only for intercepts (1a,1b) lying between them and the reference pole.

Some of these relationships can be formulated or supplemented by reference to a *tangent of demarcation* (dashed radius vector) from the reference pole to the monovular locus [The "near" intercepts are the illustrated intercepts of the tangent line with the ovals; the "far" intercepts are the intercepts that would be made if the tangent line were to be extended on the other side of the reference pole.] For rule 1, the far intercept of the tangent of demar-

Mutual Non-Focal Self-Inversion of An Annular Parabolic Cartesian & Its Non-Focal Self-Inversion Loci Referred To The h=0, 0° Focus of Self-Inversion

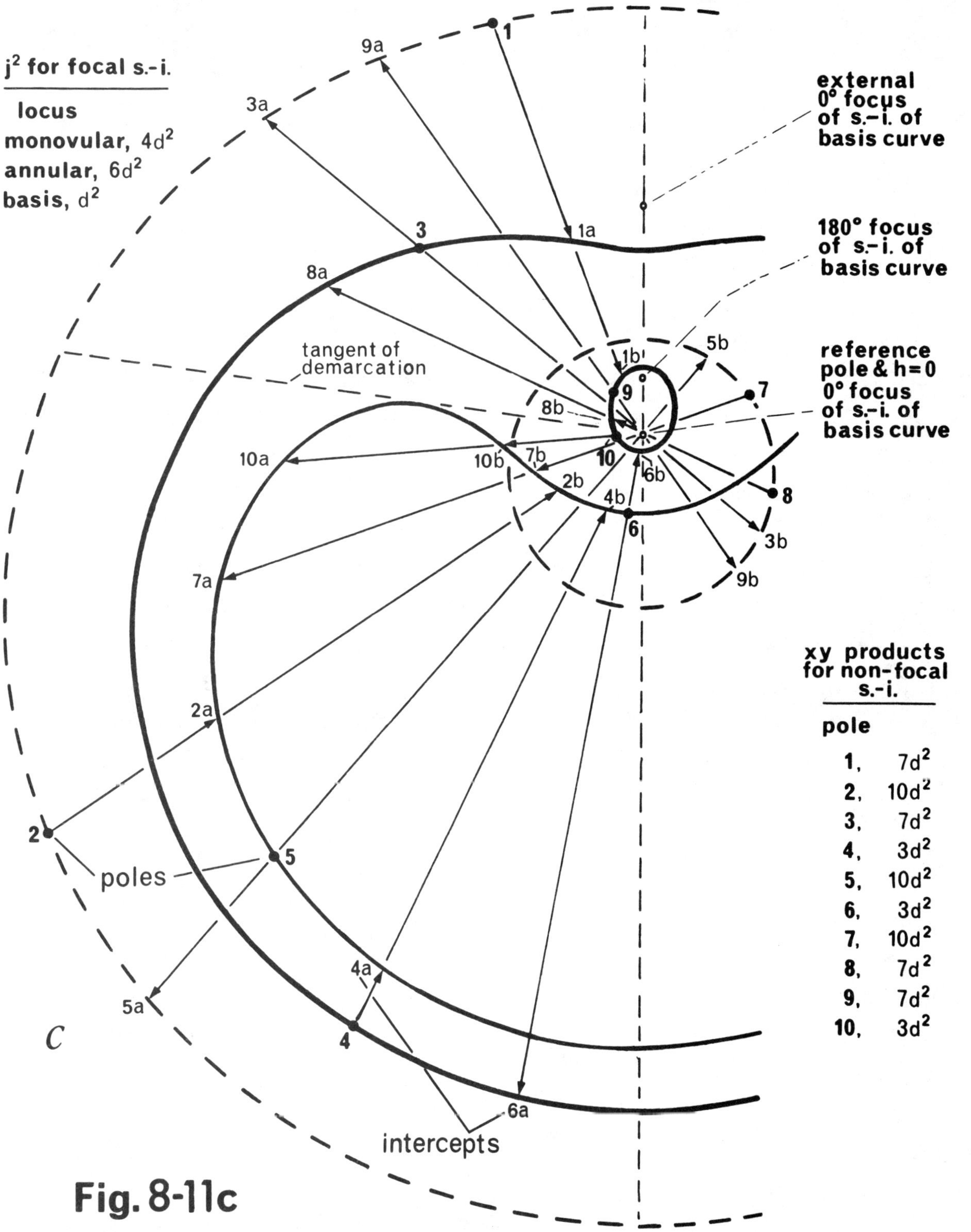

Fig. 8-11c

cation with the interior annular oval marks one terminus of this non-focal self-inversion locus of the monovular locus (the length of the segment of this tangent from the tangent point to the intercept in question is the unit for the inversion, namely $\sqrt{10}$d).

Similarly, for rule 2, the near intercept of the tangent with the interior basis oval marks one terminus of this non-focal self-inversion locus of the monovular locus (the corresponding length of the segment of the tangent is $\sqrt{3}$d); for rule 3, the near intercept of the tangent with the exterior annular oval marks one terminus of this non-focal self-inversion locus (the corresponding length of the segment of the tangent is $\sqrt{10}$d).

The same intercepts that are referred to above define the segments of the basis and annular ovals for which the monoval is a non-focal self-inversion locus (as described in rule 4). Accordingly, rules 1 through 4 can be restated in part or in toto in terms of intercepts with the tangents of demarcation rather than in terms of the relative positions of the poles and intercepts. On the other hand, the tangents of demarcation play no role with regard to rules 5 through 7.

The above rules are purely illustrative. Although similar rules hold for all of the other annular and monovular loci obtained for other values of the xy products for non-focal self-inversion of the basis curve, modifications are required for changes in the relative sizes and locations of the ovals (for example, for an annular locus with its interior oval smaller and within the basis interior oval or intersecting the basis interior oval, and for cases in which monovular loci intersect the basis interior oval, etc.).

On the other hand, points on interior annular ovals that comprise non-focal self-inversion loci of monovals may lie both interior to and exterior to the monovals. For example, if one refers to Fig. b, it is evident that the tangent of demarcation for the largest monoval intersects the annular interior oval within the monoval itself. Thus, considering a segment of the annular interior oval as a non-focal self-inversion locus of the largest monoval (rule 1), portions of the segment lie both interior and exterior to the monovular basis locus. Since the segment intersects the monovular locus, it also is evident that, for quartics, points of non-focal self-inversion loci may be incident upon the basis curve for cases in which neither of the corresponding intercepts is at infinity (as is obligatory when a point of a non-focal self-inversion locus of a hyperbola is incident upon the curve).

Another complication is introduced by the fact that the annular loci may self-invert at either 0° or 180° about the reference pole (depending upon whether the xy product for non-focal self-inversion is taken to be less than or greater than d^2 --the square of the modular value for the basis curve). While rule 7 applies to all annular loci considered as non-focal self-inversion loci of the basis ovals, the same rule applies only to annular loci considered as non-focal self-inversion loci of one another if both self-invert at 0° about the reference pole, or if one self-inverts at 180° and the other at 0°.

Contrariwise, if both annular loci self-invert at 180°, non-focal self-inversion occurs only if the intercept with the interior oval (comparable to point 1b for rule 7) is on the opposite side of the reference pole. This distinction, in terms of the nature of the self-inversion of the annular loci about the reference pole, is responsible for the difference between rules 1 and 2. Rule 1 holds for poles on interior ovals of all annular loci that self-invert about the reference pole at 180°, and rule 2 for those (like the basis curve) that do so at 0°.

With reference to rule 7, this rule holds for 180° annular ovals and the basis curve. For 0° ovals and the basis curve, the intercepts with the large and small ovals (corresponding to points 8a and 8b) of the basis curve must be on the same side of the reference pole as the pole for the inversion (corresponding to pole 8). A similar situation holds for any two 0° annular ovals. On the other hand, for two 180° annular ovals the situation is still different; in this case the intercept on the small oval (corresponding to 8b) must be on the same side of the reference pole as the pole for the inversion (corresponding to pole 8), while the intercept with the large oval (corresponding to 8a) must be on the opposite side of the reference pole.

Lastly, still another relationship exists that is not illustrated in Fig. C. When two annular curves that both self-invert at 0° about the reference pole are considered, both ovals of one of the curves are inter-ovals with respect to the other (i.e., both of the ovals of the former curve are of intermediate size). In such cases, each of the two inter-ovals of the one curve is an inter-oval non-focal self-inversion locus of the two ovals of the other annular curve. These examples sufficiently illustrate the relationships between ensembles of monovular and annular loci that are mutual non-focal self-inversion loci.

Fig. 8-12. Inter-loop and exterior non-focal self-inversion loci of the $e = 2$ limacon about the DP as reference pole for $j^2 = a^2/2$ and a^2. Fig. a illustrates exterior and interior non-focal self-inversion loci of the basis $e = 2$ limacon referred to the DP, for $j = \sqrt{2}a/2$. The basis curve is shown by the lighter solid tracings, and non-applicable continuations of these segments by the light dashed tracings. The large arrows mark the *points of demarcation* between these segments. The very light dashed line extending from the DP at the angle of a DP tangent is designated the *first chord of demarcation.* This chord intersects the illustrated half of the basis curve at the *points of demarcation.*

Three examples of radius vectors that illustrate the manner of derivation of the non-focal self-inversion loci are shown by straight dashed lines, labelled x and y, emanating from poles (marked by solid circles) on the loci to intercepts with the basis curve. A secant perpendicular to the line of symmetry at the DP intersects the large loop at a distance a from the DP, as indicated at the far right. The diameter of the small loop is (b-a), that of the large loop is (b+a). Accordingly, the axial distance from the large-loop vertex to the small-loop vertex is 2a. Fig. b represents the same basis curve and the comparable non-focal self-inversion loci for $j = a$.

As reference poles for non-focal self-inversion, foci of self-inversion fall into a unique category, inasmuch as they are the only such poles for which the non-focal self-inversion loci become coincident points for modular j. However, foci of self-inversion often qualify as circumpolar foci only by virtue of the self-inversion property itself (this applies to self-inversion at any reference angle).

In fact, poles of congruent-inversion (see Fig. 6-6, point midway between poles 1 and 18; Fig. 8-30d,e,g,h; and Fig. 8-38, pole p_3') enjoy no exceptional circumpolar focal status (circumpolar intercept transforms about these poles are not of lower degree than those about all neighboring points). Accordingly, the properties of non-focal self-inversion loci of limacons about the highest-ranking circumpolar focus, namely the DP or DP homologue, are of great interest.

To derive these loci, one substitutes the roots, $r = a\text{-}b\cos\theta$ and

Inter-Loop & Exterior Non-Focal Self-Inversion Loci of The e=2 Limaçon About The DP As Reference Pole ($j^2 = a^2/2$ & a^2)

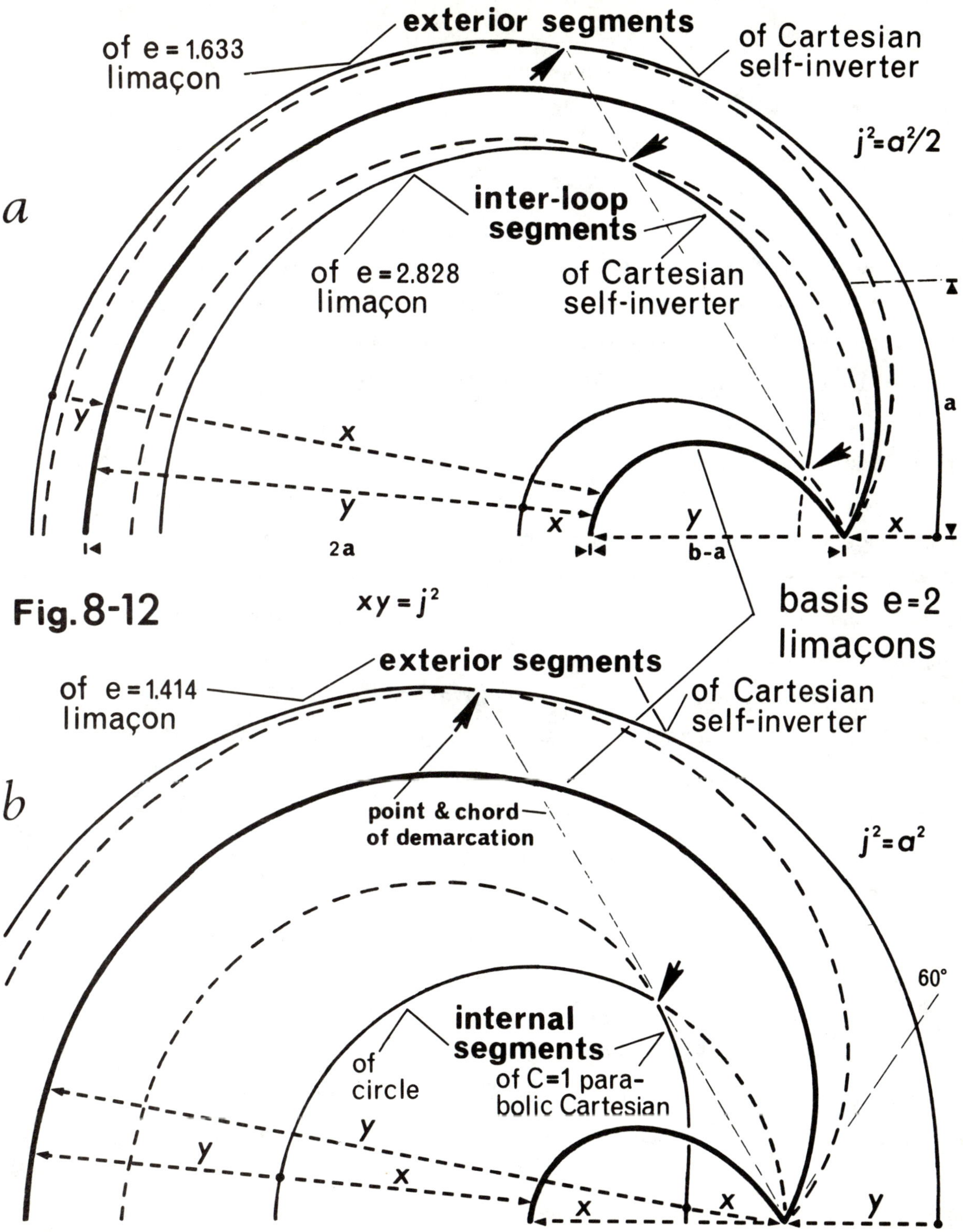

Fig. 8-12

$r = -a - b\cos\theta$ of the standard eq. of a limacon, $r = \pm a - b\cos\theta$ (in the form that yields two values of r for each value of θ) in the basis root eq. 8-9-6 for all non-focal self-inversion loci of monovular and annular biovular (or internalized two-looped) curves, leading to eq. 8-12-1, which is seen to

interior, inter-loop, and exterior non-focal self-inversion loci of limacons referred to the DP

$$r = \pm\sqrt{(a^2 \pm j^2)} - b\cos\theta \qquad (8\text{-}12\text{-}1)$$

represent other limacons of eccentricity $b/\sqrt{(a^2 \pm j^2)}$. As applied to hyperbolic limacons, the upper alternate sign within the radical represents interior and exterior non-focal self-inversion loci (and exterior loci of elliptical limacons), while the lower sign is for the inter-loop non-focal self-inversion loci (and interior loci of elliptical limacons). The external alternate sign is a true alternate.

It is clear from inspection of eq. 1 that, for the DP as reference pole, the condition analogous to taking modular j for a focus of self-inversion as reference pole is $j = a$, since for this value of j and the lower alternate sign the constant term vanishes. For a *focus of self-inversion* as reference pole these conditions yield a unique and most highly symmetrical locus among the non-focal self-inversion loci, namely a limacon, as opposed to other, less symmetrical, Cartesian self-inverters (see Table 8-12-1).

An analogous situation arises when the DP is taken to be the reference pole. In this case, though, it is limacons that play the role of the general non-focal self-inversion loci, whereas circles (or segments thereof) become the unique, most highly symmetrical subspecies and the analogues of the modular loci (see Table 8-12-1).

In the cases of elliptical limacons, the modular analogues consist of complete circles that are bisected by the line of symmetry of the basis curve and pass through the DP homologue. The situation is different for hyperbolic limacons, for which the analytical solution of eq. 1 does not yield the complete non-focal self-inversion loci as defined, but only a segment thereof.

A detailed discussion of the analytical basis for the discrepancy between eq. 1 and the complete interior and exterior non-focal self-inversion loci becomes very complicated. The essence of the factors involved is contained in the

Table 8-12-1. Non-Focal Self-Inversion Loci of Limacons Referred To The Focus of Self-Inversion And The Double Point

Basis Curve	Reference Pole	Non-Focal Self-Inversion Loci	Modular Locus (or analogue thereof)
hyperbolic limacon	focus of self-inversion	Cartesian self-inverters	*elliptical limacons* (genus of QBI curves, and genus of Cartesian self-inverters of highest circumpolar symmetry)
hyperbolic limacon	double point (highest ranking circumpolar focus)		
	a. pre-chord of demarcation	limacons (genus of QBI curves, and genus of Cartesian self-inverters of highest circumpolar symmetry)	*circle* (limacon of highest circumpolar symmetry)
	b. post-chord of demarcation	Cartesian self-inverters	Cartesian self-inverters (including internal segments consisting of *parabolic Cartesians*, one of 3 genera of Cartesian self-inverters of highest circumpolar symmetry)

section, *Analytical Transforms Versus Geometrical Expectations* on pages X-19 to 22. The present treatment of this matter is mainly descriptive. Consider first the lengths of two radius vectors directed from the DP to the vertices of the large loop (the x vector) and the small loop (the y vector). Since the x vector is of length 3a and the y vector of length a, the inter-loop axis is of length $x-y=2a$.

With the unit of linear dimension for non-focal self-inversion being $j=a$, (Fig. b), the midpoint of the inter-loop axis is the sole point of intersection with the interior non-focal self-inversion locus. For the exterior locus, the eq., $d(d+L) = a^2$ defines the distance, d (from the non-focal self-inversion locus to the basis curve), along an extension of the x radius vector from its intercept with the large loop to the point of intersection with the exterior locus. Since the inter-loop distance is 2a, $L = 2a$, and $d = (\sqrt{2}-1)a$ (the point on the exterior locus is not shown in Fig. b).

It is a well-known property of hyperbolic limacons that, though the lengths of the x and y radius vectors to the large and small loops decrease as these vectors leave the vertices, the difference between their lengths remains constant at 2a (whence the 0° transform, $x-y = 2a$). The consequences of this situation for the non-focal self-inversion loci is that the points on the interior locus continue the be at the midpoints of the segments of the x radius vectors (i.e., $d = a$) between their intersections with the two loops (the distances of these points from the origin being $b\cos\theta$), while points on the exterior locus continue to lie on the extensions of the x radius vectors at a distance $d = (\sqrt{2}-1)a$. This situation prevails until the angles of the *chords of demarcation* are reached (see Fig. b), at which angles the x radius vectors have diminished to a total length of 2a, and the y radius vectors to length 0. [The chords of demarcation are at the angles of the DP tangents; the one shown in Fig. b is referred to as the *first* chord of demarcation, while the corresponding chord for the lower half of the limacon (the extension of which is shown) is referred to as the *second* chord of demarcation.]

The points at which the chords of demarcation intersect the non-focal self-inversion loci are designated as *points of demarcation*. At these points the non-focal self-inversion loci change in nature. Until the attainment of these points, the interior locus is a segment of the circle $r = -b\cos\theta$ and the exterior locus is a segment of the hyperbolic limacon $r = a(1-\sqrt{2}\cos\theta)$, as labelled in Fig. b.

The reason for the change in nature is that the x radius vector from the DP no longer intersects the small loop. Instead, it extends directly from the DP to the large loop. Accordingly, its length begins to diminish from 2a. The consequence for the interior non-focal self-inversion locus is that $d = [-L\pm\sqrt{(L^2+4a^2)}]/2$ begins to increase (by the distance between the dashed and solid non-focal self-inversion loci beyond the point of demarcation). At an angle of 90° between the x radius vector and the coordinate reference line, this vector is of length a, and $d = [-a\pm\sqrt{(5a^2)}]/2 = 0.618a$.

The length of the exterior extension of the x radius vector reaches its maximum value $d = a$ at the angle for which the length of the vector itself is

0, namely at the angle of the *second* chord of demarcation (the extension of which is labelled "60°" in Fig. b). Beyond this angle the y radius vector intersects the small loop of the curve again. Accordingly, L, the length of the interior chord of the small loop increases from 0 to a, as the x radius vector progresses from the position of the second chord of demarcation to coincidence with the positive direction of the coordinate reference axis. Concomitantly, d decreases in value from a units to 0.618a units.

Parallel circumstances hold for $j = a/\sqrt{2}$ (Fig. a), but because the value of j is $a/\sqrt{2}$ units rather than a units, the non-focal self-inversion loci differ. The exterior locus, from the point opposite the large-loop vertex to the points of demarcation, is a segment of the $e = 1.633$ limacon; the inter-loop locus between the corresponding points consists of portions of the $e = 2\sqrt{2}$ limacon. In this instance, though, since $j < a$ and the length of the inter-loop segment is 2a, there are two points of the inter-loop non-focal self-inversion locus on each segment (solid light inter-loop tracings).

Accordingly, portions of both the small and large loops of the $e = 2\sqrt{2}$ limacon participate in the inter-loop locus and both portions terminate at *points of demarcation* determined by intersections with the chords of demarcation (since for chords of length less than 2a, the points of both the inter-loop and exterior non-focal self-inversion loci take a different course).

The nature of the non-focal self-inversion loci beyond the points of demarcation now is elucidated. These loci are not derived, as above, by considering non-focal self-inversion relative to two points of radius vectors from the DP, where one point lies on each loop of the basis curve. Instead, they are derived by considering the case where the DP, itself, is one point and the other point is incident upon one or the other of the loops. Eq. 2c, derived from eqs. 2a and 2b is the combined root eq. for this derivation.

(a)	$(r_1-r)r = j^2$	root equation for interior non-focal self-inversion locus for either loop	(8-12-2)
(b)	$(r-r_1)r = j^2$	eq. 2a modified for exterior loci	
(c)	$(r_1-r)r = \pm j^2$	combined root eq. for interior (upper alternate sign) and exterior non-focal self-inversion loci of individual loops of limacons referred to the DP	

Taking the eq. of the basis hyperbolic limacon to be $r = \pm a + b\cos\theta$, and substituting in eq. 2c gives eq. 3a for the sought-after non-focal self-inversion locus (the upper alternate sign of the j^2 term is for interior loci). The corresponding simple intercept format is eq. 3b and the rectangular eq. is 3c. Examination of these eqs. reveals that they represent a group (supergenus) of Cartesian self-inverters, restricted only by $a < b$ (hyperbolic basis subspecies) and $j < (a+b)/2$ (no locus exists unless j is less than half the diameter of the large loop).

(a) $r^2 - br\cos\theta \pm ar \pm j^2 = 0$ — polar eq. of non-focal self-inversion loci of individual loops of limacons referred to the DP (8-12-3)

(b) $\cos\theta = \dfrac{x^2 \pm ax \pm j^2}{bx}$ — simple intercept format of eq. 3a

(c) $[x^2+y^2-bx \pm j^2]^2 = a^2(x^2+y^2)$ — eq. 3a in rectangular form

$[x^2+y^2-2dx+d^2]^2 = C^2d^2(x^2+y^2)$ — rectangular eq. of bipolar parabolic Cartesians, $u^2 = Cdv$ (8-12-4)

$r = d(\pm C + 2\cos\theta)$ — basis hyperbolic limacons [segments of parabolic Cartesians comprise interior non-focal self-inversion loci of large loop $(j = d)$] (8-12-5)

Comparison of eq. 3c with the rectangular eq. for parabolic Cartesians reveals that the loci of eq. 3c are parabolic Cartesians for the case in which the upper alternate sign is employed (i.e., the interior loci); $j = d$, $b = 2d$, and $a = \pm Cd$. For these conditions the basis $r = \pm a + b\cos\theta$ limacon can be rewritten as eq. 5. It is evident from eq. 5 that interior non-focal self-inversion loci consisting of parabolic Cartesians exist (with C always less than 2) for all hyperbolic limacons--factored to give the form of eq. 5--provided that d (the factored-out constant) is taken to be the unit of linear dimension for the non-focal self-inversion.

The value of the constant term within the parentheses of the factored eq. is the value, C, of the bipolar parabolic Cartesian, $u^2 = Cdv$. Since the diameter of the small loop is $(2-C)d$, and d is the unit of linear dimension, there can be no interior parabolic Cartesian non-focal self-inversion locus of the small loop [since for any locus to exist, $(2-C)d$ must be greater than $2d$]. [These findings provide a basis for constructing parabolic Cartesians from hyperbolic limacons (for $C < 2$).]

In the case of the lower alternate sign of eq. 3c, and the same values of a, b, and j, the non-focal self-inversion loci beyond the points of demarcation are segments of *displaced parabolic Cartesians* (see analysis of Cartesian self-inverters in Fig. 8-13), while in Fig. a, for $j = a/\sqrt{2}$, the locus beyond the point of demarcation is a monovular Cartesian self-inverter. In Fig. b (for which $j = d$ in the eq. of the basis limacon satisfies the condition for a parabolic Cartesian non-focal self-inversion locus), a portion of the non-focal self-inversion locus is a segment of the $C = 1$ parabolic Cartesian. In all these cases, the non-applicable segments of the curves that contribute to the non-focal self-inversion loci (that lie to either side of the points of demarcation) are shown by dashed lines, so that the full form of each curve (or its outer loop or outer oval) can be seen.

It should be emphasized, lastly, that in all of the cases for which portions of the internal loci consist of segments of parabolic Cartesians, these segments do not, strictly speaking, apply to the basis curve as a whole. For the basis curve as a whole, the loci terminate at the points of demarcation. It is only if the basis curve is considered arbitrarily to consist only of its large loop that segments of parabolic Cartesians contribute to the non-focal self-inversion loci (of course, the construction referred to above is valid independently of the concept of non-focal self-inversion).

[In Fig. b, points on the solid segment of the circular locus are equidistant from both loops along radii through the DP. In contrast, points on the solid segment of the parabolic Cartesian locus are at the same distance from the DP as are points on its dashed continuation from the large loop (along the same radii).]

Fig. 8-13. Cartesian self-inverters: effects of signs of terms on loci and their polar generation. The generalized simple intercept format for Cartesian self-inverters is given by eq. 1. The parameters, E, F^2, and G are assumed to

$$\cos\theta = \frac{\pm x^2 \pm Ex \pm F^2}{\pm Gx} \quad \text{generalized simple intercept format of Cartesian self-inverters (cast about any of 3 foci of s.-i.)} \qquad (8\text{-}13\text{-}1)$$

$$\cos\theta = \frac{+x^2 + Cdx + d^2}{+2dx} \quad \text{example of simple intercept format for parabolic Cartesians } (x^2 \text{ and } d^2 \text{ terms must be of same sign}) \qquad (8\text{-}13\text{-}2)$$

be positive in this formulation in which signs are employed for individual terms. All combinations of signs are allowed. The representatives of all combinations of signs are illustrated here and are useful in the treatments that follow.

Instead of plotting complete curves for each combination, an example of the curve in one orientation is given at the left without indication of the manner in which it is traced. At the right the curve is not shown in its entirety. Instead, its orientation and the mode by which it is traced are indicated by the directional segments, the angle labels, and the location of the focus of self-inversion.

For these illustrations, the parameters of parabolic Cartesians are used, with eq. 2 being an example for the case in which the signs are positive for all terms. When this is the case, eq. 2 represents annular or monovular parabolic Cartesians that have the bipolar eq. $u^2 = Cdv$. If the signs of the x^2 and d^2 terms are opposite, it represents *displaced parabolic Cartesians* with the bipolar eq. $u^2 = Cdv+2d^2$. Eq. 3 is the polar solution of eq. 2 for r (the x intercept), from which it can be seen that two ovals (4 vertices) exist only for $(C-2)^2 > 4$, i.e., $C > 4$.

$$r = d\{(2\cos\theta-C)\pm\sqrt{[(2\cos\theta-C)^2-4]}\}/2$$ polar form of eq. 2 solved explicity for r (x intercept) (8-13-3)

(a) $r_1 r_2 = d^2$ product of the roots of eq. 3 for x^2 and d^2 terms of the same sign ($\theta = 0°$ or $180°$) (8-13-4)

(b) $r_1 r_2 = -d^2$ product of the roots of eq. 3 for x^2 and d^2 terms of opposite sign ($\theta = 0°$ or $180°$)

Analysis of the loci defined by eq. 2 for all combinations of signs reveals that from the point of view of congruence, the only critical condition on the signs of the terms has to do with whether those of the x^2 and d^2 terms are the same or different. For a given value of C, say $C = 2$, all 8 sign combinations in which the x^2 and d^2 terms have the same sign give congruent monovular curves (sets b and c), while all 8 combinations in which they have different signs give congruent but different curves, namely annular biovals--*displaced parabolic Cartesians* (sets d and e). The only differences between the congruent sets consist in reflections and the modes in which the curves are traced.

Cartesian Self-Inverters: Effects of Signs of Terms On Loci & Their Polar-Coordinate Generation (Parabolic & Displaced-Parabolic Bipolar Cartesian Exs.)

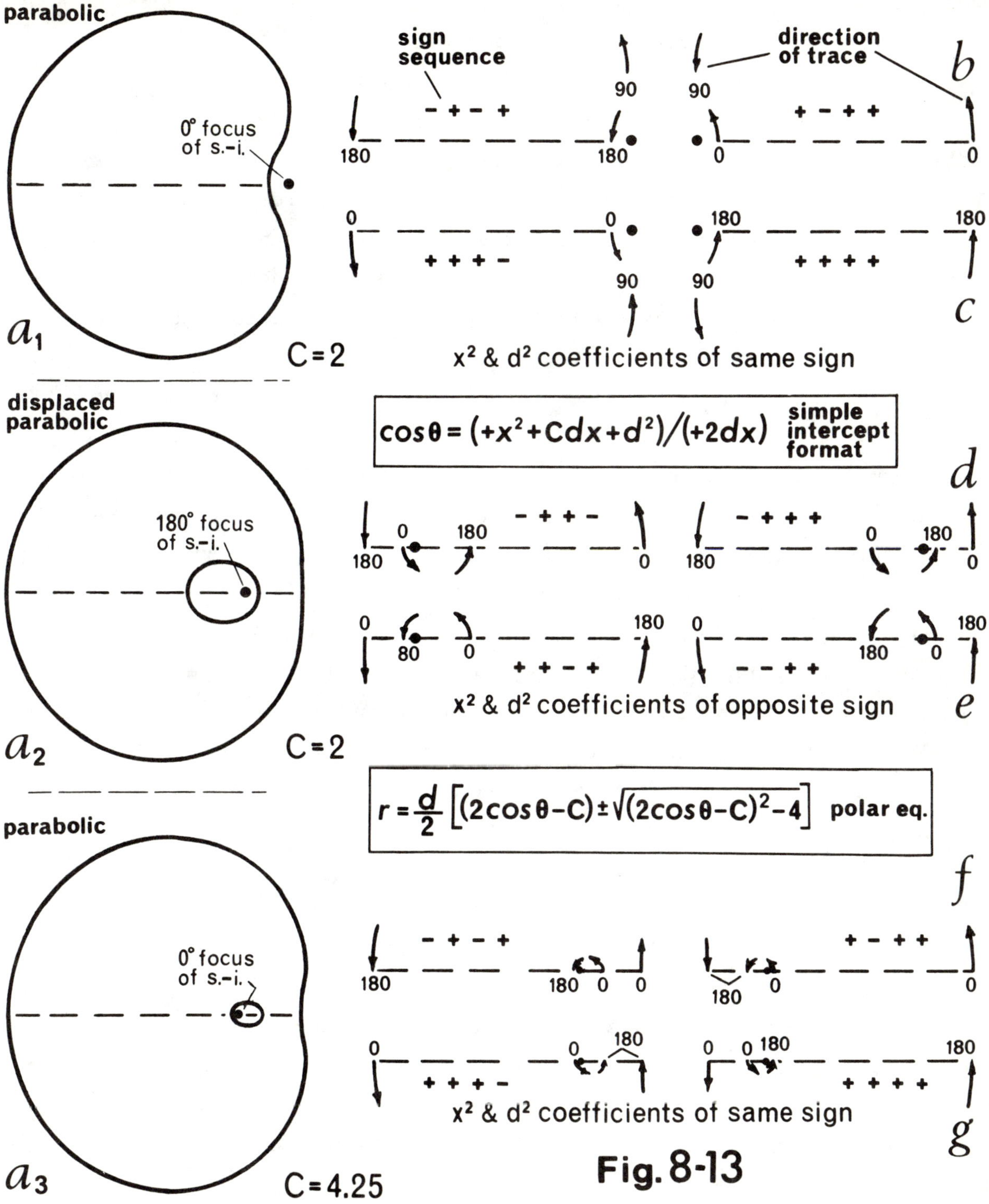

Fig. 8-13

Since the 8 combinations of signs referred to fall into equivalent complementary pairs (such as + − − + and − + + − , where the sequence is from left to right for the numerator, followed by the sign of the denominator), examples for only 4 different sets of signs illustrate completely the differences between the loci.

Consider first the 4 loci obtained for $C = 2$ with the signs of the x^2 and d^2 terms the same. For Fig. b-left (signs − + − + and + − + −), the tracing for $\theta = 0^\circ$ to 90° is imaginary. The real locus begins at 90° for both roots of eq. 3 at the upper tangent point of a line orthogonal to the axis from the focus of self-inversion. For θ progressing from 90° to 180°, one root traces the upper segment of the curve CW from the tangent point downward to the near axial vertex (i.e., near the focus of self-inversion), and the other root traces the upper segment CCW up and around the loop to the far axial vertex.

The tracing for Fig. c-left (+ + + − and − − − +) continues the pattern of Fig. b-left but for the lower half of the curve. From 0° to 90° one root traces the lower segment of the curve from the far vertex CCW around the loop to the 90° tangent point, while the other traces the lower segment CW from the near vertex to the tangent point. Together, these two 0° and 180° tracings make up a complete locus with the same orientation as that of curve a_1. Their 0° to 360° tracings, of course, each make up a complete locus.

For the combinations of signs of Fig. b-right and Fig. c-right, the curve is reflected in the focus of self-inversion (equivalent to reflection about a normal to the line of symmetry at the focus of self-inversion). Otherwise, Fig. b-right is similar to Fig. c-left except that the upper half of the curve is traced, while Fig. c-right is similar to Fig. b-left but the lower half of the curve is traced.

Consider now the 4 loci for *displaced parabolic Cartesians* obtained for $C = 2$ but for differing signs for the x^2 and d^2 terms. The orientations and modes of tracing of the annular ovals are illustrated in Figs. d and e, which are essentially self-explanatory. It should be remarked, however, that the tracing of displaced parabolic Cartesians differs notably from the tracing of annular parabolic Cartesians (Figs. f and g) in that for each set of qualifying signs the upper half of one oval is traced simultaneously with the lower half of the other, whereas for annular parabolic Cartesians the upper or lower portions of both ovals are traced simultaneously.

These differences have their basis in the fact that when the signs of the x^2 and d^2 terms differ, the $h = 0$ pole is a focus of self-inversion possessing the 180° inter-oval mode, and the product of the roots (eq. 4b) is negative (this product also is given by the ratio of the constant term of eq. 2 to the coefficient of the x^2 term, i.e., $-d^2$, when the terms are of opposite sign; see also page XIV-48). For a negative product of the roots, one root has to be negative in the ranges of θ for which the other is positive. Geometrically, this means that one half of one oval is traced CW or CCW (one chord segment of the annular curve) while the other half of the other oval is being traced in the same sense (by the other chord segment). Corresponding diagrams for displaced parabolic Cartesians with $C = 4\frac{1}{4}$ are not given, since modes of tracing of these curves are identical for all non-zero values of C. [The key condition for an intercept format to lead to a generic double-focus of self-inversion (see pages XIV-48 et seq.) is that the product of the roots of the corresponding polar eq. be constant (or that the product of appropriate pairwise groupings of the roots be constant). This is the case for the intercept formats of self-inversion of eqs. XIV-43a-e but not for the intercept formats of eqs. XIV-44.]

Some General Properties of Cartesian Self-Inverters

Members of the supergenus (or subfamily) of Cartesian self-inverters play a prominent role in subsequent treatments. Accordingly, having noted the general influences of the signs of terms on Cartesian-self-inverters, using parabolic and displaced parabolic Cartesians as examples, some of the general properties of this group are elucidated here. For the sake of simplicity, the simple intercept format for Cartesian self-inverters is represented in the form of eq. 5, with the understanding that the parameters, E, F^2, and G, may be positive or negative (the exponent of the F^2 term is employed to maintain dimensional balance but the term may be either positive or negative).

$$\cos\theta = \frac{x^2+Ex+F^2}{Gx}$$ simple intercept format of Cartesian self-inverters cast about a focus of self-inversion; E, F^2, $G \neq 0$, positive or negative (8-13-5)

$$r = (G\cos\theta-E)/2 \pm \sqrt{[(G\cos\theta-E)^2-4F^2]}/2$$ polar form of eq. 5 solved explicitly for r (x intercept) (8-13-6)

Cartesian self-inverters number among their member genera the most highly symmetrical non-central quartics. Among the known members are limacons, bipolar linear, parabolic, and displaced parabolic Cartesians, polar-circular linear Cartesians (which include limacons and some bipolar linear Cartesians), true exchange-limacons, and non-focal self-inversion loci of (1) limacons with reference to both the focus of self-inversion and the DP, and (2) parabolic and displaced parabolic Cartesians about the focus of self-inversion.

The corresponding roots of eq. 5 are given by eq. 6, whence the number of ovals represented and the products and sums of the roots (vertex locations) readily are ascertained. The conditions for 4 vertices are 7a, from which it can be seen that displaced parabolic Cartesians, for which the F^2 term is negative, always possess two ovals, but that parabolic Cartesians, for which the F^2 term is positive, do not.

(a) $(G-E)^2 > 4F^2$ and $(-G-E)^2 < 4F^2$, or (b) $F^2 < 0$ — conditions for 4 vertices (8-13-7)

(b) $r_1r_2r_3r_4 = F^4$, $r_1r_2 = F^2$ — products of 4 or 2 roots

(c) $r_1+r_2+r_3+r_4 = 2G$, $r_1+r_2 = (G-E)$ or $(G+E)$ — sums of 4 or 2 roots

(a) $(G-E)/2 \pm\sqrt{[(G-E)^2-4F^2]}/2$

$(G+E)/2 \pm\sqrt{[(G+E)^2-4F^2]}/2$ — axial locations of the 2 or 4 vertices relative to the $h=0$ focus (8-13-8)

Parabolic Cartesians possess two ovals only for $(G-E)^2 > 4F^2$ and $(-G-E)^2 < 4F^2$ [limacons are the special cases of eqs. 5 and 6 for which $(G-E)^2 = 4F^2$ or $(-G-E)^2 = 4F^2$]. The products and sums of 2 or 4 roots of monovular and biovular members of the group are given by eqs. 7,b,c (recall that for limacons eq. 5 is cast about the focus of self-inversion--eq. X-2 --not the DP). The axial locations of the 2 or 4 vertices are given by eqs. 8. The vertex locations of 8a are for $\theta = 0^\circ$ in eq. 6, while those of 8b are the negative of those for $\theta = 180^\circ$ in eq. 6.

The simple intercept eq. for the group, cast about any of the 3 foci of self-inversion (but about the 180° inter-oval-mode focus of self-inversion for the F^2 term negative) is eq. 9. It is evident from this eq. that the double-focus defined by the $\cos^2\theta$-condition is $h = \frac{1}{2}G$ for all members, whence the homology of this focus in limacons and Cartesians (for $h = \frac{1}{2}G$, the term in

$\cos\theta$ on the left of eq. 9 vanishes, with the consequence that there is no $\cos^2\theta$ term in the resulting eq.). Casting eq. 9 in the form of the simple intercept format (i.e., solving for $\cos\theta$) yields eq. 10, from which the remaining axial focal conditions for all members of the group readily follow.

simple intercept eq. for Cartesian self-inverters

$$[x^2+(2h-G)x\cos\theta+(h^2-Gh+F^2)]^2 = E^2(x^2+2hx\cos\theta+h^2) \qquad (8\text{-}13\text{-}9)$$

simple intercept format for Cartesian self-inverters (8-13-10)

$$\cos\theta = \frac{[E^2h+(G-2h)(x^2+h^2-Gh+F^2)]\pm E\sqrt{\{G(G-2h)x^2+h[2h^2G+h(E^2-G^2-4F^2)+2F^2G]\}}}{(G-2h)^2x}$$

$$h = \frac{(4F^2+G^2-E^2)\pm\sqrt{[(4F^2+G^2-E^2)^2-16F^2G^2]}}{4G}$$ 2 additional real or imaginary foci of Cartesian self-inverters (8-13-11)

$$(4F^2+G^2-E^2)^2 > 16F^2G^2$$ condition for two additional real foci for Cartesian self-inverters (8-13-12)

(a) $h(+)$ of eq. 11, (b) $h(-)$ of eq. 11, (c) $\frac{(4F^2+G^2-E^2)}{2G}$ — 3 inter-focal distances for foci of self-inversion (8-13-13)

In addition to two or 4 vertices and $h=0$ and $h=\frac{1}{2}G$ foci (for $h=0$ the radical vanishes, leaving a term $\pm Gx$), two additional real foci exist whenever the constant term of the radicand vanishes. Setting the bracketed term of the radicand equal to 0 and solving for h yields eq. 11 for the two additional real or imaginary foci of all Cartesian self-inverters. These foci are real whenever the radicand of eq. 11 is positive (eq. 12)--this condition being equivalent to the conditions for 4 vertices or two real ovals (eqs. 7). [For bipolar parabolic Cartesians, eq. 12 reduces to $C>4$, while for bipolar linear Cartesians (eqs. XIV-90 to 92), it reduces to $B\neq C$, $A\neq C$, i.e., the conditions for non-limacon loci (two real ovals).]

Accordingly, biovular Cartesian self-inverters possess 8 discrete real foci, of which the $\frac{1}{2}G$ focus is a generic double-focus (see Figs. 8-21 and 26). Like the $h=0$ focus, the two additional foci defined by eqs. 11 and

12 are foci of self-inversion. Insofar as focal rank is concerned (i.e., the reciprocals of the degrees of corresponding circumpolar intercept transforms), these foci are equivalent. In many respects, however, their properties differ. Thus, the ovals self-invert about each focus by a different mode --about one by the 0° inter-oval mode, about another by the 0° intra-oval mode, and about the 3rd by the 180° inter-oval mode, and, for $F^2 > 0$, the intercept format takes the same form for all.

The chief restriction on the identity of the foci is that if $F^2 < 0$ (i.e., if the signs of the x^2 and F^2 terms differ), the format of eq. 5 is cast about the 180° inter-oval-mode focus. The focus with the 0° intra-oval mode lies on the line of symmetry exterior to the two ovals, while the other two foci lie within the small oval (the ovals for parabolic and displaced parabolic Cartesians always are annular). Eq. 13 gives the 3 inter-focal distances for these foci.

In general, the modular values of j for the 3 foci (i.e., the units of linear dimension for self-inversion) are not inter-focal distances, the parabolic and displaced parabolic Cartesians being among the exceptions (for which $j^2 = d^2$). However, it is a known characteristic of the group that the square of the modular distance for any given focus of self-inversion is the product of the distances to the other two foci (i.e., any two foci self-invert about the third). Accordingly, values of j^2 for the 3 foci are given by eqs. 14.

	products of inter-focal distances	focus	
(a)	$j_o^2 = F^2$	$h = 0$	(8-13-14)
(b)	$j_{(+)}^2 = [A^2\surd(A^4-16F^2G^2)+A^4-16F^2G^2]/8G^2$	$h(+)$ of eq. 11	
(c)	$j_{(-)}^2 = [A^2\surd(A^4-16F^2G^2)-A^4+16F^2G^2]/8G^2$	$h(-)$ of eq. 11	
	$A^2 = (4F^2+G^2-E^2)$		

simple intercept format for Cartesian self-inverters for j_o equal to the distance between the $h = 0$ and $h = \frac{1}{2}G$ foci

$$\cos\theta = \frac{x^2+Ex+\frac{1}{4}G^2}{Gx} \qquad (8\text{-}13\text{-}15)$$

Given that the distance from the $h = 0$ focus to the focus of the $\cos^2\theta$-condition is $\frac{1}{2}G$, under what conditions does the focus of the $\cos^2\theta$-condition lie at modular distance from the $h = 0$ focus? The eq. for this condition is $F^2 = \frac{1}{4}G^2$ and the corresponding simple intercept format is eq. 15. Accordingly, the condition is satisfied by all parabolic and displaced parabolic Cartesians (compare with eq. 8-13-2). For bipolar linear Cartesians (pages XIV-87 to 89), it applies whenever $A^2(B^2+C^2) = B^2C^2$, and the same criterion can be applied for other members of the group.

Fig. 8-14. Interior and inter-loop non-focal self-inversion loci of elliptical and hyperbolic limacons referred to the focus of self-inversion, for $j^2 = (a^2-b^2)^2/4b^2$. Complete interior loci--which consist only of monovals--are illustrated for the elliptical limacon subspecies, $e = 0.415$, $e = 0.5411961$, $e = 0.75$, $e = 0.9$, and $e = 0.95$. For the $e = 0.99$ elliptical and the $e = 1.01$ and $e = 1.1$ hyperbolic subspecies, only a greatly magnified region in the neighborhood of the reference pole (the focus of self-inversion) is shown.

Complete inter-loop loci are illustrated for the hyperbolic limacon subspecies, $e = 1.306562965$, $e = 1.65$, $e = 2$, and $e = 2.4$. In the case of the locus for the $e = 2$ limacon, the basis curve also is indicated (dashed segments). The non-focal self-inversion loci that are illustrated in this figure are the basis curves for the loci of Fig. 8-15.

Reference poles, $h = 0$, exterior poles of inversion to central curves, $h = \sqrt{[(a^2-b^2)^2/2b^2]}$, and the corresponding interior poles $h = -\sqrt{[(a^2-b^2)^2/2b^2]}$, are indicated by solid circles (the latter poles are at a distance of $\sqrt{2}j$ along the axes on each side of the reference poles). In the case of Fig. b, space does not permit indicating the exterior pole. In the cases of Figs. a, f, g, and m, the magnification of the curves is so great that only the reference pole can be shown.

Eq. 1 represents the loci of this figure. It is derived from eq. 8-9-7, with the upper alternate sign and j modular for the non-focal self-inversion, i.e., $j^2 = xy = (a^2-b^2)^2/4b^2$. It differs from the eq. of limacons cast about

interior and inter-loop non-focal self-inversion loci of elliptical and hyperbolic limacons, respectively, referred to the focus of self-inversion, for modular j

$$4b^2r^2 + 4br(a^2\cos\theta-b^2) + 2(b^2-a^2)^2 = 0 \qquad (8\text{-}14\text{-}1)$$

the focus of self-inversion only in the presence of the coefficient, 2, of the $2(b^2-a^2)^2$ term. For j modular, interior loci of elliptical limacons exist only for $e \geq (\sqrt{2}-1)$; for a limiting value of $e = (\sqrt{2}-1)$, the locus is merely the midpoint of the loop diameter. For inter-loop loci of hyperbolic limacons, the corresponding point locus is attained at a value of $e = (\sqrt{2}+1)$,

The roughly circular locus of Fig. a is for an elliptical limacon whose eccentricity departs by 0.000786 units from the limiting value. The elongated ovular locus of Fig. m is for a hyperbolic limacon whose eccentricity departs by 0.0142 units from the corresponding limiting value. For Fig. a, the reference pole is at a distance of roughly 10 times the diameter of the oval; the corresponding distance ratio for Fig. m is roughly 6.

Fig. b is representative of the interior non-focal self-inversion loci of the $e = 0.5411961$ limacon. Among all elliptical limacons, this subspecies is unique in that the two central inversions of its interior non-focal self-inversion locus (for modular j) belong to the same subspecies. They thus are *curves of intersection* (see pages XI-11,12) of the two genera to which they belong. These central curves are illustrated in Fig. 8-15b. The curve of Fig. c derives from the $e = 3/4$ limacon. [Both the $e = 3/4$ limacon and its first interior non-focal self-inversion locus also are illustrated in Fig. 8-11a.]

As e increases toward 1, the concavity in the neighborhood of the reference pole of both the basis limacon and its non-focal self-inversion locus increases in depth. For $e = 0.95$ this concavity has almost the appearance of the cusp of the cardioid but the reference pole still lies just outside the curve. For $e = 0.99$ and $e = 1.01$, both the basis limacons and the non-focal self-inversion loci are indistinguishable from the cardioid at the scale of Fig. e. Accordingly, the region in the neighborhood of the reference pole has been greatly magnified and portions of both the basis limacons and the derived loci are shown (the axial diameters of the full curves are roughly 700 times as great as the sizes of the regions shown, and the distances to the poles for central inversions are roughly 5 times the sizes of the regions).

Figs. f and g illustrate the appearance of the respective curves very close to the transition point between elliptical and hyperbolic limacons. At the transition point (in the cardioid), the inner curve of Fig. f has become elongated to a cusp, the tip of which is in coincidence with the reference pole and DP homologue; similarly, by attainment of $e = 1.01$, the tip of the cusp has opened out into a loop from the DP, and the reference pole remains

Interior & Inter-Loop Non-Focal Self-Inversion Loci of Elliptical & Hyperbolic Limaçons Referred To The Focus of Self-Inversion; $j^2 = (a^2-b^2)^2/4b^2$

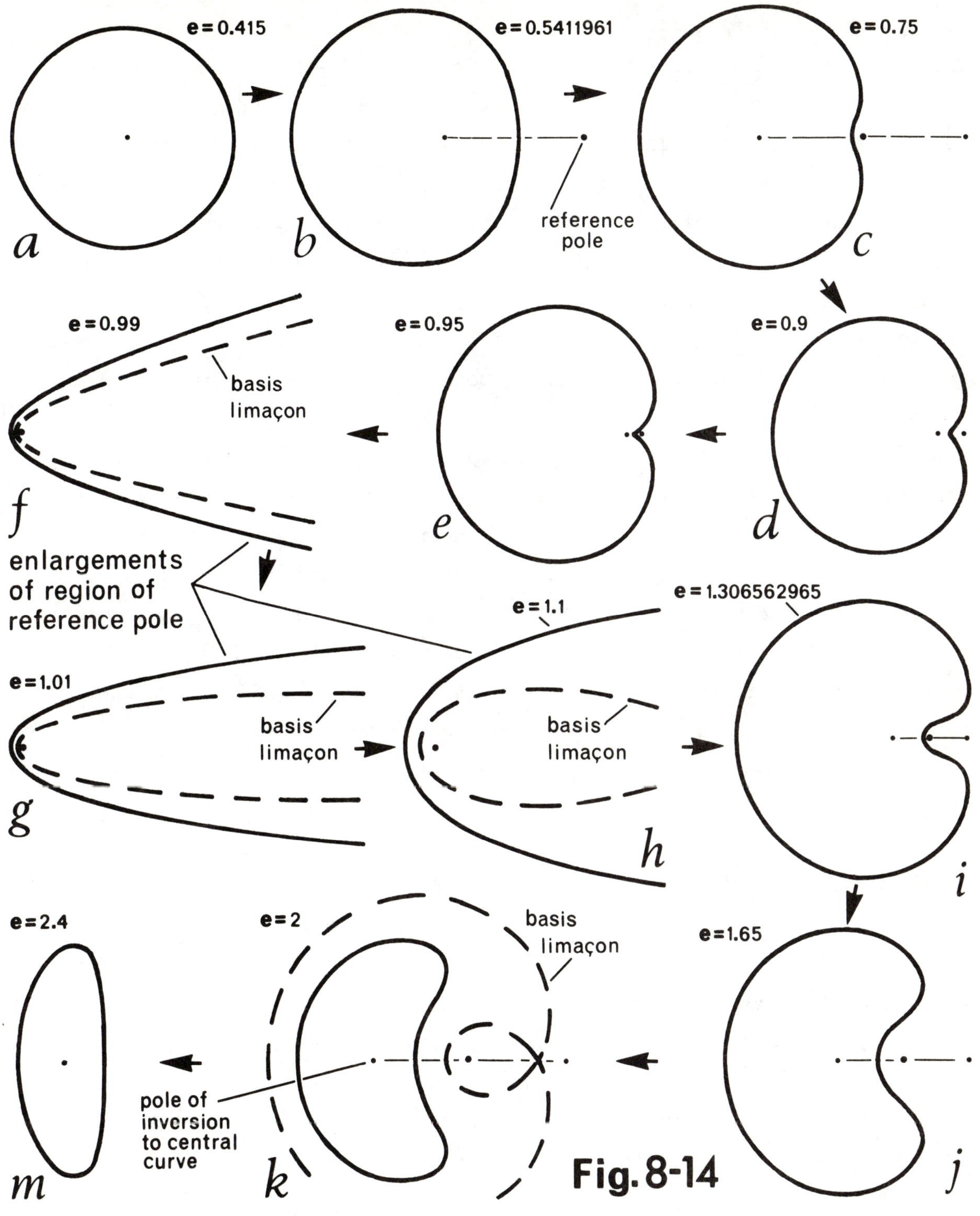

Fig. 8-14

close to the loop vertex.

Since the DP and the focus of self-inversion become coincident in the cardioid, j approaches 0 as the cardioid condition is approached. Concomitantly, the non-focal self-inversion loci become more and more nearly coincident with the basis elliptical or hyperbolic limacons. One could, in fact, regard the cardioid as its own modular non-focal self-inversion locus, since the loci of eq. 1 remain defined for $a=b$; for equality of a and b the eq. becomes degenerate and defines either the point at the origin or the cardioid.

The loop of the basis curve continues to enlarge and the inter-loop non-focal self-inversion loci continue to retract from the basis curve as e increases. The situation for $e=1.1$ also is shown in great magnification in Fig. h. The differences between Figs. f, g, and h, which are apparent only at great magnification for the basis limacons and their non-focal self-inversion loci, are readily apparent in the corresponding central inversions of Figs. 8-15f-h.

The locus of Fig. i, for $e=1.306562965$, together with the poles for its inversions to central curves, are illustrated without magnification. This locus is unique among the non-focal self-inversion loci of hyperbolic limacons (for modular j) in that both of its central inversions belong to the same subspecies. This central subspecies (Fig. 8-15i) thus also is a *curve of intersection* of the two genera of central curves (of Fig. 8-15). The loci of Figs. j and k are illustrated without magnification. For Fig. k, the basis $e=2$ limacon also is shown. Its presence serves to illustrate the location of a complete inter-loop non-focal self-inversion locus relative to the basis curve.

Recall (see legend to Fig. 8-11) that, by definition, all limacons self-invert for modular j (the distance of the focus of self-inversion from the DP or DP homologue), and that, by choice, the illustrated non-focal self-inversion loci also undergo non-focal self-inversion for modular j, but that the modulus for traditional self-inversion of the latter locus is $\sqrt{2}$ times modular j. The poles for central inversions also are at a distance of $\sqrt{2}$ times modular j from the reference pole. In other words, the distance of the exterior pole for central inversion from the reference pole (in Fig. k) is $\sqrt{2}$ times the distance of the DP from the reference pole.

Fig. 8-15. Central Cartesians: dissimilar and identical subspecific inversions of the limacon interior and inter-loop non-focal self-inversion loci of Fig. 8-14. The two central Cartesian basis subspecies for interior and inter-loop non-focal self-inversion loci of limacons referred to the focus of self-inversion consist of mutually-inverting monovals that prove, in general, to be dissimilar curves. By contrast, the biovular central Cartesians derived in the text from bipolar linear Cartesians (pages XIV-86 to 95), as well as all other biovular central Cartesians, self-invert about their centers at both 0° and 180° (see Figs. 8-19 to 28 and Tables 8-10-1 and 8-17-4).

As a concluding topic relating to non-focal self-inversion loci, this figure illustrates the two monovular central Cartesian basis subspecies for each of the limacon non-focal self-inversion loci of Fig. 8-14. The derivation and circumpolar symmetry of these curves have been treated briefly in Fig. 8-10. A general derivation of the eqs. for central Cartesians from the simple intercept format for Cartesian self-inverters (eq. 8-13-5) is given here.

Although the pole about which the Cartesian self-inverter format is cast is referred to as the $h=0$ focus, for biovular curves with $F^2>0$ this pole may be any one of the 3 foci of self-inversion, inasmuch as the simple intercept format takes the same form about all 3 poles (see, for example, eqs. XIV-90 to 92). This remarkable property of these Cartesian self-inverters--the possession of similar simple intercept formats about 3 non-equivalent point-foci (only the values of the parameters differ)--is possessed by no other known curve (not even for only two non-equivalent point-foci).

Derivation of The General Equations of Central Cartesians From The Simple Intercept Format For Cartesian Self-Inverters

The eqs. of the central Cartesians in question now are obtained from the starting curves--Cartesian self-inverters--of eqs. 8-13-5 by translating the coordinate origin to poles on either side of the $h=0$ focus at the modular distance for self-inversion about this focus, i.e. poles at $h=\pm F$. Rearranging the format and making these translations lead to eqs. 1a,b.

Inversions of 1a,b give the sought-after central Cartesians. But these are not centered; they are cast about the reciprocal inversion poles. To center them, one translates the origin of the inverse locus of eq. 1a to $H=-j^2/2F$, and the origin of the inverse locus of eq. 1b to $H=+j^2/2F$, giving eqs. 2a,b.

equation 8-13-5 with the origin translated to the modular poles

(a) $[r^2+(2F-G)r\cos\theta+F(2F-G)]^2 = E^2(r^2+2Fr\cos\theta+F^2)$	cast about interior pole of monovular species at $h = F$ ($x+F$ translation)	(8-15-1)
(b) $[r^2-(2F+G)r\cos\theta+F(2F+G)]^2 = E^2(r^2-2Fr\cos\theta+F^2)$	cast about exterior pole of monovular species at $h = -F$ ($x-F$ translation)	

centered central Cartesians derived by line-of-symmetry inversions of Cartesian self-inverters

(a) $F^2[(2F-G)^2-E^2]r^4+j^4(4F^2-G^2-E^2)r^2/2 + E^2j^4r^2\cos^2\theta+(j^8/16F^2)[(2F+G)^2-E^2] = 0$	eq. 1a inverted about interior pole (of monovular species) and translated $-j^2/2F$	(8-15-2)
(b) $F^2[(2F+G)^2-E^2]r^4+j^4(4F^2-G^2-E^2)r^2/2 + E^2j^4r^2\cos^2\theta+(j^8/16F^2)[(2F-G)^2-E^2] = 0$	eq. 1b inverted about exterior pole (of monovular species) and translated $+j^2/2F$	

[These translations are simply the reverse of those for an axial inversion of a centered central basis curve followed by a casting of the inverse locus about the focus of self-inversion that replaces the line of symmetry (see Maxims 18 and 8-15-4a).]

Inasmuch as eqs. 2a,b involve the product $2F = 2\sqrt{F^2}$, it is evident that the derivation does not hold for $F^2 < 0$, i.e., for simple intercept formats cast about 180° inter-oval-mode foci. This finding is merely an expression of the fact that inversions of Cartesian self-inverters about poles modular to the 180° foci of self-inversion do not yield central curves (the distinctive property of the inverse loci about these poles is that they are congruent for a common value of the unit of linear dimension; see Maxim 8-20-11).

Whether the biovular central curves represented by eqs. 2a,b are annular or possess opposed ovals depends upon whether the basis Cartesian self-inverter format employed is for curves cast about the 0° inter-oval-mode foci or for curves cast about the 0° intra-oval-mode foci. In the former case, the central curves are of the opposed-oval type, while in the latter case they are annular (see Maxim 8-20-9).

We now examine the circumstances under which eqs. 2a and 2b represent curves that (1) are identical (i.e., congruent and coincident), (2) belong to

Central Cartesians: Dissimilar & Identical Subspecific Inversions of The Limaçon Interior & Inter-Loop Non-Focal Self-Inversion Loci of Fig. 8-14

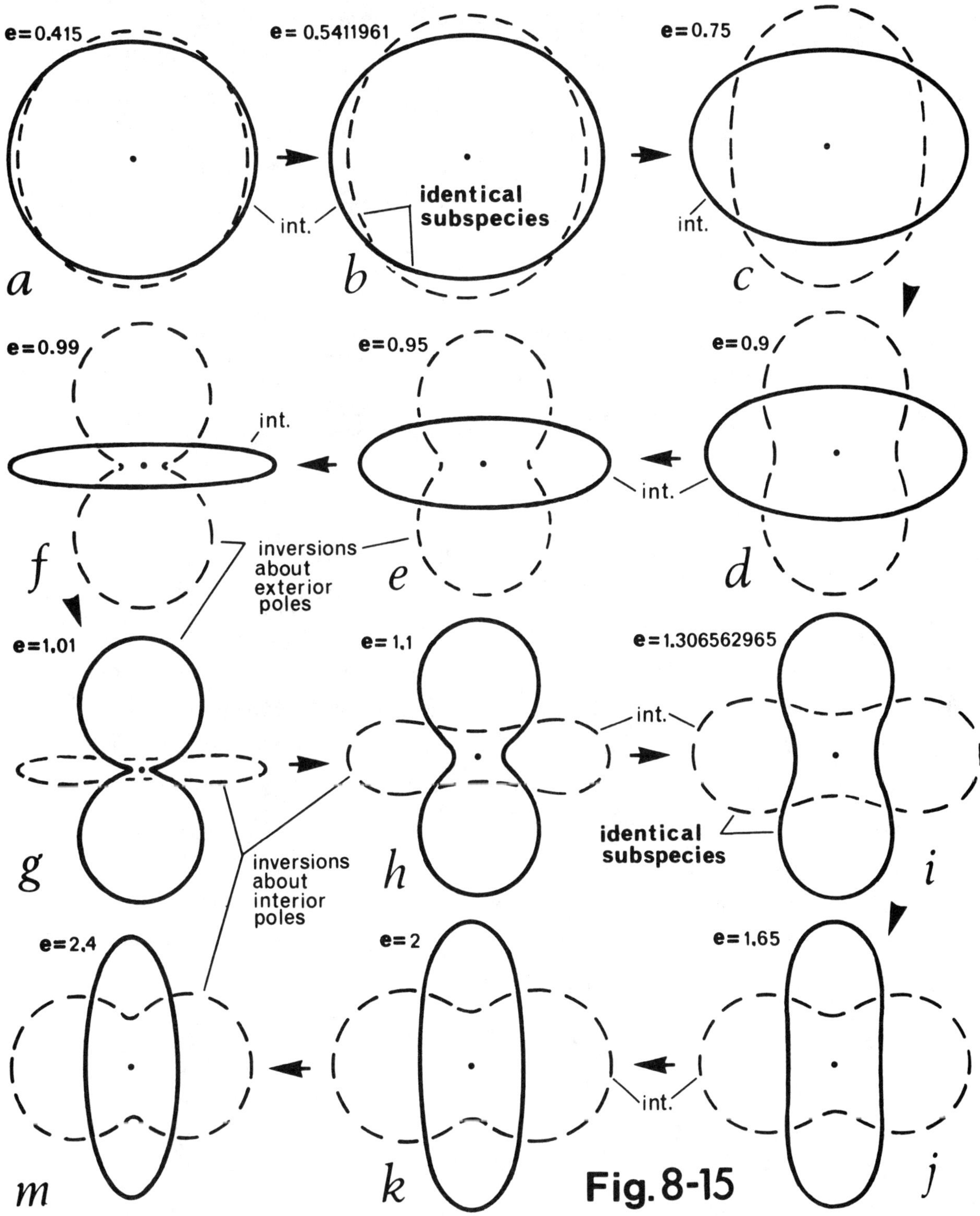

Fig. 8-15

the same subspecies and are similarly oriented, (3) belong to the same subspecies but have corresponding axes orthogonal to one another, (4) are congruent and have corresponding axes orthogonal to one another, and (5) belong to different subspecies.

Considering case (1) first, it readily is apparent that the loci of eqs. 2a,b cannot be identical, since this requires that either F or G vanish, which is not allowed. To test case (2), one merely magnifies or reduces the size of either curve by a linear factor n by substituting nr for r. Dividing the resulting eq. termwise by n^2 leaves the coefficients of the terms in r^2 and $r^2\cos^2\theta$ the same in both eqs. The coefficients of the terms in r^4 and the constant terms will be the same if eq. 3 is satisfied.

$$\text{(a)}\quad n^2[(2F-G)^2-E^2] = [(2F+G)^2-E^2] \qquad \text{condition for similarly-oriented curves of same subspecies} \qquad (8\text{-}15\text{-}3)$$

$$\text{(b)}\quad (2F-G)^2 > E^2 \quad \text{or} \quad \text{(c)}\quad (2F+G)^2 < E^2 \qquad \text{alternate conditions for similarly-oriented curves of same subspecies}$$

But an n always can be found to satisfy eq. 3, provided only that both sides of the eq. have the same sign. This leads to the alternate conditions 3b,c, either one of which is sufficient to ensure similar orientation and subspecific identity. In the case of parabolic Cartesians represented by the standard format of eq, 8-13-3, for example, condition 3b ($C^2 < 0$) cannot be fulfilled and 3c yields $C > 4$. Thus, the inverse loci of biovular parabolic Cartesians (all of which are annular) about poles at $h = \pm d$ (i.e., poles modular to the 0° inter-oval-mode focus) are central Cartesians (ovals of Cassini) that belong to the same subspecies and are similarly oriented.

To test for condition (3), in which the central Cartesians belong to the same subspecies but have corresponding axes orthogonal to one another, one first rotates 2a or 2b 90° and magnifies or reduces by a linear factor n, giving eq. 4a. Dividing by n^2, substituting $(1-\cos^2\theta)$ for $\sin^2\theta$, and

$$\text{(a)}\quad F^2[(2F-G)^2-E^2]n^4r^4+j^4(4F^2-G^2)n^2r^2/2-j^4E^2n^2r^2/2 + E^2j^4n^2r^2\sin^2\theta+(j^8/16F^2)[(2F+G)^2-E^2] = 0 \qquad \text{eq. 2a rotated } 90^\circ \text{ and magnified or reduced by a factor } n \qquad (8\text{-}15\text{-}4)$$

$$\text{(b)}\quad -F^2[(2F-G)^2-E^2]n^4r^4-j^4(4F^2-G^2)r^2/2-E^2j^4r^2/2 + E^2j^4r^2\cos^2\theta-(j^8/16F^2n^2)[(2F+G)^2-E^2] = 0 \qquad \text{eq. 4a divided by } n^2 \text{ and restated}$$

changing signs yields eq. 4b. This eq. for the rotated eq. 2a is seen to differ from eq. 2b only in the coefficients of 3 terms. In order for all the coefficients to be identical, two conditions, 5a and 5b, must be satisfied.

(a) $G^2 = 4F^2$ — conditions for subspecific identity of loci of 2b with loci of 4b — (8-15-5)

(b) $n^2[E^2-(2F-G)^2] = [(2F+G)^2-E^2]$

$$(n^2+1)E^2 = 16F^2 \qquad \text{eq. 5b for } G = +2F \qquad (8\text{-}15\text{-}6)$$

[A necessary condition for subspecific identity of two centered orthogonal central curves follows from a comparison of eqs. 2b and 4 (and from elementary considerations of the properties of polar eqs.). If the eqs. of the two curves in question possess terms in r^2 and $r^2\cos^2\theta$, then the coefficient of the r^2 term must be $-\frac{1}{2}$ times the coefficient of the term in $r^2\cos^2\theta$. Only if this obtains will a 90° rotation leave the coefficients in the same ratio ($Ar^2/2-Ar^2\cos^2\theta$ rotated 90° becomes $Ar^2/2-Ar^2\sin^2\theta =$ $Ar^2/2-Ar^2+Ar^2\cos^2\theta = -Ar^2/2+Ar^2\cos\theta$, in which the coefficients are in the same ratio, namely $-\frac{1}{2}$, as for the unrotated curve). This is the basis for the observation on page XIV-76 concerning the generalized simple intercept format for Cassinians as applied to the monovular species.]

The condition 5a, $G^2 = 4F^2$, generally does not hold for Cartesian self-inverters. For example, though it applies to parabolic and displaced parabolic Cartesians, it applies only to the $e = 0.541196100$ and $e = 1.30652965$ subspecies (Figs. 8-14b and i) of limacon interior and inter-loop non-focal self-inversion loci about the focus of self-inversion [$j^2 = (a^2-b^2)^2/4b^2$].

However, this is only a necessary condition for subspecific identity of orthogonal central Cartesians. The 2nd condition that must be fulfiled is 5b. If one takes $G = +2F$ in 5b, for genera of Cartesian self-inverters in which condition 5a applies, one obtains eq. 6. There is no loss of generality in discarding the the negative root ($G = -2F$), since corresponding loci are congruent for formats that differ only in the signs of the G terms (note in Fig. 8-13 that for given values of the first 3 signs, the loci are congruent for differing 4th signs).

Accordingly, for all formats for which $G = \pm 2F$, the central basis curves are orthogonal and belong to the same subspecies provided that an n exists such that $(n^2+1)E^2 = 16F^2$, i.e., for all subspecies for which $E^2 < 16F^2$ (since the smallest value $n+1$ can take on is 1). For parabolic Cartesians (eq. 8-13-2) this translates to $C < 4$, while for the non-focal self-inversion loci of limacons (referred to above) it becomes $e < 0.859532504$ and $e > 1.243751648$.

Accordingly, the two central Cartesians derived by inverting monovular parabolic Cartesians are similar to one another (belong to the same subspecies) and are orthogonal. For the limacon non-focal self-inversion loci, only those of eccentricity 0.541196100 and 1.306562965 satisfy both conditions and invert to similar but orthogonal central curves.

[This also had been shown to be the case for monovular Cassinians (page XIV-76), and it is shown below (Fig. 8-22 and 8-31) that central parabolic Cartesians (see eqs. 8-17-1a,b) and Cassinian ovals are the same curves.]

Considering now case (4) for orthogonal axes and congruence, conditions 7a,b must be fulfilled. For parabolic Cartesians (eq. 8-13-2), 7b yields $C = 2\sqrt{2}$,

(a) $G = 2F$ (b) $E = 2\sqrt{2}F$ conditions for congruent, orthogonal central Cartesians (8-15-7)

for which congruence and orthogonality are readily verified, while for the limacon non-focal self-inversion loci, 7b yields $e = \sqrt{2}$ and $e = \sqrt{(2/3)}$, neither of which is consistent with condition 7a. In other words, the central Cartesians derived from the limacon non-focal self-inversion loci (Fig. 8-14) never are congruent, though in two cases they are similar (Figs. 8-15b,i).

The conditions for central inversion to an orthogonal congruent curve also can be approached readily by a circumpolar symmetry analysis, since central inversion to an orthogonal congruent curve is equivalent to the existence of a 90° transform of self-inversion, $xy = \text{constant}$, for the center. Thus, if one takes eq. 2b for the centered basis central Cartesian obtained by inverting about the exterior pole, and forms the 90° transform (see Appendix I, page *vii*, *Symmetry, An Analytical Treatment*), one obtains eq. 8. This is a 90°

general equation for the 90° circumpolar intercept transforms of eq. 2b about the center

$$F^2[(2F+G)^2-E^2]x^2y^2(x^2+y^2) + j^8[(2F-G)^2-E^2](x^2+y^2)/16F^2 + j^4(4F^2-G^2)x^2y^2 = 0 \qquad (8\text{-}15\text{-}8)$$

$$(x^2+y^2)[F^2(16F^2-E^2)x^2y^2-E^2j^8/16F^2] = 0 \qquad \text{eq. 8 for } 4F^2 = G^2 \qquad (8\text{-}15\text{-}9)$$

(a) $(x^2+y^2) = 0$ (b) $xy = j^4E/4F^2\sqrt{(16F^2-E^2)}$ solutions to eq. 9 (8-15-10)

(c) $xy = Cd^2/4\sqrt{(16-C^2)}$ eq. 10b for central parabolic Cartesians (Cassinian ovals)

transform of self-inversion only when $4F^2 = G^2$, whereupon the 4th-degree term in x^2y^2 vanishes, yielding the degenerate 90° transform of eq. 9. This represents the two 90° transforms, 10a,b, the former of which has only imaginary solutions and the latter of which has real solutions for $16F^2 > E^2$, duplicating the above results. For central parabolic Cartesians, eq. 9b becomes 9c, for which real solutions exist for all $C < 4$, i.e., for the monovular species.

The conditions for self-inversion of the monovular species at 90° ($4F^2 = G^2$) also can be derived readily by finding the conditions for which the product of the roots for the monovular species of eq. 2a or 2b is constant.

In summary, the central curves derived by inverting Cartesian self-inverters about poles lying on the line of symmetry at modular distance from the $h = 0$ focus of self-inversion are similar and similarly oriented if either $(2F-G)^2 > E^2$ or $(2F+G)^2 < E^2$, i.e., for biovular species. They are similar and orthogonal if $G = 2F$ and an n exists such that $n^2[E^2-(2F-G)^2] = [(2F+G)^2-E^2]$. They are congruent and orthogonal if $G = 2F$ and $E^2 = 4F^2+G^2$, i.e., $G = 2F$ and $E = 2\sqrt{2}F$ (for the influence of the signs of terms, see Fig. 8-13). On the other hand, the curves are dissimilar if $(2F-G)^2 < E^2$, $(2F+G)^2 > E^2$, and (a) $G \neq 2F$ or (b) $G = 2F$ but $n^2[E^2-(2F-G)^2] \neq [(2F+G)^2-E^2]$, i.e., $(n^2+1)E^2 = 16F^2$.

Returning to the illustrations, both central inversions about axial poles (small solid circles in Fig. 8-14) that are at modular distance from the reference pole (the focus of self-inversion; large solid circles in Fig. 8-14) are shown for each of the 12 loci depicted in Fig. 8-14. The locus obtained by inverting about the interior axial pole is shown in heavy solid tracing for $e < 1$ and light dashed tracing for $e > 1$, and vice versa for the exterior-pole loci (see labelling of the loci; int. = interior-pole locus). Inversions about these poles are orthogonal but neither centered nor concentric and, generally, not congruent. In the present figure the curves have been centered (eqs. 2a,b), plotted about a common center, and have had their sizes adjusted to give equality of the major axes (since the curves are mutually inverse, this also leads to equality of the minor axes).

Fig. a shows inverse loci of the almost-circular locus of Fig. 8-14a. The orthogonality of the two inversions--in the sense of having major(and minor) axes orthogonal--can be understood as follows. The slight asymmetry of the non-focal self-inversion locus becomes reversed in its inverse locus about the almost-central interior pole, i.e., the inverse curve has its major and minor axes at 90° to those of the basis curve. The exterior axial pole is at a relatively great distance from the basis curve and is incident upon one of its axes. Since the basis curve departs very little from having two lines of symmetry, the locus obtained by inverting it about a distant axial pole is very similar to it and reflected in the orthogonal axis (see Figs. 8-7a,c). As a result, the two inverse loci are orthogonal. Although the two loci appear to be congruent or very nearly so, they depart significantly from this condition.

The loci of Fig. b are congruent, because the value $e = 0.541196100$ is one of the two values of the eccentricity that satisfies eq. 5 for subspecific identity of orthogonal loci, and because the sizes of the inverse loci have been adjusted for axial equality. The subspecies of Fig. b thus is a curve of intersection between the two central Cartesian monovular species--one species consisting of the inverse loci about the interior poles and the other consisting of the inverse loci about the exterior poles.

For $e = 0.75$ the loci still are very similar but can be seen to depart from congruence [In this connection, because of common optical illusions, many of the figures must be viewed from several angles.] For $e = 0.9$ the departure from congruence is marked and remains so in Figs. e-h. Figs. f and g are for values of e only 0.01 units on either side of 1, the value for the cardioid. This is the value for which the non-focal self-inversion loci convert from the interior (for $e < 1$) to the inter-loop type (for $e > 1$).

With attainment of $e = 1.1$, it is evident that the loci are approaching a second condition of congruence and, at $e = 1.306562965$, this congruence is attained. Accordingly, both the curves of Fig. b and of Fig. i are curves of intersection of the two monovular species of central curves derived by inverting about the interior and exterior poles (i.e., these curves are common members of both species).

[These intersections are for a common value of e. It has not been ruled out, however, that the two species possess other members in common for differing values of e nor that within each species there exist loci for some values of $e < 1$ that duplicate loci for values of $e > 1$.]

As e increases beyond the value for Fig. i, the locus obtained by inverting about the exterior pole gradually loses its biconcave appearance (as both concavities become more shallow). In Fig. k ($e = 2$) the exterior pole locus is very slightly convex in the formerly concave regions; the convexity is greater in Fig. m, though the regions retain a flattened appearance. Concomitantly with these changes in the exterior-pole locus, the interior-pole locus retains its biconcavity but the two lobes become broader as they more nearly approach circularity (if they were to attain circularity, of course, the slightly convex regions of the inverse locus would attain linearity).

In summary, the exterior-pole locus begins ($e = 0.415$) with an almost-round shape and elongates along the x-axis giving rise to eccentric figures with elliptical appearance that progressively flatten and elongate; eventually, for $e > 1$, these figures become concave in the regions of the minor-axis vertices. The loci broaden progressively with increasing e, becoming markedly biconcave, and with lobes that approach circularity.

The exterior-pole locus also begins with an almost-round shape and progressively flattens. It becomes biconcave, however, without first greatly elongating, and the lobes progressively approach circularity as e approaches 1. The loci are very similar for $e = 0.99$ and $e = 1.01$, but as e increases thereafter, the lobes become progressively less circular in appearance and the concavities become less pronounced. At $e = 2$ the locus is everywhere convex and, with increasing e, the loci become more elliptical in appearance and the curvature in the minor-axis vertex regions increases.

In addition to subspecific identity of at least two members of the interior-pole and exterior-pole loci, other members of the two species have very similar appearances, for example, those of Figs. e and m.

Comparative Aspects of Foci of Self-Inversion For Different Angles

Having encountered curves that self-invert at 90° and 270° (see also pages XIV-76 and 77, and Figs. 8-17,19, and 30), questions arise as to the relationships between various angles of self-inversion and the properties of the dissimilar inverse loci of curves that self-invert at various angles. These topics are dealt with briefly here.

Comparative analyses of self-inversion at (a) 0°, (b) 180°, (c) 0° *and* 180° about a single pole, and (d) θ and $180^\circ + \theta$ ($\theta \neq 0^\circ$ or 180°) can be carried out working from first principles together with the inversion-mapping Maxims already considered. The results that emerge from these analyses reveal that unique exchange relationships exist between a basis curve and its dissimilar inverse loci for each of these angle-categories of self-inversion.

Thus, if a curve possesses a 0° focus of self-inversion in the finite plane, its dissimilar inverse loci about a point in the plane either: (1) possess a corresponding 0° focus of self-inversion in the finite plane; or (2) lose the 0° focus of self-inversion in the finite plane but gain a corresponding line of symmetry (and a corresponding 0° focus of self-inversion on the ideal line at infinity in the direction orthogonal to the line of symmetry). The second alternative applies to the situation where the curve is inverted about a pole at modular distance from the focus of self-inversion. These conclusions stem from Maxim 8-15-1 (in conjunction with the properties embodied in foregoing inversion-mapping Maxims).

On the other hand, 180° foci of self-inversion are neither lost nor gained in an inverse locus about any point in the plane, which means that they do not exchange with lines of symmetry. Unlike the case for a 0° focus of self-inversion, the inverse loci about points on the modular circle for a 180° focus are unexceptional. These conclusions stem from Maxim 6-15-3.

If curves possess coincident foci of self-inversion for angles of both 0° and 180°, they are, by definition, central curves with two lines of symmetry. If, in addition, the curves are quartics, they are biovular central Cartesians (see Tables 8-10-1 and 8-17-4), a group that is the object of detailed treatments in many of the following figures. The coincident 0° and 180° foci of self-inversion of these central curves are not coincident in any dissimilar inverse locus (although the 180° focus may become coincident with another 0° focus).

Foci of self-inversion of quartics at other angles than 0° or 180° are known only for Cassinian monovals and for a few individual subspecies of monovular central Cartesians, which self-invert about the center at both 90° and 270° (see Figs. b and i). Foci with these angles of self-inversion are not conserved in any dissimilar inverse locus. In their place an axial inverse locus possesses two poles of congruent-inversion (for the appropriate value of the unit of linear dimension) in addition to the focus of self-inversion that replaces the line of symmetry.

In a non-axial inverse locus, basis foci of self-inversion at 90° and 270° are replaced by 4 poles of congruent-inversion (for the appropriate values of the units of linear dimension), in addition to two 0° foci of self-inversion that replace the lines of symmetry, and a single pole of congruent-inversion (homologous to that obtained in non-axial inverse loci of central conics) about the point inverse to the reflection of the pole of inversion in the center (see Fig. 6-6). An example of such a curve is given in Fig. 8-30d.

Foci of self-inversion at other angles than 0°, 90°, 180°, and 270° are unknown in quartics but a simple example of a composite non-central locus consisting of two concentric central QBI curves self-inverting at angles of θ and $180^{\circ}+\theta$ ($\theta \neq 0^{\circ}$, 90°, 180°, or 270°) is considered in the legend for Fig. 8-30 (*Curves That Congruent-Invert But Do Not Self-Invert*). Inversion of such a composite non-central figure about an axial pole of one of its component central curves yields a locus that possesses 3 poles of congruent-inversion, but for a different value of the unit of linear dimension about each pole.

CIRCUMPOLAR MAXIMS 8-15-

1: *Upon inversion of points in the plane about a pole P, all point-pairs inverse at 0° in a circle of inversion C_1 become inverse at 0° in a circle of inversion C_2 centered on pole P', the inverse at 0° of pole P in circle C_1, provided that poles P and P' are not coincident (i.e., provided that pole P is not incident upon circle C_1).*

2: *Upon inversion of points in the plane about a pole P incident upon a circle of inversion C, all points inverse at 0° in C become reflected in a common line.*

3: *Upon inversion of points in the plane about a pole P, all point-pairs inverse at 180° in a circle of inversion C_1 become inverse at 180° in a circle of inversion C_2 centered on pole P', the inverse at 180° of pole P in circle C_1.**

4: *In the inversion of curves to dissimilar inverse loci:*

(a) *the gain (or loss) of a 0° focus of self-inversion in the finite plane is accompanied by the loss (or gain) of a line of symmetry and a focus of self-inversion on the ideal line at infinity in the orthogonal direction* (see also Maxim 6-8-10);

(b) *180° foci of self-inversion are conserved, i.e., they are neither gained nor lost;*

(c) *0° and 180° foci of self-inversion that are coincident are conserved but are non-coincident in the inverse locus**; and*

*Since a pole of inversion cannot be coincident with its modularly-inverse pole in a 180° focus of self-inversion (unlike the situation for 0° foci of self-inversion), a 180° focus of self-inversion cannot become transferred to (nor exist at) the ideal line at infinity (see also Maxim 8-22-21).

**Since the inverses of poles that are modularly inverse to a pole of inversion in coincident 0° and 180° foci cannot also be coincident.

4: *(d) foci of self-inversion at 90° and 270° are not conserved in any dissimilar inverse locus. Those of quartics are replaced by two poles of congruent-inversion in axial inverse loci and by 4 poles of congruent-inversion in non-axial inverse loci. Together with one pole of congruent-inversion possessed by virtue of the basis curve having two lines of symmetry, this makes a total of 5 poles of congruent-inversion for the non-axial loci (for the appropriate values of the units of linear dimension).*

5: *Conservation of a focus of self-inversion by the inversion transformation implies conservation of the corresponding angle of self-inversion.*

6: *Modularly-inverse pole-pairs about foci of self-inversion at angles of 0°, 180°, 0° and 180°, and θ and 180°+θ (θ ≠ 90° or 270°, except for central curves*) are equivalent inversion poles.*

*In general, poles modularly inverse in a focus of self-inversion at an angle θ (and 360°-θ), where θ ≠ 0° or 180°, are not equivalent inversion poles.

Fig. 8-16. The equilateral limacon as the curve of demarcation of parabolic Cartesians. A common circumstance among quartic genera is that members of the congeneric species do not possess the same number of real ovals. A non-central genus with both monovular and annular biovular species--parabolic Cartesians--is presented as a first example.

The transition from the monovular to the annular species typically occurs by the gradual pinching together of the opposed segments of a single oval (Figs. g and h) with increasing values of the parameter, C, of the eq. for the genus (bipolar eq., $u^2 = Cdv$). The two-looped curve formed when the opposed segments make contact, i.e., the curve that demarcates between monovals and annular biovals (Fig. 1) is referred to as the *curve of demarcation.* [In an analogous situation in phase-plane analysis, the curve that separates regions of different qualitative behavior is referred to as the "separatrix."]

It is inherent in this transition from non-central monovals to non-central biovals that the numbers of imaginary point-foci in the demarcated species differ, and that a multiple-focus not present in either of the demarcated species exists in the curve of demarcation.

The present figure is an extension of Fig. 2-1f_1-f_3; more species are represented here and all the non-incident axial foci are marked. The arrows and the values of the parameter C, adjacent to each curve, indicate the progression of subspecies for monotonically increasing values of C. The curve of demarcation (the equilateral limacon for $C = 4$) is shown by the heavy tracing. In the case of the locus for $C = 5$, only a magnified portion of the curve is

The Equilateral Limaçon As The Curve of Demarcation of Bipolar Parabolic Cartesians

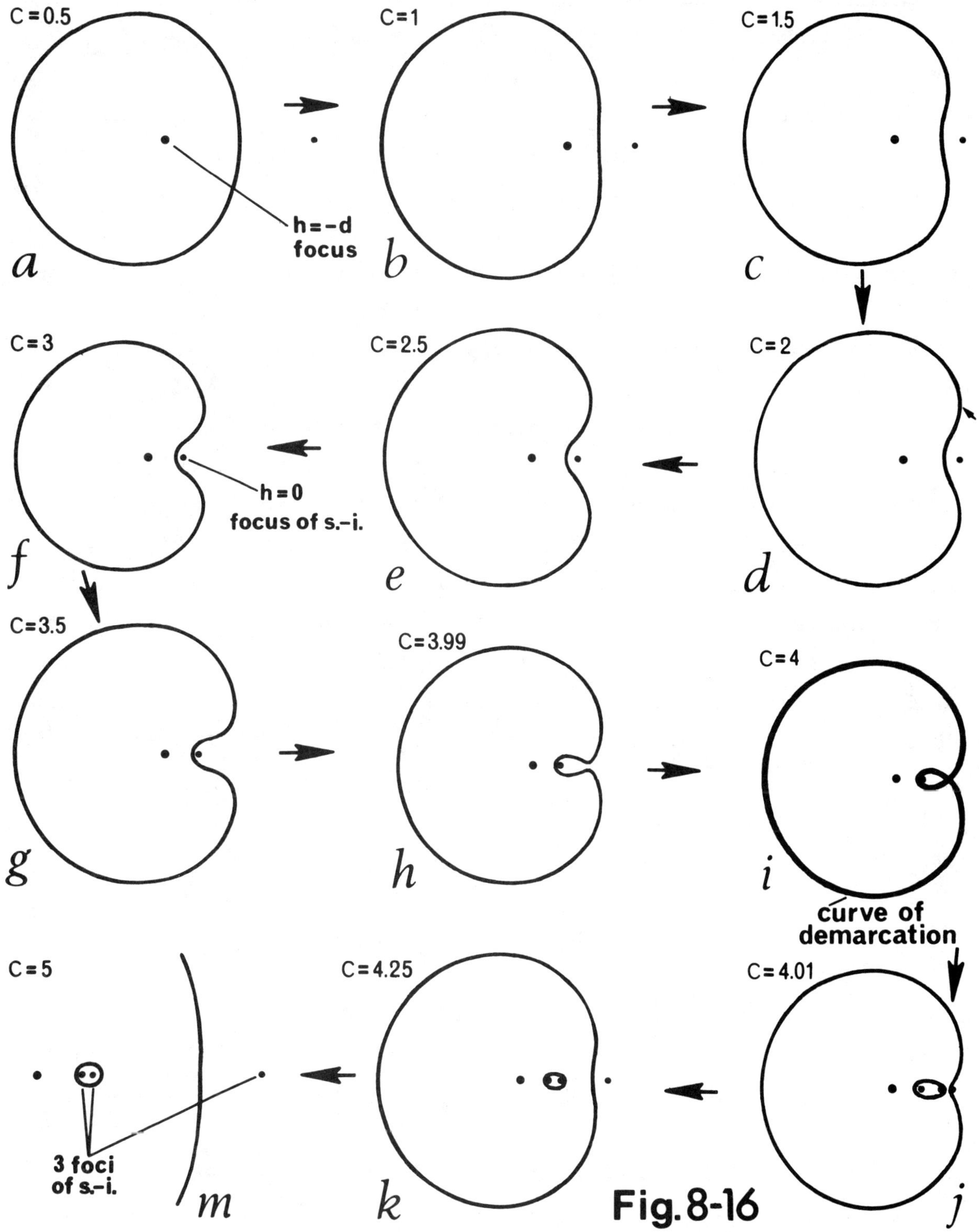

Fig. 8-16

depicted, in order that all of the foci of self-inversion can be shown. In the case of Fig. j, the 0° intra-oval-mode and 180° inter-oval-mode foci of self-inversion are very nearly incident (see Fig. 8-22e') but have been displaced slightly for the purpose of illustration.

For $C = \frac{1}{2}$, the oval is everywhere convex and possesses 4 discrete real axial foci; the $h = 0$ focus of self-inversion, the $h = -d \cos^2\theta$-condition double-focus, and the two real axial vertices. For $C = 1$, a very slight concavity is present. This gradually deepens with increasing C, and between $C = 3\frac{1}{2}$ and $C = 3.99$ its opposed rims approach the axis. These rims are in close proximity to one another for $C = 3.99$. The rims come into contact for $C = 4$, forming a two-looped curve with a 5th discrete real focus at the resulting DP (the latter is either a double-focus or a quadruple-focus, depending upon whether or not one counts the two trivial foci of self-inversion; see Table 8-16-1 and the legend to Fig. 8-20).

For C increasing from 4, the loops detach from one another, forming discrete annular ovals. In this process two additional real vertices replace the DP, bringing the number of real vertex foci to 4. The $h = 0$ and $h = -d$ foci are the only non-incident foci for values of C between 0 and 4, but for $C > 4$ two additional foci of self-inversion come into being in close proximity to the new vertices (or, one could say that, as C decreases to 4, the two foci of self-inversion of the annular species converge toward the vertices and, ultimately, come into coincidence at the DP).

The interior oval is very small compared to the exterior one, and the size disparity increases with increasing values of C. At the same time the interior oval migrates toward the center of the large oval. For very large C the diameter of the large oval approaches 2C, that of the increasingly centered small one approaches 2/C, the size disparity approaches a ratio of C^2, and both ovals approach circularity.

The scheme of Table VIII-1, in *Symmetry, An Analytical Treatment*, listed the number of discrete real axial foci and the number of lines of symmetry for several monovular and biovular species and their curves of demarcation. Subsequent studies (see legend to Fig. 20) suggest that the total number of real and imaginary axial circumpolar focal loci of monovular and biovular species and their curves of demarcation--taking into account focal multiplicity--is the same within any genus (possibly even within corresponding axial genera of an inversion superfamily). Accordingly, the curves of demarcation typically demarcate species with differing numbers of ovals or segments, while they themselves often have communicating loops, and generally possess the greatest circumpolar symmetry of any curves in their respective genera (for example, the equilateral limacon is the only parabolic Cartesian that possesses a non-trivial linear circumpolar intercept transform).

Table 8-16-1 lists the point-foci of parabolic Cartesians that are incident upon the line of symmetry, as well as the linear focal loci that intersect the line of symmetry. The Table does not take into account modes of self-inversion (i.e., an angle-independent circumpolar point-focus--a point-focus defined by a condition on the simple intercept eq. or format--that also is a pole about which a curve self-inverts is reckoned only as a single focus). The analytical bases for the Table are set forth in the legend for Fig. 8-20.

Table 8-16-1. Circumpolar Focal Loci of Bipolar Parabolic Cartesians Referred To The Line of Symmetry

Monovular ($C < 4$) →→	Curve of Demarcation ($C = 4$) The Equilateral Limacon	→→→→ Annular ($C > 4$)	
2 real vertices	2 real vertices	4 real vertices	
2 imaginary vertices	2 1 DP (double-focus)		
3 radicand constant-condition foci (two are imaginary, one is a focus of self-inversion)	3 radicand constant-condition foci of self-inversion (the two at the DP are trivial)	3 radicand constant-condition foci of self-inversion	
2 1 $\cos^2\theta$-condition double-focus	2 1 $\cos^2\theta$-condition double-focus	2 1 $\cos^2\theta$-condition double-focus	
1 covert linear focal locus orthogonal to the line of symmetry at $h = -d$	1 covert linear focal locus orthogonal to the line of symmetry at $h = -d$ ($h = -\frac{1}{2}b$)	1 covert linear focal locus orthogonal to the line of symmetry at $h = -d$	
1 line of symmetry	1 line of symmetry	1 line of symmetry	
11	11	11	total

Fig. 8-17. The equilateral hyperbola and the equilateral lemniscate as curves of demarcation of central inversions of parabolic Cartesians (eqs. 1 derived from eqs. 8-15-2a,b for basis curves 8-13-5). This figure depicts axial central inversions (i.e., axial--as opposed to non-axial--inversions that yield curves that possess a true center) of each of the parabolic Cartesian subspecies of Fig. 8-16 (with corresponding letter-labels).

centered, central parabolic Cartesians (with opposed-oval biovular species) derived by inverting parabolic Cartesians (eq. 8-13-5) about poles on the line of symmetry at modular distance from the $h = 0$ focus

(a) $r^4+d^2r^2/2-d^2r^2\cos^2\theta+d^4(1/16-1/C^2) = 0$ interior-pole inversions (of monovular species) (8-17-1)

(b) $r^4-C^2d^2r^2/2(16-C^2)+C^2d^2r^2(\cos^2\theta)/(16-C^2) - C^2d^4/16(16-C^2) = 0$ exterior-pole inversions (of monovular species)

For the monovular species, the inversions are about exterior poles at modular distance to the right of the 0° intra-oval-mode focus of self-inversion (eq. 1b). The corresponding pole for the curve of demarcation is coincident with the DP, while for the annular species, the poles are in the right inter-oval region (at modular distance to the right of the 0° inter-oval mode-focus of self-inversion; the latter focus is within the small oval at the far left--see Figs. 8-16j,k,m).

Since the central inversions of any given biovular parabolic Cartesian about both the pole in question and a pole at modular distance to the left of the 0° inter-oval-mode focus of self-inversion belong to the same subspecies and have corresponding lines of symmetry in parallel, the inverse loci about either of the poles in question represent those for both poles. For the monovular species, however, the inverse loci about the interior poles, though subspecifically identical to those about the exterior poles, are orthogonal to the latter.

The inversions of parabolic Cartesians about line-of-symmetry poles at modular distance from the 0° intra-oval-mode focus of self-inversion of the monovular species, and the corresponding inverse loci of the biovular species, are ovals of Cassini. These inverse loci are shown in larger size in the composite illustrations of Fig. 8-19.

The Equilateral Hyperbola & The Equilateral Lemniscate As Curves of Demarcation of Central Inversions of Bipolar Parabolic Cartesians

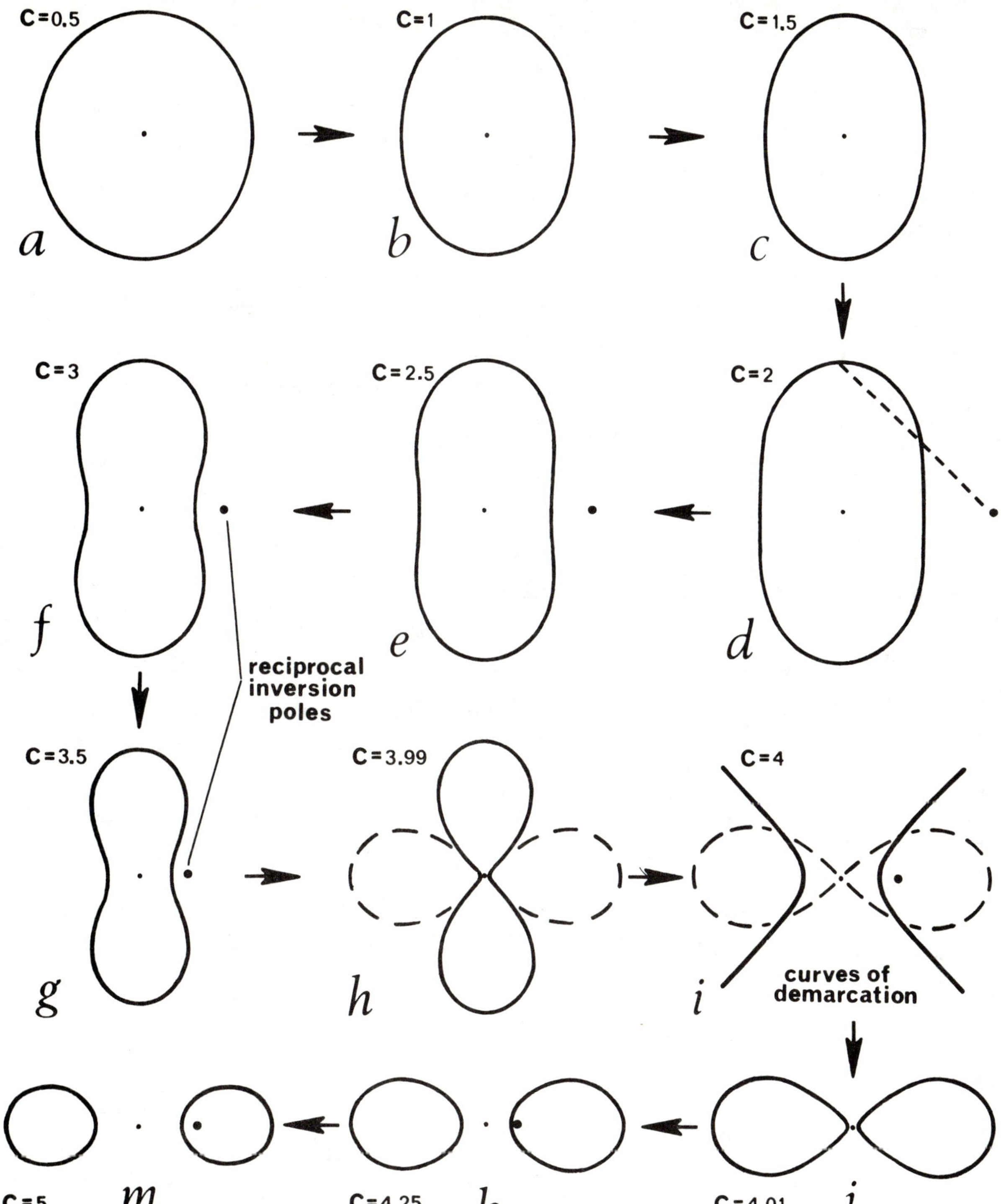

Fig. 8-17

[Eq. 1a will be recognized as the polar eq. of Cassinian ovals with the customary relative orientations of the monovular and biovular species (Fig. 8-19b and eqs. XII-11 and 8-31-5), where d is the distance between the bipolar poles (the $\cos^2\theta$-condition double-foci) and C is the reciprocal of the parameter with the same designation in the bipolar eq., $uv = Cd^2$. The relative orientations of the two species represented by eq. 1b are those of the present figure and Fig. 8-19a.]

Fig. h, for $C = 3.99$, illustrates both of the central inverse curves (adjusted to be centered, concentric, and congruent) in order to clarify the transition to the two curves of demarcation in Fig. 1. For $C > 4$, both central inversions have corresponding lines of symmetry in parallel. Accordingly, Figs. j, k, and m represent both inverse loci in proper orientation. This means that both curves of demarcation of Fig. i become transformed to similarly-oriented identical subspecies for values of C greater than 4.

The equilateral hyperbola (curve of demarcation) of Fig. i has been enlarged for the purpose of illustration. The center-focus of each curve is shown by a small solid circle. The locations of the intra-oval foci of central Cartesians (defined by $\cos^2\theta$ and $\sin^2\theta$-conditions) are discussed in the legend of Fig. 8-10 (see also Figs. 8-21, 8-26, and 8-27).

Inasmuch as the equilateral limacon is the curve of demarcation of parabolic Cartesians, it follows from principles of Inversion Taxonomy, as elucidated in preceding Circumpolar Maxims, that every equilateral QBI curve (i.e., the equilateral hyperbola and its inverse loci about all points in the plane) also is a curve of demarcation of one or another genus of inverse loci of parabolic Cartesians.

For these particular central parabolic Cartesians (Cassinian ovals), the equilateral hyperbola and the equilateral lemniscate are the curves of demarcation. For the axial-vertex inversion cubics of parabolic Cartesians (*Cassinian axial-vertex cubics*)--identical subspecies of which are obtained for each vertex for any given value of C (see Maxim 44)--the equilateral strophoid is the curve of demarcation (see Fig. 8-18), etc., for other genera.

Comparisons of the individual illustrations of Fig. 8-16 with those of the present figure reveal interesting relationships between the basis curves and their central inversions, but these are not treated in detail. In these comparisons one must bear in mind that the reciprocal inversion poles (which lie at a distance of $\frac{1}{2}j^2/d$ to the right of center) cannot be indicated for all cases because of being either too distant or in too great proximity to the curve. The reciprocal inversion poles (for inversions about exterior poles of the monovular basis curves) are indicated by the large solid circles of Figs. d,e,f,g,i,k, and m.

Curve d is taken as an illustrative example. In this case the regions of the minor-axis vertices approach linearity over a considerable extent. This means that the axial-vertex regions of the basis curve (Fig. 8-16d) must approach circularity, and that the exterior pole would be incident upon circular extensions of both of these regions (since linear regions of inverse curves have to be inverse to circular--or linear--regions of a basis curve, and only incident inversions of circles--and lines--are linear). That these relationships apply is evident from inspection of Fig. 16d.

Additionally, the near circularity of the regions of the major-axis vertices (Fig. d) means that certain other regions of the basis curve also are very nearly circular but that circular extensions of these other regions would not pass through the exterior inversion pole. The regions of the basis curve in question are those extending from the near points of intersection of radius vectors from the pole of inversion at the same angles as are made by radius vectors from the reciprocal inversion pole to the major-axis vertices of the inverse curve. One such radius vector is shown in Fig. d (dashed line) and the terminus of another at the near point of the curve is indicated by the small arrow in Fig. 8-16d.

Basis Genera of Inversion Superfamilies

It was noted above that the central curves derived by inverting Cartesian self-inverters about the two poles modular to a given 0° focus of self-inversion are central inversions of one another (Maxims 8-10-2 and 8-20-9, and eqs. 8-15-2a,b). Under certain conditions (eqs. 8-15-3,4,5, or 7), the members of each pair of central inversions belong to the same subspecies. Under other conditions (last paragraph of legend to Fig. 8-15), they generally do not.

The members of the central genera of any inversion superfamily always possess greater circumpolar symmetry than any of their (non-central) inverse loci, since all central curves possess the simplest non-trivial 180° transform, $x = y$. For this reason, and for convenience, one of the central genera often is referred to as the *basis genus* for the superfamily. In the case of

Table 8-17-1. Curves of Intersection of The QBI Superfamily and The Cassinian-Based Inversion Superfamily

	QBI SUPERFAMILY	CASSINIAN-BASED INVERSION SUPERFAMILY
basis central curves	Central Conics ($e = 0$ to ∞)	Cassinian Ovals ($C = 0$ to ∞: self-invert about their centers; monovals self-invert at 90° and 270°)
inverse central curves	Central Quartics (inversions of central conics about their centers	Cassinian Ovals
transverse- (and major) axis vertex cubics	Central Conic Axial-Vertex Cubics	Cassinian Axial-Vertex Cubics (see Fig. 8-22)
minor-axis vertex cubics	Ellipse b-Vertex Cubics (no equilateral member)	Axial-Vertex Cubics of Opposed-Oval Central Modular-Focal Quartics (see Figs. 8-22 and 8-32; no member for $C = \frac{1}{4}$)
most symmetrical non-central quartic genus	Limacons	Bipolar Parabolic Cartesians

CURVES OF INTERSECTION OF THE QBI AND CASSINIAN-BASED INVERSION SUPERFAMILIES

The ensemble of all equilateral hyperbolic species (one subspecies, $e = \sqrt{2}$, from each QBI genus)

all Curves of Demarcation (one subspecies, $C = \frac{1}{4}$, from each genus* in the Cassinian-based inversion superfamily)

*See Table 8-17-2 for the composition of the central genus of the Cassinian-based inversion superfamily.

Table 8-17-2 The Genus Central Parabolic Cartesians

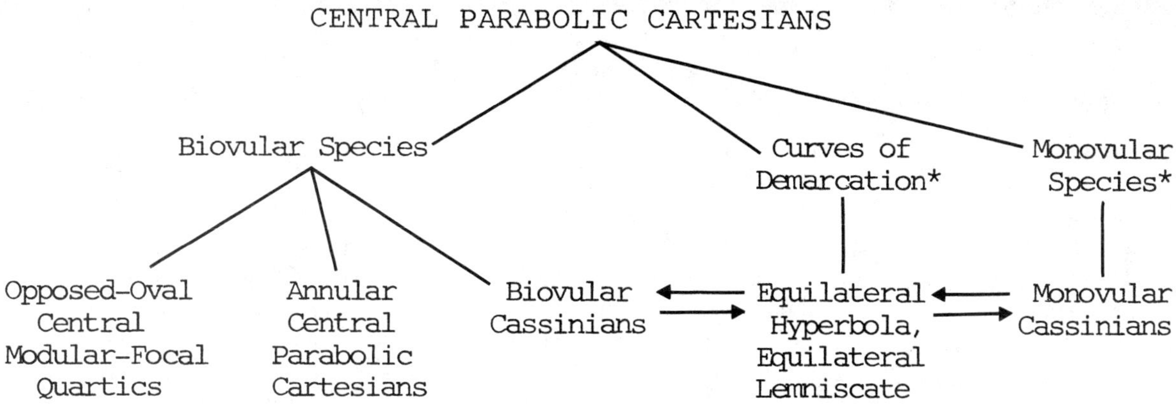

*There are no real curves of demarcation nor corresponding real monovular species for either opposed-oval central modular-focal quartics or annular central parabolic Cartesians.

the QBI superfamily, the central conics are assigned this role, since they are of lower degree than central quartics.

In the light of analyses of Cartesian self-inverters and their inverse loci, the scheme of Table 8-17-1 can be drawn up for relations between the QBI superfamily and the Cassinian-based inversion superfamily (curves derived by inverting Cassinian ovals or any of their inverse loci about all points in the plane for all values of C).

Aside from groups that possess only monovular or biovular species (see Table 8-17-4), genera of *central Cartesians* generally possess 3 biovular species, two monovular species, and two single-membered species of curves of demarcation. The central genus with the greatest circumpolar symmetry is *central parabolic Cartesians*. This group possesses two species with opposed ovals (see Fig. 8-22 and legend to Fig. 8-21), one species with annular ovals (see Fig. 8-22), only one monovular species (each member subspecies of which self-inverts at 90° and 270°), and two curves of demarcation (Table 8-17-2). The corresponding inversion superfamily is designated the *Cassinian-based inversion superfamily*.

Curves of Demarcation And Curves of Intersection

It is evident from Table 8-17-1 and from the foregoing that the curves of demarcation of the Cassinian-based inversion superfamily are *curves of intersection* with the QBI superfamily, i.e., these superfamilies possess in common one subspecies from each of their member genera, namely, the equilateral QBI subspecies and the $C = \frac{1}{4}$ subspecies of the Cassinian-based inversion superfamily.

[In the case of the QBI superfamily, the basis central genus consists of only 3 species (see pages I-22 and 23), and no subspecies of any one of the 3 species inverts to a subspecies of another species. The parabola is regarded as a single-membered species (with its center and other vertex at the axial point at infinity). It is the central conic homologue of a curve of demarcation of central Cartesians (see legend to Fig. 8-24). A *genus* of the QBI superfamily consists of the inversions of every subspecies of basis conic about corresponding poles, for example, the axial-vertex cubics, the ellipse b-vertex cubics, the LR vertex cubics, limacons (the inverse loci about the traditional foci), etc.]

In the cases of superfamilies based upon genera of *central Cartesians*--the curves represented by the eq., $(x^2+y^2)^2+B^2x^2+C^2y^2+D^4 = 0$--the central genera, as noted above, often possess 3 species of biovals, two species of monovals, and two curves of demarcation, and within each of these 3 categories the subspecies invert to one another. Accordingly, following principles of Inversion Taxonomy, the non-central genera of superfamilies based on these central curves derive from corresponding inversions of but a single species in each category, and a single curve of demarcation.

For example, all subspecies of the genus *parabolic Cartesians* are obtained by inverting all subspecies of biovular Cassinians, all subspecies of monovular Cassinians, and the equilateral lemniscate about the poles of the *Cartesian condition* (see legend of Fig. 8-20). If the corresponding inverse loci of the subspecies of the species *annular central parabolic Cartesians* were to be included, this would duplicate the inverse loci of the biovular Cassinians; inclusion of the corresponding inverse locus of the equilateral hyperbola would duplicate the inverse locus of the equilateral lemniscate, etc. These relationships are elucidated in Table 8-17-3 for the Cassinian-based inversion superfamily.

The fact that the equilateral limacon is the curve of demarcation of parabolic Cartesians raises the question of the identity of the curves of demarcation of other qualifying genera of Cartesian self-inverters. If one

Table 8-17-3. The Cassinian-Based Inversion Superfamily

	BIOVULAR		CURVES OF DEMARCATION	MONOVALS
		Annular Central Parabolic Cartesians		
CENTRAL CURVES (6 species, two of which are single-membered)	Opposed-Oval Central Modular-Focal Quartics	Biovular Cassinians	Equilateral Hyperbola, Equilateral Lemniscate	Monovular Cassinians
AXIAL CURVES (an infinite number of genera plus an infinite number of axial inversions of one curve of demarcation)	Axial Bicircular Quartic and Serpentine Circular Cubic inversions of the subspecies of any biovular genus of central parabolic Cartesians		Axial Equilateral QBI Curves	Axial Monovals and Serpentine Cubics
NON-AXIAL CURVES (an infinite number of genera plus an infinite number of non-axial inversions of the equilateral hyperbola or equilateral lemniscate or any inversion thereof)	Non-Axial Bicircular Quartic and Serpentine Circular Cubic inversions of the subspecies of any genus of central parabolic Cartesian or the subspecies of any genus of axial bicircular quartic or serpentine circular cubic		Non-Axial Equilateral QBI curves	Non-Axial Monovals and Serpentine Cubics

1. modular-pole inversions
2. from $C < \frac{1}{4}$ to $C = \frac{1}{4}$
3. from $C > \frac{1}{4}$ to $C = \frac{1}{4}$
4. axial and modular-circle inversions of any one of the 3 biovular genera
5. transverse and conjugate-axis inversions of either curve of demarcation
6. axial inversions
7. inversions about all points in the plane except points on axes, 0° circles of inversion, and modular circles (see Fig. 8-22)
8. inversions about all points in the plane except points on axes and circles of inversion
9. all non-axial inversions of either curve of demarcation
10. all non-axial inversions

considers the polar eq. of Cartesian self-inverters (eq. 8-13-6), it is evident that the conditions for 3 real intersections of the curves with the polar axis, i.e., the condition for curves of demarcation, is simply $[|G|-|E|]^2 = 4F^2$. Accordingly, if one takes G and E to be positive (which leads to no loss of generality; see Fig. 8-13), the simple intercept format for curves of demarcation of Cartesian self-inverters can be written as eq. 2a.

simple intercept format of curves of demarcation of qualifying Cartesian self-inverters cast about the focus of self-inversion

(a) $$\cos\theta = \frac{x^2+Ex+(G-E)^2/4}{Gx} \qquad G,E>0 \qquad (8\text{-}17\text{-}2)$$

simple intercept format of limacons cast about the focus of self-inversion

(b) $$\cos\theta = \frac{x^2-bx+(a^2-b^2)^2/4b^2}{-a^2x/b} \qquad (8\text{-}17\text{-}3)$$

curves of demarcation of central inversions of qualifying Cartesian self-inverters

$$\frac{E(E-G)^2[x-j^2/(E-G)]^2}{Gj^4} - \frac{E(E-G)y^2}{j^4} = 1 \qquad G,E>0$$

When this format is compared with the format for limacons (eq. 2b), it is evident that the curves of demarcation of qualifying Cartesian self-inverters are limacons. Inverting the latter about the DP or DP homologue yields the centered central conics of eq. 3.

Accordingly, the curves of demarcation of all qualifying genera of Cartesian self-inverters are limacons of eccentricity $\sqrt{(E/G)}$ [centered, central curves of demarcation also can be obtained by letting $(E-G) = 2F$ in eqs. 8-15-2a,b]. It follows that for every limacon subspecies there exists at least one genus of Cartesian self-inverter for which it is a curve of demarcation (see Fig. 8-24).

However, inasmuch as some genera of Cartesian self-inverters possess only annular ovals (for example, displaced parabolic Cartesians and linear Cartesians), they have no curve of demarcation. In the "non-qualifying" genera of Cartesian self-inverters, that have only monovals, or annular ovals that do

Table 8-17-4. Members of The Supergenus Central Curves With 4th-Degree-Maximum Inverse Loci

$A(x^2+y^2)^2+B^3x^2+C^3y^2+D^5 = 0$ — general equation in standard form, locus centered at origin, lines of symmetry in coincidence with coordinate axes

A, B^3, C^3, D^5, real

Genera: *QBI CENTRAL CURVES* (comprise the curves of demarcation of inversion superfamilies based on qualifying* genera of this group)

GENUS: 1. *Central Conics* $(A=0)$
GENUS: 2. *Central Quartics* ($D^5=0$; central inversions of central conics)

GENERA: *CENTRAL CARTESIANS*

$(x^2+y^2)^2+B^2x^2+C^2y^2+D^4=0$, $B^2, C^2, D^4 \neq 0$, real, general eq. in standard form

GENERA: *Monovular Central Cartesians* (only exceptional subspecies self-invert)

GENERA: *Biovular Central Cartesians* (*Central Bicircular Quartics***) (self-invert by two modes)

GENUS: 1. *Central Linear Cartesians* (their axial inverse loci--*Linear Cartesians*--possess linear bipolar eqs.)

GENERA: 2. all others

GENERA: *Monovular And Biovular Central Cartesians* (QBI central curves are their curves of demarcation)

GENUS: 1. *Central Parabolic Cartesians* (both monovals and biovals self-invert in multiple modes; certain axial inversions possess parabolic bipolar eqs.)

GENUS: 2. *Central True Exchange-Limacons* (see Fig. 8-26)

GENERA: 3. all others (only exceptional subspecies of monovals self-invert; see Fig. 8-15b,i)

* "Qualifying genera" are genera that possess both monovular and biovular species.

**"Bicircular quartics" are 4th-degree curves that consist of two ovals that self-invert in multiple modes.

not come into contact, the condition $(|G|-|E|)^2 = 4F^2$ cannot be fulfilled.

In the light of the above considerations, it is evident that every subspecies of central conic, together with the ensemble of its inverse loci about all points in the plane (including central quartics), are *curves of intersection* with one or another central Cartesian-based inversion superfamily. The curves of intersection with the latter superfamilies are the *curves of demarcation* of these superfamilies, and any given superfamily intersects the QBI superfamily in curves of only a single eccentricity.

The situation can be visualized in terms of Table 8-17-1 by appropriately altering the eccentricity of the ensemble of QBI curves to $\sqrt{(E/G)}$, where E and G are the parameters of the simple intercept format (eq. 8-13-5) of the Cartesian self-inverters from which the central genus of the superfamily is derived. The relationships between *central Cartesians* and *QBI central curves* are outlined in Table 8-17-4.

[The above analysis also could be carried out directly in terms of the eqs. for central Cartesians, say eq. 8-15-2b, rather than through the approach via Cartesian self-inverters and the simple intercept format thereof.]

Fig. 8-18. The equilateral strophoid as the curve of demarcation of parabolic Cartesian axial-vertex cubics. This figure illustrates the vertex cubics obtained by inverting the loci of Fig. 8-16 (plus curves h and i for $C = 3.75$ and 3.95) about the vertex lying between the focus of self-inversion and the $\cos^2\theta$-condition double-focus (in the annular curves this is the vertex of the interior oval that lies closest to the $\cos^2\theta$-condition double-focus).

A portion of the basis curve is illustrated by the dashed tracing superimposed on curve f for the $C = 3$ axial-vertex cubic, from which the length of the unit of linear dimension for the inversion is evident (the distance from the reciprocal inversion pole--or pole of inversion--to either point of intersection of the curves). The solid circles mark the reciprocal inversion poles. A single reciprocal inversion pole serves for curves with C-values of ½, 1, 1½, and 2. Corresponding C-values are given beneath the loci.

For $C = 4$, the basis curve (Fig. k) is the equilateral limacon--the curve of demarcation of parabolic Cartesians--and its vertex cubic is the equilateral strophoid--the curve of demarcation for the loci of the present figure. For $C > 4$, the basis curves are annular and the axial-vertex cubics are serpentine circular cubics. These self-invert about all 3 axial vertices, by a different mode for each.

Since the equilateral limacon is the curve of demarcation of parabolic Cartesians, a corresponding equilateral QBI curve is the curve of demarcation of every genus of inverse loci of parabolic Cartesians about a corresponding pole (see Table 8-17-3). Fig. 8-17 illustrates the series of corresponding curves

The Equilateral Strophoid As The Curve of Demarcation of Parabolic Cartesian Axial-Vertex Cubics

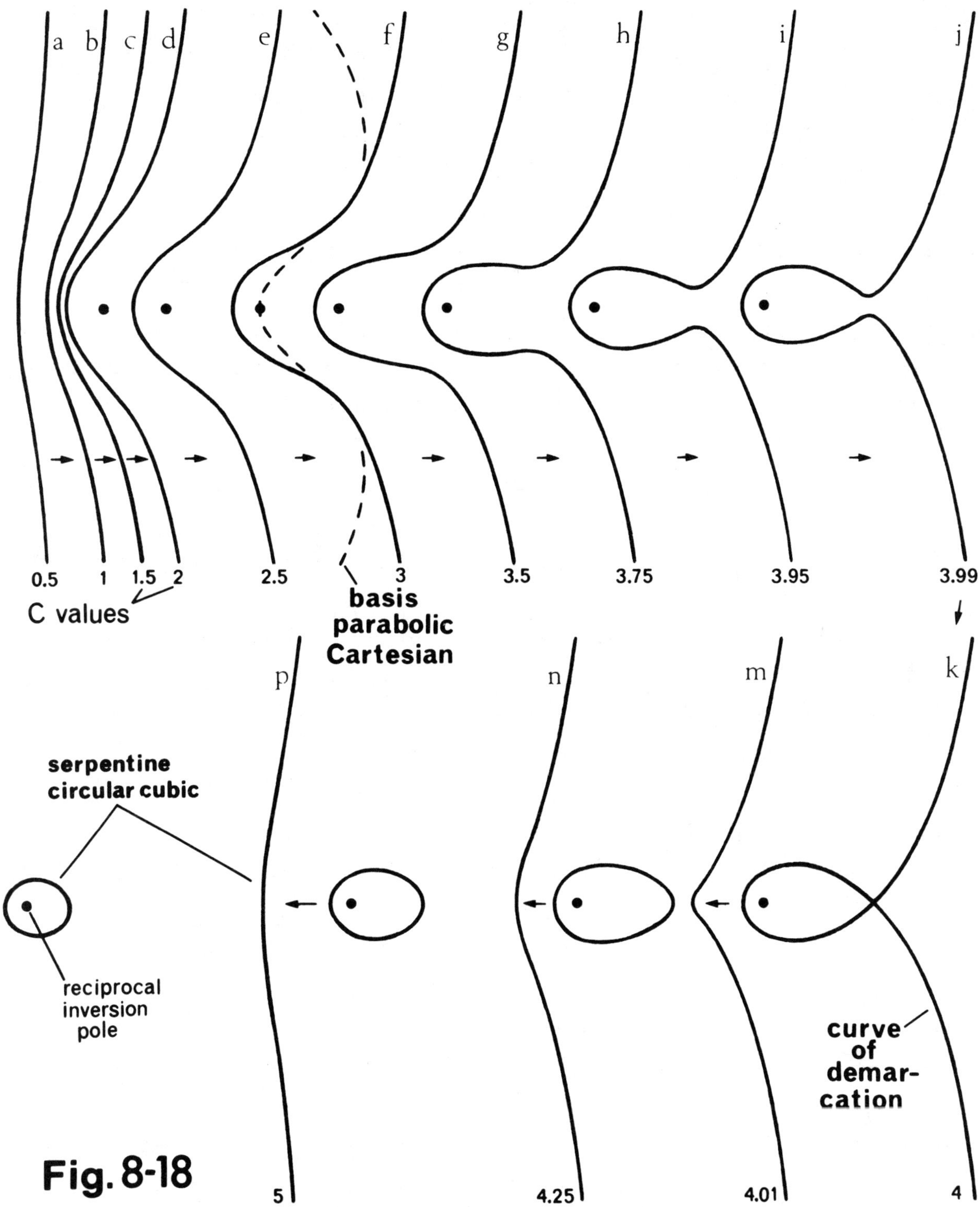

Fig. 8-18

for inversions of the monovular subspecies about the exterior poles modular to the foci of self-inversion (the reflection of the $\cos^2\theta$-condition double-focus in the focus of self-inversion), and of the biovular species about the right inter-oval modular poles.

The axial-vertex inversion subspecies of the present figure, though they were obtained by inverting the basis curves about a corresponding vertex, also are representative of the vertex cubics obtained by inverting the monovular members about the other axial vertex and by inverting the biovular members about the other 3 axial vertices (i.e., for a given value of C, all the axial-vertex cubics for corresponding axes belong to the same subspecies). In other words, the inverse cubics obtained by inverting a given axial inverse locus of a central Cartesian about its two or 4 axial vertices differ only in size and orientation (Maxim 45).

The vertex cubics that are illustrated, however, are not the only vertex cubics of the superfamily for C-values greater than 4, nor is it typical of curves of demarcation that only a single species of vertex cubics exists. Although members of the same species of vertex cubics are obtained by inverting the corresponding monovular ($C < 4$) central parabolic Cartesians of Fig. 8-17 about their y-axis vertices (as a consequence of the fact that these monovals self-invert about the center at 90° and 270°), a second species of vertex cubics is obtained by inverting the members of a congeneric central annular species (see Fig. 8-22a) about their y-axis vertices and by inverting the members of a second congeneric central opposed-oval species (see Fig. 8-22f) about their y-axis vertices (which are the only axial intersections). [Only in the annular central curves are both species of vertex cubics obtained by *axial* inversions.]

Furthermore, for inversion superfamilies for which elliptical QBI curves are the curves of demarcation, each subspecies of central curve of demarcation (ellipses and elliptical central quartics) possesses two subspecies of vertex cubics.

The two species of serpentine circular cubics, obtained by inverting the annular central curves referred to above about the vertices on each line of symmetry, are found to have characteristic but different symmetry properties. Each of these properties is unique to members of the possessing species, i.e., no other vertex cubic of the members of any other inversion superfamily based upon central Cartesians possesses either of the properties in question (see Figs. 8-21, 22, 26, and 27).

Fig. 8-19. The equilateral lemniscate and the equilateral hyperbola as curves of demarcation of monovular and biovular Cassinians and their central inversions. This figure complements Fig. 8-17, in which some of the same curves are displayed individually, rather than in centered ensembles. The C-value for a given Cassinian of this figure is the reciprocal of the C-value for the same subspecies in Fig. 8-17, inasmuch as the present values are those of the bipolar eq., $uv = Cd^2$, of the central curves, while those of Fig. 8-17 are those of the bipolar eq., $u^2 = Cdv$, for the related parabolic Cartesians (the relationship between these curves is that the former are central inversions of the latter; see also Table 8-17-2).

Fig. b is the conventionally displayed confocal ensemble of Cassinian ovals, in which the solid circles, F, mark the common loop foci of each curve in the ensemble. Corresponding C-values are adjacent to each curve. The curve of demarcation--the equilateral lemniscate--is for the value $C = \frac{1}{4}$, and is shown by the heavily traced curve with the DP. The entire curve is not shown because of its close proximity to the adjacent curve for $C = 0.2625$. Only small segments of the curves for $C = 0.255$ and 0.245 are shown for the same reason.

Inasmuch as the ensemble of curves of Fig. b have the space-filling property, any ensemble of their inverse loci about a common pole for a common value of the unit of linear dimension (in this case, $j = d$) also have the space-filling property. Since the pole of inversion of Fig. b that yields Fig. a is the center, the inverse locus of the equilateral lemniscate is the equilateral hyperbola. This curve--the curve of demarcation of the inverse ensemble--is the only inverse locus of Fig. a that does not belong to the same subspecies as the corresponding basis locus of Fig. b.

If only the sizes of corresponding loci were to change in the transformation, from Fig. b to Fig. a, the curve of demarcation would be subspecifically unchanged from one ensemble to the other. Though the changes produced by the central inversion are manifold, their key aspects are (a) that the biovals self-invert at 0° and 180°, whereas the monovals self-invert at 90° and 270°, and (b) the curve of demarcation is the only member of the ensemble upon which the pole of inversion is incident. Accordingly, the biovals maintain their orientation, the monovals become rotated 90°, and the equilateral lemniscate "opens" to an equilateral hyperbola, the arms of which demarcate monovals and biovals for C-values approaching ¼ from above and below (the portions of the loci at great distances in Fig. a are inverse to portions of the loci in the neighborhood of the DP in Fig. b).

Because the units of linear dimension $[\sqrt{\pm(C^2-1/16)}]$ for congruent inversions of the loci of Fig. b are not employed, the loci of Fig. a are not confocal. The large solid circles labelled F are the traditional foci of the equilateral hyperbola--the curve of demarcation. The two unlabelled small solid circles are the loop foci of the $C = 0.2$ biovals. The foci of the other biovals become increasingly distant from the center with increasing values of C. Segments of monovular and biovular loci for C-values only one thousandth of a unit greater and less than the demarcating value of ¼ are shown adjacent to the lateral arms of the equilateral hyperbola at the upper right and upper left. Because of the close proximity of the loci in the neighborhood of the inner vertices, only segments of most of the ovals are shown.

If either ensemble is inverted about a non-central pole, all the inverse loci are dissimilar to the basis loci, and the curve of demarcation becomes the corresponding equilateral QBI curve obtained by inverting either curve of demarcation about the pole in question. For example, for inversion of the ensemble of Fig. a about either traditional focus of the equilateral hyperbola, or of Fig. b about either confocal pole (F), the curve of demarcation of the inverse ensemble is the equilateral limacon, and the other members of the ensemble are Cartesian self-inverters, specifically monovular and biovular parabolic Cartesians.

For inversion of either ensemble about a transverse-axis vertex, the curve of demarcation is the equilateral strophoid, and the other members are Cassinian axial-vertex cubics. For an axial pole of inversion at twice the distance of the vertices of the equilateral hyperbola of Fig. a, the curve of demarcation is the equilateral pseudo-exchange-limacon (see Fig. 6-3m), etc.

[The relatedness of Cassinians to quadratics on other grounds than the fact that the $C = ¼$ subspecies of Cassinian is the equilateral lemniscate was known previously from several transformation relationships. For example, the locus of the points of intersection of two circles that have their centers at the foci of a given central conic and that touch any tangent of the conic is a Cassinian oval (see Hilton, 1920, page 325).]

The Equilateral Lemniscate & The Equilateral Hyperbola As Curves of Demarcation of Monovular & Biovular Cassinians & Their Central Inversions

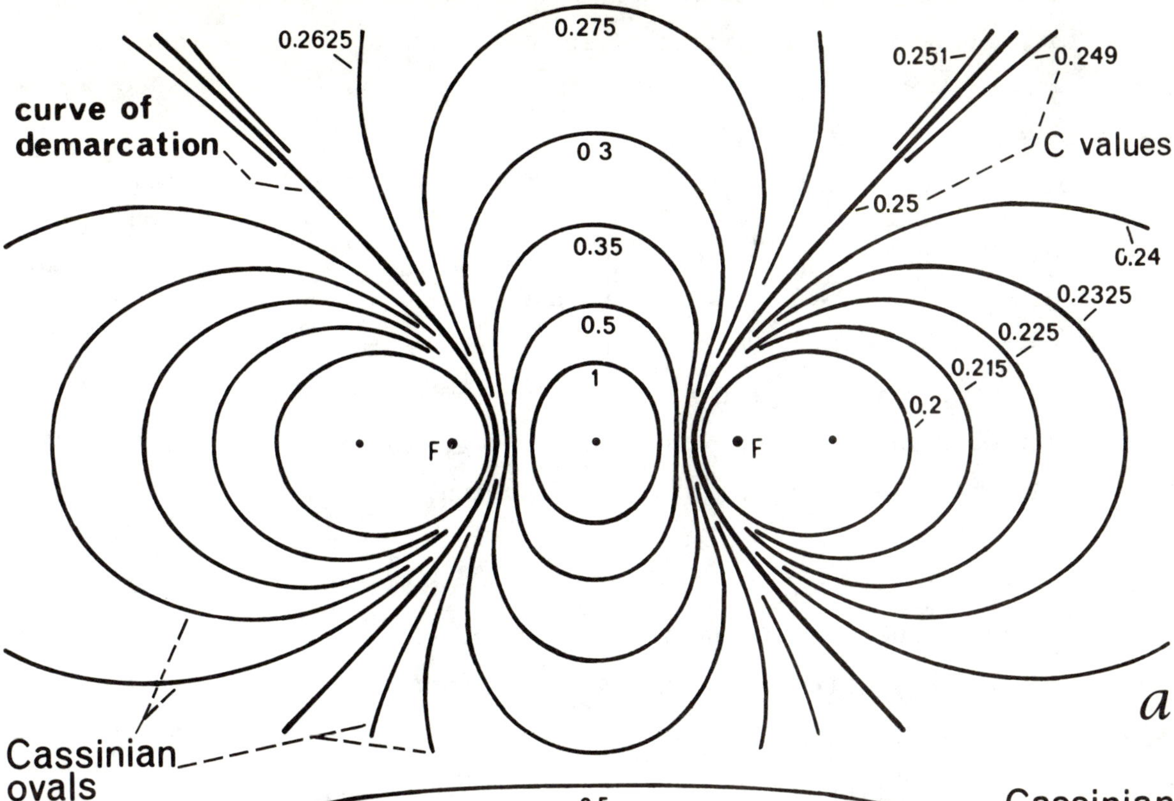

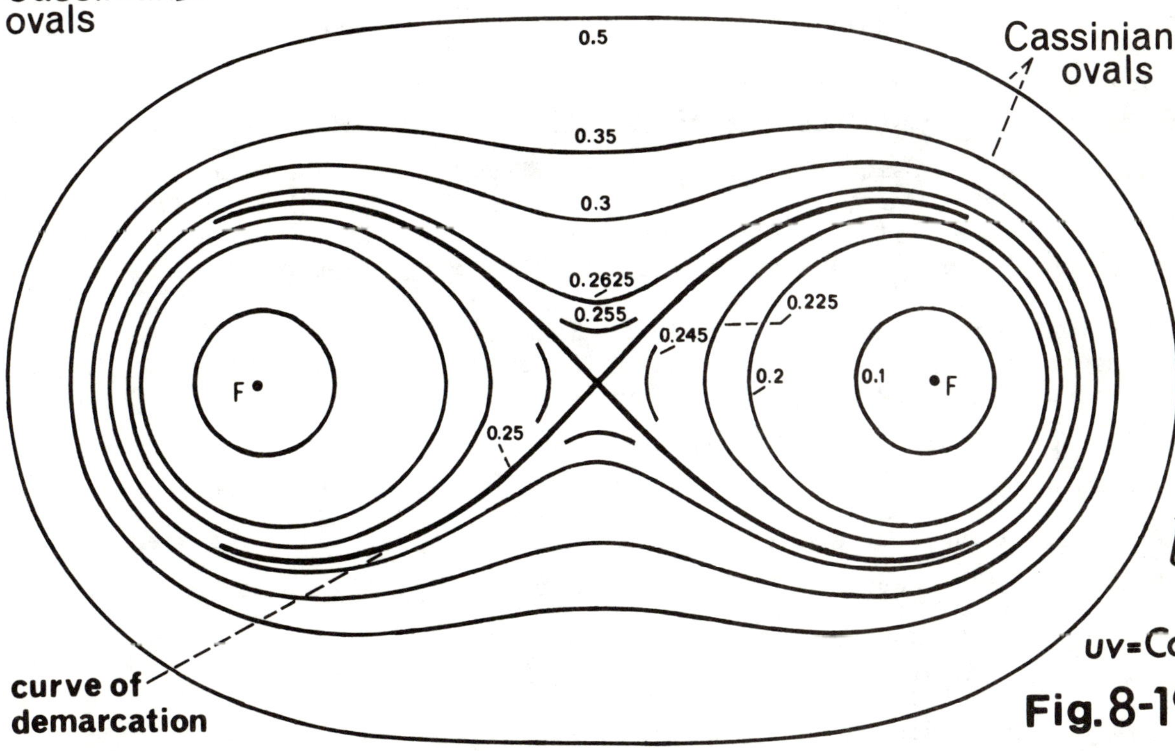

$uv = Cd^2$

Fig. 8-19

Fig. 8-20. Correspondences between axial poles in axial inversions to and from central parabolic Cartesians. Treatments of the topics of curves of demarcation and Inversion Taxonomy raise questions about the conservation of circumpolar focal loci, the locations and focal status of poles inverse to circumpolar foci, and the properties of inverse loci about point-focal as opposed to non-point-focal poles (see also Maxim 32). Some of these questions are answered here.

[The classical view, as applied to classically-defined foci (see pages xvi and I-2 of *Symmetry, An Analytical Treatment*), is that foci invert to foci (see Hilton, 1920).]

In Fig. a, the basis $C = 0.2$ Cassinian (a $C = 5$ opposed-oval central parabolic Cartesian; see Table 8-17-2) is inverted about the $\cos^2\theta$-condition focus of the right oval to the $C = 5$ parabolic Cartesian. This inverse curve then is inverted about the pole that is inverse to the center of the basis curve. The latter pole is the (modular) reflection of the reciprocal inversion pole in the 0° inter-oval-mode focus of self-inversion. This sequence of inversions yields an *inversion image* of the basis curve, i.e., a curve belonging to the same subspecies as the basis curve.

In Fig. b the same sequence of operations is carried out for the same basis Cassinian subspecies, but for a pole of inversion that is modularly inverse (i.e., inverse for modular j) to the pole that was employed in Fig. a. The inverse curve in this case is the same subspecies of parabolic Cartesian that was obtained in Fig. a (but with reversed orientation).

[This is unlike the situation for central QBI curves; although the specification of an axial pole, $h = \pm nj$ $(n \neq 0)$, for a given biovular species of central Cartesian defines a dissimilar inverse subspecies, it does so uniquely only for modular $\pm nj$, i.e., for $n = 1$, in which case the inverse subspecies is a central curve. For other values of $\pm nj$, two axial poles exist on each side of the center--one pair at $\pm nj$ and the other pair at $\pm j^2/nj = \pm j/n$ --inversions about which produce members of the same subspecies, but of different sizes and with opposed orientations (compare Figs. a_2 and b_2).]

In Fig. c the same inversion sequence is carried out for the same basis Cassinian subspecies, but for a pole of inversion lying midway between the poles employed in Figs. a and b. In this case the first inversion does not yield a parabolic Cassinian because the pole of inversion is not one of the poles specified by the *Cartesian condition* (the poles defined by equating the *Cartesian term* to 0; see eqs. 6b, 8a, and 11, below).

In Fig. d the basis curve is the corresponding $C = 0.2$ (Cassinian parameter) annular central Cartesian (see derivation in the legend to Fig. 8-31).

Correspondences Between Axial Poles In Axial Inversions To & From Central Parabolic Cartesians

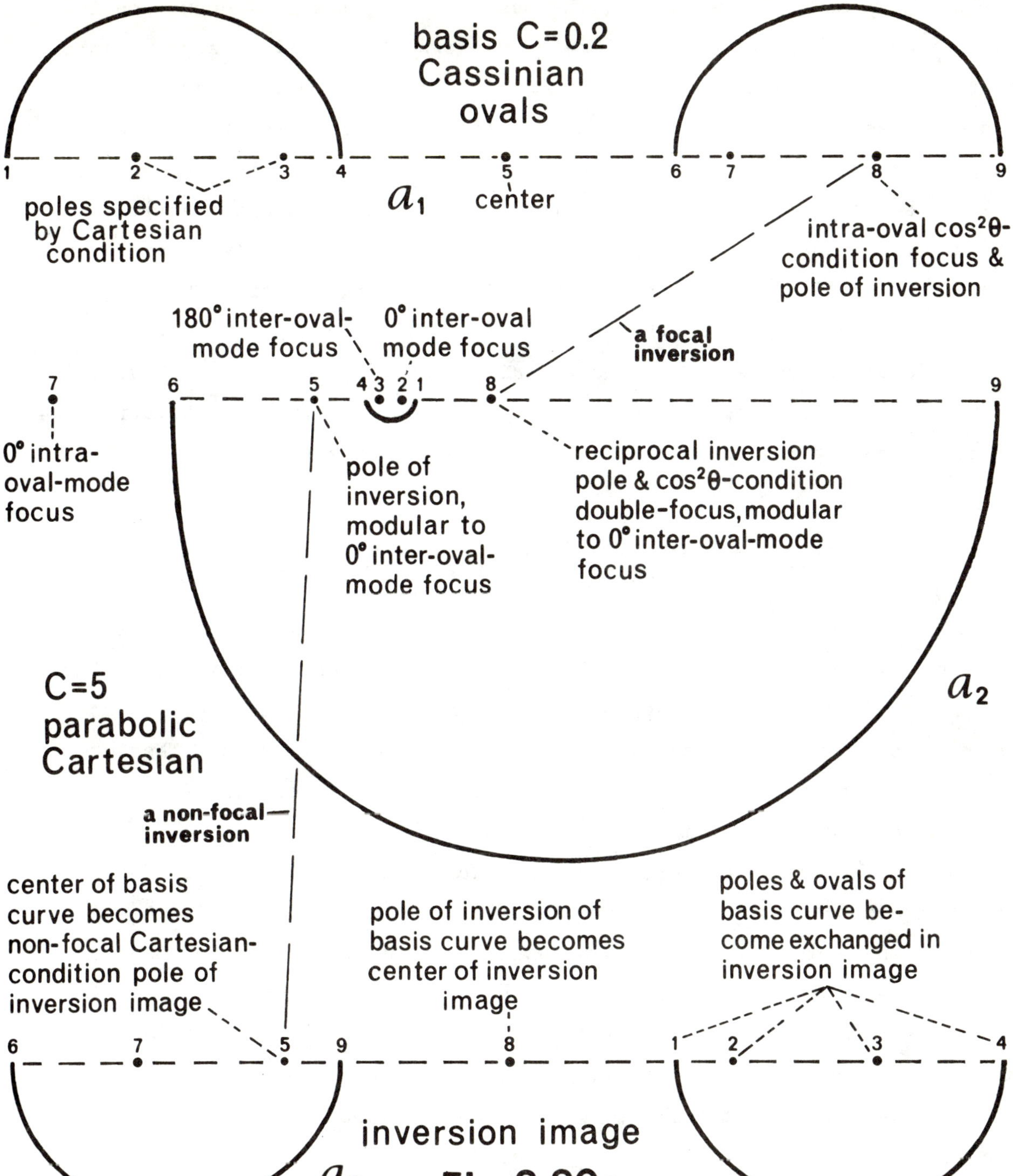

Fig. 8-20a

[This central curve is obtained by inverting the $C = 0.2$ Cassinian about an intra-oval pole at $h = \pm D$, for $D^4 > 0$.] The pole of inversion in this case is the right interior pole specified by the *Cartesian condition*. This inversion also yields the $C = 5$ parabolic Cartesian.

For the purpose of illustration, the basis curves, the 1st-generation inverse curves, and the *inversion images* are shown at convenient sizes, rather than to the same scale, and only half (or less for Fig. d_2) of each curve is depicted. Except for the vertices, all pertinent axial poles, the corresponding inverse poles in the 1st generation, and the poles inverse to these poles in the inversion images, are marked by solid circles and identified with numbers--with the same numbers being employed consistently within each triad of curves. The vertices merely are identified by numbers. The large solid circles identify poles that have circumpolar point-focal rank.

Figs. a-d are illustrative of the principles and topics treated below, and are referred to repeatedly as the occasion demands. A further systematic discussion of the figures appears after the analytical treatments.

Correlation of The Symmetry of Inverse Curves With The Focal Rank of The Pole of Inversion

General treatments of the inversion and circumpolar symmetry properties of Cartesian self-inverters and central Cartesians (the central inverse loci of Cartesian self-inverters) bring to light a wealth of relationships not encountered in previous studies, which mostly concerned QBI curves. These relationships both restrict the scope of applicability of certain previously elaborated Circumpolar Maxims and lead to numerous new Maxims.

One such relationship concerns the poles about which Cartesian self-inverters invert to central Cartesians and vice versa (see pages XIV-94 and 95). According to Maxim 3 (page VI-9)--based primarily on studies of the QBI superfamily--*for curves of even degree, the higher the circumpolar symmetry rank of the pole of inversion, the higher, in general, the circumpolar symmetry of the inverse curve*. For example, with the possible exceptions of foci of self-inversion, the DP or DP homologue of a QBI cubic or quartic is the highest-ranking focus (i.e., it possesses intercept transforms of lowest degree). Inversions about these poles yield conics--from most points of view the most highly symmetrical curves in the inversion superfamily.

The same relationships clearly do not hold for the non-QBI cubics and quartics considered here. Since central Cartesians (Figs. a_1-d_1, a_3-d_3) possess the lowest degree transforms in inversion superfamilies based upon them, they are the most highly symmetrical species. Yet the 4 axial poles about which Cartesian self-inverters invert to central Cartesians (for example, Fig. a_2, poles 5 and 8; Fig. b_2, poles 5 and 7; Fig. c_2, poles 5 and 7½; and Fig. d_2, poles 6 and 8) typically do not possess point-focal rank. The basis for this difference from the situation in the QBI superfamily is treated below (*Cartesian Self-Inverters As The Most Highly Symmetrical Axial Bicircular Quartics*). Though many cases for which Maxim 3 applies to non-QBI curves could be cited, it is clear that this Maxim has limited applicability.

Conservation of Lines of Symmetry And 0° Modes of Self-Inversion

It is evident from properties of bicircular quartics, serpentine circular cubics (see Table 8-10-1), and QBI curves that the sum of the number of 0° modes of self-inversion for finite moduli and the number of lines of symmetry is conserved by the inversion transformation. These relationships are outlined in Table 8-20-1 for biovular parabolic Cartesians and Cassinians.

Table 8-20-1. Conservation of Lines of Symmetry And 0° Modes of Self-Inversion (For Finite Moduli)

	Biovular Parabolic Cartesians (annular ovals)		⇄	Biovular Cassinians (opposed ovals)	
modes of 0° self-inversion	1	0° intra-oval mode (lateral focus)		1	0° intra-oval mode (center-focus)
	1	0° inter-oval mode (interior focus)			
lines of symmetry	1			2	
	3			3	total

CIRCUMPOLAR
MAXIM 8-20-1: *Axial inversions to and from central Cartesians conserve the sum of the number of lines of symmetry and the number of 0° modes of self-inversion for finite moduli.*

8-20-2: *The inversion transformation conserves the sum of the number of lines of symmetry and the number of 0° modes of self-inversion for finite moduli.*

8-20-2': *The inversion transformation conserves the sum of the numbers of 0° modes of self-inversion for finite and infinite moduli.*

The Central Cartesian Basis Genera For Inversion Superfamilies Possessing Bicircular Quartic Members

With knowledge of these conservation relationships, general eqs. for the basis central genera of inversion superfamilies that possess bicircular quartic members can be written as 1a-c (see also Table 8-17-4).

equation of central Cartesian basis genera for inversion superfamilies possessing bicircular quartic members

(a) $$(x^2+y^2)^2+B^2x^2+C^2y^2+D^4 = 0 \qquad (8\text{-}20\text{-}1)$$

(b) $$r^4+r^2(B^2\cos^2\theta+C^2\sin^2\theta)+D^4 = 0$$ polar form of eq. 1a

(c) $$r^2 = \frac{-(B^2\cos^2\theta+C^2\sin^2\theta)\pm[(B^2\cos^2\theta+C^2\sin^2\theta)^2-4D^4]^{\frac{1}{2}}}{2}$$ eq. 1b solved explicitly for r^2

(a) $$r^2 = [-B^2\pm(B^4-4D^4)^{\frac{1}{2}}]/2$$ x-axis roots (vertices) of 1a-c (8-20-2)

(b) $$r^2 = [-C^2\pm(C^4-4D^4)^{\frac{1}{2}}]/2$$ y-axis roots (vertices) of 1a-c

$$r_1r_2 = \sqrt{(D^4)} = \pm D^2, \quad \text{for } D^4 > 0$$ product of roots of 1c of different absolute values (8-20-3)

Expressing eq. 1a in polar form (eq. 1b) and solving explicitly for r^2 gives eq. 1c. For this eq., the product of the roots of different absolute values is constant at $\sqrt{(D^4)}$ (eq. 3). As a consequence, the biovular members of every genus self-invert about the center in both the 0° and 180° modes (for the modular value, $j = D$).

[The conditions, $B^4 = 4D^4$, $C^4 = 4D^4$, and $B^2 = C^2$, which are encountered repeatedly in the following, correspond to central curves that consist of real or imaginary circles. The first two conditions represent congruent, non-concentric circles; the last condition represents concentric circles.]

It will be noted from eq. 2 that when D^4 is negative there are two real and two imaginary vertices on each axis. For 4 real vertices to exist on both axes (annular ovals), D^4 must be positive and less than both $\frac{1}{4}B^4$ and $\frac{1}{4}C^4$, and both B^2 and C^2 must be negative. If the other conditions apply but only B^2 is negative, there are 4 real x-axis vertices but no real vertex on the y-axis (opposed ovals). The parameters B^2 and C^2 of eq. 1 play symmetrical roles. Any implications regarding axial properties of a curve based on the value of B^2 also can be drawn in relation to C^2, but for exchanged axes.

Excepting inversions about the center or the modular poles on each line of symmetry (see Maxim 8-20-3, below), inverse loci of the biovular central Cartesians of eq. 1a (Figs. a_1-d_1, a_3-d_3) about poles on either axis or the modular circle (a centered circle of modular radius) consist of axial bicircular quartics (Figs. a_2-d_2) or axial serpentine circular cubics (Figs. 8-18, 22b,d, 31 and 32) with two modes of 0° self-inversion, and one mode of 180° self-inversion; inverse loci about poles that do not lie on either axis or the modular circle consist of non-axial bicircular quartics or non-axial serpentine circular cubics with 3 modes of 0° self-inversion and one mode of 180° self-inversion.

CIRCUMPOLAR
MAXIM 8-20-3: *On each line of symmetry there are two poles about which a biovular central Cartesian inverts to other subspecies of biovular central Cartesians; these poles lie at modular distance on each side of center, i.e., at the distance of the unit of linear dimension for self-inversion. In these inversions, annular curves invert to two subspecies of opposed-oval curves, whereas opposed-oval curves invert to one annular and one opposed-oval subspecies.*

All modifications of eq. 1a, aside from those producing translations and/or rotations, destroy the property of self-inversion about the center in 0° and 180° modes. Thus, addition of a term in x^2y^2 does not alter the locus from a central curve but the self-inversion property is lost; the product of the roots no longer is constant and the curve inverts to an 8th-degree locus. Addition of terms in Axy^2, Ax^2y, $Ax(x^2+y^2)$, or $Ay(x^2+y^2)$ destroys the central self-inversion property because a line of symmetry is lost.

[Note that the curves of demarcation of members of the central genera of eq. 1a (and the central inversions of these members) must be either quartics or quadratics.]

Conservation of Circumpolar Focal Loci

A re-examination of the breadth of applicability of Circumpolar Maxim 26 (page XII-17), which is based primarily on studies of QBI curves, is in order at this juncture. According to this Maxim, the sum of the number of real and imaginary circumpolar point-foci and the number of orthogonal lines of symmetry of a basis curve is conserved by line-of-symmetry inversions. The many and diverse interrelationships that have been found to exist in the realm of non-QBI quartics have led to several new working hypotheses with regard to the application of conservation rules to circumpolar focal loci, as follows.

1. With few exceptions, the topic of conservation of circumpolar focal loci is treated most conveniently by dealing separately with (a) modes of self-inversion and poles that qualify as circumpolar foci exclusively by virtue of the self-inversion property, and (b) angle-independent circumpolar focal loci. [This separate treatment is consistent with the classical approach, in which poles of 0° and 180° self-inversion (referred to as "centers") were not reckoned as foci unless they were coincident with foci as defined in the classical sense (see pages xvi and I-2 of *Symmetry, An Analytical Treatment*, and Hilton, 1920).]

2. Treatments of conservation of circumpolar focal loci in relation to a line of symmetry should take into account all angle-independent circumpolar focal loci--real and imaginary, finite and infinite--that are incident upon or intersect that line.

3. Covert circumpolar focal loci should be taken into account.

   In this connection:

   (a) A line through the DP of limacons orthogonal to the line of symmetry is not a covert circumpolar focal locus; points on this line were assumed to possess circumpolar focal rank on the basis of indirect evidence, but this cannot be verified analytically (see page X-17).

   (b) Nor are lines parallel to the line(s) of symmetry and passing through the reciprocal inversion pole of either axial or non-axial QBI curves covert circumpolar axial focal loci. Nor, either, do the axial foci of the basis curves come to lie on these lines in the inverse loci (see the treatment on pages XII-52 and 53).

4. The presence of only a single discrete circumpolar focus defined by the $\cos^2\theta$-condition signifies that two $\cos^2\theta$-condition foci are in coincidence as a double-focus (that should be reckoned as two foci). This situation occurs in the most highly symmetrical even-degree axial genus in inversion superfamilies based on central Cartesians (see analytical treatments below, in legend for Fig. 8-31, and on pages XII-6 and 11).

5. In the cases of some curves of high circumpolar symmetry, such as curves of demarcation, a complete specification of circumpolar focal loci often depends upon a circumpolar symmetry analysis of the general eq. for all members of the ensemble under consideration. A complete representation of the multiplicity of

foci may not be evident from analyses of subspecific or generic eqs. alone. For example, considered as curves of demarcation of Cartesian self-inverters, limacons are most appropriately analyzed as special cases of Cartesian self-inverters, rather than as QBI curves; similarly, central conics viewed as special cases of axial inversions of central quartics should be analyzed in that framework, rather than as quadratics.

6. In the reckoning of circumpolar focal loci, each focus must be assigned the full multiplicity of the analytical treatment. For example, a focal condition, $(h-d)=0$ defines a single point-focus at $h=d$, a focal condition $(h-d)^2=0$ defines a double-focus at $h=d$, a focal condition $(h-d)^3=0$ defines a triple-focus, etc. Six explicit examples are considered.

limacon line-of-symmetry simple intercept eq.

(a) $$x^2(b+2h)^2\cos^2\theta+2x[x^2(b+2h)+h(2h^2+3bh+b^2-a^2)]\cos\theta + x^4 + [2h(b+h)-a^2]x^2 + h^2[(b+h)^2-a^2] = 0 \qquad (8\text{-}20\text{-}4)$$

(b) $$\cos\theta = \frac{a^2h-(b+2h)(x^2+h^2+bh)\pm a[b(b+2h)x^2+h^2(a^2-2bh-b^2)]^{\frac{1}{2}}}{x(b+2h)^2}$$ simple intercept format of eq. 4a

central quartic simple intercept eq. (conic parameters)

(a) $$[j^4(b^2\pm a^2)-4a^2b^2h^2]x^2\cos^2\theta+2b^2h(j^4-2a^2h^2-2a^2x^2)x\cos\theta - a^2b^2x^4+(\mp j^4-2a^2b^2)a^2x^2+h^2b^2(j^4-a^2h^2) = 0 \qquad (8\text{-}20\text{-}5)$$

simple intercept format of eq. 5a

(b) $$\cos\theta = \frac{-b^2h(j^4-2a^2h^2-2a^2x^2)\pm aj^2(b^2\pm a^2)^{\frac{1}{2}}[b^2x^4+(\mp j^4-2b^2h^2)x^2+b^2h^4]^{\frac{1}{2}}}{[j^4(b^2\pm a^2)-4a^2b^2h^2]x}$$

simple intercept eq. of central Cartesians

(a) $$(4h^2+B^2-C^2)x^2\cos^2\theta+2hx[2(x^2+h^2)+B^2]\cos\theta + x^4 + (C^2+2h^2)x^2 + (h^4+h^2B^2+D^4) = 0 \qquad (8\text{-}20\text{-}6)$$

(b) $$\cos\theta = \frac{-h[2(x^2+h^2)+B^2]\pm[(C^2-B^2)x^4+(C^2-B^2)(C^2-2h^2)x^2+(CT)]^{\frac{1}{2}}}{(4h^2+B^2-C^2)x}$$ simple intercept format of 6a

$$(CT) = h^4(C^2-B^2)+h^2(B^2C^2-4D^4)+D^4(C^2-B^2)$$ the *Cartesian term*

In eq. 4a, the $\cos^2\theta$-condition, $(b+2h)^2=0$, defines a double-focus at $h=-\frac{1}{2}b$, a property unique to the genus *limacons* among all QBI curves. The constant-condition $h^2=0$ defines the DP or DP homologue as a double-focus, while the constant-condition $a^2=(b+h)^2$ defines the two vertices. These 6 foci are the only angle-independent circumpolar foci defined by eq. 4a.

Solving eq. 4a for $\cos\theta$ yields the simple intercept format 4b. The term $(b+2h)^2$ of the denominator is the redundant expression of the existence of the double-focus defined by the $\cos^2\theta$-condition, since the term in the denominator is a multiplicative component of the $\cos^2\theta$ term. In the numerator, the $(b+2h) = 0$ condition also is expressed twice, redundantly, i.e., with the force of a double-focus, once as the radicand x^2-condition and once as the external x^2-condition. Either condition alone leads to the reduction of the degree of intercept transforms.

There are two radicand constant-conditions: $a^2-2bh-b^2 = 0$ and $h^2 = 0$. The former defines the focus of self-inversion at $h = (a^2-b^2)/2b$. The latter condition would appear to be merely the redundant expression of the double-focus at the DP. Indeed, without the perspective provided by an analysis of limacons as curves of demarcation of Cartesian self-inverters, it would be interpreted solely in this manner.

On the other hand, real or imaginary intersections of curves with lines of symmetry are not otherwise known to be defined by radicand conditions. To appreciate the full significance of the radicand constant-condition $h^2 = 0$, one must refer to the general simple intercept format for Cartesian self-inverters (eq. 8-13-10) cast about a focus of self-inversion, or to the general format for all axial bicircular quartics cast about a point on the line of symmetry (see eq. 10, below).

Inspection of the radicands of these formats reveals that the angle-independent foci that are in coincidence with the 3 foci of self-inversion (and their imaginary counterparts in monovular species) are defined by radicand constant-conditions. Accordingly, the significance of the $h^2 = 0$ radicand constant-condition of eq. 4b for limacons, considered as curves of demarcation of Cartesian self-inverters, is that it defines two additional but trivial foci of self-inversion that are in coincidence with the DP (see Figs. 8-16i and j, in which the foci of self-inversion of Fig. j for $C = 4.01$ merge into the DP for $C = 4$).

In the case of eq. 5a for central quartics (which are QBI curves), the $\cos^2\theta$-condition, $h^2 = j^4(b^2\pm a^2)/4a^2b^2$, defines the two "loop foci," and the $\cos\theta$-condition, $h = 0$, defines the center. The constant-condition, $h^2 = 0$, defines the self-intersection of the hyperbolic species at the DP, as well as the DP homologue of the elliptical species, which consists of two coincident isolated points of the locus that lie at the true center (see legend of Fig. 8-21).

This condition is not redundant; it emphasizes the fact that a center considered only as the point of intersection of two lines of symmetry has the force of a single focus, but a center that also is a DP must be reckoned as a triple-focus (with single force as the center plus double force as the DP). Lastly, the constant-condition, $h^2 = j^4/a^2$, defines the two vertices.

In the case of the corresponding simple intercept format 5b, the $h = 0$ condition redundantly defines the loop foci. As is typical of central curves of greater than 2nd degree, no single condition on the radicand is focal (the condition, $h^4 = 0$, allows an x to be extracted from the radical, but this alone is not focal, i.e., alone it does not lead to the reduction of the degrees of transforms).

Eq. 6a is the $\cos\theta$ simple intercept eq. for central Cartesians. For it, the $\cos^2\theta$-condition, $h^2 = (C^2-B^2)/4$, defines the intra-oval foci, the $\cos\theta$-condition defines the center, while the constant-condition (in the brackets) defines the 4 vertices. In the simple intercept format, 6b, the center is defined redundantly by the external condition, $h = 0$, and equating the denominator to 0 redundantly defines the intra-oval foci. Again, no single condition on the radicand is focal.

The radicand constant-condition, however, is most significant. Though it is not focal, it identifies the axial poles (for example, in Fig. a_1, it defines poles 2,3,7, and 8, and in Fig. d_1, poles 1,4,6, and 9) about which inversions yield the most highly symmetrical axial quartic genus in each superfamily (each superfamily being based upon a species of monovular and/or biovular central basis curves, such as Cassinians). Vanishing of both the radicand constant term and the radicand x^2 term, for the condition $(C^2-D^2)(B^2+C^2) = 0$, is not allowed, inasmuch as it leads to an eq. of the form $\cos\theta = (x^2+M^2)/Nx$, for a non-centered circle.

Inasmuch as it is likely that definitive rules regarding the conservation of angle-independent circumpolar focal properties of curves will be as complex as Plückers eqs., which relate the degree, n, class, n', and numbers of nodes, cusps, inflections, etc., the formulation of such rules is deferred. Table 8-20-2 lists the known angle-independent circumpolar focal loci of (a) 3 highly symmetrical equilateral QBI curves, (b) monovular central Cartesians and Cartesian self-inverters, and (c) biovular central Cartesians and Cartesian self-inverters. These focal loci are referred to the transverse axis or the single line of symmetry. As is evident from the discrepancy between Cartesian self-inverters and central Cartesians (11 versus 9 known circumpolar focal loci), knowledge concerning angle-independent circumpolar focal loci is incomplete.

In connection with the contents of this table, it should be pointed out that the arms of hyperbolas do not have axial vertices at infinity. The points inverse to inversion poles of hyperbolic central quartics, as these poles approach the DP, recede to infinity in directions approaching the directions of the DP tangents. For elliptical central quartics, the direction of the inverse of the DP homologue is undefined (the inverse locus thus is the ideal line at infinity).

When central conics are analyzed as a genus of axial inverse loci of central quartics, it is found that, as the pole of inversion (eq. 7) approaches the center, the $\cos^2\theta$-condition foci of hyperbolic species recede along the axis to infinity; those of the elliptical species are complex, with both the real and imaginary components also becoming infinite as the pole of inversion approaches the center.

abscissae of the $\cos^2\theta$-condition foci of inverse loci of central quartics, $a^2b^2(x^2+y^2)^2 = j^4(b^2x^2\mp a^2y^2)$, about an axial pole, $x = h$

$$H = \frac{a^2b(a^2-2h^2)\pm a^4[\mp h^2+(b^2\pm a^2)]^{\frac{1}{2}}}{2bh(h^2-a^2)} \qquad (8\text{-}20\text{-}7)$$

The 3 equilateral QBI curves, considered in the general context of curves of demarcation of genera of central Cartesian-based inversion superfamilies, are seen to have a total of 9 circumpolar focal loci referred to the transverse axis. All have two vertices and two $\cos^2\theta$-condition foci. For both the equilateral limacon and the equilateral hyperbola the latter are in coincidence as a double-focus [for the equilateral hyperbola they are in coincidence at the axial point of the ideal line at infinity (see eq. 7, for $a = b$ and h tending to 0)]. Only the limacon has an angle-independent covert linear focal locus.

Table 8-20-2. Angle-Independent Axial Circumpolar Focal Loci

EQUILATERAL QBI CURVES

Equilateral Limacon	Equilateral Hyperbola	Equilateral Lemniscate
2 vertices	2 vertices	2 vertices
2 1 $\cos^2\theta$-condition double-focus	2 1 $\cos^2\theta$-condition double-focus (at the axial point at ∞)	2 $\cos^2\theta$-condition foci ("loop foci")
1 1 radicand constant-condition focus of self-inversion*	2 radicand constant-condition foci (traditional foci)	
1 covert linear focal locus		
2 1 DP (double-focus)		2 1 DP(double-focus)
	1 center	1 center
1 line of symmetry	2 lines of symmetry	2 lines of symmetry
9	9	9

CARTESIAN SELF-INVERTERS	CENTRAL CARTESIANS
Monovular	
4 vertices (2 real, 2 imaginary)	4 vertices (2 real, 2 imaginary)
2 1 $\cos^2\theta$-condition double-focus	2 $\cos^2\theta$-condition foci
3 radicand constant-condition foci (2 imaginary, 1 focus of self-inversion)	1 center ($\cos\theta$-condition)
1 covert linear focal locus (orthogonal to line of symmetry at $\cos^2\theta$-condition focus)	
1 line of symmetry	2 lines of symmetry
11	9
Biovular	
4 real vertices	4 real vertices
2 1 $\cos^2\theta$-condition double-focus	2 $\cos^2\theta$-condition foci
3 radicand constant-condition foci of self-inversion	1 center ($\cos\theta$-condition), a double-focus of self-inversion
1 covert linear focal locus	
1 line of symmetry	2 lines of symmetry
11	9

*Treated in the context of a QBI curve the equilateral limacon has only 1 radicand constant-condition focus of self-inversion.

For the limacon, the DP is a double-focus by virtue of two intersections of the curve with the line of symmetry. For the hyperbola, the center is a single-focus, whereas for the lemniscate it is a triple-focus by virtue of also being a DP. The limacon also possesses a focus of self-inversion, defined uniquely (among QBI curves) by a radicand constant-condition. The corresponding radicand constant-condition foci of the hyperbola are the traditional foci. Being a 4th-degree central curve, the lemniscate has no radicand constant-condition foci (Maxim 8-20-4, below). Lastly, the limacon has one line of symmetry, while the hyperbola and lemniscate have two.

Both groups of monovular Cartesians have two real vertices, two imaginary vertices, and two $\cos^2\theta$-condition foci, the latter of which are in coincidence in Cartesian self-inverters. Monovular Cartesian self-inverters have 3 radicand constant-condition foci; two of these are imaginary and the 3rd is a focus of self-inversion. Monovular central Cartesians have no radicand constant-condition foci(Maxim 8-20-4) but they have a center. Monovular Cartesian self-inverters have a covert linear focal locus orthogonal to the line of symmetry at the position of the double-focus of the $\cos^2\theta$-condition, but they have only one line of symmetry, as opposed to two for monovular central Cartesians.

For the biovular species (Figs. a-d) the principal differences from the circumstances cited above for the monovular species are that all 4 vertices are real and that all the radicand constant-condition foci (of self-inversion) of the Cartesian self-inverters (Figs. a_2,b_2,d_2) are real. The discrepancy between the total numbers of angle-independent circumpolar focal loci for Cartesian self-inverters and central Cartesians is as yet unresolved.

Cartesian Self-Inverters As The Most Highly Symmetrical Axial Bicircular Quartics

In analyses of the circumpolar symmetry of bicircular quartics, derived as inversions of eq. 1, one repeatedly encounters a term--the *Cartesian term* (CT)--which, when equated to 0--the *Cartesian condition* (eq. 8a)--leads to the simplification of eqs. (for example, see eq. 6b). These simplifications closely parallel those that occur in the axial eqs. of QBI curves when the *limacon term* (8b) is equated to 0 (the *limacon condition*). It will be noted that eq. 8a defines axial poles of the basis 4th-degree central curves (eq. 1), just as eq. 8b for the homologous *limacon condition* defines axial poles of central conics. [The terms and conditions are said to be *homologous* because limacons also comprise a genus of Cartesian self-inverters.]

(a)	$h^4(C^2-B^2)+h^2(B^2C^2-4D^4)+D^4(C^2-B^2) = 0$	the *Cartesian condition* (*Cartesian term* = 0)	(8-20-8)
(b)	$b^2\mp(h^2-a^2)$ or $\mp h^2+(b^2\pm a^2) = 0$	the *limacon condition* (*limacon term* = 0)	
(c)	$[(a^2\pm b^2)x^2\mp b^2h^2+b^2(b^2\pm a^2)]^{\frac{1}{2}}$	radicand for central conics	

The key difference between the two situations is that equating the limacon term to 0 defines *focal* poles (the degrees of the corresponding intercept transforms reduce), whereas equating the Cartesian term to 0 generally does not (for example, poles 1,4,6, and 9 of Fig. d_1 are not circumpolar point-foci). The basis for this difference (which underlies the limitation of the scope of general applicability of Circumpolar Maxim 3 to QBI curves) is that the radicand of the simple intercept format of central Cartesians (eq. 6b) contains terms in x^4, x^2, and a constant, whereas the corresponding radicand for central conics (term 8c) contains only terms in x^2 and a constant.

Accordingly, the vanishing of the constant term in the latter case is a *focal* condition (because the radicand becomes a perfect square), whereas in the former case it is not (because, for the radicand to become a perfect square, two of the 3 terms must vanish). [In this connection, by definition, the radicand of a $\cos\theta$ simple intercept format always becomes a perfect square for the $\cos^2\theta$-condition.]

CIRCUMPOLAR
MAXIM 8-20-4: *Single radicand conditions for simple intercept formats of central curves of greater than 2nd degree do not define additional circumpolar focal loci beyond those specified by conditions on the corresponding simple intercept equations.*

8-20-5: *The most symmetrical axial quartics in a central Cartesian-based inversion superfamily are subspecies of Cartesian self-inverters, i.e., they are subspecies of the genera in which the two foci defined by the $\cos^2\theta$-condition are in coincidence as a double-focus.*

Axial inversions of biovular central Cartesians fall into 6 categories: (1) self-inversions about the center; (2) inversions to complementary central curves--annular (Fig. d_1) to opposed (Fig. a_1-c_1) or opposed to annular; (3) inversions of opposed-oval curves to other opposed-oval central curves (see Fig. 8-22); (4) inversions to opposed-oval axial bicircular quartics (inter-oval and lateral inversions of opposed-oval curves and inter-oval inversions of annular curves); (5) inversions to annular axial bicircular quartics (Figs. a_2-d_2; intra-oval inversions of opposed-oval curves and exterior and interior inversions of annular curves); and (6) inversions to axial serpentine circular cubics.

Consider first the general simple intercept eq. for axial inversion genera in superfamilies based upon central Cartesians (eq. 1). The genera in question possess representatives of the above-mentioned 6 categories plus interior, exterior, and incident axial inversions of the monovular species and the curves of demarcation (when the latter exist). Eq. 9 is the rectangular form of the

x+H translation of axial inversion genera (about pole x = h) of inversion superfamilies based upon the central curves of eq. 1

$$[j^4+2hj^2(x+H)+h^2K^2]^2 + B^2[j^4(x+H)^2+2hj^2(x+H)K^2+h^2K^4] + C^2j^4y^2 + D^4K^4 = 0 \qquad (8\text{-}20\text{-}9)$$

$$K^2 = (x^2+y^2+2Hx+H^2)$$

standard form of simple intercept eq. for curves of eq. 9

$$[4M^4H^2+4hHj^2N^2+j^4(4h^2+B^2)-C^2j^4]x^2\cos^2\theta + \qquad (8\text{-}20\text{-}10)$$

$$[2x^2(2M^4H+hj^2N^2)+4M^4H^3+6hH^2j^2N^2+2Hj^4(6h^2+B^2)+4hj^6]x\cos\theta +$$

$$M^4x^4+x^2[2M^4H^2+2hHj^2N^2+j^4(2h^2+C^2)]+M^4H^4+2hH^3j^2N^2+H^2j^4(6h^2+B^2)+4hHj^6+j^8 = 0$$

$$M^4 = (h^4+h^2B^2+D^4), \quad N^2 = (2h^2+B^2)$$

eq. for this group, obtained by inverting the central basis curves of eq. 1 about an axial pole h, followed by an x+H translation of the inverse loci. Formulated in standard form this becomes eq. 10, for which the $\cos^2\theta$-condition is eq. 11a.

By homology with the situation for limacons among axial QBI curves, the following question is asked. Which inverse loci comprise the most symmetrical axial quartic genus in each superfamily (based upon a given genus of central basis curve of eq. 1)? To answer this question, one examines the same focal condition which, for axial QBI curves, yields limacons. In other words, one determines the axial inverse loci for which the $\cos^2\theta$-condition defines coincident foci, i.e., double-foci.

(a) $H = \dfrac{-hj^2(2h^2+B^2)\pm j^2[h^4(C^2-B^2)+h^2(B^2C^2-4D^4)+D^4(C^2-B^2)]^{\frac{1}{2}}}{2(h^4+h^2B^2+D^4)}$ $\cos^2\theta$-condition of eq. 10 (8-20-11)

(b) $(CT) = h^4(C^2-B^2)+h^2(B^2C^2-4D^4)+D^4(C^2-B^2)$ the *Cartesian term* (radicand of eq. 11a)

(b') $h^2 = \dfrac{(4D^4-B^2C^2)\pm[(4D^4-B^2C^2)^2-4D^4(B^2-C^2)^2]^{\frac{1}{2}}}{2(C^2-B^2)}$ solution of *Cartesian condition*, for which radicand of eq. 11a vanishes

(b") $h_1h_2 = \sqrt{(D^4)}$ product of roots of *Cartesian condition* of different absolute values

(c) $h^2 = d^2/4$ (c') $h^2 = 4d^2(1/16-C^2)$* solution to eq. 11b' for Cassinians (parameter C of bipolar eq., $uv = Cd^2$)

Reference to eq. 11a reveals that the radicand of the $\cos^2\theta$-condition is simply the Cartesian term, namely 11b. Since the condition for a double-focus, i.e., the two foci being in coincidence, is the vanishing of the radicand, the poles of inversion to the most symmetrical curves are obtained--as in the case of the *limacon term* (8b)--by equating the Cartesian term to 0 and solving for h^2, yielding 11b' (a solution $h=0$ is not consistent with the Cartesian condition, for then either $D^4=0$, which is not allowed, or $B^2-C^2=0$, the latter being the condition for concentric circles).

It will be noted that the denominator of eq. 11b' may be made positive or negative merely by rotating the curves 90° (exchanging B^2 and C^2) and that the signs of the other terms do not depend upon the identity of the line of symmetry upon which the poles of inversion are incident. Accordingly, whether the poles of the Cartesian condition are real or complex depends only upon the sign of the radicand. It is readily ascertained that the existence of real Cartesian-condition poles depends upon having either $4D^4$ less than both B^4 and C^4 or $4D^4$ greater than both B^4 and C^4.

These conditions are fulfilled in all monovular and annular curves repre-

*[For monovular Cassinians, which also have 4 Cartesian-condition poles, the two poles on the orthogonal axis lie at $k^2 = 4d^2(C^2-1/16)$.]

sented by eq. 1. Since every genus of eq. 1 that possesses biovular species possesses an annular species, it follows that every genus of the curves of eq. 1 that possesses either or both monovular or biovular species inverts to a genus of Cartesian self-inverter (see Table 8-17-3 and Fig. 8-27).

Since the product of roots of different absolute magnitude that are defined by the Cartesian condition is $\sqrt{(D^4)}$ (eq. 11b"), the poles defined thereby always are inverse to one another for modular j (see eq. 3). In fact, these poles must be inverse to one another for modular j if all biovular Cartesian self-inverters are to be annular. Thus, under these conditions, if one set of poles is within opposed central ovals, the other set also will be within them, while if one set of poles is interior to both annular central ovals, the other set will be exterior to both.

Accordingly, for a given central basis genus (eq. 1) with parameters B^2, C^2, and D^4 [for Cassinians $D^4 = d^4(C^2-1/16)$, where C is the parameter of the bipolar eq., $uv = Cd^2$; see page II-7], inversions about poles h (for example, poles 2,3,7, and 8 of Figs. a_1-c_1), defined by eq. 11b', yield members of the most symmetrical genus of non-central quartic curves in the superfamily (for example, Figs. a_2,b_2, and d_2). Eq. 11b' must define all Cartesian self-inverters because it includes all genera with coincident $\cos^2\theta$-condition foci, and all Cartesian self-inverters (as defined by eq. 8-13-5) have this property.

Moreover, though eq. 11b' defines 4 real poles for biovular species (such as poles 2,3,7, and 8 of Fig. a_1), the inverse loci about these poles must be members of the same subspecies (i.e., the loci of Figs. a_2 and b_2, for example, must belong to the same subspecies).

[For example, there is only one parabolic Cartesian for any given value of the bipolar parameter C (Fig. a_2), and this locus is inverse to 3 different central curves--a Cassinian with the reciprocal value of the bipolar parameter C (Figs. a_1-c_1) and its central inversions to an annular central curve (Fig. d_1 and Fig. 8-22a) and a dissimilar opposed-oval central curve (Fig. 8-22f).]

Converting eq. 11b' for the poles in question to Cassinian parameters yields eqs. 11c,c' (poles 2 and 8 of Figs. a_1-c_1 are those of eq. 11c, while poles 3 and 7 of the same figures are specified by eq. 11c'). [It is of interest to note that in the ensemble of Fig. 8-19a, the traditional focus of the equilateral hyperbola is in coincidence with the focus of eq. 11c' for the $C = 0.2$ biovals.]

CIRCUMPOLAR

MAXIM 8-20-6: *On a given line of symmetry, the poles about which a given subspecies of central Cartesian inverts to a given subspecies of axial bicircular quartic or serpentine circular cubic occur in quartets. The poles in each quartet lie in modularly-inverse pairs on each side of center, i.e., if two poles lie at* $\pm nj$*, the other two lie at* $\pm j/n$*, where* n *can take on any real value except 0 or 1, and* j *is the modular value (the unit of linear dimension for self-inversion).*

8-20-7: *All axial inverse loci of biovular central Cartesians are obtained by inverting about poles on 3 loci: all axial poles from* $h=0$ *to* $h=j$*; all axial poles from* $k=0$ *to* $k=j$*; and all poles incident upon one quadrant of the modular circle.*

Maxim 8-20-7 follows from Maxim 8-20-6 but is formulated to emphasize one of the remarkable differences between the Inversion Taxonomy of (1) biovular Central Cartesians, and (2) central conics, central quartics, and non-self-inverting monovular central Cartesians. To obtain a given axial inversion ensemble of, for example, a central conic of given eccentricity, the central curve must be inverted about all poles on one of the lines of symmetry from the center to plus (or minus) infinity.

For biovular central Cartesians, the same goal is accomplished by inverting about all poles on an axial segment of finite length. In both cases the inverse locus at one limit is subspecifically identical with the basis curve (see Maxims 6-7-5 and 6-8-3) while at the other limit it is a different central curve.

[For monovular central Cartesians, the inverse locus at the center may be either a different, or merely a differently oriented but subspecifically identical, central curve (as with monovular Cassinians).]

We next derive the general eq. for axial bicircular quartics and serpentine circular cubics, cast about the focus of self-inversion that replaces the line of symmetry, by letting $H=-\frac{1}{2}j^2/h$ in eq. 10, yielding eq. 12. The latter is cast in the form of a $\cos\theta$ simple intercept format to facilitate comparisons.

general eq. of axial bicircular quartics and serpentine circular cubics obtained by letting $H=-\frac{1}{2}j^2/h$ in eq. 10 (8-20-12)

$$\cos\theta = \frac{(D^4-h^4)(4h^2x^2+j^4)\pm h[16h^4x^4(CT)+8h^2j^4r^2\{2h^2(C^4-4D^4)-(CT)\}+j^8(CT)]^{\frac{1}{2}}}{4h^3j^2r(h^2-C^2+D^4/h^2)}$$

general equation of Cartesian self-inverters obtained for the Cartesian condition

(a) $$\cos\theta = \frac{(D^4-h^4)(4h^2x^2+j^4)\pm 4h^3j^2x\sqrt{(C^4-4D^4)}}{4h^3j^2x(h^2-C^2+D^4/h^2)}, \quad D^4 \neq h^4 \qquad (8\text{-}20\text{-}13)$$

radicand of equation 13a expressed in terms of parameters E, F^2, and G of equation 8-13-5 for Cartesian self-inverters

(b) $$(C^4-4D^4) = E^2G^2j^8/F^4[(2F\pm G)^2-E^2]^2$$

Examination of eq. 12 confirms that the curves are Cartesian self-inverters only if the Cartesian term vanishes, because only then does the radicand become a perfect square (as it must for the $\cos^2\theta$-condition) and the format become that of Cartesian self-inverters (eq. 8-13-5). Eq. 12 also serves to emphasize the homology of the Cartesian condition (for inversions of 4th-degree central curves to bicircular quartics and serpentine circular cubics cast about a focus of self-inversion) with the *limacon condition* for the simple intercept format of axial inversions of central conics to QBI quartics (eqs. XII-16b and 8-20-8b).

In both cases, the vanishing of the term in question leads to the vanishing of the 4th-degree term of the radicand and the simplification of the 2nd-degree term. In the case of bicircular quartics, the constant term of the radicand also vanishes, whereas for axial inversions of central conics, the constant term only simplifies. The homology is complete, however, if the comparison is made with the general simple intercept format of bicircular quartics (eq. 14, below), rather than with the format cast about a focus of self-inversion.

The condition $h^4 = D^4$ leads to the vanishing of the external term in the numerator. This condition, in fact, is the condition for inversion of the loci of eq. 1 to other central loci of the same or complementary type. When the external term of the numerator vanishes, squaring the eq. to eliminate the radical leads to an eq. of the same form as that for the basis central curves of eq. 1.

If now, the Cartesian term of eq. 12 is set equal to 0, eq. 13a results. When terms in h are replaced by the solutions of eq. 11b', this gives the general eq. for Cartesian self-inverters in terms of B^2, C^2, and D^4, the parameters of the central curves of eq. 1. It readily is ascertained that the radicand of eq. 13 always is positive for Cartesian self-inverters, since conversion to the parameters of eq. 8-13-5 yields the expression of eq. 13b (recall that $C^4 = 4D^4$ is a condition for congruent circles). In eq. 13a, h^4 may not equal D^4, because this condition, together with the Cartesian condition, is the condition for congruent intersecting circles (mutually-inverting quartic circles) inverted about a point of intersection to give intersecting lines. In other words, these conditions are inconsistent with one another.

Accordingly, it has been ascertained that the bicircular quartic species with the highest degree of circumpolar symmetry belong to genera of the supergenus *Cartesian self-inverters*. These species occupy the homologous position in central Cartesian-based inversion superfamilies that limacons occupy among non-central genera of QBI quartics.

The Poles of The Cartesian Condition

The inquiry now is made as to when the poles of inversion of the basis central genera that yield Cartesian self-inverters (eq. 11b') are coincident with the foci of the $\cos^2\theta$-condition. To answer this question one need merely solve $h^2 = (C^2-B^2)/4$ --the $\cos^2\theta$-condition of the simple intercept eq. 6a for the basis central curves--simultaneously with eq. 8 for the vanishing of the Cartesian term. This leads to the result, $B^2 = -C^2$, for which the only real loci are Cassinians. Accordingly, only in Cassinians are two of the poles for inversion to Cartesian self-inverters (namely, parabolic Cartesians) coincident with the foci specified by the $\cos^2\theta$-condition.

Since the only other circumpolar axial point-foci of central Cartesians are the center and the vertices, it is evident that none of the poles about which other central Cartesians invert to Cartesian self-inverters is an angle-independent circumpolar point-focus.

[The condition $B^2 = -C^2$ never applies to annular central Cartesians, since both B^2 and C^2 must be negative if 4 real vertices are to exist on both axes.]

The Foci of Self-Inversion of Axial Bicircular Quartics And Serpentine Circular Cubics

The next query concerns the conditions under which the foci of self-inversion of bicircular quartics (Fig. a_2, poles 2,3,7; Fig. b_2, poles 2,3,8; Fig. c_2, poles $2\frac{1}{4}, 2\frac{1}{2}, 7\frac{1}{2}$; Fig. d_2, poles 1,4,9) and serpentine circular cubics (Figs. 8-18m,n,p; 8-22b,d; 8-31j; 8-32c,h) are generic double-foci. In other words, under what conditions are these poles angle-independent foci specified by conditions on simple intercept eqs. and formats? To answer this question, the simple intercept format of eqs. 14 is obtained from eq. 10 by solving for $\cos\theta$.

general solution for the simple intercept format of inverse loci of eq. 1 about a pole $x = h$, recast about a pole $x = H$

(a) (8-20-14)

$$\cos\theta = \frac{-[x^2\{4A^4H+2hj^2M^2\}+\{4A^4H^3 + 6hH^2j^2M^2+2Hj^4(6h^2+B^2)+4hj^6\}]\pm\sqrt{[\text{radicand}]}}{x[4A^4H^2+4hHj^2M^2+j^4(4h^2+B^2)-C^2j^4]}$$

$$A^4 = (h^4+h^2B^2+D^4), \quad M^2 = (2h^2+B^2)$$

(b) radicand of equation 14a

$$4j^4x^4(CT)+4j^4x^2[-2H^2(CT)-2hHj^2\{2h^2(C^2-B^2)+(B^2C^2-4D^4)\}-j^4(C^2-B^2)(2h^2-C^2)]$$
$$+ 4j^4(H^4(CT)+2hH^3j^2\{2h^2(C^2-B^2)+(B^2C^2-4D^4)\}+H^2j^4\{6h^2(C^2-B^2)+(B^2C^2-4D^4)\} +$$
$$4hHj^6(C^2-B^2) + j^8(C^2-B^2))$$

radicand 14b for $(CT) = 0$

(c) $$4j^4x^2[-2hHj^2\{2h^2(C^2-B^2)+(B^2C^2-4D^4)\}-j^4(C^2-B^2)(2h^2-C^2)] +$$
$$4j^4(2hH^3j^2\{2h^2(C^2-B^2)+(B^2C^2-4D^4)\}+H^2j^4\{6h^2(C^2-B^2)+(B^2C^2-4D^4)\}+$$
$$4hHj^6(C^2-B^2) + j^8(C^2-B^2))$$

It is evident from inspection of eq. 14a that any radicand condition for which the radical vanishes, leaving a term in x and no term of greater degree than x^2, will be focal and will yield a Cartesian self-inverter. Accordingly, the only possibility for a focal condition on the radicand that does not yield a Cartesian self-inverter is for the radical to vanish, leaving no term in x. Since this would leave the simple intercept format of a circle, and since the foci of self-inversion of non-central curves are not known to be specified by conditions other than radicand conditions, it is evident that the only foci of self-inversion that could be specified by radicand conditions on eq. 14a are those of Cartesian self-inverters. This constitutes yet another parallel between the properties of these curves and the properties of limacons.

This situation is confirmed by examining the radicand 14b. The 4th-degree term, the H^2 portion of the x^2 term, and the H^4 portion of the constant term all vanish for the Cartesian condition (setting the Cartesian term equal to 0), leaving 14c. Vanishing of both the x^2 and constant terms of 14c (the condition for which is $C^4 = 4D^4$) or of the x^2 terms alone cannot be focal conditions for the foci of self-inversion, as either of these would leave the format for a circle.

The only possible focal condition (other than the entire radicand 14b being a perfect square, for the $\cos^2\theta$-condition) is the vanishing of the constant term (eq. 15a), which is a polynomial in H^3, H^2, H, and a constant. The cubic eq., 15a, is solved readily since the location of one of the foci of self-inversion already is known (the focus of self-inversion that replaces the line of symmetry lies at $H = -\frac{1}{2}j^2/h$).

radicand constant-condition of 14b for (CT) = 0

(a) $$2hH^3[2h^2(C^2-B^2)+(B^2C^2-4D^4)] + H^2j^2[6h^2(C^2-B^2)+(B^2C^2-4D^4)] + 4hHj^4(C^2-B^2) + j^6(C^2-B^2) = 0 \qquad (8\text{-}20\text{-}15)$$

locations of the foci of self-inversion of Cartesian self-inverters, where h is defined by equation 11b'

(b) $$H = -\tfrac{1}{2}j^2/h, \quad \frac{hj^2(B^2-C^2)\pm[(B^2-C^2)(B^2C^2-4D^4)-h^2(B^2-C^2)^2]^{\frac{1}{2}}j^2}{[(B^2C^2-4D^4)-2h^2(B^2-C^2)]}$$

15b for parabolic Cartesians, where $h = \frac{1}{2}d$ (the $\cos^2\theta$-condition focus) and C is the Cassinian C of the bi-polar eq., $uv = Cd^2$

(c) $$H = -\tfrac{1}{2}j^2/2h, \quad \frac{-hj^2\pm j^2[(\frac{1}{2}d^2-h^2-4C^2d^2)]^{\frac{1}{2}}}{(2h^2-\frac{1}{2}d^2+4C^2d^2)}$$

(a) $j_1^2 = j^4/4h^2$

(b) $j_2^2 = \dfrac{A^3j^4(4D^4-B^2C^2-2A^3h)}{h[(B^2C^2-4D^4)-2h^2(B^2-C^2)]^2}$

the squares of the moduli for self-inversion for two of the foci of 15b (8-20-16)

Accordingly, the locations of the 3 foci of self-inversion for Cartesian self-inverters are given by eqs. 15b. Those of parabolic Cartesians (Figs. a_2,b_2, and d_2) are obtained by substituting for B^2, C^2, and D^4 their values from the eq. for Cassinians, $(x^2+y^2)^2-d^2(x^2-y^2)/2=d^4(C^2-1/16)$, since Cassinians (Figs. a_1-c_1) belong to the central basis genus for parabolic Cartesians (Fig. d_1 is a subspecies of the annular basis species).
[There exist 3 central biovular basis species for biovular parabolic Cartesians (see Fig. 8-22). However, a monovular basis species and curves of demarcation exist only for Cassinian ovals (see Fig. 8-19 and Table 8-17-2).]

Knowing the locations of the 3 foci of self-inversion, one readily derives the squares of the moduli for self-inversion about each focus, since the product of the distances from any given focus to the other two is the square of its modulus for self-inversion. In other words, in the self-inversion of the loci about any of the 3 foci of self-inversion, the other two foci of self-inversion are inverse to one another (see Maxim 8-15-6). Two examples of these squared moduli, derived from 15b, are given by eqs. 16a,b.

In summary, virtually all the basic circumpolar symmetry properties of Cartesian self-inverters in relation to the basis central Cartesians are found to parallel those of limacons in relation to central conics, placing Cartesian self-inverters in the position of being the quartic non-central genera with the highest degree of circumpolar symmetry in their respective superfamilies.

1. They are the only genera in which the foci defined by the $\cos^2\theta$-condition are generic double-foci, i.e., the two foci are in coincidence in all subspecies of each genus.

2. They are the only genera for which the x^4 term vanishes from the radicand of the simple intercept format.

3. They are the only genera for which there is great simplification of the x^2 and constant terms of the radicand of the simple intercept format.

4. They are the only genera in which the foci of self-inversion are specified by radicand conditions, specifically, radicand-constant conditions.

5. They are the only genera in which the foci of self-inversion are generic double-foci, i.e., these foci are angle-independent circumpolar point foci as well as being foci of self-inversion.

Correspondences Between Axial Poles In Axial Inversions To & From Central Parabolic Cartesians

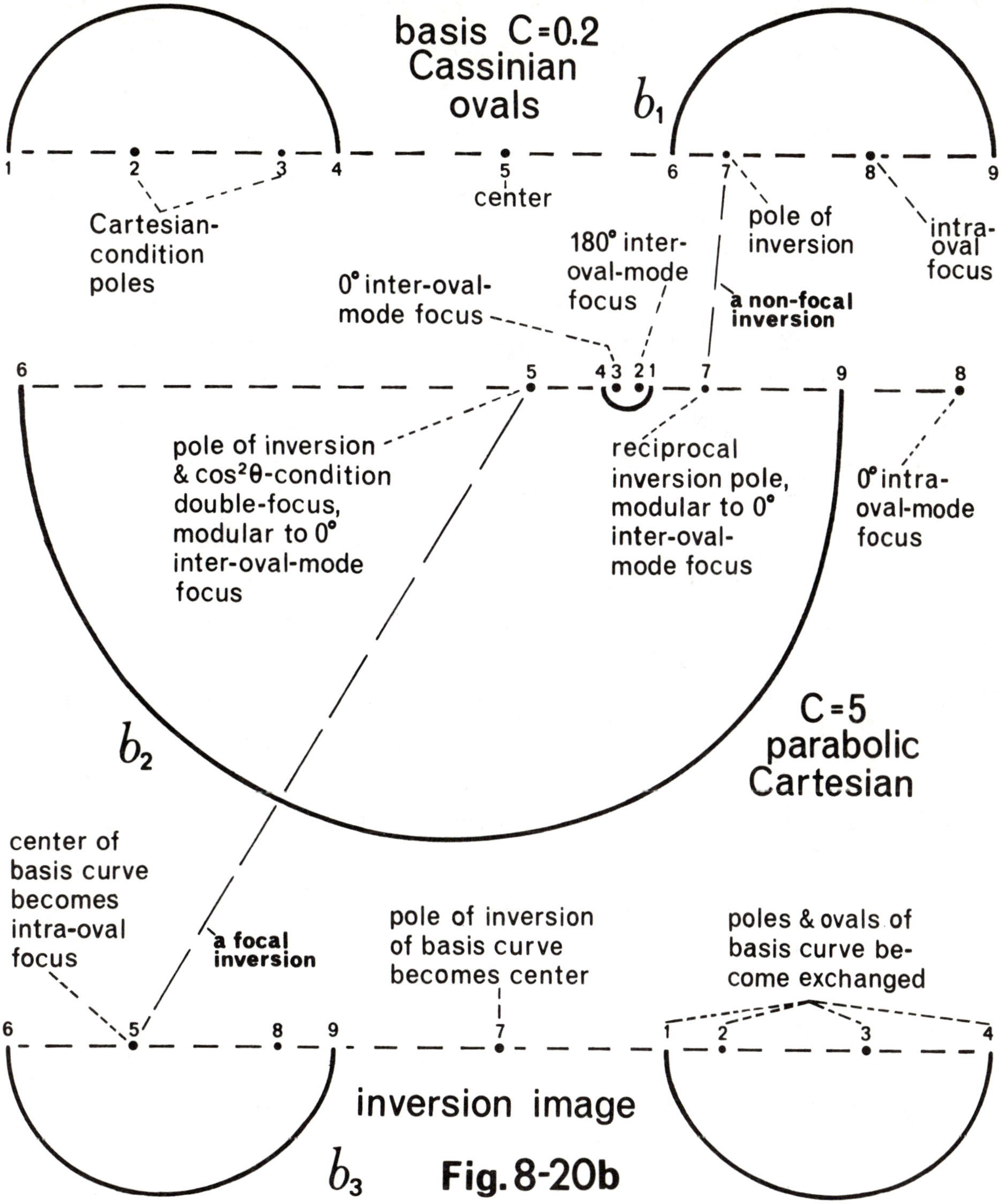

Fig. 8-20b

6. They are the only genera in which the cosθ simple intercept formats are expressible in polynomial form.

In consequence of the above findings, though other axial bicircular quartics and serpentine circular cubics also self-invert about 3 different poles, these poles generally are not angle-independent circumpolar foci, just as the poles for inversion of these curves to and from the basis central Cartesians generally are not angle-independent circumpolar foci (not even for inversions to and from Cartesian self-inverters).

Axial Inverse Loci of Central Cartesians And Recovery of Inversion Images

The figures now are considered individually. The $C = 0.2$ Cassinian of Fig. a_1 is inverted about the right intra-oval focus (pole 8), the reciprocal inversion pole of which in the inverse $C = 5$ parabolic Cartesian is the $\cos^2\theta$-condition double-focus (for all inversions in Figs. a-d the unit of linear dimension is taken to be the modulus for self-inversion of the curve about the pertinent focus). The pole inverse to the reflection of the pole of inversion in the line of symmetry (i.e., the pole inverse to pole 2) becomes the 0° inter-oval-mode focus of self-inversion.

In the self-inversion of the basis curve a_1 about the center, the non-focal poles 3 and 7--defined by the Cartesian condition--are inverse to the focal poles 2 and 8, which are defined by both the Cartesian condition and the $\cos^2\theta$-condition. The poles inverse to poles 3 and 7, when the basis curve is inverted about pole 8, are the 180° inter-oval-mode and 0° intra-oval-mode foci of self-inversion of the parabolic Cartesian of Fig. a_2 (poles 3 and 7). The center (pole 5), on the other hand, inverts to a non-focal location in the inverse curve; it becomes the reflection of the $\cos^2\theta$-condition focus in the 0° inter-oval-mode focus. The vertices, of course, invert to one another.

A locus belonging to the same subspecies as the basis curve (an *inversion image*) is recovered by inverting the parabolic Cartesian of Fig. a_2 about the aforementioned reflection of the reciprocal inversion pole in the 0° inter-oval-mode focus of self-inversion (the focus of self-inversion that replaces the line of symmetry). Although the basis curve and its *inversion image* are depicted as being congruent in the figure, in fact, a linear reduction by a factor of ¼ occurs (Maxim 8-20-14).

Correspondences Between Axial Poles In Axial Inversions To & From Central Parabolic Cartesians

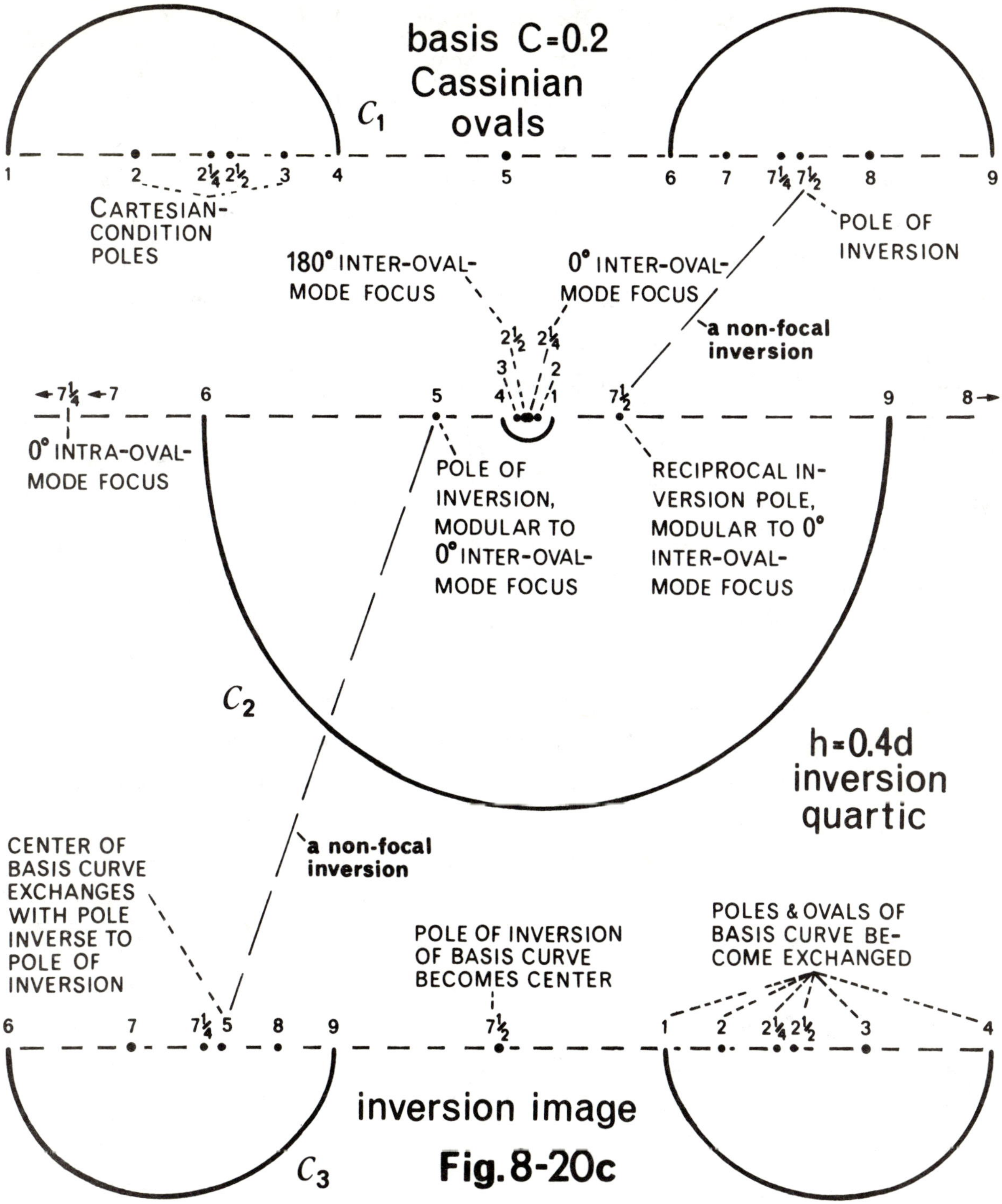

Fig. 8-20c

The positions of the poles in the inversion image are related systematically to their positions in the basis curve. Thus, both the vertices and the two intra-oval poles exchange positions within each oval (as do other inverse pole-pairs), and the ovals also exchange positions. The only exception is that the pole of inversion of the basis curve (pole 8) comes to lie at the center of the inversion image (rather than exchanging positions with pole 7). Within the corresponding oval of the inversion image this pole (pole 8 of the basis curve) is replaced by pole 5--the center--of the basis curve.

Corresponding exchanges occur in the inversions depicted in Figs. b and d, both of which involve initial inversions about poles defined by the Cartesian condition. The exchanges of Fig. c, on the other hand, are somewhat modified, inasmuch as the initial inversion of this figure is about an unexceptional pole. [All of the inversion exchanges, however, follow the same rules, which are listed at the end of this section.]

In addition to poles 5 and 8, there are two other poles on the line of symmetry (as well as two other poles not incident upon it; see Fig. 8-22) about which the parabolic Cartesian of Fig. a_2 inverts to central curves, namely, unmarked poles on each side of the 0° intra-oval-mode focus at modular distance from it ($j = 3.872983345d$). The annular central curve obtained by inverting about either of these poles is the locus of Figs. d_1 and d_3. The poles of inversion of the basis central curve a_1 to this annular central curve lie at $h = \pm D$, namely at $h = \pm 0.387298335d$. Since these poles are only $0.012701665d$ units from the pole of inversion employed in Fig. c_1, the locus of Fig. c_2 more resembles a central curve than do the loci of Figs. a_2 and b_2 (both of which belong to the $C = 5$ parabolic Cartesian subspecies).

The pole of inversion of the basis curve of Fig. b_1 is the non-focal Cartesian condition pole 7, which is modularly inverse to pole 8 about the center. In this case, also, the other 3 intra-oval poles, 2,3, and 8, become the foci of self-inversion of the inverse $C = 5$ parabolic Cartesian. The orientation of this locus, however, is reversed (from that in Fig. a_2), and the roles of the inverse poles 2 and 3 are exchanged. Pole 3 now is the 0° inter-oval-mode focus and pole 2 is the 180° inter-oval-mode focus.

Since the pole of inversion (pole 7) is not defined by the $\cos^2\theta$-condition, it does not invert to the $\cos^2\theta$-condition focus of the inverse curve. Instead, the reciprocal inversion pole is the reflection of this focus

Correspondences Between Axial Poles In Axial Inversions To & From Central Parabolic Cartesians

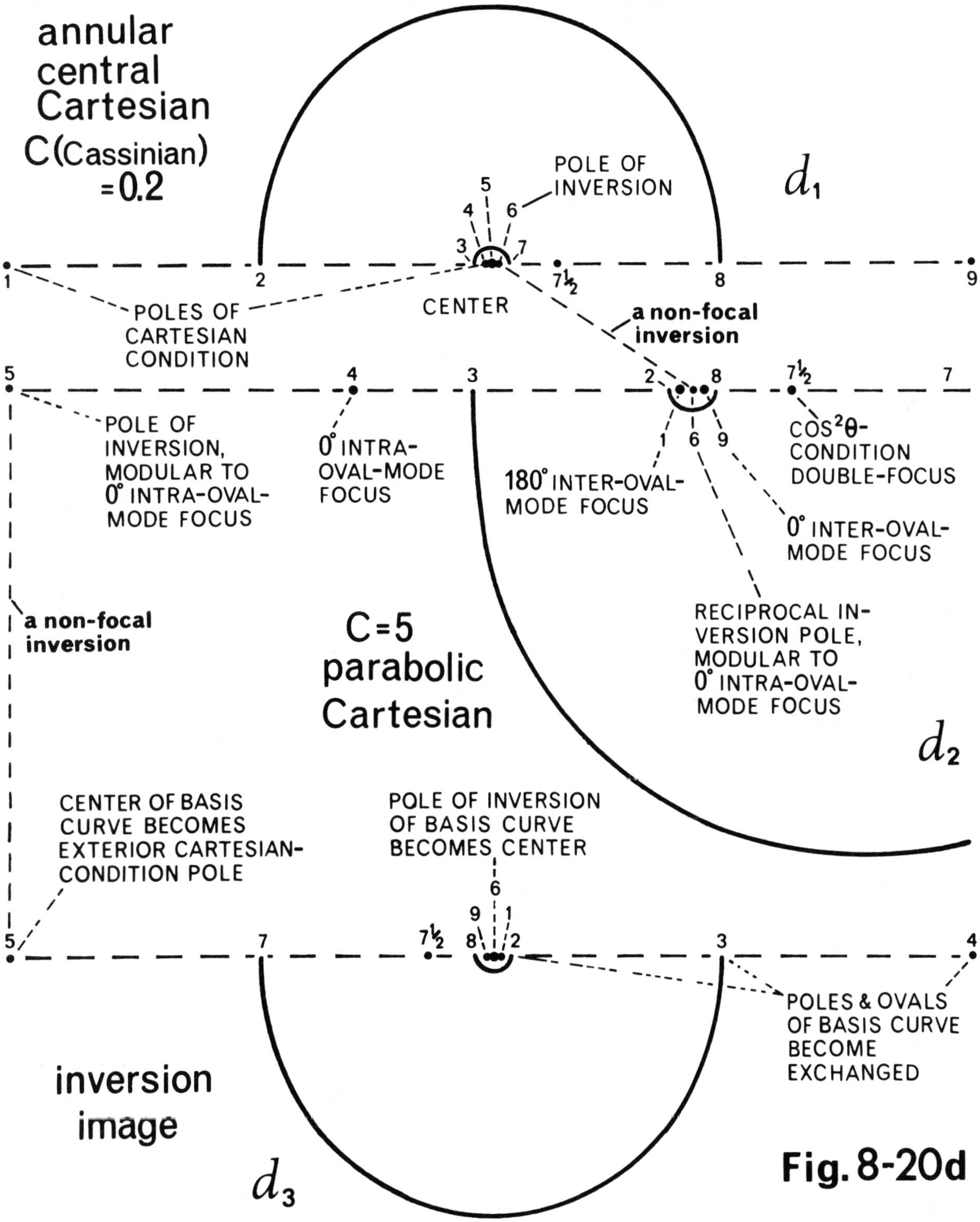

Fig. 8-20d

in the 0° inter-oval-mode focus. The relationships in the inversion image--again reduced by a factor of ¼--follow the pattern described for Fig. a. The sequence of poles from left to right now is 6,5,8,9,7(center),1,2,3,4, instead of 6,5,7,9,8(center),1,2,3,4, reflecting the altered status of poles 7 and 8 (as a consequence of which pole 5 replaces pole 7 in Fig. b_3 rather than pole 8, as in Fig. a_3).

The same operations are followed in Fig. c, but now the pole of inversion (pole 7½) is chosen arbitrarily to be midway between poles 7 and 8. The chief consequences of this change are that the inverse curve is not a Cartesian self-inverter, its $\cos^2\theta$-condition foci are imaginary (see Fig. 8-21), and none of its foci of self-inversion (poles 2¼,2½, and 7¼ of Fig. c_2) is an angle-independent circumpolar focus (i.e., none of these foci is specified by a radicand condition on the simple intercept format). Nor are these foci inverse to any of the intra-oval poles, 2,3,7, or 8, of Fig. c_1 as they are in Figs. a_2 and b_2. Instead, the foci of self-inversion are inverse to unexceptional poles of the basis curve and its inversion image (poles 2¼,2½, and 7¼ of Fig. c_1)

As in the cases of Figs. a and b, the inversion image is ¼ the size of the basis curve (which is a characteristic property of this inversion sequence), the pole of inversion of the basis curve (pole 7½) becomes the center of its inversion image, the intra-oval poles exchange relative positions, and the ovals exchange positions (sequence 6,7,8,9,7½,1,2,3,4, as opposed to 1,2,3,4,5,6,7,8,9 in the basis curve).

The basis curve of Fig. d_1 is the annular central Cartesian [eq., $(x^2+y^2)^2-70d^2x^2/3-30d^2y^2+25d^4/9=0$; derived in the legend of Fig. 8-31] obtained by inverting the $C=0.2$ Cassinian of Figs. a_1-c_1 about poles at $h=\pm D=\pm(1/16-C^2)^{\frac{1}{4}}d$. The 4 poles for inversion to Cartesian self-inverters are interior and exterior to both ovals at distances specified by the Cartesian condition. Unlike the situation for the basis Cassinian ovals of Figs. a_1-c_1, none of these poles has circumpolar point-focal rank, inasmuch as the $\cos^2\theta$-condition foci on its x-axis are imaginary.

[This is characteristic of the x-axis $\cos^2\theta$-condition foci of annular central parabolic Cartesians derived by transverse-axis inversions of Cassinians (since the $\cos^2\theta$-condition foci lie on the y-axis, their x-axis locations are imaginary). On the other hand, the opposed-oval central curves derived by inversions of Cassinians about the modular poles on the y-axis possess $\cos^2\theta$-condition foci on the x-axis (see Figs. 8-22 and 8-23).]

Since the pole of inversion of Fig. c_1 is one of those specified by the Cartesian condition, the poles inverse to the other 3 poles are the foci of self-inversion of the inverse $C = 5$ parabolic Cartesian, as in Figs. a and b (Maxim 8-22-11). The reciprocal inversion pole (pole 6) occupies an unexceptional location in the small oval of the parabolic Cartesian, and the $\cos^2\theta$-condition double-focus of the latter (pole $7\frac{1}{2}$) corresponds to an unexceptional inverse pole (pole $7\frac{1}{2}$) of the basis curve. The poles for inversion of the basis curve to the $C = 0.2$ Cassinian are inter-oval, lying at $h = \pm D = (25/9)^{\frac{1}{4}}d = \pm 1.2909945d$.

The locations of thé poles in the inversion image follow the pattern already noted in Figs. a-c. The pairs of poles specified by the Cartesian condition exchange locations and the image itself is reflected in the orthogonal line of symmetry. In addition, the inversion pole of the basis curve becomes the center of the inversion image and the center of the basis curve takes the position in the inversion image of one of the poles specified by the Cartesian condition, namely the one employed as the pole of inversion.

In summary, in the axial inversions of basis central Cartesians to bicircular quartics, and recovery of inversion images by inverting the bicircular quartics about an axial pole inverse to the center of the basis curve, the following exchanges occur.

In the 1st-generation axial bicircular quartic:

1. For opposed-oval basis curves, the pole of the basis curve that is modularly inverse to the pole of inversion inverts to the 0° intra-oval-mode focus of self-inversion of the inverse locus. For annular basis curves, it inverts to the 0° inter-oval-mode focus (see Maxims 8-30-6 and 10).
2. For both annular and opposed-oval basis curves, the reflection (in the line of symmetry) of the pole of the basis curve that is modularly inverse to the pole of inversion inverts to the 180° inter-oval-mode focus (Maxim 8-30-6).
3. For opposed-oval basis curves, the reflection of the pole of inversion of the basis curve in the line of symmetry inverts to the 0° inter-oval-mode focus. For annular basis curves, it inverts to the 0° intra-oval-mode focus (Maxim 8-30-9).
4. For opposed-oval basis curves, the 1st-generation pole inverse to the center of the basis curve, and the 1st-generation reciprocal inversion pole are reflections of one another in the 0° inter-oval-mode focus, and are the two poles about which inversion yields the basis curve or its 2nd-generation inversion image (see Maxim 8-20-9).
5. If the pole of inversion of the basis curve is focal (a $\cos^2\theta$-condition focus or a vertex), the reciprocal inversion pole of the 1st-generation bicircular quartic also is focal, and both poles are defined by the same types of focal conditions.

In the inversion image of the central Cartesian:

1. Ovals of the basis curve become exchanged.
2. The vertices of the exchanged ovals become exchanged.
3. Modularly-inverse poles of the basis curve become exchanged (excluding the participation of the pole of inversion of the basis curve).
4. The center of the basis curve comes to occupy a position in the inversion image that is modularly inverse to the location of the pole of inversion in the basis curve, but on the opposite side of center.
5. The pole of inversion of the basis curve becomes the center of the inversion image.
6. For modular inversions, the inversion image becomes reduced to ¼ the linear dimensions of the basis curve (Maxim 8-20-14).

The inversions depicted in Figs. a and b have implications for Circumpolar Maxim 32, which was formulated exclusively on the basis of findings for QBI curves. This Maxim asserts that:

Excluding self-inversions and inversions about true centers, inversion loci about focal poles are distinguished from those about non-focal poles by the increase or appearance (or both) of: (a) the focal multiplicity of the reciprocal inversion pole; (b) generic multiple-foci; (c) foci at infinity; or (d) an orthogonal line of symmetry.

The inversions of Figs. 20a and 20b provide two instances in which two basis curves belonging to the same subspecies are inverted to two inverse curves belonging to the same subspecies, one by a focal inversion and the other by a non-focal inversion. Since the basis and inverse loci in each set belong to the same subspecies, the applicability of Maxim 32 to this situation can be evaluated only in relation to the focal multiplicity of the inversion poles and reciprocal inversion poles.

In one case, the focal multiplicity of the reciprocal inversion pole is greater than that of the pole of inversion for the focal inversion (Fig. a_1 to Fig. a_2) but unchanged for the non-focal inversion (Fig. b_1 to Fig. b_2), consonant with Maxim 32. In the other case, however, the focal multiplicity of the reciprocal inversion pole is less than that of the pole of inversion for the focal inversion (Fig. b_2 to b_3) but unchanged for the non-focal inversion (Fig. a_2 to a_3).

Accordingly, it is evident that the domain of applicability of Maxim 32 must be restricted further by excluding inversions between pairs of inverse loci that can invert to one another about more than one non-equivalent pole (i.e., about equivalent inversion poles that are not also equivalent poles).

CIRCUMPOLAR MAXIMS 8-20-

8: *Axial serpentine circular cubics and axial bicircular quartics invert to two subspecies of biovular central Cartesians about the members of two pole-pairs on their (single) lines of symmetry; they invert to one subspecies about the members of one pole-pair, and to the second subspecies about the members of the other pole-pair.*

9: *An axial bicircular quartic with a non-intersecting line of symmetry possesses two axial 0° intra-oval-mode foci of self-inversion; the inverse loci about the axial modular poles of these foci belong to different subspecies of opposed-oval central Cartesians, one subspecies for the poles of each focus.*

10: *An axial serpentine circular cubic or axial bicircular quartic with an intersecting line of symmetry inverts to a single subspecies of axial-vertex serpentine circular cubic about its vertices, to a single subspecies of opposed-oval central Cartesian about the two axial poles at modular distance from the 0° inter-oval-mode focus of self-inversion, and to a single subspecies of annular central Cartesian about the two axial poles at modular distance from the 0° intra-oval-mode focus of self-inversion.*

11: *Axial inverse loci of a given biovular central Cartesian possess only two axial poles about which inversions yield curves that are congruent with one another for a common value of the unit of linear dimension. These are the poles at modular distance from the 180° inter-oval-mode focus of self-inversion.*

12: *An axial serpentine circular cubic and an axial bicircular quartic whose line of symmetry intersects the curve possess 5 axial multimember sets of exceptional equivalent inversion poles. For any given basis curve, all of the inverse loci about the poles of any given set are members of the same subspecies. Three of these sets consist of two poles each, and the pole-pairs lie at modular distances from the foci of self-inversion; a fourth set consists of the 3 foci of self-inversion, and the fifth set consists of the 3 or 4 vertices. An axial bicircular quartic whose line of symmetry does not intersect the curve has only 4 such exceptional multimember sets of equivalent inversion poles, since it possesses no axial vertices.*

13: *There are 4 generally non-focal poles about which biovular central Cartesians invert to Cartesian self-inverters. For annular central curves these lie on a line of symmetry in pairs interior and exterior to both ovals. For opposed-oval central curves they are located in pairs within each oval; in the members of one opposed-oval species they lie on the line of symmetry that intersects the curve, while in the members of the other species they are concyclic.*

14: *The inversion image obtained by first inverting a biovular central Cartesian about an axial pole for modular j (j for self-inversion about the center), and then inverting the resulting axial bicircular quartic or serpentine circular cubic about the pole inverse to the center of the basis curve, also for modular j (j for self-inversion about the focus of self-inversion that replaces the line of symmetry), is ¼ the linear size of the basis curve.*

[In connection with Maxims 8-20-10 and 11, the different results obtained by inverting about axial poles modular to 0° and 180° foci of self-inversion can be understood in terms of Maxims 8-15-2,3, and 8-22-21. A pole of inversion modular to a 0° focus of self-inversion is coincident with its modularly-inverse pole in the 0° focus, whereas this is not the case for a pole of inversion modular to a 180° focus. As a consequence, inversion transfers the 0° focus to the ideal line at infinity, with the concomitant gain of a line of symmetry, whereas the inverse of the pole modularly inverse to the pole of inversion at 180° always remains in the finite plane.]

Fig. 8-21. Tracking the axial foci of 0° axial inversions of the $C = 0.2$ ovals of Cassini. This figure complements Figs. 20a-c. The loci of the latter figures are larger (but not to the same scale), they include inversion images, and pertinent axial poles are identified individually. On the other hand, only 3 levels of axial inverse loci of the basis Cassinian ovals are depicted. The present figure gives the locations of all pertinent axial poles, represents all loci to the same scale, and represents 30 levels of axial inversions for modular j. It does not, however, include inversion images.

The advantages of the present figure are that: (a) it depicts the nature of the inverse loci for a continuous sequence of inversion poles progressing from the center to the neighborhood of the far vertices; and (b) it enables one to track the corresponding poles from one inverse locus to the next at increments of only 0.02d (where d is the distance between the $\cos^2\theta$-condition foci, i.e., the outer poles of the Cartesian condition).

[By taking the orthogonal line of symmetry of the basis curve as the reference line (vertical line consisting of short light dashes), the distances of the axial poles of inversion from the center of the basis curve readily are apparent, since they are the same as the distances of the reciprocal inversion poles from the reference line.]

At each of the 30 levels, which are numbered in the left margin, the location of each pole is marked by individual identification symbols:

x's.....................................poles inverse to the center
squares.................................0° intra-oval-mode foci
triangles...............................180° inter-oval-mode foci
open circles............................0° inter-oval-mode foci
small solid circles.....................$\cos^2\theta$-condition foci
large solid circles.....................reciprocal inversion poles
heavy tracings of long dashes...............vertices
a light tracing of short dashes.............center reference line

In cases of poles that recede to or approach from the ideal line at infinity, the locations can be shown only in the limited regions in which they lie within the confines of the figure. In addition, the levels at which inverse loci are exceptional are marked and the loci are identified.

Tracking The Axial Foci of 0° Axial Inversions of The C=0.2 Ovals of Cassini

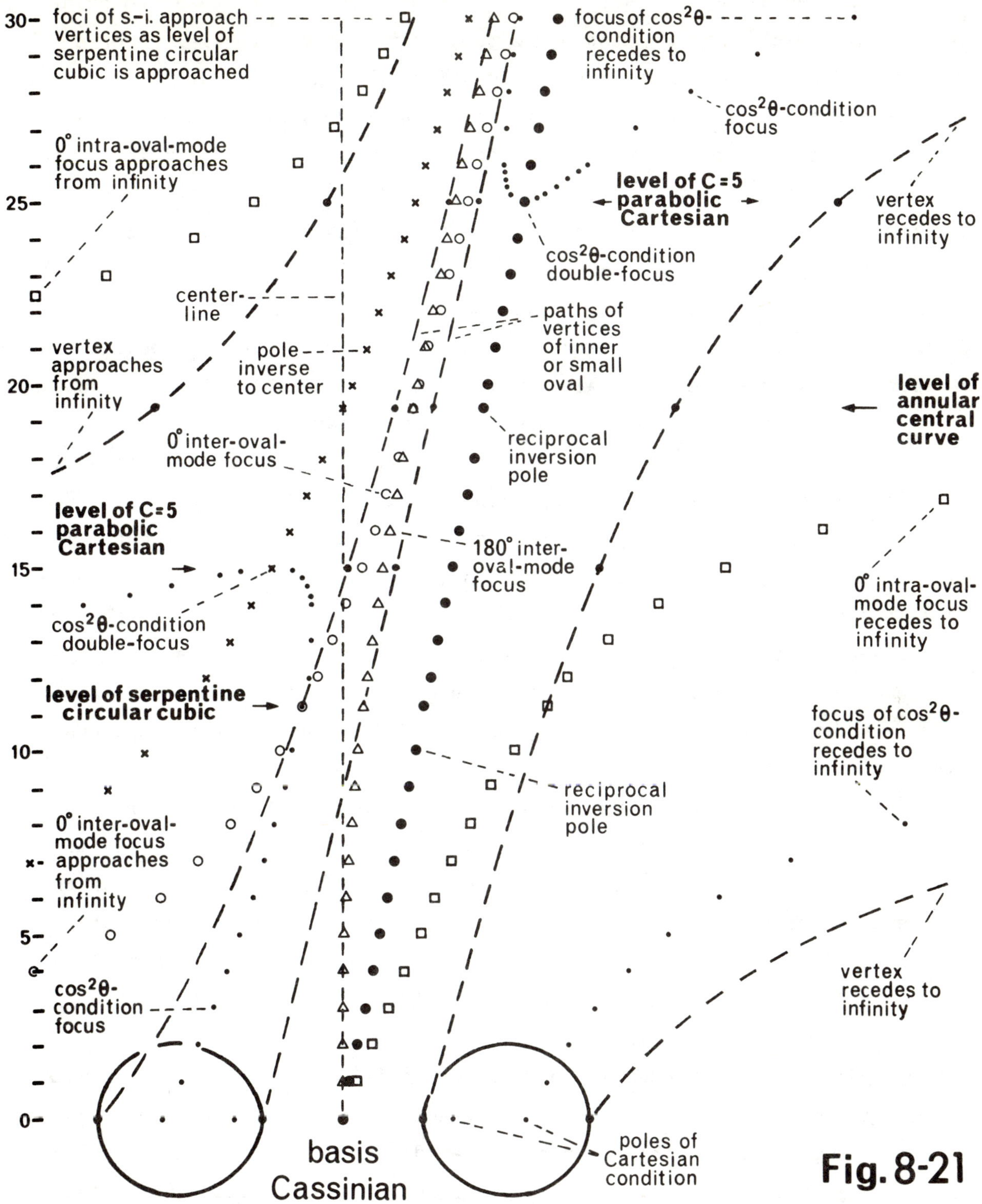

Fig. 8-21

Beginning at level 0, for self-inversion about the center, the vertices of each oval exchange positions. Similar exchanges occur for the two intra-oval poles specified by the Cartesian condition and for all other points that are modularly inverse to one another. As the pole of inversion is taken to be at locations progressively more distant to the right of center, axial inverse points depart progressively from their locations in the coincident inverse of the basis curve. The closer a point of the basis curve is to a pole of inversion, the more distant is the corresponding inverse point. Accordingly, for a central inversion, the point inverse to the center (marked by x's at near locations) is the most distant inverse pole (the ideal line at infinity); among exceptional points, the point inverse to the right near vertex (rightmost heavy dashed line) is next most distant.

Coincidence of the pole of inversion with the near right vertex leads to formation of the first exceptional dissimilar inversion locus, which is a serpentine circular cubic designated as the *Cassinian axial-vertex cubic*. This is the first position at which the 3 foci of self-inversion of the inverse locus (in the finite plane) achieve angle-independent circumpolar focal rank.

For a given value of C, only a single subspecies of axial-vertex cubic is formed by inverting any axial inverse locus of a member of the Cassinian-based inversion superfamily about the corresponding axial vertices; only a total of two axial-vertex inversion subspecies exist for all axial vertices (see Fig. 8-22). Progressive approximation of the pole of inversion to the near right vertex is accompanied by the following progressive changes in the locations of inverse axial poles.

The 180° inter-oval-mode focus, initially coincident with the center, departs very little from the centerline at first but becomes progressively more proximate to the near left vertex (the inverse of the far left vertex) as the latter approaches the centerline. At the level of the serpentine circular cubic ($x = 0.2236067978d$), just beyond level 11 ($x = 0.22d$), these two poles come into coincidence and the inverse locus self-inverts about the double-focus so formed in the 180° inter-oval mode.

Accompanying these progressive changes, the position of the far left vertex moves progressively to the right, as does the position of the left $\cos^2\theta$-condition focus, and that of the 0° inter-oval-mode focus, which is approaching from infinity at the left. These 3 poles also converge into coincidence just beyond level 11 in the serpentine circular cubic, forming a triple-focus at the far vertex of the oval, about which the curve self-inverts in the 0° inter-oval mode (the oval and the serpentine arm invert to one another).

The pole inverse to the center of the basis curve does not have focal rank in dissimilar inverse curves. This pole lies at the same distance to the left of the 0° inter-oval-mode focus as the reciprocal inversion pole lies to its right, and it is the pole about which each inverse locus inverts to an inversion image of the basis curve.

Both the far right vertex and the $\cos^2\theta$-condition intra-oval focus of the right oval recede to infinity as the pole of inversion approaches the near right vertex. At the same time, both the 0° intra-oval-mode focus and the near right vertex recede increasingly from the centerline. The former recedes at a greater rate than the latter, with the result that these two poles come into coincidence in the serpentine circular cubic, forming a generic double-focus at the vertex of the serpentine arm, about which both segments of the curve self-invert in the 0° intra-oval mode.

[Plots of the inverse positions of the vertices and the center were derived by modular inversion of these poles about the pole of inversion. The positions of the foci of self-inversion were derived making use of the relationships embodied in Maxims 8-22-11 and 8-30-6. The locations of the $\cos^2\theta$-condition foci were derived from eq. 8-20-11.]

[Note that circumpolar foci generally do not invert to circumpolar foci. For example, the poles inverse to the two highest-ranking foci of Cassinians--the center and the $\cos^2\theta$-condition foci--do not, in general, possess circumpolar point-focal rank in the inverse curves (nor do poles inverse to the centers or traditional foci of central conics). The $\cos^2\theta$-condition foci of the inverse curves of the present figure, for example, are not inverse to the $\cos^2\theta$-condition foci of the basis curve (if they were, the pole inverse to the right $\cos^2\theta$-condition focus would not recede to infinity until the pole of inversion came into coincidence with this focus in the basis curve.)]

The 4 cubics obtained by inverting central Cartesians about the vertices on a given axis (as well as the corresponding vertex cubics of its axial inverse loci) belong to the same subspecies (Maxim 44). In consequence, the serpentine circular cubic formed when the pole of inversion comes into coincidence with the far right vertex (between levels 33 and 34) belongs to the same subspecies as the vertex cubic formed between levels 11 and 12, though its size is different (see Maxim 8-22-2) and its orientation is opposed. This means that the $\cos^2\theta$-condition focus also comes into coincidence with the 0° inter-oval-mode focus between levels 33 and 34 (at $h = 0.6708203933$).

The Locations of The $\cos^2\theta$-Condition Foci In Axial-Vertex Serpentine Circular Cubics

The fact that the $\cos^2\theta$-condition focus comes into coincidence with the far vertex of the oval and the 0° inter-oval-mode focus to form a generic triple-focus in Cassinian axial-vertex cubics raises the following question. Is this phenomenon unique to the axial-vertex cubics of central parabolic Cartesians (of which Cassinians are one of two opposed-oval species; see Table 8-17-2 and Figs. 8-22c,f) or does it occur in other genera of serpentine circular cubics? To answer this question, the $\cos^2\theta$-condition of eq. 8-20-10 is formed by equating the coefficient of the $\cos^2\theta$ term to 0 (eq. 1a)

$\cos^2\theta$-condition for axial bicircular quartics and serpentine circular cubics

(a) $4A^4H^2+4hHj^2(2h^2+B^2)+j^4(4h^2+B^2)-C^2j^4 = 0, \qquad A^4 = (h^4+h^2B^2+D^4)$ (8-21-1)

(b) $H = \dfrac{j^2(C^2-B^2-4h^2)}{4h(2h^2+B^2)} = \dfrac{j^2[B^2+C^2\mp 2(B^4-4D^4)^{\frac{1}{2}}]}{\pm 4h(B^4-4D^4)^{\frac{1}{2}}}$ — $\cos^2\theta$-condition for axial serpentine circular cubics

(c) $H = -\frac{1}{2}j^2/h$ — $\cos^2\theta$-condition for Cassinian axial-vertex cubics (identical with the condition defining the location of their 0° inter-oval-mode focus)

It will be noted that for $A^4 = 0$, one root, H_1, of eq. 1 goes off to infinity, leaving a second root, H_2, at the pole defined by eq. 1b. Since $A^4 = 0$ defines the x-axis vertices of central Cartesians (eq. 8-20-1), this means that in all axial-vertex inversions of the central basis curves to serpentine circular cubics, one $\cos^2\theta$-condition focus goes off to infinity and the other assumes the location defined by eq. 1b.

It follows from eq. 1b that biovular Cassinians (for which $B^2+C^2 = 0$) are the only species of central Cartesian for which the $\cos^2\theta$-condition focus of the axial-vertex cubic comes to lie at the pole defined by eq. 1c. For a vertex inversion of the central basis curve, eq. 1c defines the location of the cubic vertex inverse to the contralateral basis vertex (for example, in Fig. 20a$_1$, if pole 6 were the pole of inversion, the $\cos^2\theta$-condition focus would be in coincidence with the pole inverse to pole 4).

For opposed-oval basis curves, the pole inverse to the contralateral vertex is the far vertex of the cubic oval, but for annular basis curves this pole is the axial vertex of the serpentine arm. But biovular Cassinians comprise the only central species in which the $\cos^2\theta$-condition focus comes into coincidence with the pole inverse to the contralateral vertex. Accordingly, these results show that, for the curves in question, the $\cos^2\theta$-condition focus does not come into coincidence with the vertex of the serpentine arm (and the 0° intra-oval-mode focus), i.e., that the vertex of the serpentine arm never is a generic triple-focus.

The question now is considered of whether the $\cos^2\theta$-condition focus of a serpentine circular cubic ever comes into coincidence with the *near* vertex of the cubic oval. To investigate this matter, one need merely make eq. 1b specific for one of the 4 vertices of the basis curve, say the far right vertex (eq. 2), and replace H of eq. 1b by the corresponding known inverse location of the near vertex of the cubic oval. For the far right vertex of an annular or opposed-oval basis curve as the pole of inversion, the location in question is the near left vertex.

$h = \sqrt{[-B^2+\sqrt{(B^4-4D^4)}]}/\sqrt{2}$	far right x-axis vertex of the basis central Cartesian (eq. 8-20-1)	(8-21-2)
(a) $B^2 = 2D^2$ (b) $(C^2-B^2) = 4D^2,\quad D^2>0$	conditions for the $\cos^2\theta$-condition focus being in coincidence with the cubic vertex inverse to the near left vertex of the basis central Cartesian	(8-21-3)
$(x^2+y^2)^2+B^2x^2+C^2y^2+(C^2-B^2)^2/16 = 0,\quad C^2-B^2>0$	central basis loci defined by eq. 3b	(8-21-4)
(a) $h^2 = (C^2-B^2)/4$	$\cos^2\theta$-condition for the loci of eq. 4	(8-21-5)
(b) $h = \pm D,\quad D^4>0$	x-axis locations of poles for inversion of one subspecies of biovular central Cartesian to another subspecies (see Figs. 8-22,32,33)	

[Members of the two opposed-oval species invert to members of the annular species about modular poles on the line of symmetry that intersects the curve, but to each other about modular poles on the orthogonal line of symmetry. Members of the annular species invert to members of one opposed-oval species about modular poles on one line of symmetry, and to members of the other opposed-oval species about modular poles on the other line of symmetry. Accordingly, all inversions between central subspecies are between complementary types (annular to-and-from opposed-oval types) except those involving opposed ovals and modular poles that do not lie on the line of symmetry that intersects the curve.]

Making this replacement leads to the conditions of eqs. 3a,b. Condition 3a is for basis congruent, non-concentric circles, and is disregarded. However, condition 3b (which requires that D^4 of eq. 8-20-1 be positive) is valid and leads to the central basis curves of eq. 4. Since the 180° inter-oval-mode focus also is in coincidence with the near vertex of the cubic oval, a generic triple-focus is formed in the corresponding serpentine circular cubics. [Condition 3b also holds for the near right vertex of the basis curve. For opposed-oval curves this vertex would invert to the vertex of the serpentine arm, an alternative excluded above. But for annular basis curves, the near right vertex inverts to the far vertex of the cubic oval. Since the $\cos^2\theta$-condition focus is in coincidence with this vertex only in Cassinian axial-vertex cubics, the annular curves in question are central parabolic Cartesians (see Table 8-17-2 and Fig. 8-22). This is confirmed by inverting Cassinians about modular transverse-axis poles to the corresponding annular curves (employing the conversion eq. 7, below). When the coefficient of the y^2 term is subtracted from that of the x^2 term, and the difference is equated to 4 times the positive square root of the constant term, one obtains the condition $B^2+C^2=0$, the relationship for Cassinians.]

Central Modular-Focal Quartics

The identification and naming of the curves of eq. 4 hinges on the fact that the $\cos^2\theta$-condition of the central curves of eq. 8-20-1 is specified by eq. 5a (see eq. 8-20-6a). Combining eq. 5a with eq. 3b leads to eq. 5b. In other words, for the central curves of eq. 4, the $\cos^2\theta$-condition foci lie at $\pm D$. But $h=\pm D$ identifies the poles about which one central Cartesian inverts to another (see Maxim 8-20-3 and eq. 8-20-3). Accordingly, the curves in question --the *central modular-focal quartics*--possess the distinctive property that their $\cos^2\theta$-condition foci lie at modular distance from center, corresponding to a pole for inversion to another central curve.

Since eq. 4 represents both annular and opposed-oval central curves, it is evident that it represents two species of central modular-focal quartics (see Figs. 8-32a,e). This finding raises the question of whether two species of serpentine circular cubics also are represented by the corresponding x-axis vertex-inversion cubics.

If the opposed-oval curves in question (Fig. 8-32e) were biovular Cassinians (Fig. 8-22c), the answer to this question would be affirmative, even if the annular curves were annular central parabolic Cartesians (i.e., annular central inversions of Cassinians; Fig. 8-32a). The reason for this is that the real

$\cos^2\theta$-condition foci of annular central inversions of Cassinians lie on the y-axis (Fig. 8-22a). The annular curves of eq. 4 have their real $\cos^2\theta$-condition foci on the x-axis, so that if they were subspecifically identical to annular central inversions of Cassinians they would have to be rotated 90°, in which case the axial-vertex cubics in question would be the y-axis serpentine circular vertex cubics (Fig. 8-32c,h), which belong to a different species from *Cassinian axial-vertex serpentine circular cubics* (Fig. 8-22b). However, the opposed-oval curves are not Cassinians, since no Cassinian satisfies the condition, $C^2-B^2=4D^2$.

Eq. 4 represents all central modular-focal quartics; thus, to establish that the x-axis serpentine circular vertex cubics of both the annular and opposed-oval central curves of eq. 4 belong to the same species, it is necessary only to show either (1) that the annular central inverse loci of opposed-oval curves that satisfy the condition $C^2-B^2=4D^2$ have their $\cos^2\theta$-condition foci at $h=\pm D$, or vice versa, (2) that the opposed-oval central inverse loci of the annular curves satisfying the same condition have their $\cos^2\theta$-condition foci at $h=\pm D$. This would establish that the central curves in question belong to the same inversion superfamily, and that all the x-axis serpentine circular vertex cubics belong to the same species.

The general eq. for the annular central curves obtained by inverting the opposed-oval loci of eq. 8-20-1 about poles at $\pm D$ (or vice versa) is eq. 6 (derived in the legend of Fig. 8-31), for which the $\cos^2\theta$-condition is eq. 7.

$$(x^2+y^2)^2 + j^4(3D^2-\tfrac{1}{2}B^2)x^2/(2D^2+B^2)D^2 + j^4(C^2+D^2-\tfrac{1}{2}B^2)y^2/D^2(2D^2+B^2) + j^8/16D^4 = 0 \qquad (8\text{-}21\text{-}6)$$

eq. 8-20-1 inverted about a pole at $h=D$

$$h^2 = \frac{j^4(C^2-2D^2)}{4D^2(B^2+2D^2)} \qquad (8\text{-}21\text{-}7)$$

locations of the $\cos^2\theta$-condition foci of eq. 6

$$(x^2+y^2)^2 + 2j^4(3C^2-5B^2)x^2/(C^4-B^4) + 2j^4(5C^2-3B^2)y^2/(C^4-B^4) + j^8/(C^2-B^2)^2 = 0 \qquad (8\text{-}21\text{-}8)$$

eq. 6 for central modular-focal quartics

In order for real $\cos^2\theta$-condition foci to exist for both the basis curves (eq. 8-20-1) and the loci of eq. 6, the inequalities of eq. 10 must hold. Inequalities 10b do not hold for Cassinian parameters [$B^2=-\tfrac{1}{2}d^2$, $C^2=\tfrac{1}{2}d^2$,

$D^2 = d^2\sqrt{(1/16-C^2)}$] but both 10a and 10b hold for the ranges of parameters for which central modular-focal quartics exist ($C^2-B^2 = 4D^2 > 0$) [For example, annular central modular-focal quartics exist for $B^2 = -2d^2$, $C^2 = -d^2$, while opposed-oval curves exist for $B^2 = -2d^2$ and $C^2 = d^2$].

$$h^2 = j^4/(C^2-B^2) = [j^8/(C^2-B^2)^2]^{\frac{1}{2}}$$ locations of $\cos^2\theta$-condition foci of eq. 8 (8-21-9)

(a) $C^2-B^2 > 0$

(b) $C^2 \gtreqless 2D^2$, $B^2+2D^2 \gtreqless 0$

conditions for real x-axis $\cos^2\theta$-foci of the loci of both eqs. 8-20-1 and eq. 6 (8-21-10)

If one eliminates D^2 from eq. 6 by substituting $B^2-C^2 = 4D^2$, one obtains eq. 8 for inverse loci of eq. 4 about a pole at $h = D$. Thus, if eq. 4 represents opposed-oval central modular-focal quartics, eq. 8 represents their annular inverse loci (and vice versa). Eq. 9 gives the locations of the $\cos^2\theta$-condition foci of the loci of eq. 8. Since these locations are seen to be modular (i.e., h is the 4th-root of the constant term), we have established what we set out to demonstrate, namely, that the x-axis serpentine circular cubics of all central modular-focal quartics belong to a single species (since all central modular-focal quartics invert to one another about an axial pole at modular distance from the center, and have their real $\cos^2\theta$-condition foci on the x-axis, i.e., on the axis that intersects the opposed-oval curve and the corresponding axis of the annular curve).

Examples of a basis annular central modular-focal quartic, the corresponding opposed-oval central curve obtained by inverting it about a pole at $h = D$, and the axial-vertex serpentine circular cubics thereof, are given in Figs. 8-22 and 8-32 (but in the former figure curves a, d, and f have to be rotated 90°).

Returning to the figure, the pre-serpentine circular cubic loci considered hereto may be characterized as Phase I opposed ovals. In these, the $\cos^2\theta$-condition foci are intra-oval, two of the foci of self-inversion are inter-oval, and one of them is lateral. None of them is an angle-independent circumpolar focus.

In progression to levels beyond that of the serpentine circular cubic, the foci of self-inversion that were coincident with the oval vertices enter the small oval, which progressively diminishes in size. The vertex of the right oval of the basis curve, which had receded to infinity at the right, now approaches from infinity at the left, appearing in the Phase II annular curves. The 0° intra-oval-mode focus, which had been in coincidence with the vertex of the serpentine arm, exits to the right to become lateral (i.e., exterior to both ovals).

The $\cos^2\theta$-condition focus that participated in the triple-focus at the far vertex of the cubic oval moves to the left into the inter-oval region. At the same time the $\cos^2\theta$-condition focus of the right basis oval, which had receded to infinity at the right, now approaches from infinity at the left, whereupon both $\cos^2\theta$-condition foci are in the same inter-oval region, though widely separated.

As the pole of inversion closely approaches the near pole of the Cartesian condition, the $\cos^2\theta$-condition foci also closely approach one another. The second exceptional locus is formed when the pole of inversion comes into coincidence with the pole of the Cartesian condition, whereupon the two $\cos^2\theta$-condition foci also merge to a double-focus. At this level the locus consists of the $C = 5$ parabolic Cartesian, the most symmetrical axial quartic in the Cassinian-based inversion superfamily. This locus marks the second position at which all 3 foci of self-inversion achieve angle-independent circumpolar focal rank.

As the pole of inversion progresses beyond the near Cartesian-condition pole, the Phase III annular curves, characterized by complex $\cos^2\theta$-condition foci, are formed. During the early portion of this phase the lateral 0° intra-oval-mode focus recedes rapidly to the right, while the two foci of self-inversion within the small oval approach one another. The $\cos^2\theta$-condition foci depart from coincidence but now are, as mentioned, complex.

As the pole of inversion approaches toward the center and achieves modular distance from it [$h = \sqrt{(0.15)}d$], the two foci of self-inversion within the small oval come into coincidence, forming a true center. Simultaneously, the lateral 0° intra-oval-mode focus recedes to infinity. The locus now is the $C = 5$ (or $C = 0.2$ in Cassinian parameters) annular central Cartesian (Fig. 8-20d_1). As in the basis curve itself, only the two poles of self-inversion in coincidence at the center have angle-independent circumpolar focal rank.

For all poles of inversion beyond the pole at modular distance from center, the subspecies of Phases I-III and the demarcating exceptional subspecies are duplicated, with two principal differences. One respect in which they differ is that the loci have opposed left-to-right orientation; the other is that they all are of smaller size (see Maxim 8-22-2). As a consequence of the reversed orientation, the relative positions of the reciprocal inversion pole and the pole inverse to the center of the basis curve are exchanged, though, of course, both are at modular distance from the 0° inter-oval-mode focus.

As the pole of inversion progresses beyond modular distance from center, Phase IV annular ovals are formed. The foci of self-inversion that were in coincidence at the center diverge and now have their relative positions with respect to the centerline reversed. At the same time, the 0° intra-oval-mode focus, having receded to infinity at the right, begins to approach from infinity at the left. These reversals of position, of course, are consonant with the reversed orientation of the loci. All the annular subspecies of Phase III are duplicated in Phase IV.

When the pole of inversion achieves coincidence with the intra-oval $\cos^2\theta$-condition foci of the basis curve (the far pole of the Cartesian condition), a second $C = 5$ parabolic Cartesian is formed, the 3 foci of self-inversion again achieve angle-independent circumpolar focal rank, and real, coincident $\cos^2\theta$-condition foci reappear. Similarly, the Phase II annular subspecies are duplicated by the Phase V annular loci, as the intra-oval foci of self-inversion again approach the vertices of the small oval and the right vertex of the large oval, and the right $\cos^2\theta$-condition focus recedes to the right.

Although loci are not depicted beyond level 30, the $C = 0.2$ Cassinian axial-vertex cubic is formed again between levels 33 and 34, and all the foci of self-inversion again achieve coincidence with the vertices and angle-independent circumpolar focal rank. The finite $\cos^2\theta$-condition focus, of course, becomes coincident with the far vertex of the cubic oval to form a triple-focus.

Beyond the level of the second vertex cubic, the opposed ovals of Phase VI form, as the pole of inversion progresses to the right of the far right vertex of the basis curve. The changes in vertices and foci that occur in this Phase are the reverse of those of Phase I; the subspecies that are formed, as the pole of inversion recedes along the x-axis to the right, duplicate in reverse sequence those formed as the pole passed from the center to the near right vertex.

If one continues to employ the modular values of j, the loci become ever smaller but, nonetheless, subspecifically identical with those of Phase I. If one lets j increase commensurately with the distance of the pole of inversion from the center, the locus in the limit is coincident with the basis curve itself (self-inversion), i.e., with the inverse locus obtained with the pole of inversion at the center of the basis curve (Maxims 6-7-3 and 6-8-5).

Tracking The Exceptional Poles of An Opposed-Oval Central Modular-Focal Quartic

A similar tracking of the axial foci and other pertinent poles of axial inverse loci (i.e., inverse loci about poles on the axis that passes through the $\cos^2\theta$-condition foci) of an opposed-oval central modular-focal quartic shows many resemblances but certain significant differences. Among the differences are the persistence of two discrete real $\cos^2\theta$-condition foci for all positions of the pole of inversion (see also the similar situation in Fig. 8-27), and the formation of serpentine circular cubics with a generic triple-focus at the near vertex of the cubic oval (Figs. 8-22d and 8-32c,h). Additionally, no Cartesian self-inverters (which have $\cos^2\theta$-condition double-foci) are formed, inasmuch as there are no real solutions to the Cartesian condition for $C^2-B^2=4D^2$. Accordingly, the 3 foci of self-inversion achieve angle-independent circumpolar focal rank only in the serpentine circular cubics.

The differences relating to the $\cos^2\theta$-condition foci come about in the following manner (for the opposed-oval curve upon which the example is based, $B^2=-0.220170459d^2$ and $C^2=-0.063920456d^2$; this locus is shown in Fig. 8-22f). As in axial inversions of Cassinian ovals, the far right vertex and the right $\cos^2\theta$-condition focus recede to infinity as the pole of inversion approaches the near right vertex. The course of the left $\cos^2\theta$-condition focus, however, is not toward the far left vertex, but toward the near left vertex, with which it comes into coincidence in the serpentine circular cubic.

For poles of inversion at positions beyond the near right vertex and progressing toward the modular pole, the finite $\cos^2\theta$-condition focus leaves coincidence with the near vertex of the serpentine circular cubic and enters the right inter-oval region of the annular inverse locus, rather than leaving coincidence with the far vertex and entering the left inter-oval region, as in the present figure.

The next exceptional locations of the $\cos^2\theta$-condition foci occur at the level of the inverse annular central locus, for which the pole of inversion is at modular distance from center. At this level, the $\cos^2\theta$-condition focus that originated in the left basis oval--the *original left focus*--is coincident with the reciprocal inversion pole (see Maxim 8-20-3); correspondingly, the focus originating in the right basis oval--the *original right focus*--is in coincidence with the basis centerline, and the two are reflections of one another in the center of the annular central locus. Accordingly, the $\cos^2\theta$-condition foci do not come into coincidence to form a double-focus as in Cartesian self-inverters.

For poles of inversion beyond modular distance from center and progressing toward the far right basis vertex, the original left focus recedes to infinity in parallel with the near right vertex, while the original right focus progressively approaches the left vertex of the small oval, with which it comes into coincidence in the serpentine circular vertex cubic--the *axial-vertex serpentine circular cubic of an opposed-oval central modular-focal quartic* (as a component of the triple-focus of the near vertex).

For inversion poles beyond the far right basis vertex, the original right focus enters the small oval, while the original left focus approaches from infinity at the left, within the large oval. These changes duplicate in reverse sequence, reverse orientation, and at much smaller size, those that occurred in the inverse curves as the pole of inversion approached the near right vertex from center.

Fig. 8-22. Relations between highly symmetrical cubics and quartics that self-invert in multiple modes. In the foregoing, 7 groups of curves with exceptional symmetry properties have been encountered in central Cartesian-based inversion superfamilies (superfamilies based upon inversions of the central curves of eq. 1).

$$(x^2+y^2)^2 + B^2x^2 + C^2y^2 + D^4 = 0 \qquad \text{central Cartesians} \qquad (8\text{-}22\text{-}1)$$

The 7 groups referred to are:

1. The genus *parabolic Cartesians*. In the Cassinian-based inversion superfamily these curves are the homologues of limacons in the QBI superfamily. Hence they are the non-central quartics with the highest degree of circumpolar symmetry in the superfamily. They are the only Cartesian self-inverters for which a pole of inversion to a central curve possesses point-focal rank, and one of only two groups of curves in the Cassinian-based inversion superfamily with a 2nd-degree bipolar eq.
2. The monovular and biovular species of *Cassinian ovals*. This is the other group of curves with a 2nd-degree bipolar eq. It is the only known group for which the members of the monovular species self-invert at 90° and 270° (and, in consequence, possess 4 Cartesian-condition poles), and the group providing one of two central basis species hereto encountered for biovular members of the Cassinian-based inversion superfamily.
3. The species *annular central parabolic Cartesians*, which is complementary to biovular Cassinians. This species comprises the second hereto encountered basis species for biovular members of the Cassinian-based inversion superfamily.
4. The species *Cassinian axial-vertex serpentine circular cubics*, the only species in the Cassinian-based inversion superfamily in which the finite focus of the $\cos^2\theta$-condition is in coincidence with the *far* vertex of the oval, which is a generic triple-focus. [All 4 axial-vertex serpentine cubics of a given monovular Cassinian belong to the same subspecies and can be presumed to have other exceptional symmetry properties compared to the axial-vertex serpentine cubics of other monovular species.]
5. The species *opposed-oval central modular-focal quartics*, the only opposed-oval central basis species for bicircular quartics and serpentine circular cubics (and for a complementary species of annular central modular-focal quartics) in which the poles of inversion to the complementary central species are focal (defined by the $\cos^2\theta$-condition).
6. The species *annular central modular-focal quartics*, the complementary species referred to above and the only annular central basis species in which the poles of inversion to the complementary central curves are focal (defined by the $\cos^2\theta$-condition).
7. The species *axial-vertex serpentine circular cubics of opposed-oval central modular-focal quartics*, the only species of serpentine circular cubics in which the $\cos^2\theta$-condition focus is in coincidence with the *near* vertex of the oval, which is a generic triple-focus.

It is shown below that *annular central parabolic Cartesians* and *annular central modular-focal quartics* are members of the same species (a fact recognized in Table 8-17-2 but not yet proved), with the consequences that (a) the hereto encountered groups of cubics and quartics with the noted exceptional symmetry properties number only 6, and (b) all of these groups belong to the Cassinian-based inversion superfamily. Although the serpentine circular cubics of species 7 possess a line of symmetry and are incident inversions of Cassinian ovals, they are not axial-vertex inversions (see below).

On the other hand, though the member subspecies of both have opposed ovals, the species *opposed-oval central modular-focal quartics* is not identical with the species *biovular Cassinians*. As a consequence there are 3 central basis species for the biovular members of the superfamily, a finding with important consequences for the Inversion Taxonomy of all central Cartesian-based inversion superfamilies (see Table 8-17-2).

[Since, of these central basis species, (a) only for biovular Cassinians is there a corresponding monovular species, and (b) only Cassinians previously were known and named, the superfamily to which these curves belong is designated as the *Cassinian based-inversion superfamily*.]

A corresponding subspecies from each of the 6 groups in question is represented in the figure. Since all 6 groups belong to the Cassinian-based inversion superfamily, each of the 6 represented subspecies inverts to the other 5 about two or more poles. In the case of the basis curve, a, all of the non-vertex poles are marked. Unmarked poles are located readily by reference to Maxims 6-4-4, 8-10-4, 8-23-9,10, and 8-22-7,18. The method of locating non-axial poles is illustrated for curves b, d, and f.

[The annular central parabolic Cartesian was taken to be the basis curve for the figure because it possesses the property, uniquely, that the two serpentine circular cubics (curves b and d) are its line-of-symmetry vertex-inversion cubics. In the cases of the two opposed-oval central curves, c and f, only one of the serpentine circular cubics is obtained by a line-of-symmetry vertex inversion (curve b from curve c, and curve d from curve f).]

The $\cos^2\theta$-condition foci of the curves are shown by the largest solid circles and labelled F. Each curve has two such foci but those of the parabolic Cartesian, e (and e'), are in coincidence as a double-focus, while one of these foci is at infinity for each of the serpentine circular cubics. Exceptional but non-focal poles are indicated by the smallest solid circles and identified by labels (these consist chiefly of Cartesian-condition poles, and poles at modular distance from center). Solid circles of intermediate size identify centers, reciprocal inversion poles, vertices, and poles of inversion to axial-vertex serpentine circular cubics.

The basis central curve, a, possesses 5 poles about which it inverts to central curves. It self-inverts about the center; it inverts to opposed-oval central modular-focal quartics, f, about the two $\cos^2\theta$-condition foci (on the y-axis at modular distance from center); and it inverts to the Cassinian ovals, c, about the two poles at modular distance from center on the x-axis. It possesses 4 poles--all on the x-axis--about which it inverts to the para-

Relations Between Highly Symmetrical Cubics & Quartics That Self-Invert In Multiple Modes

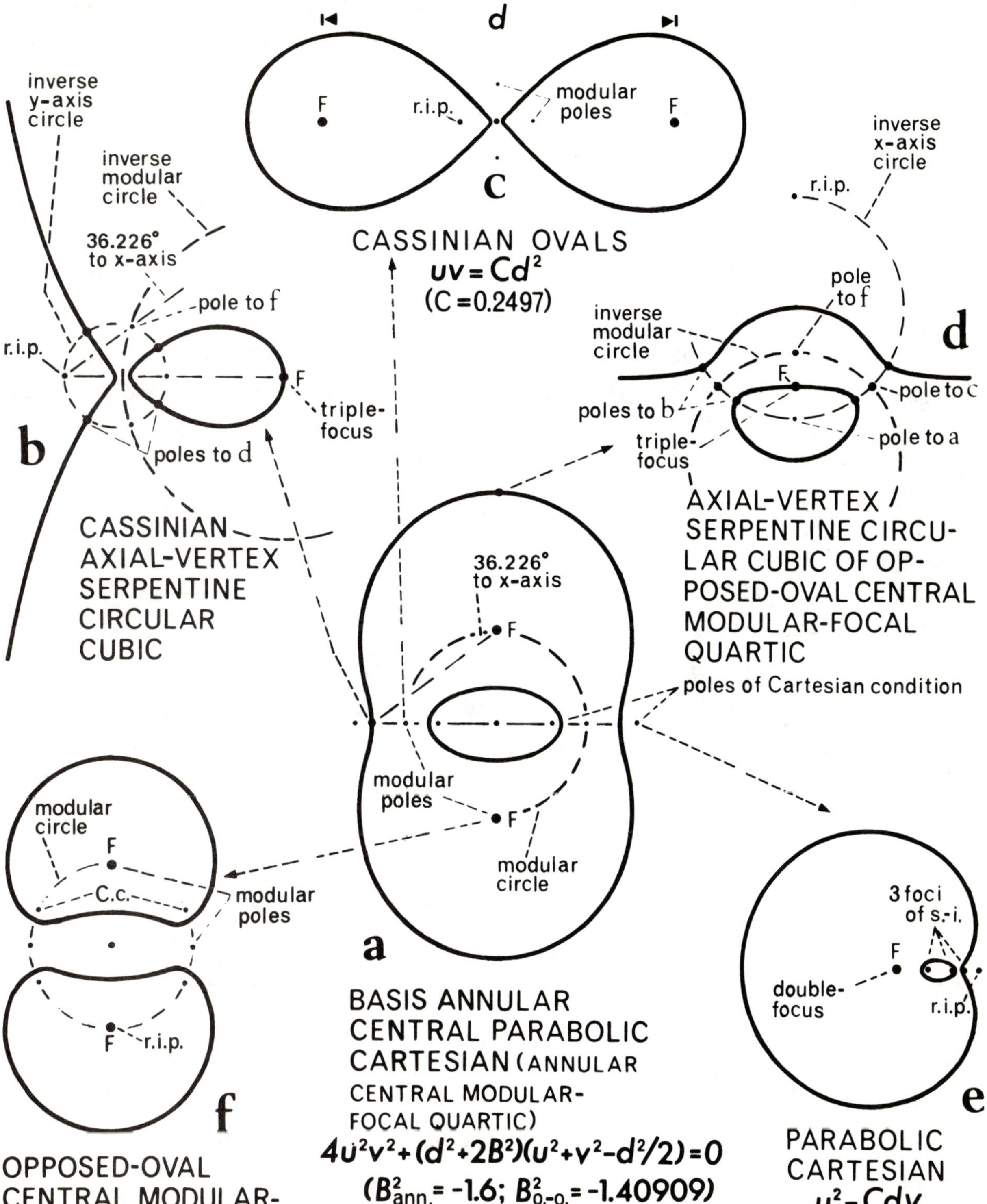

CASSINIAN OVALS
$uv = Cd^2$
(C = 0.2497)

CASSINIAN AXIAL-VERTEX SERPENTINE CIRCULAR CUBIC

AXIAL-VERTEX SERPENTINE CIRCULAR CUBIC OF OPPOSED-OVAL CENTRAL MODULAR-FOCAL QUARTIC

BASIS ANNULAR CENTRAL PARABOLIC CARTESIAN (ANNULAR CENTRAL MODULAR-FOCAL QUARTIC)
$$4u^2v^2 + (d^2 + 2B^2)(u^2 + v^2 - d^2/2) = 0$$
$(B^2_{ann.} = -1.6;\ B^2_{o.-o.} = -1.40909)$

OPPOSED-OVAL CENTRAL MODULAR-FOCAL QUARTIC

PARABOLIC CARTESIAN
$u^2 = Cdv$

Fig. 8-22a-f

bolic Cartesian, e (and e'). These poles are defined by the Cartesian condition (eq. 8-20-8). It inverts to Cassinian axial-vertex serpentine circular cubics, b, about all 4 x-axis vertices, and to the axial-vertex serpentine circular cubics of opposed-oval central modular-focal quartics, d, about all 4 y-axis vertices. It possesses only 3 non-incident exceptional poles--the modular $\cos^2\theta$-condition foci on the y-axis (largest solid circles labelled F) and the center.

It follows from Maxim 6-4-4 that each of the 5 inverse loci, b-f, also inverts to the curves mentioned above about the corresponding inverse poles. Since all of the poles in question lie on one or the other line of symmetry of curve a, it follows that for inversion of curve a about a pole on one line of symmetry, poles inverse to the other poles on the same line of symmetry lie on the corresponding line of symmetry of the inverse curve. On the other hand, the poles on the other line of symmetry of the basis curve invert to concyclic positions on the *inverse x-axis circle* or the *inverse y-axis circle* in the plane of the inverse curve. In other words, poles inverse to one set of poles collinear on one line of symmetry of the basis curve also are collinear in the inverse curve, but the poles inverse to the other set on the other line of symmetry become concyclic on the circle inverse to the other line of symmetry.

The *inverse-axis circles* upon which the concyclic poles lie are illustrated in the plane of the two axial-vertex cubics, b and d, and the opposed-oval central modular-focal quartic, f (labelled "modular circle"). In the case of the x-axis vertex inversion of curve a to curve b, the *inverse y-axis circle* is centered at the vertex of the serpentine arm (since the other vertex of the large oval of curve a, being at twice the distance of the line of symmetry from the pole of inversion, comes to lie midway between the reciprocal inversion pole and the pole inverse to the basis center).

Poles inverse to poles on the y-axis of curve a lie on the *inverse y-axis circle* of curve b, along radii at the same angle to the x-axis as the corresponding radii to the basis poles of curve a. Thus, the inner vertex lies along a radius at an angle of 16.345°, the $\cos^2\theta$-condition focus at an angle of 36.226°, and the outer vertex at an angle 61.345°. The two poles inverse to these vertices (labelled "poles to d") are, of course, incident upon the inverse curve at its points of intersection with the inverse y-axis circle (as indicated by the 4 medium-sized solid circles).

On the other hand, the pole inverse to the upper $\cos^2\theta$-condition focus, F, of the basis curve lies at the point where the radius vector from the reciprocal inversion pole at an angle of 36.226° to the x-axis intersects the circle (labelled "pole to f"). Curve b inverts to curve f about this pole, just as the basis curve does about the pole inverse to it. The other pole for inversion of curve b to the basis curve is the reflection of the reciprocal inversion pole in the vertex of the serpentine arm (illustrated for curve b but not labelled).

Inversion of the basis curve about either the left or right x-axis pole at modular distance from center (labelled "modular poles" in the figure) yields the Cassinian ovals, c. For the latter curve, only the $\cos^2\theta$-condition foci, the center, and the modular poles on the x-axis and the y-axis are marked. Of the latter, the left pole on the x-axis is the reciprocal inversion pole, the right pole is the second pole for inversion to the basis curve, and the y-axis poles are the poles for inversion to the opposed-oval central modular-focal quartic, f.

Curve c inverts to the parabolic Cartesian, e (and e'), about two pairs of x-axis poles. One pair consists of the $\cos^2\theta$-condition foci (labelled F); the other pair is modularly inverse to these foci about the center, lying between the near vertices and the modular poles (see Fig. 8-21). Curve c inverts to Cassinian axial-vertex serpentine circular cubics, b, about the 4 x-axis vertices. The inverse y-axis circle (not shown) of modular radius, centered at the center of curve c, intersects the two ovals at the 4 points about which c inverts to the serpentine circular cubics, d.

Inversion of the basis curve about each of the 4 y-axis vertices yields the subspecies of serpentine circular cubic, d, for which the inverse x-axis and inverse modular circles are shown. The poles inverse to the 11 exceptional poles on the x-axis of the basis curve are accounted for as follows in the plane of curve d. The poles inverse to the 4 poles for inversion to the Cassinian axial-vertex serpentine circular cubic, b, are the intersections of the inverse x-axis circle with curve d (labelled "poles to b"). The poles inverse to the 4 poles for inversion to the parabolic Cartesian, e, are incident upon this circle just within the oval and just outside of (on the side opposite from the oval) the serpentine arm. The poles inverse to the two poles for inversion to Cassinians are incident upon the same circle in the region between the oval and the serpentine arm (labelled "pole to c"). Lastly, the second pole for inversion to the basis curve (the pole inverse to the center) lies upon the inverse x-axis circle at a position diametrically opposed to the reciprocal inversion pole (pole to a). Thus, of the 11 exceptional x-axis poles of curve a, 4 of the inverse poles are incident upon the axial-vertex cubic, 3 are intra-oval, two are inter-segment, and two are on the opposite side of the serpentine arm.

The basis curve a inverts to a single subspecies of parabolic Cartesian about the 4 poles defined by the Cartesian condition. The parabolic Cartesian e (and e') is obtained by inverting about the exterior right pole. Since there are 4 equivalent inversion poles for inversion to this subspecies, there also must be 4 equivalent inversion poles in the plane of any inverse curve about which that curve also inverts to this same subspecies of parabolic Cartesian. In the present case, as in the case of the vertex cubics, b and d, 3 of these poles are the foci of self-inversion in the finite plane; the 4th is the ideal line at infinity, including a focus of self-inversion on this line in the direction orthogonal to the line of symmetry (see Maxim 6-4-6).

The reflection of the pole of inversion of curve a to curve e (and e') in the line of symmetry (the y-axis) inverts to the 0° intra-oval-mode focus of curve e (and e'), incident upon the x-axis between the curve and the reciprocal inversion pole; the near (right) interior Cartesian-condition pole inverts to the 0° inter-oval-mode focus (see Maxims 8-22-11 and 8-30-6). The two modular poles invert to axial poles lying at modular distance on each side of the 0° inter-oval-mode focus. Together with the poles inverse to the basis vertices, which are the vertices of curve e (and e'), these points account for the 11 exceptional x-axis poles of the basis curve.

The exceptional y-axis poles, of course, invert to positions on the inverse y-axis circle (shown only for curve e'), centered on the 0° intra-oval-mode focus and passing through the reciprocal inversion pole and the pole inverse to the center. This circle intersects the curve at the points about which the latter inverts to the serpentine circular cubic, d. The basis modular poles (F), defined by the $\cos^2\theta$-condition, invert to positions on this circle between the ovals, on radius vectors emanating from the reciprocal inversion pole at $\pm 147.310^{\circ}$ to the x-axis ("poles to f" of curve e'; however, these inverse poles are not $\cos^2\theta$-condition foci).

It will be recalled that axial bicircular quartics and serpentine circular cubics invert to curves congruent with one another, but otherwise unexceptional, about axial poles lying at modular distance on each side of the 180° inter-oval-mode focus. The poles of the basis curve that are inverse to these 1st-generation axial poles are found to be modularly inverse about the center, one interior, to the right, and nearer to the center than the Cartesian-condition pole, and the other exterior, to the left, and more distant than the far Cartesian-condition pole. They, of course, lie on opposite sides of both the center and the Cartesian-condition pole inverse to the 180° inter-oval-mode focus.

A characteristic relationship involves the pole of the central basis curve inverse to the 180° inter-oval-mode focus and the poles of the central basis curve (poles A) inverse to the poles modular to this focus (poles A') in the 1st-generation curve. In general, for any pair of inverse axial poles of a biovular central Cartesian, the ratio of the sizes of the inverse loci about these poles is the reciprocal of the ratio of the distances of the poles from the center. Thus, although the inverse loci about the modular poles (poles A') in the 1st-generation curve are congruent to one another (Maxim 8-20-11), the inverse loci about the inverse poles (poles A) in the plane of the central basis curve are not. They are subspecifically identical but incongruent in definite size ratio.

The metric relationship referred to above states that the ratio of the distances of poles A of the basis curve from the pole inverse to the 180° inter-oval-mode focus is the reciprocal of the square-root of the ratio of their distances from the center. In other words, the pole of the central basis curve inverse to the 180° focus of self-inversion is so situated that the ratio of the distances of poles A from it is the square-root of the ratio of their distances from the center.

This relationship applies regardless of whether or not the pole inverse to the 180° focus of self-inversion is specified by the Cartesian condition, and regardless of whether or not the central basis curve belongs to the Cassinian-based inversion superfamily (which are the only central curves that invert to central parabolic Cartesians). [A detailed treatment of the locations and characteristics of the poles of inversion of the $C = 4.005$ parabolic Cartesian to its exceptional inverse loci is given below (for curve e').]

CIRCUMPOLAR
MAXIM 8-22-1: *Biovular central Cartesians invert to subspecifically different central Cartesians about modular poles on the lines of symmetry. In these inversions, the modular distance becomes halved, i.e., if j is modular for self-inversion of the basis curve, $\frac{1}{2}j$ is modular for self-inversion of the inverse central curve.*

8-22-2: *The linear size ratio of subspecifically identical inverse loci about modularly-inverse poles on the lines of symmetry of a central Cartesian is the inverse ratio of the distances of these poles from the center.*

Curve f is the opposed-oval central modular-focal quartic obtained by inverting the basis curve about the lower $\cos^2\theta$-condition focus (which is at modular distance from center). The same curve is obtained by inverting the Cassinian ovals, c, about the modular poles on the y-axis. The poles inverse to the 11 exceptional x-axis poles of curve a come to lie on the illustrated inverse x-axis circle of curve f, which also is its modular circle. The poles are: the vertices, at the points where this circle intersects the inverse locus (unmarked); the Cartesian-condition poles within the ovals in pairs (marked by small solid circles and labelled "C.c" in the upper oval); the two modular poles (about which the curve inverts to Cassinians) at modular distance on the new x-axis (the inverse of the modular circle of curve a); and the center, at the point inverse to the contralateral $\cos^2\theta$-condition focus.

Central Modular-Focal Quartics As Species of The Cassinian-Based Inversion Superfamily

It now is shown that central modular-focal quartics comprise two species of the Cassinian-based inversion superfamily, and that members of the annular species are identical with annular central parabolic Cartesians. Since one of the 3 parameters of the general eq. (1) for central Cartesians is eliminated in the modular-focal species of eq. 2, one can express the eq. for the group in terms of but a single parameter if one denotes the distance between the two $\cos^2\theta$-condition foci as d, as is done in bipolar eqs. (see Chapter II).

In this event, the constant term of eq. 2 can be represented as in eq. 3, and (either B^2 or) C^2 eliminated, leading to eq. 4a. If, now, one lets $B^2 = nd^2$, eq. 4a becomes 4b. It is evident from eq. 4b that the coefficient of the y^2

term for this curve is simply the square of distance, d, greater than the coefficient of the x^2 term.

$(x^2+y^2)^2+B^2x^2+C^2y^2+(C^2-B^2)^2/16 = 0$	central modular-focal quartics	(8-22-2)
$(C^2-B^2)^2/16 = d^4/16$	equality of constant term of eq. 2 to the 4th power of the modular distance	(8-22-3)
(a) $(x^2+y^2)^2+B^2x^2+(B^2+d^2)y^2+d^4/16 = 0$	eq. 2 re-expressed in terms of the distance between the $\cos^2\theta$-condition foci	(8-22-4)
(b) $(x^2+y^2)^2+nd^2x^2+(n+1)d^2y^2+d^4/16 = 0$	eq. 4a with $B^2 = nd^2$	
$4u^2v^2+(d^2+2B^2)(u^2+v^2-\frac{1}{2}d^2) = 0$	eq. 4a cast about the center in bipolar form	(8-22-5)
$(x^2+y^2)^2+(3-2n)d^2x^2/(4n+2) + (2n+5)d^2y^2/(4n+2)+d^4/16 = 0$	central modular-focal quartic complementary to eq. 4b obtained by inversion about a $\cos^2\theta$-condition focus	(8-22-6)

The bipolar eq., 5, of these curves is derived readily (see Chapter II). Since eq. 5 is of 4th degree, it is evident that, from the point of view of bipolar coordinates, central modular-focal quartics are less symmetrical than either Cassinians or parabolic Cartesians, both of which have only 2nd-degree bipolar eqs. (given in the figure under the respective curves).

The eq. of the central modular-focal quartic complementary to that of eq. 4a or 4b can be derived by making use of eq. 8-21-7 or 8-31-3 (provided that the modular poles lie on a line of symmetry that intersects the curve; see note following eq. 8-21-5). Letting $B^2 = nd^2$, $C^2 = (n+1)d^2$, and $D^2 = j^4/d^2$, one obtains eq. 6. Accordingly, since B^2 for the basis curve a is $-1.6d^2$, its value for the complementary central curve, f, is $[3-2(-1.6)]d^2/[4(-1.6)+2] = 6.2d^2/(-4.4) = -1.4090909d^2$.

The standard eq. (7) for Cassinians with the transverse axis incident upon the x-axis now is taken as the basis curve. It will be noted at the outset that, since d is the distance between the $\cos^2\theta$-condition foci, this eq. differs from the eq. of a central modular-focal quartic with $B^2 = -\frac{1}{2}d^2$ only in that the constant term is not the 4th power of the modular distance [instead of being $(\frac{1}{2}d)^4 = d^4/16$, it is $d^4(1/16-C^2)$].

standard equation of Cassinian ovals (transverse axis incident upon x-axis)

$$(x^2+y^2)^2-\tfrac{1}{2}d^2x^2+\tfrac{1}{2}d^2y^2+d^4(1/16-C^2) = 0 \qquad (8\text{-}22\text{-}7)$$

central curves obtained by inverting Cassinians about modular poles on the x-axis

$$(x^2+y^2)^2 + \frac{(3\alpha+\frac{1}{4})j^4x^2}{\alpha d^2(2\alpha-\frac{1}{2})} + \frac{(\alpha+3/4)j^4y^2}{\alpha d^2(2\alpha-\frac{1}{2})} + \frac{j^8}{16\alpha^2d^4} = 0, \quad \alpha = \sqrt{(1/16-C^2)} \qquad (8\text{-}22\text{-}8)$$

(a) $\dfrac{\delta^4}{16} = \dfrac{j^8}{16\alpha^2d^4}$ constant term of eq. 8 regarded as the 4th-power of a modular distance, $\tfrac{1}{2}\delta$ (8-22-9)

(b) $j^4 = \alpha\delta^2d^2$ square-root of eq. 9 rearranged

(a) $(x^2+y^2)^2 + (3\alpha+\tfrac{1}{4})\delta^2x^2/(2\alpha-\tfrac{1}{2}) + (\alpha+3/4)\delta^2y^2/(2\alpha-\tfrac{1}{2}) + \delta^4/16 = 0$ eq. 8 with the substitution 9b (8-22-10)

(b) $(x^2+y^2)^2+n\delta^2x^2+(n-1)\delta^2y^2+\delta^4/16 = 0$ eq. 10a with the coefficient of the δ^2x^2 term set equal to n

(c) $(x^2+y^2)^2+(n-1)\delta^2x^2+n\delta^2y^2+\delta^4/16 = 0$ eq. 10b rotated 90°

Next, the central curves complementary to those of eq. 8 are derived with the use of eq. 8-21-7 or 8-31-3 by letting $B^2 = -\tfrac{1}{2}d^2$, $C^2 = \tfrac{1}{2}d^2$, and $D^4 = j^8/16d^4(1/16-C^2)$. If the constant term of eq. 8 is regarded as the 4th-power of a modular distance $\tfrac{1}{2}\delta$, with δ the distance between the $\cos^2\theta$-condition foci, one obtains the identity of eq. 9. Making the substitution of eq. 9b in eq. 8 yields eq. 10a. If the coefficient of the δ^2x^2 term is denoted by n, this becomes eq. 10b.

Eq. 10b differs from the eq. of a central modular-focal quartic only in that the coefficient of the y^2 term is d^2 less than that of the x^2 term, rather than d^2 greater. However, it is only the possession of a difference of d^2 (the square of the distance between the $\cos^2\theta$-condition foci) between the magnitude of the two coefficients that is significant in characterizing these curves, not the sign of the difference. Thus, if the locus of eq. 10b is rotated 90°, the coefficient of the y^2 term becomes d^2 greater than that of the x^2 term (eq. 10c).

The above results show that the species of central curves obtained by inverting biovular Cassinians about the transverse-axis poles at modular distance from center are annular central modular-focal quartics. Since there is only one species of annular central inversions of Cassinians, namely annular central Cartesians, these two groups are identical. Since annular central modular-focal quartics invert to the opposed-oval species, it also follows that both species belong to the Cassinian-based inversion superfamily. Accordingly, of the hereto encountered axial bicircular quartics and serpentine circular cubics, that have exceptional symmetry properties, all are members of the Cassinian-based inversion superfamily (see Table 8-17-2).

CIRCUMPOLAR MAXIMS 8-22-

3: *A given axial bicircular quartic or serpentine circular cubic inverts to a different subspecies of central Cartesian about each of 3 pairs of poles. One of these central subspecies is annular, the other two possess opposed ovals.*

4: *A monovular central Cartesian inverts to axial monovals or axial serpentine cubics about poles on its lines of symmetry. It inverts to a generally dissimilar central curve about its center.*

5: *An axial inverse locus of a monovular central Cartesian inverts to axial monovals about poles on its line of symmetry and on its inverse-axis circle (the circle inverse to the orthogonal line of symmetry of the basis curve). It inverts to central Cartesians about the two points of intersection of these loci, and self-inverts about the center of the inverse-axis circle. The central Cartesian subspecies obtained in this manner invert to one another through their centers and generally are dissimilar.*

6: *Non-axial inverse loci of monovular central Cartesians invert to axial monovals about points on the inverse-axis circles (circles inverse to the lines of symmetry of the central basis curves); they invert to the basis central Cartesians about the two points of intersection of these circles. The basis central Cartesian subspecies obtained in this manner invert to one another through their centers and generally are dissimilar.*

7: *A given biovular central Cartesian inverts to both subspecifically identical and different biovular central Cartesians about the 5 points of intersection of the lines of symmetry and the modular circle (the centered circle of inversion). The 0° and 180° inverse loci about the center are self-inversions, whereas the pairs of inverse loci about the two points of intersection of each line of symmetry with the modular circle are dissimilar to each other and to the basis curve. If the basis curve possesses opposed ovals, one pair of inverse loci also possess opposed ovals, whereas the other are annular; if the basis curve is annular, both pairs of inverse loci possess opposed ovals* (see also Maxim 8-20-3).

8: *Biovular central Cartesians invert to axial bicircular quartics or serpentine circular cubics about points on one or the other line of symmetry or the modular circle, excluding the points of intersection of these 3 loci (they invert to central Cartesians about the points of intersection). About other points in the plane, they invert to non-axial curves.*

A Detailed Account of The Poles of Inversion of A Parabolic Cartesian To Its Exceptional Inverse Loci

Curve e'--a magnification of curve e--illustrates in detail the locations of poles of inversion to exceptional inverse loci in the plane of the parabolic Cartesian subspecies, $C = 1/0.2497 = 4.0048$. The 3 foci of self-inversion in the finite plane are marked by large open circles, while the 5 poles of inversion to central curves--labelled "pole to a," "pole to c," and "pole to f" --are marked by large solid circles (the 6th pole, modular to the 0° inter-oval-mode focus, lies off the figure to the left).

The poles of inversion to axial serpentine circular cubics--labelled "pole to b" and "pole to d"--are marked by the 7 intermediate-sized solid circles (the 8th, a "pole to c," lies off the figure to the left), while the smallest solid circles mark the positions of two poles of inversion to unexceptional axial curves--at the points where the circle of inversion of the 180° focus of self-inversion intersects the line of symmetry.

The 3 circles of inversion, centered on the 3 foci of self-inversion (and of corresponding modular radii) are shown by long dashes. Radii and chord segments from the 3 foci of self-inversion to points on the curve are illustrated with short dashes.

The pole of inversion of the basis curve a to curve e' is the exterior right Cartesian-condition pole. Inasmuch as this pole is incident upon the x-axis, the inverse x-axis is the line of symmetry of the present curve. Accordingly, the ensemble of inverse loci about poles on the line of symmetry of curve e' possesses the same members as the ensemble of inverse loci of curve a about poles on the x-axis (see Maxim 6-4-6).

The reflection of the pole of inversion in the line of symmetry, which is the exterior left Cartesian-condition pole, inverts to the 0° intra-oval-mode focus of self-inversion of curve e'. The center of curve a inverts to the modular pole on the side of the latter focus opposite the reciprocal inversion pole (which also is at modular distance). Curve e' inverts to curve a about the latter two modular poles. The y-axis of curve a inverts to the circle of inversion of the 0° intra-oval-mode focus, and all the axial inverse loci obtained by inverting curve a about poles on the y-axis are obtained by inverting curve e' about poles on this circle (Maxim 6-4-5).

The interior right Cartesian-condition pole of curve a inverts to the 0°

Relations Between Highly Symmetrical Cubics & Quartics That Self-Invert In Multiple Modes

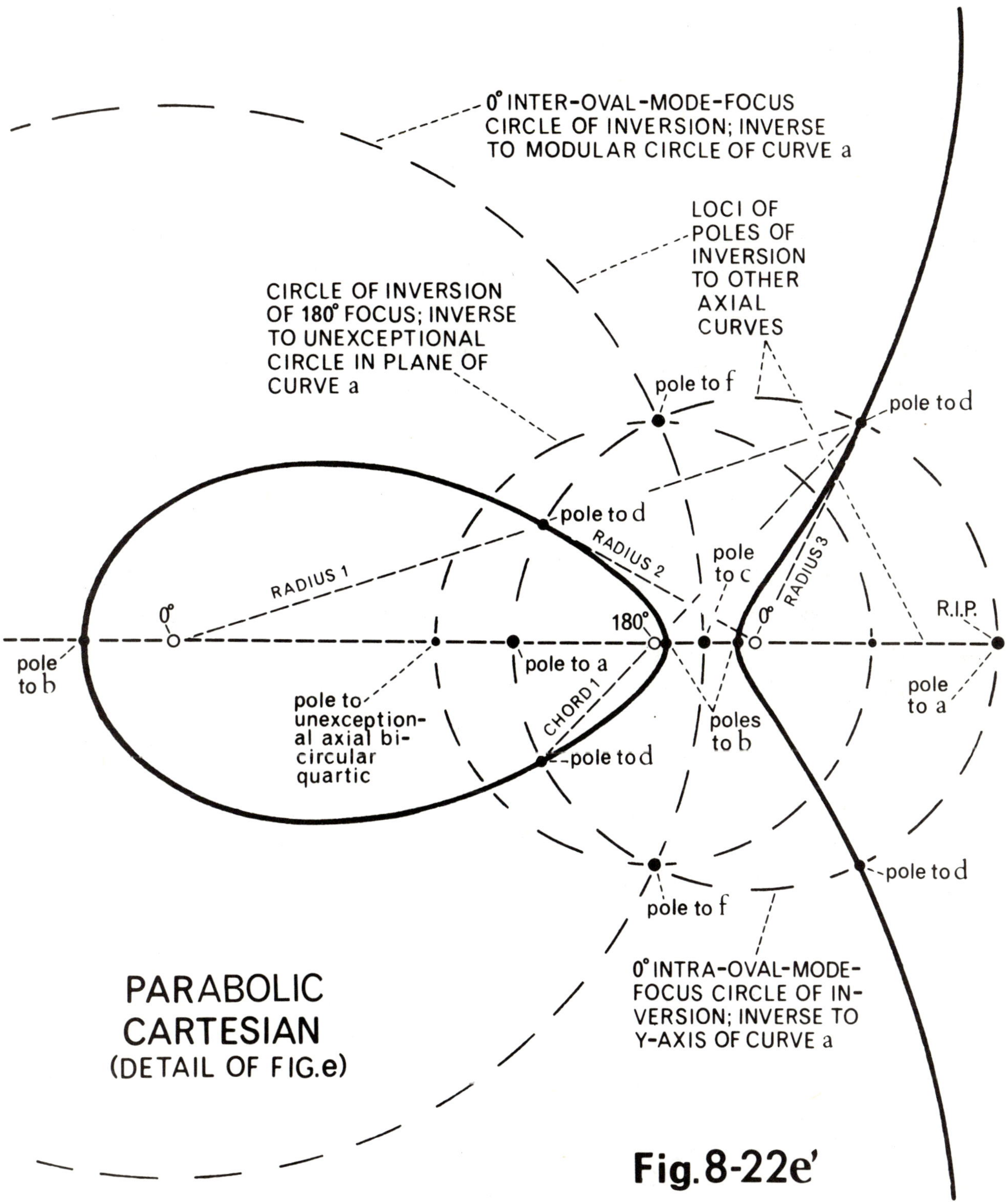

Fig.8-22e'

inter-oval-mode focus of self-inversion of curve e' (see Maxims 8-22-11 and 8-30-6), and the corresponding circle of inversion of this focus in the plane of curve e' is inverse to the modular circle of curve a. Accordingly, all the axial loci obtained by inverting curve a about poles on its modular circle also are obtained by inverting curve e' about poles on the circle of inversion of the 0° inter-oval-mode focus, including the 4 central curves obtained by inverting curve a about the 4 modular axial poles.

Inasmuch as the modular poles of curve a lie at the intersections of its modular circle with its lines of symmetry, the corresponding poles for inversions of curve e' to the same central loci occur at the intersections of the circle of inversion of the 0° inter-oval-mode focus with the loci inverse to the lines of symmetry of curve a in the plane of curve e'. The latter inverse loci are (a) the inverse y-axis circle (the circle of inversion of the 0° intra-oval-mode focus), the intersections with which are labelled "poles to f," and (b) the line of symmetry, the illustrated intersection with which is labelled "pole to c" (the other intersection of the inverse modular circle with the line of symmetry lies off the figure).

The interior left Cartesian-condition pole of curve a inverts to the 180° inter-oval-mode focus of self-inversion of curve e' (see Maxims 8-22-11 and 8-30-6). Thus, the 4 poles about which e' self-inverts, namely 3 foci of self-inversion in the finite plane and one point on the ideal line at infinity, are poles in the plane of e' inverse to the 4 poles about which curve a inverts to curve e' (Maxims 6-4-4 and 6-8-3). Unlike the circles of inversion of the two 0° foci of self-inversion, the circle of inversion of the 180° focus is inverse to an unexceptional circle in the plane of curve a, with the exception that the latter circle intersects the y-axis at the modular poles.

Accordingly, there are only 4 poles on the circle of inversion of the 180° focus in the plane of curve e' about which the inverse loci are axial curves: the inverse loci about the two poles where the 180° circle intersects the line of symmetry of e' are congruent with one another but otherwise are unexceptional axial bicircular quartics (Maxim 8-20-11); while the inverse loci about common points of intersection of the 180° circle with the other two circles of inversion (corresponding to the common intersections with the modular circle and the y-axis in the plane of curve a) yield curve f.

[If a centered annular or opposed-oval central Cartesian intersects the x-axis at points $\pm a$ and $\pm b$ $(b > a)$, and is inverted about a pole at $x = h$, then a circle, C', in the plane of the basis curve, centered at $2abh/(ab-h^2)$ of radius $\sqrt{(ab)(h^2+ab)/(h^2-ab)}$, inverts to the 180° circle of inversion, C, of the inverse axial $x = h$ locus. The pole in the plane of the basis curve that is inverse to the center of the circle C is the reflection of the modularly-inverse pole to h in the y-axis, i.e., $-j^2/h$, where j is the modular distance. The circle C' intersects the y-axis at $y = \pm j$.]

CIRCUMPOLAR MAXIMS 8-22-

9: *An axial bicircular quartic or serpentine circular cubic possesses one 180° and two 0° axial foci of self-inversion in the finite plane. The 3 circles of inversion centered on these foci (circles A and B at 0°; circle C at 180°) intersect in two common points at the axial level of the 180° focus, which lies midway between the points of intersection.*

10: *There are 3 loci in the plane of an axial bicircular quartic or serpentine circular cubic about which it inverts to other axial curves. These are the two circles of inversion, A and B, centered on the two 0° foci of self-inversion, and the line of symmetry, L. Inverse curves about points on the 3rd circle of inversion, C, centered on the 180° focus of self-inversion (except for the points where C intersects A, B, or L) are non-axial.*

11: *In the inversion of a central Cartesian about any member of a quartet of axial poles consisting of reflected modularly-inverse pole-pairs, the poles inverse to the other 3 members of the quartet are the foci of self-inversion in the finite plane of the inverse curve* (see also Maxim 8-30-6).

12: *The 180° circle of inversion, C, of an axial bicircular quartic or serpentine circular cubic inverts to an unexceptional (and uncentered) circle, C', in the plane of each of its 3 basis central curves. The inverse circle, C', intersects both lines of symmetry and the modular circle of each basis central curve; the two points of intersection of circle C' with the modular circle are coincident with the two points where the modular circle intersects one of the lines of symmetry.*

13: *An axial bicircular quartic or serpentine circular cubic inverts to central Cartesians about the 6 points of intersection of the line of symmetry, L, and the two circles of inversion, A and B, centered on the 0° foci of self-inversion (taking L, A, and B pairwise). The inverse central loci about the two points of intersection of circle A with line L have lines of symmetry inverse to A and L and a modular circle inverse to circle B; the inverse central loci about the two points of intersection of circle B with line L have lines of symmetry inverse to B and L, and a modular circle inverse to circle A; the inverse central loci about the two points of intersection of circles A and B have lines of symmetry inverse to A and B, and a modular circle inverse to line L.*

14: *An axial bicircular quartic or serpentine circular cubic inverts to non-axial curves about points in the plane that are not incident upon its line of symmetry or either of its circles of inversion, A or B, centered on the two (axial) 0° foci of self-inversion.*

15: *A non-axial bicircular quartic or serpentine circular cubic inverts to 3 pairs of subspecifically different central Cartesians about the 6 points of intersection of 3 mutually orthogonal 0° circles of inversion, A, B, and D. In a given inversion to a central curve, if a point of intersection of circles A and B provides the pole of inversion, the circles A and B invert to the two lines of symmetry; the 3rd circle, D, inverts to the modular circle. If a point of intersection of circles B and D provides the pole of inversion, then circles B and D invert to the two lines of symmetry and circle A inverts to the modular circle, etc.*

Inversions about the points of intersection of curve e' with its 0° circles of inversion and line of symmetry yield inverse loci that are subspecifically identical with those obtained by inversions of curve a about the points of intersection of the corresponding inverse loci with curve a (Maxims 6-4-4 and 8-22-18). For all 4 points of intersection of the line of symmetry with curve e', the inverse locus is serpentine circular cubic b, while for all 4 points of intersection of e' with the circle of inversion of the 0° intra-oval-mode focus (inverse to the y-axis of curve a), the inverse locus is the serpentine circular cubic d.

On the other hand, inverse loci about points of intersection of the circle of inversion of the 180° focus with curve e' yield non-axial and unexceptional serpentine circular cubics, the characteristics of which vary with the location of the pole of inversion of curve a.

It is a simple exercise to show that, given an incident pole on one oval of a non-central bicircular quartic, all the other incident poles on the ovals, about which inversions yield the same subspecies of inversion cubic as obtained by inverting about the given pole--i.e., all the *equivalent inversion poles*--can be located making use of the inversion-mapping Maxim 6-5-1.

For example, one first draws a radius vector through the given pole from the 0° inter-oval-mode focus, yielding the location of the equivalent inversion pole on the other oval (at the pole inverse to the given pole). A radius through each of the two poles from the 0° intra-oval-mode focus now yields an additional equivalent inversion pole on each oval (at the poles inverse to the first two poles). Appropriate chords from the 180° inter-oval-mode focus now yield additional equivalent inversion poles on the two ovals, etc.

In the case of incident poles for inversion to non-axial serpentine circular cubics, there are 4 poles on each basis oval (i.e., a total of 8 poles) for inversion to a given subspecies of inversion cubic. If the basis curve in question is axial, these poles can be located by means of radii emanating solely from the 0° foci of self-inversion. However, in the cases of incident poles for inversion to the axial-vertex inversion cubics (Figs. b and f), there are only two *equivalent inversion poles* on each oval. This means that only radii from the 0° inter-oval-mode focus and chord segments from the 180° focus intersect both ovals at the poles of inversion to axial-vertex cubics, and that the radii from the 0° intra-oval-mode focus are tangent to the ovals at the poles of inversion to these cubics (see Maxims 8-22-17, 8-30-4, and 8-38-1).

The situation is illustrated using the upper half of the parabolic Cartesian, e', as an example. Radius 1 from the 0° inter-oval-mode focus intersects the "pole to d" (i.e., the pole of inversion to curve d) on each oval, as do the segments of chord 1 from the 180° inter-oval-mode focus. On the other hand, radii 2 and 3 from the 0° intra-oval-mode focus are tangent to the ovals at each "pole to d."

CIRCUMPOLAR MAXIMS 8-22-

16: *The line of symmetry of the inverse axial curve obtained by inverting a biovular central Cartesian about a non-axial pole on its modular circle is inverse to the modular circle and passes through the two poles inverse to the reflections of the pole of inversion in the lines of symmetry (these poles become the 0° foci of self-inversion in the finite plane).*

17: *The single points of tangency with the curve, of radii from 0° intra-oval-mode foci of self-inversion of axial bicircular quartics, are the poles of inversion to axial-vertex serpentine circular cubics.*

18: *The poles inverse to the 8 vertices of annular central Cartesians lie at the points where the inverse curve intersects the loci inverse to the basis lines of symmetry.*

19: *A pair of equivalent inversion poles of a given central QBI curve, that are reflected in a line of symmetry, are modularly inverse in a focus of self-inversion in the finite or extended plane in all of its inverse loci.*

20: *A pair of equivalent inversion poles that are modularly inverse in a focus of self-inversion of a given mode (180° inter-oval, 0° inter-oval and intra-oval) in the finite or extended plane, in a given subspecies of curve, are modularly inverse in a focus of self-inversion of the same mode in all inverse loci of that subspecies.*

21: *A non-axial curve, B, that self-inverts at 0° and/or 180° about foci Q, R, etc., is inverted about a pole, P. The foci of self-inversion of the inverse locus, B', lie at poles, $P_Q^{*}{}'$, $P_R^{*}{}'$, etc., in the plane of B'. The latter poles are inverse to the modularly-inverse poles, P_Q^{*}, P_R^{*}, etc., of P in the foci of self-inversion, Q, R, etc., in the plane of B. 0° foci of self-inversion of B' may lie in the finite plane or on the ideal line at infinity; a 180° focus of self-inversion of B' lies only in the finite plane.*

Fig. 8-23. Mutually-inverting quartic circles as limiting complementary, confocal, central modular-focal quartics. Mutually-inverting quartic circles play an analogous role to curves of demarcation for complementary, confocal, central modular-focal quartics. The situation differs from that which generally applies when curves of demarcation are involved, in that the number of ovals is unchanged in making the transition from pre-limiting to post-limiting loci, and in that the eq. of the limiting curve is degenerate. It resembles the situation that generally applies when curves of demarcation are involved in that the limiting curve (or transition curve) is self-intersecting.

When complementary central modular-focal quartics (eqs.1a,b) are plotted as a confocal ensemble for a common value of d --the distance between the $\cos^2\theta$-condition foci--the ovals of the opposed-oval curve are located to either side of the interior oval and within the exterior oval of the annular curve. For very large negative values of n, the exterior oval approximates a circle (Fig. a) and the interior oval approximates a point-circle coincident with the latter's center.

general eq. of annular and opposed-oval central modular-focal quartics

(a) $$(x^2+y^2)^2 + nd^2x^2 + (n+1)d^2y^2 + d^4/16 = 0 \qquad (8\text{-}23\text{-}1)$$

(b) $$(x^2+y^2)^2+(3-2n)d^2x^2/(4n+2) + (2n+5)d^2y^2/(4n+2)+d^4/16 = 0$$

central modular-focal quartics complementary to those of eq. 1a

For example, for $n=-4{,}999.5$ (Fig. a), the axes of the annular exterior and interior ovals differ in length by only 0.01%, and the diameter of the former is 20,000 times as great as that of the latter. The nearly-circular ovals of the complementary opposed-oval curve ($n=-0.5002$) have 1.972 times the diameter of the interior annular oval, and they are "centered" on the $\cos^2\theta$-condition foci. The sizes of these 3 ovals relative to the size of the exterior annular oval are greatly increased in Fig. a (if shown at the same scale as the exterior annular oval, they would be invisible). In all other figures, the ovals of the complementary curves are to the same scale.

As the absolute magnitude of n (or B^2/d^2) decreases for the annular

Mutually-Inverting Quartic Circles As Limiting Complementary, Confocal, Central Modular-Focal Quartics

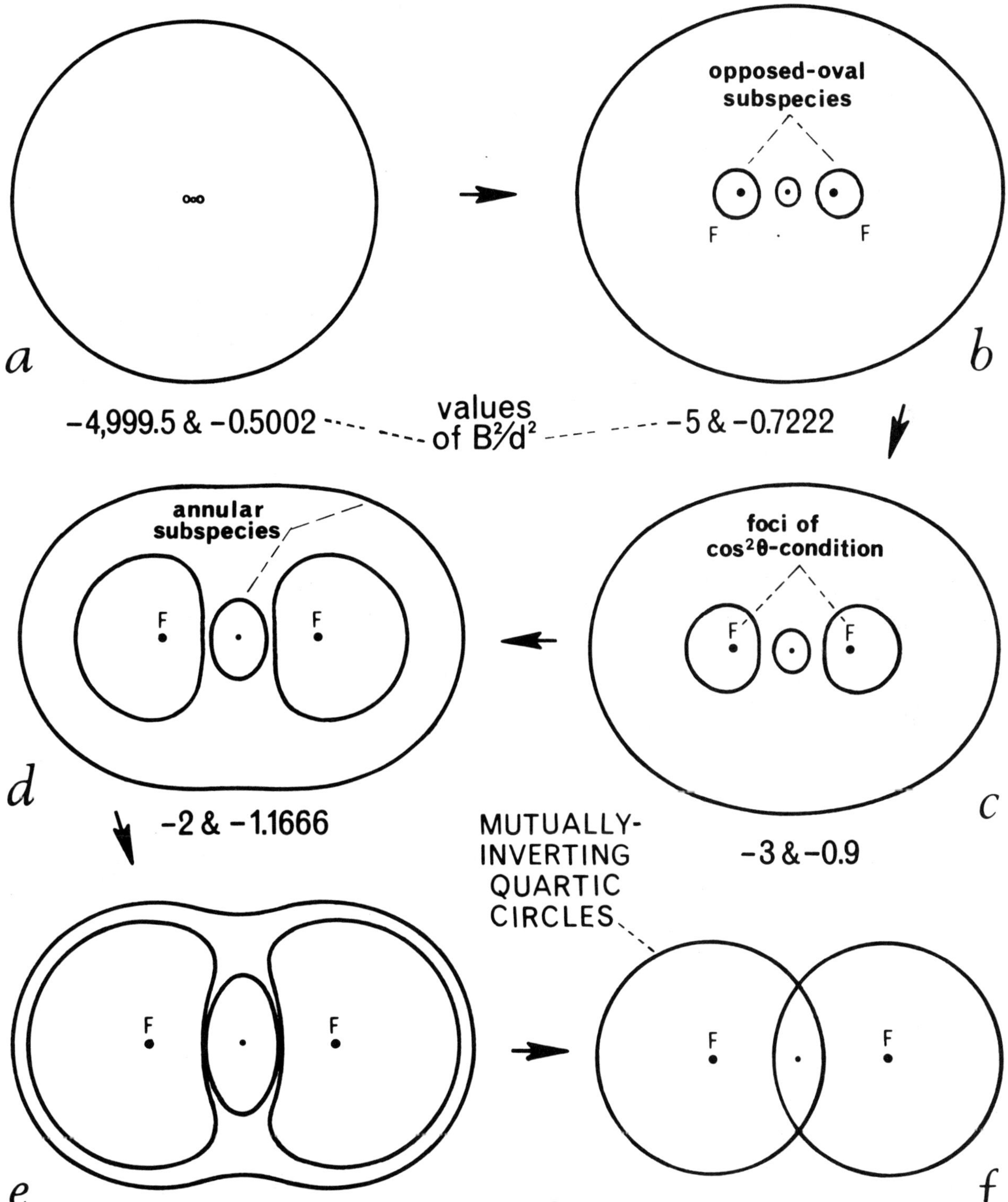

Fig. 8-23

curve (first value of each pair of values in Figs. a-f) and increases for the opposed-oval curve (second value of each pair of values), the size of the opposed ovals increases relative to the size of the exterior annular oval. Concomitantly, the relative size of the interior annular oval increases and both annular ovals elongate to elliptical form. The opposed ovals, on the other hand, flatten in their inner regions, where they face one another. The closer the values of n approximate to one another, the more the ovals of the two curves tend to nestle with one another, as best illustrated in Fig. e.

With increase of n from -1.6 to $-1\frac{1}{2}$ for the annular curve, and decrease from -1.409090909 toward $-1\frac{1}{2}$ for the opposed-oval curve, the ovals of the two loci come into increasingly close approximation. Concomitantly, the tips of the opposed ovals and the near regions of the annular ovals approach one another in the y-axis region, where they ultimately come into contact.

For the value $n = -1\frac{1}{2}$ the two loci are identical mutually-inverting quartic circles (eq. 2a) of radius $\sqrt{2}d/2$, and the eq. can be expressed in degenerate form as the product of two quadratic terms equated to 0 (eq. 2b). With increase of n from $-1\frac{1}{2}$, the former annular representative becomes an opposed-oval locus, while for decrease of n from $-1\frac{1}{2}$, the former opposed-oval locus becomes annular. The ensembles of complementary curves now progress in the reverse sequence from Fig. f to Fig. a.

mutually-inverting quartic circles as a special case of central modular-focal quartics

(a) $$(x^2+y^2)^2 - 3d^2x^2/2 - d^2y^2/2 + d^4/16 = 0 \qquad (8\text{-}23\text{-}2)$$

equation 2a expressed in degenerate form as the product of two quadratic factors which, when equated to 0, each represents one circle

(b) $$[(x-\tfrac{1}{2}d)^2 + y^2 - \tfrac{1}{2}d^2][(x+\tfrac{1}{2}d)^2 + y^2 - \tfrac{1}{2}d^2] = 0$$

If members of the annular and opposed-oval species are considered for all values of n, the opposed-oval subspecies nestle with one another and fill the central regions of the mutually-inverting quartic circles (Fig. f), the interior annular ovals nestle and fill the space in the region of overlap of the circles, and the exterior annular ovals nestle and fill the plane exterior to the two circles.

Thus, circles of one description or another are the limiting curves for both annular central modular-focal quartics (annular central parabolic Cartesians) and opposed-oval central modular-focal quartics. At one extreme, for $n = -1\frac{1}{2}$, both types of curves become a particular subspecies ($R = \sqrt{2}d/2$) of mutually-inverting quartic circles. The other extreme for the opposed-oval curves is achieved for $n = -\frac{1}{2}$, for which the loci are two point-circles, d units apart; for the annular curves, the loci approach a circle of infinite radius, with a point circle at its center, as n approaches negative infinity.

Fig. 8-24. Dual curves of demarcation, including an axial elliptical ($e = \frac{1}{2}$) QBI curve as one member. In the legend to Fig. 8-17 it was ascertained that the curves of demarcation of qualifying Cartesian self-inverters (genera of Cartesian self-inverters that possess both monovular and biovular species) are limacons (eqs. 1a,b), and it was asserted that a genus of Cartesian self-inverters exists for which a limacon of any given eccentricity, b/a, is a curve of demarcation. The formation of such genera is illustrated here, employing as a specific example a genus of Cartesian self-inverters (Fig. a) for which the $e = \frac{1}{2}$ elliptical limacon is a curve of demarcation.

simple intercept format of curves of demarcation of qualifying Cartesian self-inverters, cast about the focus of self-inversion

(a) $$\cos\theta = \frac{x^2+Ex+(G-E)^2/4}{Gx} \qquad (8\text{-}24\text{-}1)$$

(b) $$\cos\theta = \frac{x^2-bx+(a^2-b^2)^2/4b^2}{-a^2x/b}$$ simple intercept format of limacons cast about the focus of self-inversion

simple intercept format of a genus of confocal Cartesian self-inverters with a limacon curve of demarcation of eccentricity b/a (for the case $n = 1$, $b \neq a$)

(c) $$\cos\theta = \frac{x^2-nbx+(a^2-b^2)^2/4b^2}{-a^2x/b}$$

(d) $$\cos\theta = \frac{x^2-\frac{1}{2}nax+9a^2/16}{-2ax}$$ equation 2c for $b = \frac{1}{2}a$

(a) $n = (2a^2-b^2)/b^2$, 1 — values of n for which eq. 1c represents a limacon (8-24-2)

(b) $e = b\sqrt{n}/a = \sqrt{(2a^2-b^2)}/a$ — value of e for the second limacon curve of demarcation of eq. 1c, $b \leqslant \sqrt{2}a$

Inversions of these exemplified Cartesian self-inverters about the two axial poles at modular distance from the common focus of self-inversion yield two monovular and two biovular species of central Cartesians (Figs. b and c); the $e = \frac{1}{2}$ ellipse is a curve of demarcation of one central ensemble (Fig. c) and the $e = \frac{1}{2}$ elliptical central quartic is a curve of demarcation of the other (Fig. b). Both central ensembles (each consisting of both of the biovular species and one of the monovular species) also possess a hyperbolic ($e = \sqrt{7}/2$) central QBI curve of demarcation, and both ensembles invert to one another about their centers.

Eq. 1a has the general form of the simple intercept format for Cartesian self-inverters (eq. 8-13-5), but with the constant term having the particular value (in terms of the parameters E and G) that yields the genus *limacons.* In limacon parameters (limacon eq., $r = a - b\cos\theta$) this becomes eq. 1b. Any alteration of eq. 1b (or 1a) that changes the relationships between the coefficients, changes the loci represented from limacons to other Cartesian self-inverters.

One can produce an alteration that yields a genus of Cartesian self-inverters with a limacon of eccentricity b/a as a curve of demarcation simply by appending a dimensionless multiplicative coefficient n (the number adjacent to each locus in Figs. a-c) to the coefficient of any of the 4 terms in the expressions 1a or 1b, where n can range over all real values. Since the general form of the format is retained, the genus produced, say by altering eq. 1b in this manner, has a curve of demarcation consisting of a limacon of eccentricity b/a as its subspecies, $n = 1$.

This is illustrated by appending a multiplicative factor n to the coefficient of the linear term in the numerator of eq. 1b, yielding eq. 1c. For the purpose of illustration, eq. 1c is made specific for a genus of Cartesian self-inverters that has the $e = \frac{1}{2}$ elliptical limacon as a curve of demarcation (eq. 1d)--just as the genus *parabolic Cartesians* has the $e = \sqrt{2}$ hyperbolic limacon as its curve of demarcation. Representative subspecies of the genus are plotted in Fig. a. As in cases of hyperbolic curves of demarcation,

Dual Curves of Demarcation, With An Axial Elliptical (e =0.50) QBI Curve As One Member

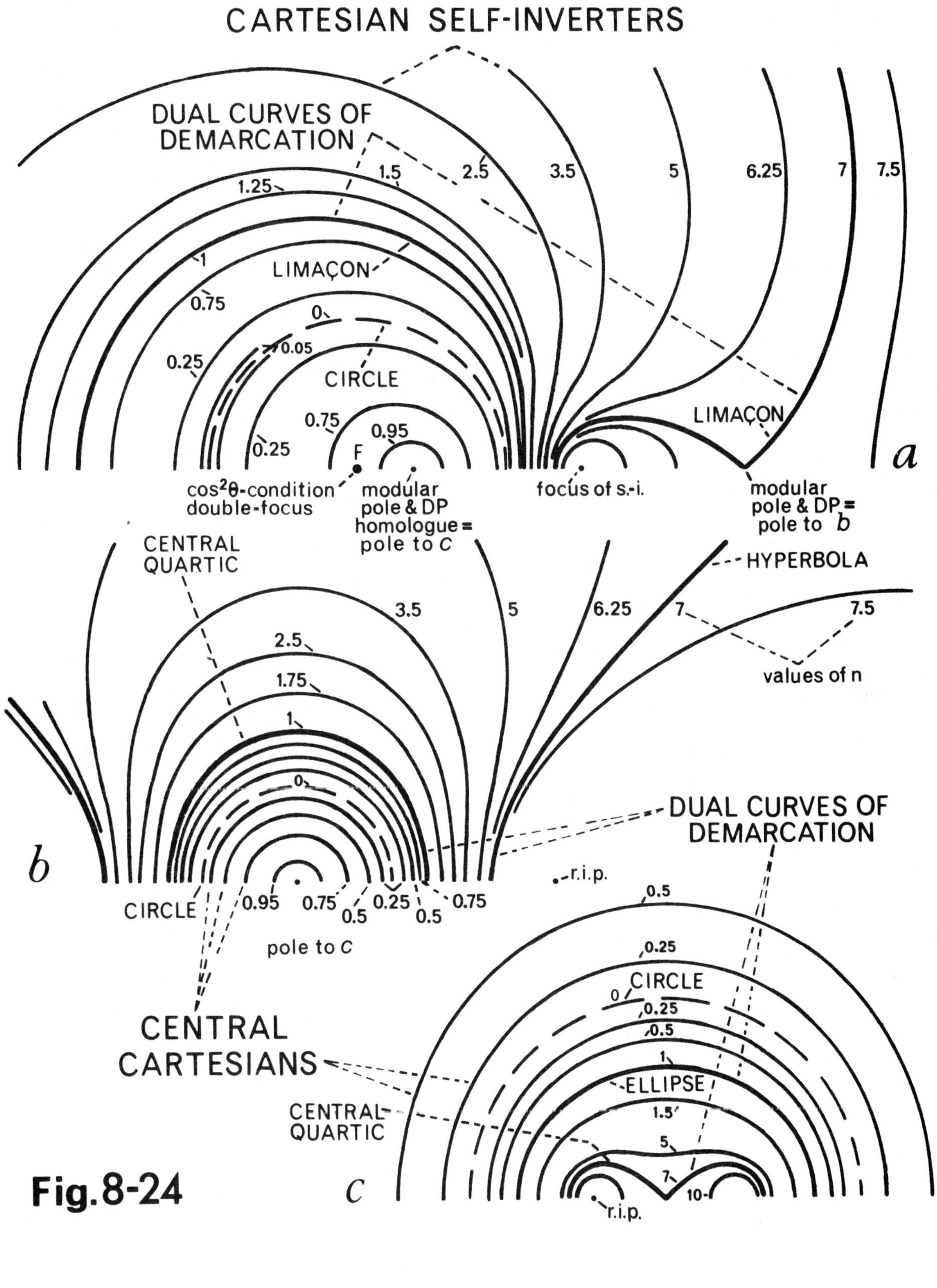

Fig.8-24

the $e=\frac{1}{2}$ elliptical limacon demarcates a monovular and an annular biovular species, but it does so in a distinctively different, though not unrelated, manner.

For n increasing progressively from the value 1 (the value for the curve of demarcation), the monovular species consists of subspecies of progressively increasing size, lying at progressively greater distances from the curve of demarcation, each enclosing all the subspecies for lesser values of n. Conversely, for values of n decreasing progressively from 1, the subspecies consist of pairs of annular biovals, with one member of the pair being of progressively increasing size and the other being of progressively decreasing size. The largest members, of progressively decreasing size, lie just within the perimeter of the curve of demarcation at progressively increasing distances therefrom; the smallest members, of progressively increasing size, encircle the DP homologue--which is an isolated DP of the limacon locus--at progressively greater distances.

[In this connection, it should be noted that elliptical QBI cubics and quartics possess two incident but isolated points at the DP homologue, though these generally are neither recognized nor illustrated. This is evident from an examination of the rectangular eqs. of limacons, $(x^2+y^2+bx)^2=a^2(x^2+y^2)$, central quartics, $(x^2+y^2)^2=j^4(x^2/a^2\mp y^2/b^2)$, and central conic axial-vertex cubics, $y^2(x\pm aj^2/2b^2)=x^2(j^2/2a-x)$. In polar coordinates these become $r^2\{(r+b\cos\theta)^2-a^2\}=0$, $r^2\{r^2-j^4(\cos^2\theta/a^2\mp\sin^2\theta/b^2)\}=0$, and $r^2\{\sin^2\theta(r\cos\theta\pm aj^2/2b^2)-\cos^2\theta(j^2/2a-r\cos\theta)\}=0$, respectively, all of which are degenerate and possess a double-root $r^2=0$, which applies for both the hyperbolic and elliptical species.]

In the case of hyperbolic curves of demarcation, one can view the DP as being the point of contact of the two ovals of the biovular species at the transition point between biovular and monovular curves. In the present instance, one can view the DP homologue as being the point to which the inner oval of the annular biovular pair "collapses" in the same type of transition (thereafter being represented by two imaginary vertices).

The loci of Fig. a are confocal about two foci--the 0° intra-oval-mode focus of self-inversion (labelled "focus of s.-i.") and the $\cos^2\theta$-condition double-focus (so identified, and labelled F). The latter focus is not at modular distance from the former one, as it is in parabolic Cartesians; rather, it is the DP homologue of the $e=\frac{1}{2}$ limacon that is at modular distance.

As n approaches the value 0, the interior oval of the annular pair progressively enlarges and the exterior oval progressively diminishes in size.

Concomitantly, both approach circularity and become more centralized within the curve of demarcation. At $n = 0$, the *limiting locus* consists of a single circle of radius $\sqrt{(a^2-j^2)} = \sqrt{(1-0.5625)}a = 0.661437828a$, centered at the $\cos^2\theta$-condition double-focus. This phenomenon is a consequence of the presence of the multiplicative coefficient n as a component of the coefficient of the linear term in the numerator since, when this term vanishes, the simple intercept format becomes that of a circle.

Had the coefficient been appended to the x^2 term, the limiting locus would have been a hyperbolic limacon, while had it been appended to the constant term, the limiting locus would have been a hyperbola. In the latter cases, the unit of linear dimension for self-inversion would have been different for each subspecies of Cartesian self-inverter (see eqs. XIV-43c,c') and some of the loci would have intersected, whereas in the present case the modular distance is constant $(9a^2/16)$ and the curves are space-filling.

Consider now the changes that occur as n increases progressively toward the value 7. In this case, the loci undergo the same type of sequence of progressive changes that parabolic Cartesians undergo as C increases progressively toward the value 4 (Fig. 8-16), and a second curve of demarcation is formed for $n = 7$. Accordingly, this genus of Cartesian self-inverters possesses *dual curves of demarcation*. Furthermore, it is characteristic of the genera of Cartesian self-inverters formed in this manner that when dual curves of demarcation exist, they occur in elliptical-hyperbolic pairs.

The existence of a second curve of demarcation depends upon the fact that for all genera for which $b < \sqrt{2}$, i.e., for all genera for which the eccentricity of the QBI curve of demarcation is less than $\sqrt{2}$, there is a second value of n, in addition to $n = 1$, for which eq. 1c represents a limacon, namely the first value given by eq. 2a [or, for eq. 1a, $n = (2G-E)/E$]. The eccentricity of the second curve of demarcation in each genus of eq. 1c is given by eq. 2b, whereupon it is evident that for $b/a < 1$ (where $e = b/a$ for the first curve of demarcation), $e > 1$, and for $b/a > 1$, $e < 1$. For eq. 1d, the value of e for the second curve of demarcation is $\sqrt{7}/2$.

It follows from principles of Inversion Taxonomy that if the $e = \frac{1}{2}$ and $e = \sqrt{7}/2$ limacons are curves of demarcation of the genus represented by eq. 1d, then a pair of corresponding QBI curves of eccentricities $\frac{1}{2}$ and $\sqrt{7}/2$ will be curves of demarcation for every genus in the superfamily obtained by inverting the ensemble of eq. 1d, i.e., the ensemble of Fig. a, about all points in the plane. Moreover, it also follows that a circle will be a *limiting locus* for every genus in the superfamily obtained by inverting about points in the plane not incident upon the limiting circle of Fig. a, and that a line will be a limiting locus for genera obtained by inverting about all points in the plane incident upon the limiting circle.

As examples of other QBI curves of demarcation for other monovular and biovular species in the superfamily obtained by inverting the loci of Fig. a about all points in the plane, the ensembles of Figs. b and c are obtained by inverting about the two poles modular to the common focus of self-inversion of the curves in the ensemble of Fig. a (labelled "pole to b" and "pole to c"). In the case of the curve of demarcation that is a hyperbolic limacon, one of these modular poles is the DP, inversion about which yields a hyperbola (Maxim 13) of the same eccentricity ($\sqrt{7}/2$), while the other modular pole is the pole for inversion to a hyperbolic central quartic (also of eccentricity $\sqrt{7}/2$; Maxim 8-10-4).

On the other hand, in the case of the curve of demarcation that is an elliptical limacon, the first modular pole mentioned above is the pole for inversion to an elliptical central quartic of the same eccentricity ($\frac{1}{2}$), while the second modular pole mentioned above is the DP homologue, inversion about which yields the $e = \frac{1}{2}$ ellipse. Accordingly, when both limacons are inverted about one or the other common modular pole, the two QBI curves obtained belong to different genera; one is a *central conic* and the other a *central quartic* (see Figs. b and c).

The situation is different for the Cartesian self-inverters. For the single monovular species ($1 < n < 7$), inversion about the two modular poles yields two different, mutually-inverse monovular central species. This ensues because the only genus of Cartesian self-inverts with monovals that invert to subspecifically-identical central curves about both modular poles is the genus *parabolic Cartesians* (see legend to Fig. 8-15).

On the other hand, in the case of the (two) biovular species, it has been shown (see Maxims 8-20-9,10) that the loci obtained by inverting a given subspecies about the two poles modular to any 0° focus of self-inversion are subspecifically-identical central curves. Accordingly, it follows that the same two biovular species occur in both Figs. b and c.

In the case of inversions of one of the annular biovular species of Fig.

a (the member subspecies of which lie within the elliptical limacon) about the two modular poles, since one pole is interior to both ovals and the other is exterior to both ovals, all the inverse loci also are annular; this is the origin of the annular loci of Fig. b (DP inversions) and Fig. c (DP-homologue inversions). On the other hand, in the case of inversions of members of the other annular species of Fig. a about the two modular poles, both poles occupy inter-oval locations, with the result that the inverse loci consist of central opposed ovals. This is the origin of the members of the single species of central curves with opposed ovals in Figs. b and c.

Accordingly, the essential differences between Figs. b and c lie in the facts that: (1) each figure represents members of a different monovular species; (2) the members of the annular species that lie interior to the monovals of Fig. b lie exterior to them in Fig. c; (3) the members of the opposed-oval species that lie exterior to the monovals of Fig. b lie interior to them in Fig. c; and (4) the members of the annular species lie between the opposed ovals of members of the other biovular species in Fig. b, but exterior to them in Fig. c.

As a result of these differences, the members of each pair of curves of demarcation in Figs. b and c are central inversions of one another that occupy "complementary" locations. The homologies of the present Figs. b and c with Figs. a and b, respectively of Fig. 8-19 are evident, as are the differences. The differences consist in the fact that: (1) the curves are members of different inversion superfamiles; (2) QBI curves of different eccentricities are the demarcating subspecies; (3) the monovular subspecies in Figs. 8-19a,b belong to a single species but those in Figs. b and c belong to two different species; (4) members of 3 species are demarcated by *dual curves of demarcation* in both Figs. b and c, but members of only two species are demarcated by single curves of demarcation in Fig. 8-19; and (5) both hyperbolic and elliptical curves of demarcation are present in both Figs. b and c, but only hyperbolic curves are present in Fig. 8-19.

QBI Curves of Unit Eccentricity As The Curves of Demarcation of Elliptical And Hyperbolic QBI Species

Lastly, following the thesis developed in the legend of Fig. 8-17, it should be pointed out that every curve of given non-unit eccentricity (b/a) in the QBI superfamily is a curve of demarcation for one or another of the superfamilies derived from the genera of Cartesian self-inverters of eq. 1c (defined by the specific value of b/a). This means that at least one member of the QBI superfamily intersects every genus of every superfamily derived from inversions of the loci of eq. 1c (for different values of b/a), and that the entire QBI superfamily, excepting only curves of unit eccentricity, intersects the ensemble of superfamilies based on the loci of eq. 1c (see also a related discussion in the legend to Fig. 8-17, and Table 8-17-1.)

In the exceptional case $b/a=1$, the cardioid becomes the curve of demarcation, limacons become the genus of demarcated Cartesian self-inverters, and the superfamily to which the curves being demarcated belong is the QBI superfamily. In other words, in every genus in the QBI superfamily, the subspecies of unit eccentricity is the curve of demarcation of members of the elliptical and hyperbolic species. Thus, for example, the parabola is the curve of demarcation of ellipses and hyperbolas.

Fig. 8-25. The $e=2$ hyperbolic limacon as the curve of demarcation of true exchange-limacons. *True exchange-limacons* are related equationally to limacons by the fact that their rectangular eq. (1a) differs from that of limacons (1b) only in that the exponents of the expressions in parentheses are exchanged (the constant a^2 also must be transposed to maintain dimensional balance; see page XI-10). Except for the $e=2$ limacon curve of demarcation, however, this genus includes no limacon. Another equationally-related genus--*pseudo-exchange-limacons*--intersects this genus in its curve of demarcation (the $e=2$ limacon) and belongs to the QBI superfamily (see page XI-10).

(a) $a^2(x^2+y^2+bx) = (x^2+y^2)^2$ true exchange-limacons (8-25-1)

(b) $(x^2+y^2+bx)^2 = a^2(x^2+y^2)$ limacons, $r=a-b\cos\theta$

(a) $\cos\theta = x(x^2-a^2)/a^2b$ simple intercept format of eq. 1a cast about a vertex (8-25-2)

(b) $4h^2x^2\cos^2\theta+[4h(x^2+h^2)-a^2(2h+b)]x\cos\theta + (x^2+h^2)^2-a^2x^2-a^2h(b+h) = 0$ simple intercept eq. of 2a

(c) $8bh^3 + 4a^2h^2 + 4a^2bh + a^2b^2 = 0$ radicand condition of 2b, defining 1 or 3 foci of self-inversion

The e=2 Hyperbolic Limacon As The Curve of Demarcation of True Exchange-Limaçons

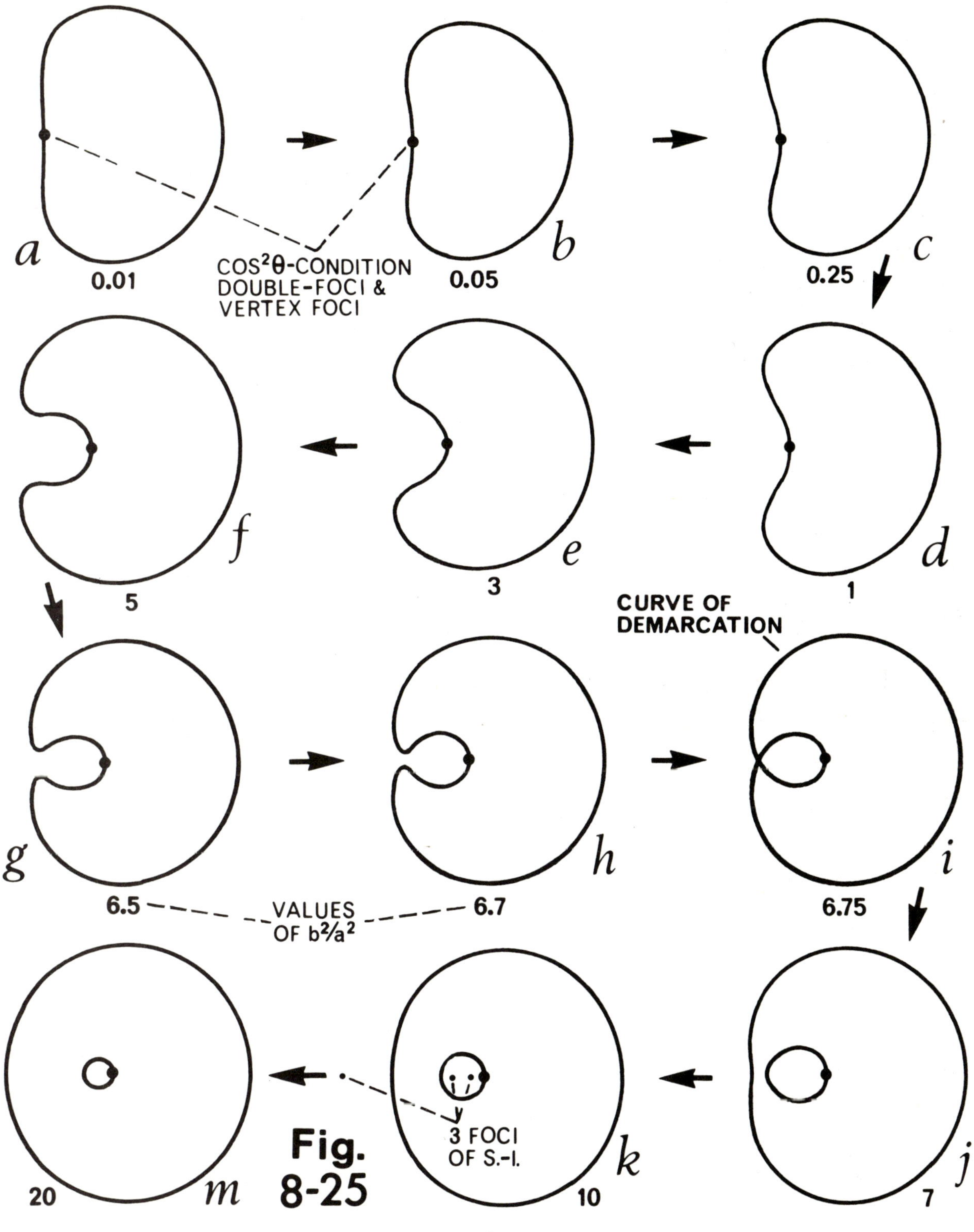

Fig. 8-25

One expects true exchange-limacons to be curves of high circumpolar symmetry, inasmuch as their eq. possesses the same expressions--"building blocks"--as occur in the eqs. of limacons, but with different exponents (see page XIV-47, *Design And Synthesis of Highly Symmetrical Curves, The Building-Block Approach*). Representative curves of this genus are illustrated here and their taxonomic affinities elucidated. One can surmise from their appearance and the fact that the $e=2$ limacon is their curve of demarcation (see also Fig. 8-16) that the curves belong to a genus of Cartesian self-inverters or some other genus of axial monovular and bicircular quartics. In fact, these curves belong to a genus of Cartesian self-inverters with a very high degree of circumpolar symmetry, a genus characterized by the fact that the $\cos^2\theta$-condition double-focus is in coincidence with a vertex.

Twelve representative subspecies ($a^2 = 0.01b^2$ to $20b^2$) of true exchange-limacons are illustrated in the figure, including the $e=2$ limacon curve of demarcation ($a^2 = 6.75b^2$). The coordinate origin of eqs. 1a and 2a, which also is a vertex, and the positions of the $\cos^2\theta$-condition double-foci, are marked by the large solid circles. Values of b^2/a^2 are given beneath or adjacent to each curve. The positions of the 3 foci of self-inversion (the monovular species have only one real focus of self-inversion) are marked by the 3 small solid circles in Fig. k.

The curves of this figure closely parallel those of Fig. 8-16, which illustrates the equilateral limacon as the curve of demarcation of parabolic Cartesians. Aside from the locations of the $\cos^2\theta$-condition foci, they differ notably in the fact that the small ovals of the annular subspecies and the concavities of the monovular subspecies are relatively greater in size than those of parabolic Cartesians.

The taxonomic affinities of these curves are elucidated by obtaining the simple intercept eq. and simple intercept format. The coordinate origin of the curve cast in the form 1a and 2a is at the vertex of the concave region of the monovals (i.e., the vertex in the region that is concave in some subspecies) and at a vertex of the small oval of the annular subspecies. Accordingly, the format (2a) is not a format of self-inversion, since a closed curve cannot self-invert about an incident pole.

Inspection of the simple intercept eq. 2b reveals the nature of the unique symmetry property of true exchange-limacons, namely that they possess a generic $\cos^2\theta$-condition double-focus ($h^2 = 0$) that is incident upon the curve. Accordingly, if the curves were axial bicircular quartics, thay would have to be Cartesian self-inverters, since only the latter among all bicircular quartics possess $\cos^2\theta$-condition double-foci.

[Possession of a $\cos^2\theta$-condition double-focus that is incident upon a vertex is not unique to true exchange-limacons. This situation also occurs in some subspecies of other genera. For example, this is the *vertex-coincidence symmetry condition* of some subspecies of linear Cartesians (see pages X-24, XIV-40 and 41). In true exchange-limacons, however, the $\cos^2\theta$-condition double-focus is *generic*, i.e., it is a property of all subspecies. In the case of linear Cartesians, no curve of demarcation exists, inasmuch as there are no monovular species (this is the case, for example, whenever F^2 of eq. 8-13-5 is negative). The *limiting curves* (the curves formed in the limit when the annular ovals come into contact) are limacons.]

Verification that true exchange-limacons are axial bicircular quartics is achieved readily by general analyses or by tests of the inverse loci about specific poles. If eq. 2b is solved for $\cos\theta$ (the $\cos\theta$ simple intercept format), the cubic radicand condition 2c is obtained. For all h satisfying 2c, the radical vanishes and the circumpolar intercept transforms reduce. Thus, eq. 2c defines 3 axial circumpolar foci, one of which is real in all species, but all 3 of which are real in the annular species (i.e., for $a^2 > 6.75b^2$). It already has been noted that foci of self-inversion generally are defined by radicand conditions, and it readily is demonstrated that the foci in question, when real, are foci of self-inversion.

If one inverts the true exchange-limacons of eq. 2a about a pole h and then translates the origin of the resulting curves to H ($x+H$ translation), one obtains the simple intercept eq. for the inverse loci of true exchange-limacons about an axial pole h (a complicated expression that need not be reproduced here). It is known from the analyses of preceding figures that if the annular curves are bicircular quartics, 4 axial poles exist about which they invert to central curves, and that two axial poles exist about which the monovular subspecies invert to central curves.

If one applies the conditions for a centered central locus to the simple intercept eq. referred to above (namely, that the coefficients of the $r\cos\theta$ term and the $r^3\cos\theta$ term both vanish), one obtains the eqs. 3a,b for H and h. Though only 4 *axial* poles exist for inversion of bicircular quartics to central curves, eq. 3b, which defines these poles, is of 6th degree. For the monovular species, $a^2 > 6.75b^2$, only two of the poles are real, whereas for the biovular species 4 of them are real. It readily is ascertained that the real solutions represent the poles modular to the one or two 0° foci of self-inversion defined by eq. 2c.

location of center of central true exchange-limacons obtained by inverting the curves of eq. 2a about poles h defined by eq. 3b

(a) $$H = \frac{j^2[4h^3-a^2(2h+b)]}{4h[a^2(b+h)-h^3]} \qquad (8\text{-}25\text{-}3)$$

(b) $8bh^6 + 8a^2h^5 + 20a^2bh^4 + 20a^2b^2h^3 - 2a^4b^2h - a^4b^3 = 0$ — eq. defining the locations of poles h about which eq. 2a inverts to central curves

(c) $(2h+3b)^3(h+3b/2)^2(h-3b/4) = 0$ — solution to eq. 3b for the curve of demarcation, $a^2 = 6.75b^2$

In the case of the monovals, the 4 non-real roots define poles of inversion to central curves in the complex domain. On the other hand, for the biovular species the two non-real roots correspond to two additional real but *non-axial* poles for inversion to central curves, namely, the off-axis poles of intersection of the circles of inversion of the two 0° foci of self-inversion (see Fig. 22e', "poles to f"). In other words, only non-real solutions for *axial* locations for two of the poles exist because these two poles are *non-axial*.

In the case of the curve of demarcation, 6 real roots exist for axial poles of inversion to central curves because all 6 poles have axial locations. Thus, there is a quintuple root at the DP, $h = -3b/2$, which is a modular pole of the 0° inter-loop-mode focus, and a single root at $h = 3b/4$, the other modular pole of the same focus. The other 4 roots at the DP exist by virtue of the fact that the 0° intra-oval-mode focus merges with the DP in the curve of demarcation (see Fig. 8-16, sequence k,j,i) and has a modular value of 0 for self-inversion. Accordingly, it is coincident with both of its axial modular poles, and with both of the intersections of its circle of inversion (of radius 0) with the circle of inversion of the 0° inter-loop-mode focus.

The situation is visualized readily by reference to Fig. 8-22e'. When the two ovals merge to form the curve of demarcation, the radius of the 0° intra-oval-mode circle of inversion shrinks to 0 (as does that of the 180° focus), carrying both "poles to f" and both "poles to a" into coincidence with the DP and one "pole to c," accounting for the quintuple root for that location.

It is evident that a general expression for true exchange-limacons would involve the general solution to the 6th-degree eq. 3b, which is not achieved readily. One can, nonetheless, derive the eqs. of the central curves by a procedure which places restrictions on the values of the parameters, E, F^2, and G.* Thus, one notes from eq. 8-30-9 that the $\cos^2\theta$-condition double-focus of Cartesian self-inverters lies at $h = \frac{1}{2}G$. For the orientation of the true exchange-limacons of the present figure, this means that G must be positive. But for positive G and monovals with 0° intercepts to the right of the origin, E must be negative (see Fig. 8-13).

It follows from eqs. 8-13-8 that the location of the vertex with which the $\cos^2\theta$-condition focus is in coincidence (the near vertex) is given by eq. 4a. Cancelling $\frac{1}{2}G$ and rearranging yields eq. 4b. Since, for the far vertex of the monovals, the eq. would be the same but with a positive sign before the radical, it is evident that the double-focus of the $\cos^2\theta$-condition comes into coincidence only with the near vertex of the monovals--since E is negative--and with the corresponding vertex of the interior oval of the annular curves.

location of vertex at coordinate origin of eqs. 1a and 2a equated to location of $\cos^2\theta$-condition double-focus of eq. 8-13-9, in terms of general parameters for Cartesian self-inverters

(a) $$h = \frac{(G-E)-[(G-E)^2-4F^2]^{\frac{1}{2}}}{2} = \frac{G}{2}, \quad G>0,\ E<0 \qquad (8\text{-}25\text{-}4)$$

(b) $E = -[(G-E)^2-4F^2]^{\frac{1}{2}}$ — 4a after cancelling $\frac{1}{2}G$ and rearranging

(c) $F^2 = G(G-2E)/4$ — solution to eq. 4b for F^2

simple intercept format of true exchange-limacons cast about the 0° inter-oval-mode focus of the annular curves and the 0° focus of the monovular subspecies

$$\cos\theta = \frac{x^2+Ex+G(G-2E)/4}{Gx}, \quad G>0,\ E<0 \qquad (8\text{-}25\text{-}5)$$

Since it already has been shown that vertex coincidence of the focus in question occurs only with the vertices of the latter oval (pages XIV-40,41),

*[and on the orientation of the loci]

it is evident that for Cartesian self-inverters the $\cos^2\theta$-condition double-focus comes into coincidence only with the vertices illustrated in the figure (large solid circles).

Solving eq. 4b for F^2 leads to eq. 4c. Substituting for F^2 from eq. 4c in eq. 8-13-5 leads to the simple intercept format of eq. 5 for true exchange-limacons cast about the 0° focus of self-inversion of the monovular species and the 0° inter-oval-mode focus of the annular species. The eq. for limacons in the same orientation as the loci of eq. 5 (eq. X-2) is eq. 6, from which it is evident that the limacon condition for eq. 5 is eq. 7b, i.e., when $E = -4G$, eq. 5 is the curve of demarcation.

e = 2 limacon curve of demarcation (Fig. i) cast about the focus of self-inversion (a and b, limacon parameters)

$$\cos\theta = \frac{x^2-bx+(a^2-b^2)^2/4b^2}{a^2x/b} \qquad (8\text{-}25\text{-}6)$$

limacon condition of eq. 6 translated into Cartesian self-inverter parameters ($E < 0$, $G > 0$)

(a) $$\frac{(a^2/b-b)^2}{4} = \frac{(a^2-b^2)^2}{4b^2} = \frac{(E+G)^2}{4} = \frac{G(G-2E)}{4} \qquad (8\text{-}25\text{-}7)$$

(b) $E = -4G$ limacon condition for eq. 5

Having obtained the simple intercept format for true exchange-limacons cast about the 0° focus of self-inversion of the monovals (and the 0° inter-oval-mode focus of the biovals) the eqs. of the corresponding opposed-oval central curves (for example, the basis curve of Fig. 8-26) can be obtained readily by substituting in eq. 8-15-2b, where one makes the substitutions of eq. 4c, namely replacing F^2 by $G(G-2E)/4$ and F by $[G(G-2E)/4]^{\frac{1}{2}}$. For example, the corresponding resulting eqs., taking the locus of Fig. k ($a^2 = 10b^2$) as the basis curve, are eqs. 8a and 9a.

Having obtained the eqs. for opposed-oval central curves, the annular central curves can be obtained by making the substitutions of eq. 8-21-7 or 8-31-3. The second species of opposed-oval central curve is obtained by rotating the annular curve 90° (exchanging the coefficients of the x^2 and y^2 terms) and repeating the substitutions of eq. 8-21-7. In other words, the opposed-oval central curves are obtained by inverting the annular central curves about the poles at modular distance on both lines of symmetry.

simple intercept format of curve k ($a^2 = 10b^2$) cast
about the 0° inter-oval-mode focus of self-inversion

(a) $$\cos\theta = \frac{x^2-3.577089346bx+1.550478045b^2}{0.781520546bx} \qquad (8\text{-}25\text{-}8)$$

(b) vertices; h = 0.3907602733b, 3.967849717b, -0.762707032b, -2.032861866b
0° inter-oval-mode focus = origin, h = 0, modulus = 1.245181931b
0° intra-oval-mode focus, h = -3.367262662b, modulus = 3.128574722b
180° inter-oval-mode focus, h = -0.460456518b, modulus = 1.155917385b
$\cos^2\theta$-condition focus, h = 0.3907602733b

opposed-oval central true exchange-limacon (Fig. 8-26)
obtained by inverting curve k about poles modular
to the 0° inter-oval-mode focus of self-inversion

(a) $$r^4+1.111442244j^4r^2/b^2-3.947997771(j^4r^2\cos^2\theta)/b^2 + 0.122825181j^8/b^4 = 0 \qquad (8\text{-}25\text{-}9)$$

(b) vertices; h = ±1.671099196b, ±0.209720787b
modular distance = 0.59200020b
$\cos^2\theta$-condition foci, h = ±0.993478456b
poles of Cartesian condition, h = ±0.401547747b, ±0.872783471b
poles modularly-inverse to the $\cos^2\theta$-condition foci, h = ±0.352764808b

Aside from limacons, 3 genera of Cartesian self-inverters with exception-symmetry properties now have been considered:

1. linear Cartesians (Fig.2-1e, pages II-5,6, X-32 to 37, XIV-38 to 40), which have the greatest bipolar symmetry;
2. parabolic Cartesians, which have the 2nd-greatest bipolar symmetry, are the only Cartesian self-inverters for which a pole of inversion to a central curve has circumpolar point-focal rank, and invert to several axial and central genera with exceptional circumpolar symmetry properties (see Fig. 8-22 and legend);
3. true exchange-limacons, which have the exceptional circumpolar symmetry property that the $\cos^2\theta$-condition double-foci are coincident with a vertex of the interior oval of the biovular species.

[Multiple-foci convey exceptional circumpolar symmetry properties by virtue of the fact that the corresponding circumpolar intercept transforms reduce in degree or simplify, as compared to those of subspecies, species, or genera in which the corresponding components of the multiple-foci are not in coincidence.]

Fig. 8-26. Tracking the vertices and $\cos^2\theta$-condition foci of axial inversions of an opposed-oval central Cartesian (the $a^2 = 10b^2$ central true exchange-limacon). The present figure illustrates the characteristic difference between biovular Cartesian self-inverters that belong to the Cassinian-based inversion superfamily (Fig. 8-21) and those that belong to the inversion superfamily based on central true exchange-limacons, namely, that the $\cos^2\theta$-condition double-foci of the biovular species of the latter are in coincidence with a vertex of the interior oval, whereas those of the former are not. It also reveals that, though the $\cos^2\theta$-condition double-foci are in coincidence with an oval vertex of the corresponding serpentine circular cubics of the former curves, this is not the case for any vertex of the serpentine circular cubics of central true exchange-limacons.

The format of the present figure follows that of Fig. 8-21 closely, with the exception that the locations of the 3 foci of self-inversion are not illustrated. One reason for this omission is that the locations are superfluous to the purposes of the present figure; the other is that these locations stand in a closely circumscribed relationship to the locations of the pole of inversion (and the vertices), which is adequately and representatively depicted in Fig. 8-21 (see Maxims 8-22-11 and 8-30-6).

The basis curve is the central true exchange-limacon of eq. 8-25-9a, for which the locations of all exceptional axial poles are listed in 8-25-9b (following eq. 8-25-9a). In the figure, the locations of the vertices, center, and $\cos^2\theta$-condition foci of the basis curve (labelled F) are marked by large solid circles. Solid circles of the same size also are employed to mark the locations of the reciprocal inversion poles.

Positions of the Cartesian-condition poles of the basis curve are marked by the solid circles of intermediate size, while small solid circles mark the locations of the $\cos^2\theta$-condition foci of the inverse loci, which are represented for intervals of 0.05b along the line of symmetry that intersects the curve. These levels are numbered from 0 (self-inversion about the center) to 37 (inversion about a lateral pole at 1.85b) in the left margin. The modular poles ($h = \pm 0.59200020b$) are not marked but are close to the axial poles that correspond to level 12 (0.60b).

The level of the pole of inversion to the serpentine circular cubic of the inner vertex lies just beyond level 4, as indicated. At this level, as in Fig. 8-21, both the outer right vertex and the $\cos^2\theta$-condition focus of the right

Tracking The Vertices & $Cos^2\theta$-Condition Foci of Axial Inversions of An Opposed-Oval Central Cartesian ($a^2 = 10b^2$ True Exchange-Limaçon)

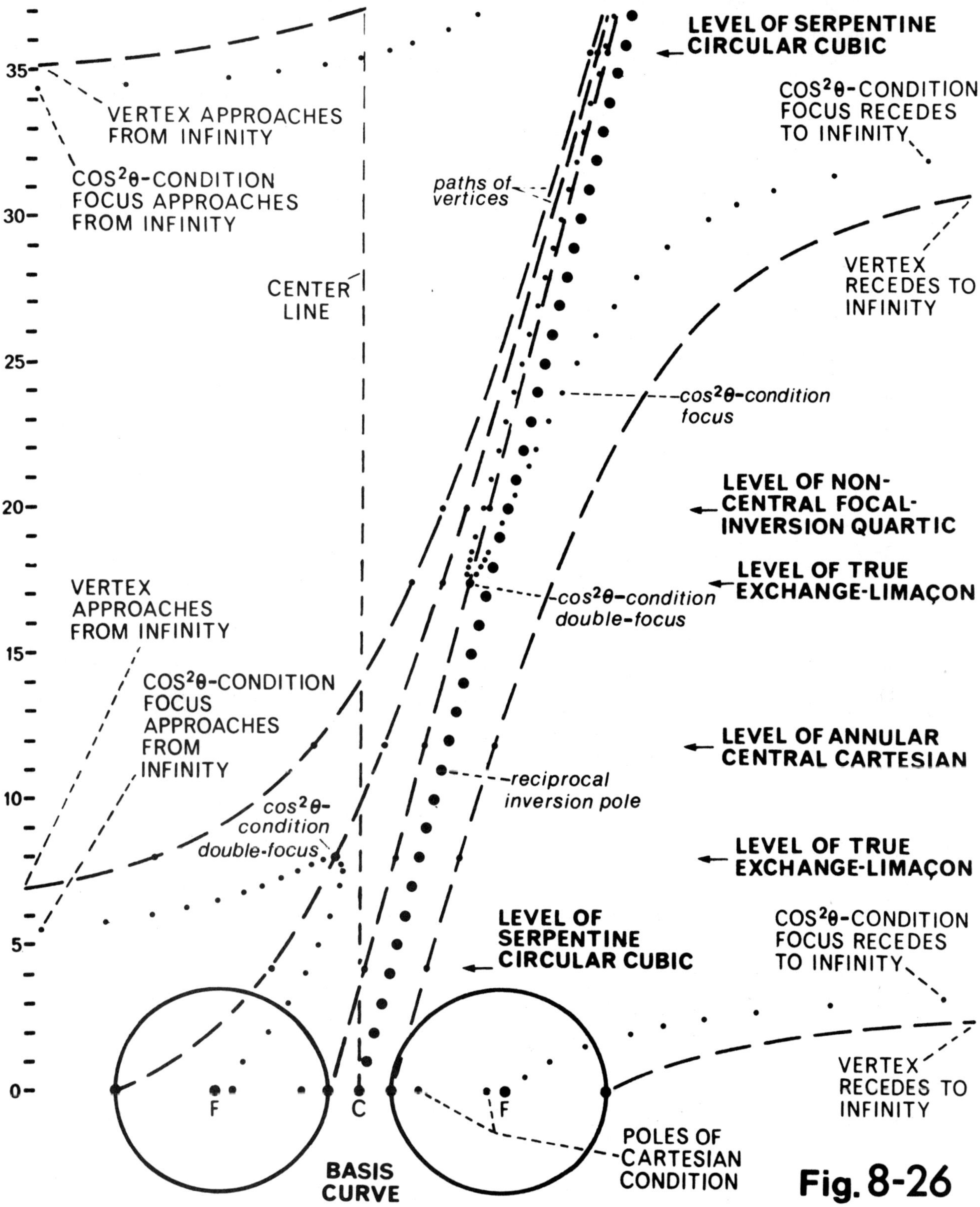

Fig. 8-26

oval lie at the axial point at infinity. The $\cos^2\theta$-condition focus of the left oval occupies an unexceptional interior location, whereas for the corresponding Cassinian axial-vertex serpentine circular cubic of Fig. 8-21 this focus is incident upon the far vertex of the oval.

On the other hand, at the level of the Cartesian self-inverter (labelled "level of true exchange-limacon"), both $\cos^2\theta$-condition foci have merged into coincidence with the far vertex of the interior oval--the unique property of these Cartesian self-inverters--whereas in Fig. 8-21 the corresponding location is not incident upon the curve but is in the left inter-oval region at the position of the pole inverse to the center (which is modular to the 0° inter-oval-mode focus)--a location unique to parabolic Cartesians.

The next exceptional location of the axial inversion pole is at a modular pole of inversion (corresponding roughly to level 12; see listings 8-25-9b), which yields an annular central inverse locus (labelled "level of annular central Cartesian"). As the pole of inversion progresses beyond the modular pole, identical subspecies of inverse loci are encountered in reverse sequence, reverse orientation, and smaller size, as compared to the inverse loci obtained about poles from the center to the modular pole (see legend to Fig. 8-21 and Maxims 8-20-6,7). Thus, the next exceptional pole is the second pole of the Cartesian condition, inversion of the basis curve about which yields true exchange-limacons.

The inverse locus obtained when the pole of inversion comes into coincidence with the $\cos^2\theta$-condition focus is designated as the *non-central focal-inversion quartic*, since it is the only non-central quartic obtained by a focal inversion (i.e., by an inversion about a point-focus). This curve possesses the distinction that the reciprocal inversion pole is in coincidence with the inter-oval $\cos^2\theta$-condition focus (the other $\cos^2\theta$-condition focus is within the interior oval).

The same curve (i.e., an identical subspecies) is obtained by an inversion about the pole modularly inverse to the $\cos^2\theta$-condition focus of the basis curve (see listing under 8-25-9b), corresponding roughly to level 7, but this locus is not labelled. In the latter case, it is the pole inverse to the center that is in coincidence with the inter-oval $\cos^2\theta$-condition focus.

The last exceptional pole of inversion is attained at the far right vertex, for which the inverse locus is a serpentine circular cubic belonging to the same subspecies as the one obtained by inverting the basis curve about the near right vertex. Inversions about poles beyond the far right vertex produce opposed-oval curves that increasingly resemble the central basis curve and, in the limit, at the axial point at infinity, achieve coincidence with it [for the appropriate (infinite) value of the unit of linear dimension] (Maxim 6-7-3).

Fig. 8-27. Tracking the vertices and $\cos^2\theta$-condition foci of axial inversions of an opposed-oval central Cartesian with non-real axial Cartesian-condition poles. Eq. 8-20-11a defines locations of the $\cos^2\theta$-condition foci of axial inverse loci of central Cartesians. The latter come into coincidence as double-foci when the radicand (the *Cartesian term*) vanishes. Equating the radicand to 0 yields the *Cartesian condition*, namely the condition for which the inverse loci have generic $\cos^2\theta$-condition double-foci.

The solution to this Cartesian condition is eq. 8-20-11b', repeated here for convenience as eq. 1a. Since the denominator of eq. 1a can be made positive or negative by rotating the curve 90°, whereas the signs of the other terms are independent of the line of symmetry of the curve upon which the poles of inversion lie, it is evident that the only requirement for real solutions to eq. 1a (i.e., for poles h about which inversion of the basis central curve yields loci in which the foci of the $\cos^2\theta$-condition are in coincidence) is that the radicand be positive (condition 1b). For the inequality 1b to hold, either inequality 1c or 1c' must apply.

solution of the Cartesian condition for which the radicand of eq. 8-20-11a vanishes

(a) $$h^2 = \frac{(4D^4-B^2C^2)\pm[(4D^4-B^2C^2)^2-4D^4(B^2-C^2)^2]^{\frac{1}{2}}}{2(C^2-B^2)} \qquad (8\text{-}27\text{-}1)$$

(b) $4D^4(4D^4-C^4) > B^4(4D^4-C^4)$ — condition for the radicand of eq. 1a to be positive

(c) $4D^4 > C^4$ and $4D^4 > B^4$

(c') $4D^4 < C^4$ and $4D^4 < B^4$

conditions for the inequality of 1b to apply

Inequality 1c', of course, holds whenever D^4 is negative, which is the condition for monovals, while the same inequality with D^4 positive holds for all annular central Cartesians (for which both B^2 and C^2 must be negative) and for opposed-oval central Cartesians, for which B^2 or C^2, but not both, is negative (see legend to Fig. 8-20). Accordingly, to illustrate the locations of the $\cos^2\theta$-condition foci of axial inversions of a central Cartesian for a case in which these foci do not come into coincidence, one need only select a curve for which D^4 is positive and neither inequality 1c nor 1c' applies.

The curves in question are opposed-oval curves in which the poles of the Cartesian condition lie on the modular circle at points at which neither line of symmetry is intersected (see Fig. 8-22f for a comparable locus). This being the case, no axial inversion pole comes into coincidence with a pole of the Cartesian condition, and no inverse locus has its $\cos^2\theta$-condition foci in coincidence. An example of such a locus is the basis curve of the present figure (eq. 2), for which $4D^4$ is greater than C^4 but less than B^4.

opposed-oval central Cartesian with real $\cos^2\theta$-condition foci on the x-axis, no axial-inversion species of which has its $\cos^2\theta$-condition foci in coincidence

$$(x^2+y^2)^2 - 2d^2x^2 - \tfrac{1}{2}d^2y^2 + \tfrac{1}{2}d^4 = 0 \qquad (8\text{-}27\text{-}2)$$

The format of this figure follows closely that of Fig. 8-26, with the positions of the vertices and center of the basis curve, and the reciprocal inversion poles marked by large solid circles, the positions of the $\cos^2\theta$-condition foci of the basis curve by intermediate-sized solid circles, and the positions of the $\cos^2\theta$-condition foci of the inverse loci marked by small solid circles. In addition, the vertices at the levels of exceptional inverse loci are marked by small solid circles. As with Fig. 8-26, the locations of the 3 foci of self-inversion are unmarked (in the cases of the serpentine circular cubics they are in coincidence with the vertices).

Although one of the $\cos^2\theta$-condition foci recedes to the axial point at infinity as a pole of inversion approaches a vertex (about which inversion yields a serpentine circular cubic) and approaches from the axial point at infinity as a pole of inversion departs from a vertex, the behavior of the $\cos^2\theta$-condition foci is characteristically different in other respects from those of Figs. 8-21 and 8-26; they differ most notably in that real $\cos^2\theta$-condition foci exist in all axial inversion representatives.

Tracking The Vertices & $\text{Cos}^2\theta$-Condition Foci of Axial Inversions of An Opposed-Oval Central Cartesian With Non-Real Axial Cartesian-Condition Poles

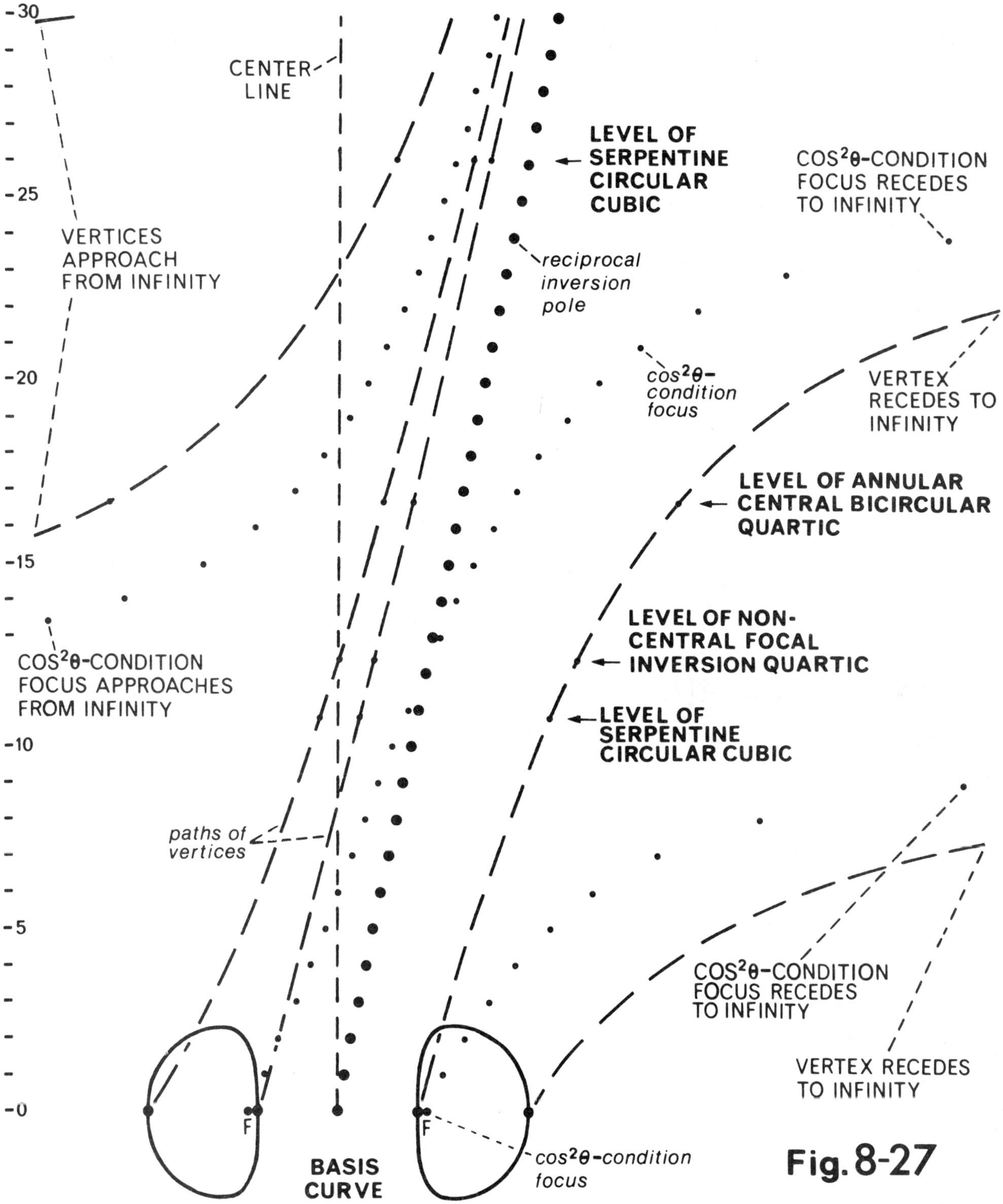

Fig. 8-27

It will be noted that there are two levels at which a single $\cos^2\theta$-condition focus comes into coincidence with a near vertex of an opposed-oval subspecies, forming a double-focus (only one of these poles of coincidence is illustrated), and that the inter-oval locations of the $\cos^2\theta$-condition foci are more nearly equidistant from the adjacent vertices than in the basis central curve.

Since the $\cos^2\theta$-condition foci of the basis curve are pre-modular, so too is the pole for inversion to the *non-central focal-inversion quartic*, for which the reciprocal inversion pole is in coincidence with a $\cos^2\theta$-condition focus. The second occurrence of this latter curve, of course, is at the level corresponding to the pole of inversion being modularly inverse to the $\cos^2\theta$-condition focus of the basis curve (unmarked, but close to level 23), for which locus the $\cos^2\theta$-condition focus is in coincidence, not with the reciprocal inversion pole, but with the pole inverse to the center.

Fig. 8-28. LCQ sextics and a sextic curve of demarcation, including circumpolar intercept transform equations. LCQ (linear-circular-quartic) sextics are so named because, for the simple intercept format in the form of eq. 1a, the 0°, 180°, and 90° transforms are linear, circular, and quartic eqs. (eqs. 1,4, and 5 of the figure). Eq. 1a does not lead to algebraically complete transforms (see Chapter VII, *Negative Intercepts, Trivial Transforms, The Degree of Transforms, And The Duality of 0° and 180° Transforms*), since for any given angle, θ, all positive and negative intercepts are not represented (and θ must progress from 0° to 360° to generate the entire locus).

simple intercept formats of LCQ sextics

(a) $\cos\theta = x(x-a)/b^2$	entire locus generated for θ progressing from 0° to 360°	(8-28-1)
(b) $\cos\theta = x(\pm x-a)/b^2$	entire locus generated for θ progressing from 0° to 180°	

$$(x-y)^2[x^2+y^2-a(x-y)][x^2+y^2+a(x-y)]\cdot(x+y-a)(x+y+a) = 0 \qquad (8\text{-}28\text{-}2)$$

algebraically complete 0° transform equation

LCQ Sextics & A Sextic Curve of Demarcation, Including Circumpolar Intercept Transform Equations

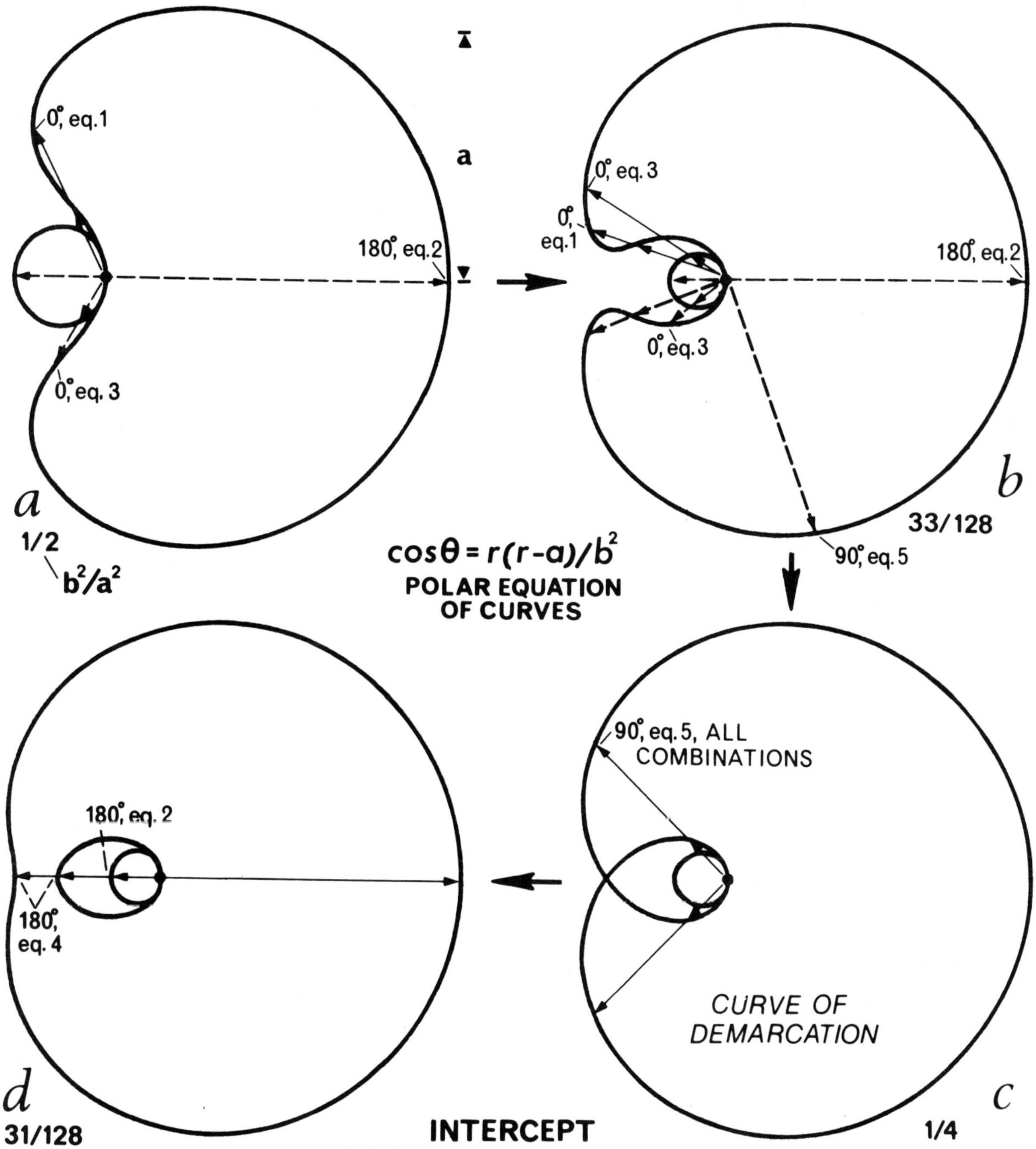

INTERCEPT TRANFORM EQUATIONS

eq. 1, $x + y = a$; eq. 3, $x^2 + y^2 = a(x - y)$; eq. 5, $x^2(x-a)^2 + y^2(y-a)^2 = b^4$

eq. 2, $x - y = a$; eq. 4, $x^2 + y^2 = a(x + y)$

Fig. 8-28

The simple intercept format 1b, derived from the corresponding rectangular eq. (eq. 3, below), represents all intercepts (with the entire locus being generated for θ progressing from 0° to 180°) and leads to algebraically complete transform eqs. Eq. 2, for example, is the algebraically complete 0° transform, from which it is evident that both linear and circular transforms exist at transformation angles of both 0° and 180° when positive and negative intercepts are taken into account.

$(x^2+y^2)^3 = a^2(x^2+y^2+b^2x/a)^2$	LCQ sextics cast about the vertex tangent-point of the medium and small ovals	(8-28-3)
$(x^2+y^2)^2 = 4(x^2+y^2+ax)^3/27a^2$	Cayley's sextic cast about the small-loop vertex	(8-28-4)
$(x^2+y^2)^2 = a^2(x^2+y^2+bx)$	true exchange-limacons cast about the $\cos^2\theta$-condition double-focus	(8-28-5)
$(x^2+y^2) = (x^2+y^2+bx)^2/a^2$	limacons cast about the DP	(8-28-6)

The present example of a sextic curve of demarcation is significant for several reasons. First, it illustrates the fact that the same phenomena, of genera possessing species of curves with different numbers of ovals or segments and a corresponding looped curve of demarcation, that are observed for curves belonging to inversion superfamilies with members of 4th degree and lower, also exist for curves belonging to inversion superfamilies with members of 4th, 5th, and 6th degree.

Secondly, these curves are related equationally to Cayley's sextic, true exchange-limacons, and limacons. The only significant differences between the eqs. of all 4 curves (eqs. 3-6) are that the exponents of the, in essence, identical parenthetical terms on the left and right differ for each curve. The true exchange-limacon differs from the limacon in having the exponents exchanged, the LCQ sextic differs from the true exchange-limacon in having the exponents of both terms increased by one unit, while Cayley's sextic differs from the LCQ sextic in having the exponents exchanged.
[Cayley's sextic, of course, is only a *subspecies* of the genus $A^2(x^2+y^2)^2 = (x^2+y^2+Bx)^3$, which has yet to be analyzed in detail, whereas eqs. 3,5, and 6 represent *genera*.]

Since limacons are curves with very high circumpolar symmetry--having linear 0° and 180° transforms and circular 90° transforms--it follows that Cayley's sextic, true exchange-limacons, LCQ sextics, and all other curves derived from the eq. of limacons by altering the exponents of the terms should have relatively high circumpolar symmetry (see Chapter XIV, *Design And Synthesis of Highly Symmetrical Curves, The Building-Block Approach*).

In keeping with this expectation, true exchange-limacons have linear and circular 0° and 180° transforms and sextic 90° transforms about the vertex of the small oval, while, as already mentioned, LCQ sextics have linear and circular 0° and 180° transforms and quartic 90° transforms about the tangent-point of the medium and small ovals (the algebraically complete 90° transform is a degenerate octic eq. consisting of the product of two quartic terms equated to 0). The circumpolar symmetry of Cayley's sextic has not been studied in detail; its 0° and 180° transforms about the small-loop vertex are of 6th degree (see pages IX-32,33).

While limacons and true exchange-limacons are closely related to one another, in that they are both Cartesian self-inverters and in that the $e = 2$ limacon is the curve of demarcation of the latter genus, the structural similarities of true exchange-limacons to LCQ sextics are greater than to limacons. Not only do LCQ sextics and true exchange-limacons greatly resemble one another if the small oval of the former is overlooked, but both curves, as expressed by eqs. 3 and 5, are cast about a homologous pole and possess their similar circumpolar symmetry properties about this pole. On the other hand, neither limacons nor LCQ sextics (nor Cayley's sextic) possess the extraordinary symmetry property that true exchange-limacons share with certain other axial-inversion genera of central Cartesians (see Figs. 8-22,26,27).

The third significant feature of LCQ sextics is that they belong to the inversion superfamily based upon a genus of (quartic) conchoids possessing high polar-linear symmetry, namely the equilateral hyperbolic polar-linear conchoids (see Figs. 3-7 and 8-42).

Considering the illustrations, curve a, for a value of $a^2 = 2b^2$, is far from the condition at which the curve of demarcation forms, with a large bi-lobed oval that is very similar to the corresponding monoval of true exchange-limacons (Figs. 8-25d,e) and a small oval that is only slightly elongated as compared to a circle. The coming together of the lobes as b decreases relative to a occurs very slowly until b^2 is very nearly equal to $\frac{1}{4}a^2$. Even for $b^2 = 33a^2/128$ (Fig. b) the lobes are relatively far apart. For this value, aside from the presence of the small oval, the curve greatly resembles the true exchange-limacon of Fig. 8-25g, and the small oval departs even less from circularity than in Fig. a.

For only a relatively minute decrease in the value of b relative to a, the approaching lobes make contact to form the curve of demarcation of Fig. c. In this curve, the looped locus greatly resembles a hyperbolic limacon and the small oval a circle (see also Fig. 8-42 for axial inverse loci of this curve). At roughly the same relatively minute increase in the value of b relative to a, the looped portion of the limacon-like locus becomes detached, with the formation of the triovular curve of Fig. d. In the transition from curve b through the curve of demarcation to curve d, the major changes in the loci occur in the region of the joining lobes and detaching ovals; the remaining portions of the 3 loci are very nearly congruent.

Aside from the transforms about the vertex of tangency (marked by the solid circles), the circumpolar symmetry of this genus of curves has been little studied. With regard to these transforms, even for transformation angles of 0°, 180°, and 90°, the situation is exceedingly complicated. The following treatment deals only with positive intercepts and geometrically valid transforms (see the discussion in Chapter X, *Analytical Transforms Versus Geometrical Expectations*), with the larger of the two intercepts of a given intercept pair being referred to in the eqs. as the x intercept. In addition, the first (smaller) intercept with the large oval of curves a and b is referred to conventionally as the intercept with "the oval of medium size," even though only two ovals are present (since these intercepts correspond to those with the oval of medium size in the triovular curves, or with the interior loop of the curve of demarcation).

With these conventions, the intercepts for a number of representative transforms are indicated by the radii terminating in arrows in all 4 curves, and the angle of the transformation and the corresponding eq. of the transform are indicated by adjacent labels. The number and identity of the transforms represented for each curve are essentially arbitrary. It is noteworthy that: (1) the 90° transforms occur only for intercepts with the large and medium ovals--the small oval does not participate; (2) there are no 180° transforms for combinations of intercepts between the small and medium ovals, since the pole of the transformation is the tangent point of these ovals;

and (3) the large oval has only 90° transforms with itself, i.e., for two radius vectors of lengths determined by intercepts with the large oval (recall that the first point of intersection with the large oval of Figs. a and b is reckoned as an intersection with the medium-sized oval); and (4) transforms for overlapping radius vectors are valid--for example, the intercept with the long radius vector extending to the right in Fig. d has a valid 180° transform with each of the 3 intercepts of the radius vectors extending to the left, and similarly for all combinations of intercepts of orthogonal radius vectors in Fig. c.

Fig. 8-29. The invariance of inverseness under inversion. The title of this figure expresses a classical property of the inversion transformation (for the finite plane), namely, that if two curves are inverse to one another with respect to a given pole, a, then their inverse loci about a pole in the plane, p, also are inverse to one another about a pole C'_C. The relationship between the invariance of inverseness under inversion (hereafter referred to as *the property*) and Maxim 6-4-6 is examined in this figure and Fig. 8-30. It also is shown by means of simple counterexamples that the classical concept of *the property* is invalid unless one extends the classically finite domain of pole C'_C to include points on the ideal line at infinity. In this broader context of "the extended plane" *the property* is valid.

As mentioned in the PREFACE, many geometrical proofs are facilitated greatly by employing the properties of the inverse loci of lines and circles and ensembles thereof (in relation to the basis ensembles). An ensemble of circles is employed in Fig. a to follow the path of the classical proof of *the property* (after Daus, 1941).

A and B are inverse points with respect to a circle C with center a and radius j (the unit of linear dimension). The circle C and points A and B are inverted about a pole p (unit of linear dimension, j) at distance L, giving the circle C' and points A' and B'. The circle C_1 is orthogonal to C and passes through point A, while the circle C_2 intersects C_1, is orthogonal to C, and passes through point B. Inasmuch as both C_1 and C_2 are orthogonal to C, C_1 also passes through point B, and C_2 through point A, after the well-known property that *if two intersecting circles are orthogonal to a circle of inverion, they intersect in a pair of inverse points.*

Corresponding to the ensemble of 3 intersecting basis circles, C, C_1, and C_2, there exists an analogous intersecting inverse ensemble, with C_1' and C_2' orthogonal to C'. The points of intersection of C_1' and C_2' are the inverse points A' and B' (of A and B with respect to p), which also are inverse with respect to the inverse circle C' [for the unit of linear dimension, $j^3/\pm(L^2-j^2)$]. Accordingly, since A and B are any pair of points inverse in C, loci consisting of ensembles of pairs of points A and B inverse in C invert to loci consisting of ensembles of pairs of points A' and B' inverse in C'.

If, now, for comparison with Maxim 6-4-4, one lets A be the basis curve, *the property* prescribes the following relations. Given a basis curve A, its inverse locus B about pole a, and the inverse loci A' and B' of A and B about a second pole p (for the unit of linear dimension, j), then loci A' and B' are inverse about the pole C_C' (the center of circle C'), the location of which is known.

Maxim 6-4-4 makes a statement of a different nature, which is more specific, and has much more far reaching consequences. This Maxim differs, first, in not being concerned with the curve B' and, second, and most essentially, in treating of the inverse locus of A' about the pole a', inverse to a, rather than its inverse locus about the pole C_C', the center of the circle C' inverse to C. The latter pole (C_C') is at a distance $\pm Lj^2/(L^2-j^2)$ from p, compared to a distance j^2/L for pole a' (the upper sign is for the case in which pole p is external to the circle C, the lower sign for the case in which p is internal).

Considerations of inverse loci of the curves A' and B' (or points A' and B') about pole C_C' follow naturally from the classical approaches to the properties of the inversion transformation in terms of the inverse loci of circles and lines and their ensembles. When, instead, inverse loci of the curves A' (inverse to A about any point in the finite plane, p) about pole a', inverse to a, are considered, it emerges that any inverse curve A' inverts about any corresponding inverse pole a' to the same curve (i.e., subspecifically identical or similar) B, that is obtained by inverting the basis curve A about pole a. In other words, all the corresponding curves in the inversion superfamily are obtained in the 1st generation (*immediate closure*; see Maxim 6-4-7), because the ensemble of curves B obtained by inverting the basis curve A about all poles a contains all the members to be found in the ensemble of curves A''

"The Invariance of Inverseness Under Inversion"

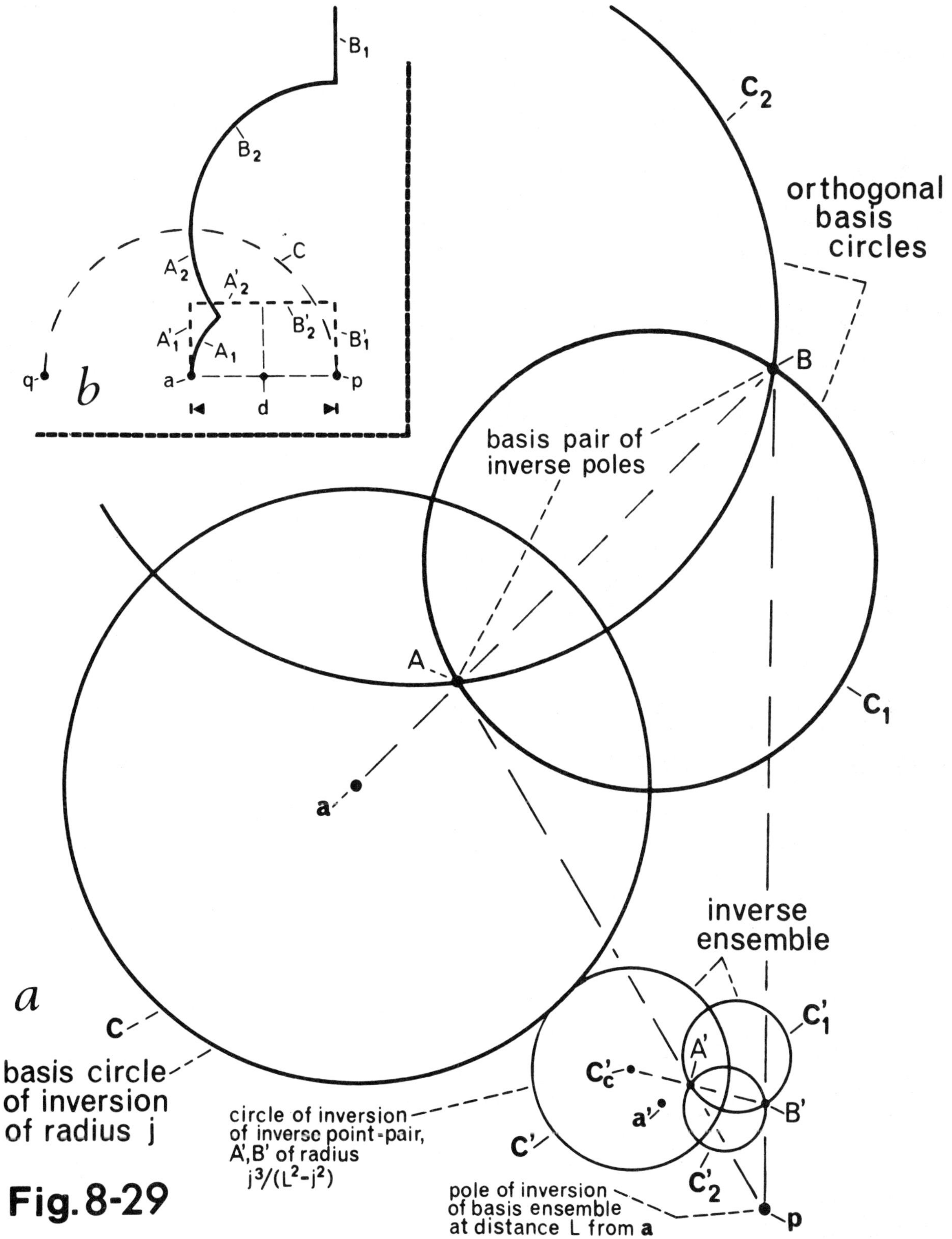

Fig. 8-29

obtained by inverting A' about all poles a' inverse to poles a. Ultimate conservation of degree (see page VIII-18) follows directly from these findings.

The invalidity of the classical concept of *the property* for poles C'_c in the finite plane is illustrated by means of two simple counterexamples. For the first of these we make use of the powerful Maxim 8-29-1, already referred to in connection with Maxim 6-5-1. [The employment of this Maxim is not essential; it merely facilitates the use of a familiar example.]

CIRCUMPOLAR
MAXIM 8-29-1: *Any unexceptional inversion of a locus C to a locus D can be regarded as a self-inversion; conversely, any self-inversion can be regarded as an unexceptional inversion of a locus A to a locus B. In the former case, one regards the union of C and D as the self-inverting curve; in the latter case, one regards the segments A and B of the self-inverting curve as the individual loci that are the inverse of one another.*

For the first counterexample consider a hyperbolic limacon and, following Maxim 8-29-1, regard it as two individual curves. The small loop is taken to be curve A and the large loop curve B, which are inverse in the pole a [the $(a^2-b^2)/2b$ focus of self-inversion within the small loop]. According to *the property*, if the limacon now is inverted about any pole p, the inverse loci A' and B' (inverse to the small and large loops, respectively) are inverse about the pole C'_c, the center of the inverse circle C' in the plane of A' and B'.

We now let p be the DP of the limacon, in which case the inverse locus is a hyperbola (Maxims 13 and 6-4-16). One arm thereof is the inverse locus A', the other is the inverse locus B'. But contrary to the assertion made by *the property*, there is no pole in the finite plane about which one arm of a hyperbola inverts to the other. But there is a pole of the ideal line at infinity for which this is the case, namely, the point at infinity in the direction of the transverse axis.

This failure of *the property* for points C'_c in the finite plane also is evident from the above algebraic expressions for the unit of linear dimension (radius) of the inverse circle C' and the distance of the pole C'_c (the center of circle C' inverse to C) from pole p, both of which have the expression (L^2-j^2) in their denominators. Both of these distances become infinite for $L = j$, i.e., when the pole of inversion p is incident upon the circle of inversion C (geometrically, this means that pole C'_c is on the ideal line at infinity).

This result is readily understood. Returning to the counterexample of the hyperbolic limacon, the circle of inversion is simply the inverse y-axis circle of the basis hyperbola. The closer p approaches to C, the larger the radius of the inverse circle C' and the more remote its center C'_c. When p becomes incident upon C, the radius of C' becomes infinite and its center recedes to infinity. In this case, the inverse circle C', of infinite radius, is a line of symmetry of the union of the inverse loci A' and B', and the pole C'_c is the point on the ideal line at infinity in the direction orthogonal to this line of symmetry.

If the point of incidence of pole p on circle C is not at one of the intersections of C with the line of symmetry of A and B, the inverse curve regains the "y-axis" line of symmetry but loses the x-axis line of symmetry of the limacon. In this event, there exists another finite pole about which A' and B' are inverse, namely, the focus of self-inversion that replaces the line of symmetry of the limacon. In that case, *the property* is valid, though its validity is "saved" only fortuitously (by virtue of the loss of the line of symmetry of the limacon).

But if the point of incidence of pole p on circle C is at one of the intersections of C with the line of symmetry of the limacon: (a) the union of the inverse loci A' and B' possesses two lines of symmetry; and (b) the pole C'_c about which A' (either an arm of a hyperbola or a loop of a hyperbolic central quartic) inverts to B' (the other arm or loop) lies on the ideal line at infinity in the direction of the transverse axis. Accordingly, *the property* is invalid in the finite plane but valid for a pole C'_c on the ideal line at infinity in a specific direction.

As a second counterexample, consider the figure consisting of the basis segments $A_1A_2B_2B_1$ of Fig. b, comprised of a circular arc A_1, a circular arc (A_2+B_2), and a linear segment B_1 that extends to infinity. The pole a is the center of the circle of inversion C. Segments A_1 and B_1 are inverse to one another about pole a, as are also segments A_2 and B_2 (which are segments of a common circle). Since the basis segments are all linear or circular, considerations of the property of inverse loci are not as clear-cut as in the example employing the loops of a hyperbolic limacon. Accordingly, attention is confined to a single inverse locus as counterexample, namely, the inverse hemi-square locus (short dashes) $A'_1A'_2B'_2B'_1$ about pole p on the circle of inversion C (for j = D)

Although the combination of two segments A_1A_2 inverts about pole a to the combination of two segments B_1B_2 of the basis figure, there is no pole in the finite plane about which this is true for the combination of segments $A_1'A_2'$ and $B_1'B_2'$ of the hemi-square inverse locus. Accordingly, *the property* also is invalid for these inverse figures. On the other hand, segments $A_1'A_2'$ invert to segments $B_1'B_2'$ about a pole on the ideal line at infinity in the direction of a line through poles a and p, whence *the property* is valid if its domain of applicability is extended to include the ideal line at infinity (the extended plane).

Fig. 8-30. Inversions of inverse curves about corresponding inverse poles belong to the same subspecies. Specific examples of the invariance of inverseness under inversion in the extended plane and the subspecific identity of inverse loci of inverse curves about corresponding inverse poles are given here. In addition, the topics of identical subspecific-inversion and equivalent inversion poles are dealt with generally, with applications to central Cartesians, a dual Cassinian with 4 lines of symmetry, and a figure with 3 lines of symmetry.

Fig a illustrates two concentric inverse loci, consisting of an equilateral hyperbola, EH, and an equilateral lemniscate, EL, inverted about their common center a, for a unit of linear dimension, j, equal to the length of the common semi-transverse axis of both curves. Pole $p = F_R$, at a distance $\sqrt{2}j$ from a, is the right traditional focus of the equilateral hyperbola, which for convenience is referred to as the basis curve, and pole F_L is the corresponding left traditional focus. Poles f_L and f_R, inverse to poles F_L and F_R, are the loop foci of curve EL.

Loci EH and EL are inverted about the pole $p = F_R$, for the same unit of linear dimension, to yield the inverse curves EH' and EL' of Fig. b. Curve EH' is the equilateral limacon, and its DP, $p = F_R$, is the reciprocal inversion pole. Pole F_L' is the inverse of the left traditional focus of the basis curve, which becomes a common 0° focus of self-inversion of both EH' and EL'. Curve EL' is the *equilateral $2^{\frac{1}{2}}a/2$ QBI inter-vertex quartic* (defined in terms of the location of the axial pole about which the equilateral hyperbola inverts to it; see page XII-1). Its DP is the pole a', inverse to pole a about pole p. Pole $C_C' = f_R'$ is the center of the circle C' inverse to circle C of Fig. a, while pole f_L' is inverse to the loop focus f_L of curve EL.

Inversions of Inverse Curves About Corresponding Inverse Poles Belong To The Same Subspecies

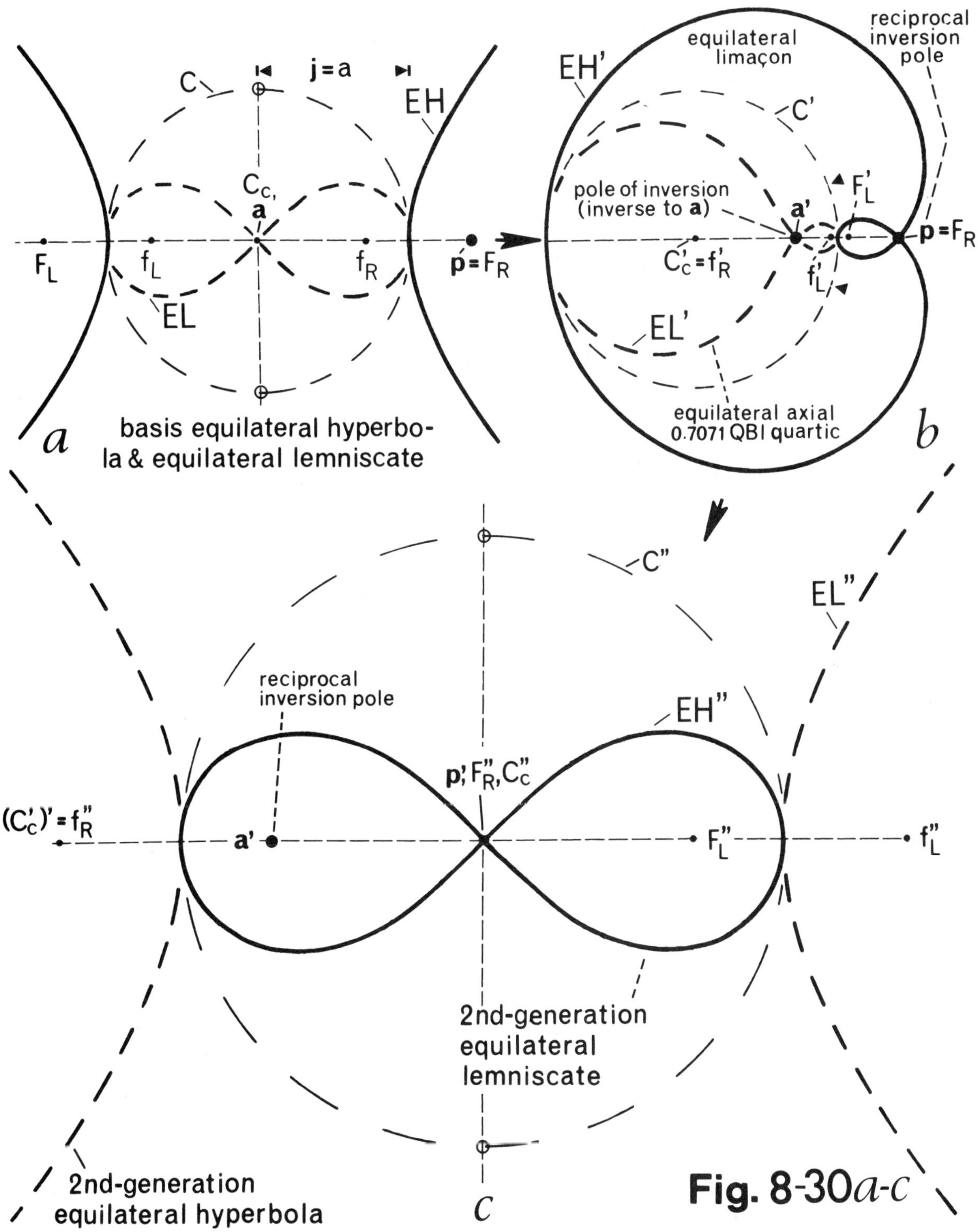

Fig. 8-30*a-c*

[Note that pole a' is the reflection of pole p in pole F'_L (the 0° focus of self-inversion of both curves). Since both poles are at modular distance from pole F'_L, they are the two poles about which both the equilateral limacon and the equilateral $\sqrt{2}a/2$ QBI inter-vertex quartic invert to central curves (the inverse locus about pole p is Fig. a, while that about pole a' is Fig. c).]

After the property of invariance of inverseness under inversion in the extended plane, curves EH' and EL' are inverse to one another about pole C'_c for a unit of linear dimension, $j^3/(L^2-j^2) = j^3/(2j^2-j^2) = j$ (see Fig. 8-29), which for this particular case is seen to have the same value as the unit of linear dimension for inversion of curves EH and EL to EH' and EL'. After Maxim 6-4-4, EH' inverts to the subspecies EL, and EL' to the subspecies EH, about pole a'.

Fig. c illustrates the latter inverse loci of the curves of Fig. b about pole a' for $j = a$. Although this inversion yields an *inversion image* of the curves of Fig. a (see Fig. 8-20), each curve is of double the linear dimension of the basis curve. Following Maxim 6-4-4, the curve inverse to EL' about pole a' is curve EL'', which belongs to the same subspecies as curve EH of Fig. a, namely the equilateral hyperbola. Or, if curve EL is considered to be the basis curve in Fig. a, then the curve inverse to EH' is EH'', which belongs to the same subspecies as curve EL, namely the equilateral lemniscate.

Accordingly, inversions about the modular poles p and a'--the two poles about which the curves of Fig. b (excluding the dashed circle) invert to central curves--give similar figures, but the origins of the two curves of Fig. a from Fig. b are the reverse of their origins for Fig. c. Analogous ensembles of inverse curves could be given for any basis pair of inverse central QBI curves.

The distinction between the relationships of invariance of inverseness under inversion and Maxim 6-4-4 is quite clear from this figure. The former asserts that if curves EH and EL are mutually inverse about pole a, their inverse loci EH' and EL', about pole p, are mutually inverse about pole C'_c. The latter asserts that the same subspecies of curve, EL, obtained by inverting a basis curve EH about a pole in the plane, a, also is obtained by inverting any inverse locus of curve EH about the corresponding inverse pole a'.

It is instructive to compare the foci of the basis curve and its central inverse locus of Fig. a with those of the similar ensemble of Fig. c. The loop foci of the equilateral lemniscate, f_L and f_R, become the traditional foci, f''_R and f''_L, of the 2nd-generation equilateral hyperbola, while the left traditional focus of the basis hyperbola becomes the right loop-focus, F''_L, of the equilateral lemniscate. The left traditional focus becomes the center of the 2nd-generation ensemble, but this is a consequence of the fact that this focus was taken to be the pole of inversion, p. In general, the pole p becomes the center of the inverse ensemble but the right traditional focus F_R becomes the left loop-focus (position of a' in Fig. c) of the 2nd-generation equilateral lemniscate (see also Fig. 8-20, and below).

The location of the inverse of pole a (the center of the basis ensemble) in the 2nd generation is variable, depending upon the selection of p. For $p = F_R$, as in the present instance, a' (which also is the reciprocal inversion pole of Fig. c), comes to lie at the position of the left loop-focus of the 2nd generation equilateral lemniscate, but when p is taken to be at 2a, a' in the 2nd generation does not have point-focal rank but lies between the DP and the left loop-focus (the left loop-focus becomes F''_R, i.e., the inverse about a' of the inverse of F_R about p).

If one follows the centers of the inverse circles, the center of circle C'' is the center of the 2nd-generation inverse ensemble, just as the center of the circle C is the center of the basis ensemble. The center of circle C', of course, following the invariance of inverseness under inversion in the extended plane, is the pole about which EH' and EL' are mutually inverse. The right common vertex of the ensemble of Fig a becomes the left common vertex of Fig. c, while the left common vertex becomes the right common vertex.

[The classical view that "the inverses of the foci of a curve are the foci of the inverse curve" (for example, Hilton, 1920, page 74) is not generallly applicable to circumpolar focal loci. In contrast with the point-focal correspondences between the basis and 2nd-generation ensembles (Figs. a and c), poles in Fig. b that are inverse to the non-incident foci of the curves of Fig. a do not, in general, possess point-focal rank for the individual inverse curves.

F'_L becomes the focus of self-inversion of both curves of Fig. b only because it is the reflection of the pole of inversion in the line of symmetry, not because it is the left traditional focus of the basis hyperbola (see Maxim 8-22-11 as applied to Fig. a, regarded as a central figure that self-inverts at 0° and 180°, as treated below). For example, in the inversion to equilateral strophoids about a pole incident upon one of the common vertices, the traditional foci of the curves of Fig. a take non-focal positions on the transverse line of symmetry of the equilateral strophoids.]

Pairs of Inverse Central QBI Curves As Figures That Self-Invert In Multiple Modes

All of the general inversion properties of biovular central Cartesians and the Maxims pertaining thereto apply with equal force to composite figures that consist of any given monovular central curve and the inverse locus about its center (since this combination also comprises a central figure that self-inverts at 0° and 180°). Figs. a,b, and c of the present illustration provide a convenient example with very close homologies with the corresponding 3 curves of Fig. 8-20a (*Correspondences Between Axial Poles In Axial Inversions To And From Central Parabolic Cartesians*). Although this example involves a basis locus (Fig. a) that consists of inverse hyperbolic, rather than elliptical, QBI subspecies, it serves to illustrate most of the homologies.

In Fig. a, the two pairs of inverse poles, (f_R,F_R) and (f_L,F_L), are the homologues of the two pairs of poles of the Cartesian condition of Fig. 8-20a, even to the point of being the only 4 poles in the plane about which one or the other of the basis curves inverts to a Cartesian self-inverter (in the present case, this curve is the equilateral limacon). [If both EH' and EL' were Cartesian self-inverters, rather than only EH', the homology would be complete.]

As for the loci of Figs. 8-20a$_1$ and 8-22a,c, and f, there are 5 poles in the finite plane about which Fig. a inverts to central curves (see Maxims 8-20-3 and 8-21-7): (a) the center, which is a pole of both 0° and 180° self-inversion; (b) the two modular poles at the x-axis vertices, about which the inverse loci are two congruent opposed equilateral strophoids, tangent at their loop vertices; and (c) the two modular poles on the y-axis (open circles), about which the inverse loci are two congruent opposed, doubly-tangent axial lemniscate-like curves (*equilateral* $k=a$ *QBI quartics*), each similar to Fig. 6-3h.

The inverse loci about all other poles incident upon the modular circle are axial curves consisting of two congruent non-axial QBI quartics (such as the curve of Fig. 6-3i) reflected in the line of symmetry (the inverse locus of the modular circle). In addition, all the axial inverse loci of this ensemble are obtained by inverting about transverse-axis poles from 0 to a, about conjugate-axis poles from 0 to a, and about poles on one quadrant of the modular circle C (see Maxim 8-20-7).

Inversions of the composite Fig. a about all 4 of the "focal" poles, f_R,

F_R, f_L, and F_L (focal for one or the other of the component basis curves), yield composite inverse loci similar to Fig. b (see Maxims 8-20-6,13). These composite loci consist of the equilateral limacon and the *equilateral* $\sqrt{2}a/2$ *inter-vertex quartic*, but they differ from one another in size and/or orientation (for example, compare Figs. 8-20a_2 and 8-20b_2).

As in the case of Fig. 8-20a_2, the inverse composite locus of Fig. b self-inverts in 3 modes--the 0° inter-oval mode about pole f'_R, the 0° "intra-segment mode" about pole F'_L (i.e., one segment of each curve inverts into its own other segment), and the 180° inter-oval mode about f'_L (a segment of one curve inverts into a segment of the other curve). As in the cases of the curves of Fig. 8-20a_2 and 8-22b,d,e, and e', there are 6 poles in the plane of Fig. b about which it inverts to a central composite locus. These are a', $p = F_R$, the two pairs of tangent vertices (yielding tangent equilateral strophoids), and the two intersections of the inverse y-axis and inverse modular circles (marked by the triangular arrowheads adjacent to the inverse modular circle C', about which the inverse loci are tangent *equilateral* $k = a$ *QBI quartics*).

Since no y-axis vertex-inversion cubic exists for Fig. a, one cannot draw homologies with the points of intersection and points of tangency of radii and chords, as in Fig. 8-22e' (poles to d), about which the inverse loci are the y-axis vertex cubic (but see Fig. e and Maxim 8-30-4, below). However, if one were to employ elliptical central QBI curves as basis curves, this homology also would exist.

In the inversion of Fig. b to Fig. c, the latter being an *inversion image*, there exist all the correspondences between the poles of the basis curve and its inversion image that were illustrated for the curves of Figs. 8-20a_1 and a_3. For example, the pole of inversion of the basis curve inverts to the center of the inversion image, the modularly-inverse poles, F_L and f_L, of the basis figure invert to exchanged positions and opposite sides in the inversion image, etc. If modular j for the 0° intra-segment-mode focus, F_L, had been employed in inverting Fig. b about pole a', the inversion image would have been ¼ the size of the basis figure, as per Maxim 8-20-14.

If one allows for the fact that biovular central Cartesians consist of two ovals, rather than two self-intersecting two-looped segments, by also taking into account a composite figure of an elliptical central QBI curve and its inverse locus about the center, all of the general Maxims for biovular members of central Cartesian-based inversion superfamilies (including axial and non-axial bicircular quartics and serpentine circular cubics) apply to composite central QBI figures. In fact, they apply in general to biovular central curves that self-invert about the center at 0° and 180°--even to the point of existence of 3 basis central biovular species (two with opposed ovals and one with annular ovals). The only exception relates to Maxim 8-20-13, which concerns inversions to Cartesian self-inverters, and even this Maxim applies in part for the traditional foci of a composite figure of central elliptical QBI curves.

Curves That Congruent-Invert But Do Not Self-Invert

It was noted in Fig. 6-6 that the non-axial inverse loci of a central basis curve B that does not self-invert possess two foci of self-inversion and one pole of congruent-inversion(see also Fig. 8-38). This is understood readily in terms of symmetry considerations and inversion-mapping Maxims. Thus, for any given non-axial pole P_1 of inversion of such a basis curve, there exist 3 equivalent poles, P_2,P_3, and P_4, namely, its reflections in the two lines of symmetry and the center. In other words, non-axial equivalent poles of central basis curves with two lines of symmetry occur in concyclic quartets --at the vertices of rectangles.

Since equivalent poles are symmetrically located and are equivalent inversion poles, the inverse loci of B about these poles are congruent. Moreover, the 2nd-generation inverse loci about corresponding poles, P_1'-P_4', inverse to these poles in all inverse loci, B', are subspecifically identical (belong to the same subspecies). If one of the poles, P_1, is the pole of inversion, then pole P_1', in the extended plane of the inverse curve B' is the ideal line at infinity. Accordingly, the 2nd-generation inverse locus of B' about P_1' (i.e., about any point of P_1') is congruent to B' [for the appropriate (infinite) value of the unit of linear dimension]. This being the case, the three 2nd-generation inverse loci about poles P_{2-4}' also are congruent to B'. But since two lines of symmetry are lost in a non-axial inversion of the central curve, two of these congruent loci also must be coincident with the basis curve (i.e., self-inversions). It follows that two of the 3 poles P_{2-4}' of congruent inversion are foci of self-inversion (the poles inverse to the reflections of pole P_1 in the lines of symmetry of B).

The basis mutually-inverse ($j = a$), concentric ensemble of an equilateral hyperbola (EH) and an equilateral lemniscate (EL) of Fig. a provides a convenient reference figure for analyzing the properties of loci inverse to curves that self-invert at other angles than 0° or 180°. Thus, if one considers EL to be rotated CCW through an angle θ ($\theta \neq 0^\circ, 90^\circ, 180^\circ$, or 270°), then the resulting ensemble EH-EL$_\theta$ self-inverts at angles of θ and $\theta+180^\circ$.

The question of concern is, what properties of the dissimilar inverse loci EH'-EL$'_\theta$ correspond to the property of self-inversion of the basis curve at the angles θ and $\theta+180^\circ$. Unlike the situation for foci or modes of self-inversion at 0° and 180° (Maxim 8-15-4), the sum of the number of lines of symmetry and the number of modes of self-inversion for the inverse loci of curves that self-invert at other angles is not conserved. In fact, it can be asserted categorically that modes of self-inversion at other angles than 0° and 180° are lost in all dissimilar inverse loci.

Since one expects the sought-after corresponding property of EH'-EL$'_\theta$ to be some characteristic inversion property, it is reasonable to inquire as to the existence and properties of the congruent inverse loci of this figure (since the composite basis curve does not self-invert at 0° or 180° and possesses no line of symmetry, no inverse locus self-inverts). Accordingly, consider that the basis ensemble EH-EL$_\theta$, is inverted about a pole, P_h, on the transverse axis of the equilateral hyperbola at $x = h$ ($h < a$). If poles exist in the plane of EH-EL$_\theta$ that are equivalent inversion poles to P_h, then poles also will exist in the plane of EH'-EL$'_\theta$ about which it undergoes congruent-inversion (for the appropriate values of the units of linear dimension). [Self-inversion already has been ruled out for the composite locus, although the individual inverse loci EH' and EL$'_\theta$ self-invert about one or two poles, depending upon the location of P_h.]

Although the basis curve EH-EL$_\theta$ consists of concentric (about the pole $C_c = a$) central curves, it is not itself a central curve (except for $\theta = 0^\circ$, 90°, 180°, and 270°, which have been excluded). Accordingly, equivalent non-axial poles do not occur in "rectangular quartets." Instead, they occur in pairs reflected in the pole of concentricity (C_c). Thus, the pole equivalent to P_h is P_{-h} on the transverse axis of the hyperbola at $x = -h$. But, although P_{-h} is the only pole of EH-EL$_\theta$ that is equivalent to P_h, it is not the only *equivalent inversion pole*. There exist two additional equivalent inversion poles, P^*_h and P^*_{-h}, on lines through C_c at angles θ and $\theta+180^\circ$ to the positive x-axis, at distances $\pm j^2/h = \pm a^2/h$, namely the poles that are modularly inverse to the poles P_h and P_{-h}.

Accordingly, the inverse loci of a non-central basis curve that self-inverts at angles θ and $\theta+180^\circ$ ($\theta \neq 0^\circ, 90^\circ, 180^\circ$, or 270°) possess 3 poles of congruent-inversion. The possession of these characteristic features is one of the properties of the inverse locus that "replaces" the property of self-inversion of the basis curve at angles of θ and $\theta+180^\circ$. One can draw homologies between these poles and the corresponding poles of biovular central Cartesians by tracking their locations for various values of the angle of rotation θ, as θ varies from 0° to 180° since, as noted above, at the limits $\theta = 0^\circ$ and 180°, the curve EH-EL$_\theta$ takes on virtually all the properties of biovular central Cartesians.

The position of the pole P_{-h} does not change with θ; it always is at the point of reflection of P_h in C_c, the pole of concentricity. However, the locations of the poles P^*_h and P^*_{-h} change with θ according to eqs. 1a,b, which are the parametric eqs. of a circle centered at $x = a^2h^3/(a^4-h^4)$ of radius $R = a^4h/(a^4-h^4)$.

(a) $$x = \frac{\pm(a^2\cos\theta \mp h^2)a^2h}{(a^4+h^4 \mp 2a^2h^2\cos\theta)}$$ x coordinates of poles P^*_h and P^*_{-h} (8-30-1)

(b) $$y = \frac{\pm a^4 h\sin\theta}{(a^4+h^4 \mp 2a^2h^2\cos\theta)}$$ y coordinates of poles P^*_h and P^*_{-h}

At $\theta = 0^\circ$ the pole $P^{*\prime}_h$ originates as the pole $C'_c = f'_R$, which is the 0° inter-loop-mode focus of curve EH'-EL' of Fig. b. [These tracking results are for $h = \sqrt{2}a$ as in Fig. a, rather than for $h < a$, as per above.] As θ increases from 0°, $P^{*\prime}_h$ travels CW along a hemi-circular arc toward pole f'_L, at which, for $\theta = 180^\circ$, it becomes the 180° inter-loop-mode focus of EH'-EL'. Pole $P^{*\prime}_{-h}$ originates as pole f'_L and travels CW along the other hemi-circular arc of the same circle, becoming the pole $C'_c = f'_R$. Two line segments connecting pole $P^{*\prime}_h$ (for $h < a$) to P_h and the pole C'_c, inverse to C_c, intersect at an angle of θ, while those to pole $P^{*\prime}_{-h}$ intersect at an angle $180^\circ-\theta$.

Accordingly, the two poles, $P^{*\prime}_h$ and $P^{*\prime}_{-h}$, of identical subspecific-inversion that are inverse to the modularly-inverse poles, P^*_h and P^*_{-h} (of P_h and P_{-h}), are the homologues of the 0° and 180° foci of self-inversion of axial bicircular quartics. On the other hand, the pole P'_{-h} (F'_L of Fig. b) of congruent-inversion, that is inverse to P_{-h} (F_L of Fig. a), the location of which does not change, is the homologue of the 0° focus of self-inversion of axial bicircular quartics that "replaces" a line of symmetry of the central basis curve.

Special cases arise for $\theta = 90^\circ$ and 270°. For these angles: (a) the locus EH'-EL' acquires two lines of symmetry; (b) the two poles of congruent-inversion, $P^{*\prime}_h$ and $P^{*\prime}_{-h}$, assume symmetrical locations in the inverse locus, whereupon the inverse loci of EH'-EL' about them are congruent with one another for a common value of the unit of linear dimension; (c) the line segments from both $P^{*\prime}_h$ and $P^{*\prime}_{-h}$ to P_h and C'_c intersect at 90°; and (d) pole P'_{-h} becomes a focus of self-inversion.

The axial inverse locus thus possesses two poles of congruent-inversion (for the appropriate single value of the unit of linear dimension) and one focus of self-inversion. If pole P_h is non-axial, the inverse locus also is non-axial and possesses 5 poles of congruent-inversion and two foci of self-inversion (see Fig. d and discussion below).

In summary, modes of self-inversion at angles of θ and $\theta+180^\circ$ ($\theta \neq 0^\circ$, 90°, 180°, or 270°) are not conserved in the dissimilar inverse loci of a non-central locus. In their place the dissimilar inverse loci possess 3 poles of congruent-inversion (for appropriate values of the units of linear dimension that are different for each pole) that are homologues of the 3 foci of self-inversion of axial bicircular quartics.

For $\theta = 90^\circ$ or 270°, the basis curve gains two lines of symmetry and acquires the inversion properties of monovular Cassinians. One line of symmetry is lost in axial inversions, whereupon one of the poles of congruent-inversion becomes a focus of self-inversion. The other two poles of congruent-inversion assume symmetrical locations and yield congruent loci for a common value of the unit of linear dimension. As θ approaches 0° or 180°, the properties of the basis curve approach those of biovular central Cartesians and the 3 poles of congruent inversion of the inverse loci approach the locations of the 3 foci of self-inversion of axial bicircular quartics.

CIRCUMPOLAR
MAXIM 8-30-1: *The equivalent inversion poles of a pole of inversion, P_h, of a basis curve that self-inverts about a single pole, C, at angles θ and $\theta+180^\circ$ are:*

(a) the reflection P_{-h} of P_h in C,
(b) poles P^*_h and P^*_{-h}, modularly inverse to poles P_h and P_{-h} (at distances $\pm j^2/h$ along lines through C at the angles θ and θ+180°), and
(c) the reflections of poles P_h, P_{-h}, P^*_h, and P^*_{-h} in lines of symmetry.

8-30-2: *Poles Q', inverse to equivalent inversion poles Q, of the pole of inversion P, of a basis curve B, are either poles of congruent-inversion or foci of self-inversion of the inverse locus B'. Poles Q' are foci of self-inversion if the corresponding poles Q are:*

(a) reflections of P in lines of symmetry,
(b) modularly inverse to P in 0° or 180° foci of self-inversion, or
(c) modularly inverse to P in a 0° and 180° focus of self-inversion.

An Inverse Monovular Cassinian Analogue With A Collinear Ensemble of 5 Discrete Poles of Congruent-Inversion And Two Foci of Self-Inversion

The concluding sections of this figure address the general topics of the numbers and locations of poles of identical subspecific-inversion in the inverse loci of central curves with two and 4 lines of symmetry, a non-central figure with 3 lines of symmetry, and non-central curves with an infinite number of parallel lines of symmetry.

Cassinian monovals are analogues of the concentric ensemble of an equilateral hyperbola and an equilateral lemniscate oriented with their transverse axes orthogonal to one another. It already has been noted that Cassinian monovals self-invert about the center at angles of 90° and 270° and that the axial inverse loci of such a curve possess two poles of congruent-inversion (for the appropriate single value of the unit of linear dimension) and one focus of self-inversion.

The basis locus of Fig. d (point quartet B) is representative of a monovular Cassinian with a C-value of 17/60ths (see hemi-outlines in Fig. f). For the present purposes it suffices to represent this monovular Cassinian by only 4 points (the axial vertices) making up a central quartet that self-inverts at 90° and 270°.

The quartet B is such an array, with the 4 points (solid squares) located at the vertices of a diamond-shaped figure with an axial ratio of 4 (points at $x = \pm\frac{1}{2}, \pm 2$). The "diamond" is centered at the origin with its long axis coincident with the y-axis. The partial outline of this figure is

Loci With Collinear Ensembles of 5 Poles of Congruent-Inversion & Two Foci of Self-Inversion (Illustrated With Point Quartets)

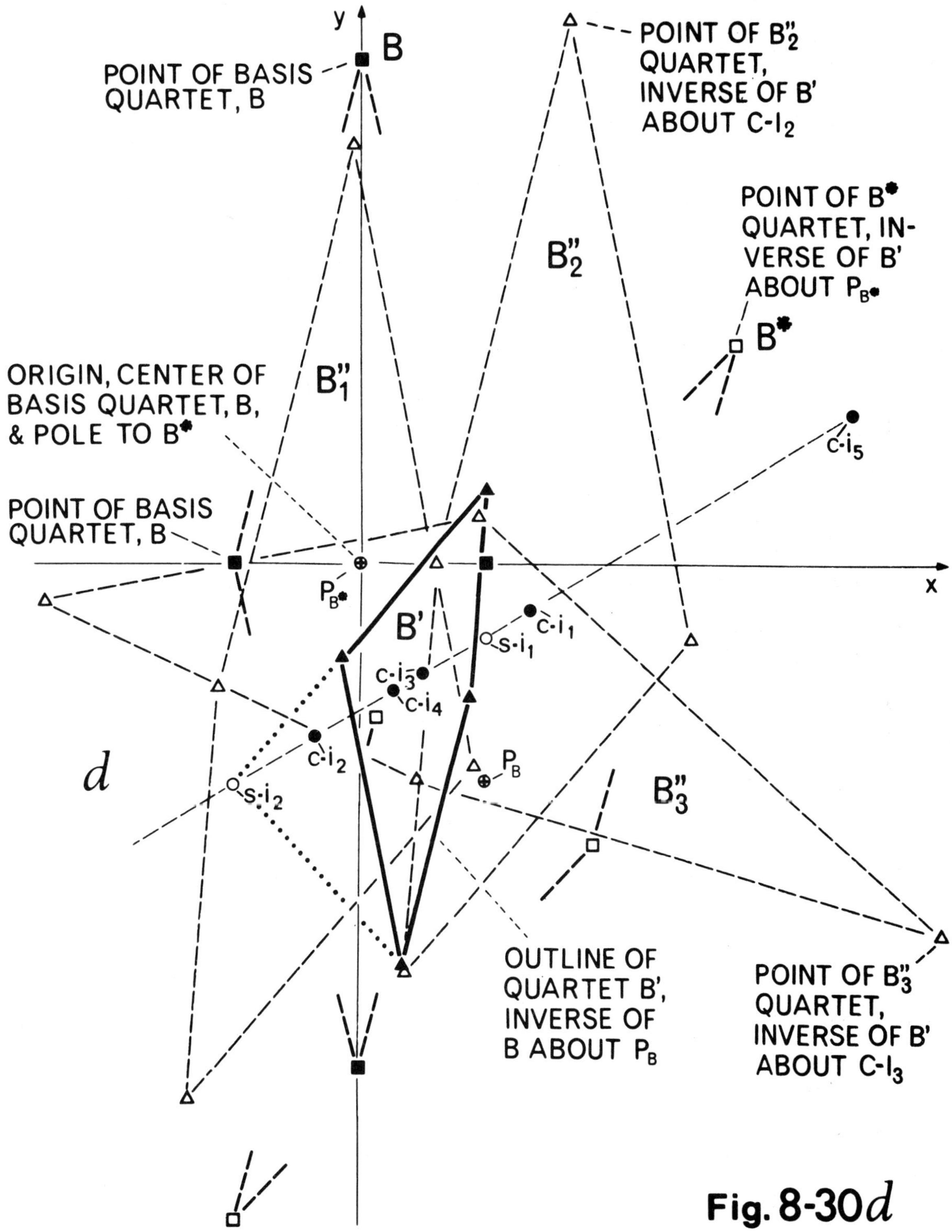

Fig. 8-30*d*

indicated by the sets of 3 collinear heavy dashes emanating from 3 of the 4 points. It is to be emphasized that the dashed linear outlines of this quartet and quartets B*, B_1'', B_2'', and B_3'' are merely convenient indications of the positions of the component points, and do not indicate the course of loci between the points (as do the dashed hemi-outlines in Figs. e and f).

The present figure illustrates the properties of the inverse loci of central curves that self-invert at angles of 90° and 270°, for non-axial poles of inversion, as opposed to the case for an axial pole of inversion treated in the preceding section. The pole of inversion, P_B, is taken to be at modular distance ($j = 1$) on a radius from the center at -60° to the x-axis. Accordingly, 3 equivalent poles lie at the points of reflection of P_B in the lines of symmetry and the center. In addition, there exist 4 equivalent inversion poles that are modularly inverse to the first 4 poles at angles of -30°, -150°, -210° and -330° to the x-axis (Maxim 8-30-1).

In consequence, there are 7 equivalent inversion poles to the pole of inversion, P_B, in the plane of the basis quartet B. While these are not shown in Fig. d, another case with 8 equivalent inversion poles may be seen in Fig. e (small solid circles) for biovular central Cartesians (which self-invert at 0° and 180°). Following Maxim 8-30-2, the inverse locus, quartet B', possesses 5 poles of congruent-inversion ($c\text{-}i_{1-5}$) and two foci of self-inversion ($s\text{-}i_{1,2}$), the latter being inverse to points of reflection of P_B in the x-axis ($s\text{-}i_1$) and y-axis ($s\text{-}i_2$).

By virtue of having taken pole P_B at modular distance from center, all of its equivalent inversion poles also lie at modular distance from center (unlike the case for Fig. e). As a result all 7 equivalent inversion poles are concyclic in the plane of quartet B, and all 7 poles of identical subspecific-inversion of quartet B' are collinear (line shown by light dashes).

CIRCUMPOLAR
MAXIM 8-30-3: *If a basis curve that self-inverts about a single pole in multiple modes is inverted about a pole on its modular circle, all of the poles of identical subspecific-inversion of the inverse locus will be collinear.*

All of the point quartets of Fig. d are 1st-generation inverse loci of quartet B' about the various indicated poles (all circular poles). Inversion of B' about the reciprocal inversion pole P_B yields the basis quartet B, while inversion about the center (which is self-inverse in P_B) yields the quartet B* (open squares), a 0° central inverse locus of the basis curve B. Note that B* is congruent with B and rotated 90°, a property of the inverse loci of all Cassinian monovals about their centers (see legend to Fig. 8-15, following eq. 8-15-6).

Three examples of the inverse loci about poles of congruent-inversion are quartets B_1'', B_2'', and B_3'' (open triangles), for the poles of congruent-inversion (large solid circles) $c\text{-}i_1$, $c\text{-}i_2$, and $c\text{-}i_3$, respectively. These loci are shown at larger than congruent size to avoid congesting the figure. Poles P_{B*}, P_B, $c\text{-}i_1$, and $c\text{-}i_2$ are concyclic about pole $c\text{-}i_3$ and lie on orthogonal diameters. Correspondingly, the loci B_1'' and B_2'' are congruent and B and B* are congruent.

The locus B_3'' is a reflection of B' in the line segment $P_B\text{-}P_{B*}$ (see Maxim 8-38-3). Loci B_1'' and B_2'' are congruent for a common value of j and are rotated 180° about poles $c\text{-}i_1$ and $c\text{-}i_2$, relative to B', and translated along the line of collinear poles of identical subspecific-inversion by amounts equal to the distance between the poles $c\text{-}i_1$ and $c\text{-}i_3$ (or $c\text{-}i_2$ and $c\text{-}i_3$).

It will be noted that in the case of a focus of self-inversion that replaces a given line of symmetry, points inverse to those points of the basis locus that are incident upon the given line of symmetry (corresponding to vertices) become points of tangency of radii from the pole of self-inversion to the inverse locus. This is illustrated for pole $s\text{-}i_2$. The only pair of points at which the dotted line extending to the upper right from $s\text{-}i_2$ intersects the quartet (solid triangles) is the pair inverse to the points of the basis quartet at $x = \pm\frac{1}{2}$, i.e., at the points reflected in the y-axis.

On the other hand, the dotted line extending to the lower right from $s\text{-}i_2$ intersects the quartet at only the single point (solid triangle), which is inverse to the basis point at $y = -2$ and corresponds to a y-axis vertex of the basis figure. It follows that such single points of tangency are poles of inversion to the corresponding axial-vertex cubics of the basis curve (see also Maxim 8-38-1).

CIRCUMPOLAR
MAXIM 8-30-4: *If radii from a 0° intra-oval-mode focus of self-inversion that replaces a line of symmetry, L, of a central Cartesian are drawn to points p_n' of the inverse locus that are inverse to points p_n, incident upon L, these radii are tangent to the inverse locus at the points p_n'. Such points, p_n', are poles of inversion to axial-vertex cubics.*

A Non-Axial Bicircular Quartic Analogue With
4 Foci of Self-Inversion and 3 Poles of Congruent-
Inversion In The Finite Plane

Inversion properties of axial bicircular quartics have been treated extensively in previous figures (particularly Fig. 8-22e') and also are dealt with in Figs. 8-31 to 33. Fig. e illustrates a non-axial bicircular quartic analogue derived by inverting a point-octet analogue of an annular central Cartesian (indicated in hemi-outline by the dashed tracing) about a non-axial pole. Accordingly, Fig. e is the counterpart of Fig. d for a conventional biovular central Cartesian that self-inverts at 0° and 180°, rather than for the very uncommon case of monovals that self-invert at 90° and 270°.

The basis octet (solid squares) consists of 4 points on the x-axis at $x = \pm\frac{1}{2}$, ± 2 and 4 points on the y-axis at $y = \pm 2/3$, $\pm 3/2$, corresponding to the vertices of the annular central Cartesian. The pole of inversion (large open circle containing an x) is at the point $(\frac{1}{2},\frac{1}{4})$, at a distance of $\sqrt{(5/16)}$ from center on a radius at an angle of 26.565° to the x-axis. The poles equivalent to the pole of inversion are marked by 3 small solid circles, the reflections of the pole of inversion in the lines of symmetry and the center. Since the distance $\sqrt{(5/16)}$ is non-modular ($j = 1$ for self-inversion about the center), 4 modularly-inverse poles exist at distances of $4\sqrt{5}/5$ on radii from the center through the pole of inversion and its 3 equivalent poles.

Thus, there are 7 equivalent inversion poles to the pole of inversion; the points inverse to these 7 poles are the poles of identical subspecific-inversion of the inverse locus. It will be noted that when the pole of inversion is included, the 8 equivalent inversion poles comprise the vertices of two concentric, parallel rectangular quartets. Accordingly, 4 poles of identical subspecific-inversion of the inverse locus are concyclic [foci 3 and 4 of self-inversion (numbered open circles) and poles 2 and 3 of congruent-inversion (numbered solid circles)] and 3 are collinear (foci 1 and 2 of self-inversion and pole 1 of congruent-inversion).

CIRCUMPOLAR
MAXIM 8-30-5: *If a basis curve that self-inverts about a single pole in multiple modes is inverted about a non-modular pole, the poles of identical subspecific-inversion in the finite plane of the inverse locus comprise two ensembles, one concyclic and the other collinear. The concyclic ensemble possesses one more pole than the collinear ensemble.*

Foci of Self-Inversion & Poles of Congruent-Inversion of Inverse Loci of A Biovular Central Cartesian & Centered, Dual, Orthogonal Monovular Cassinians (Illustrated With Point Octets)

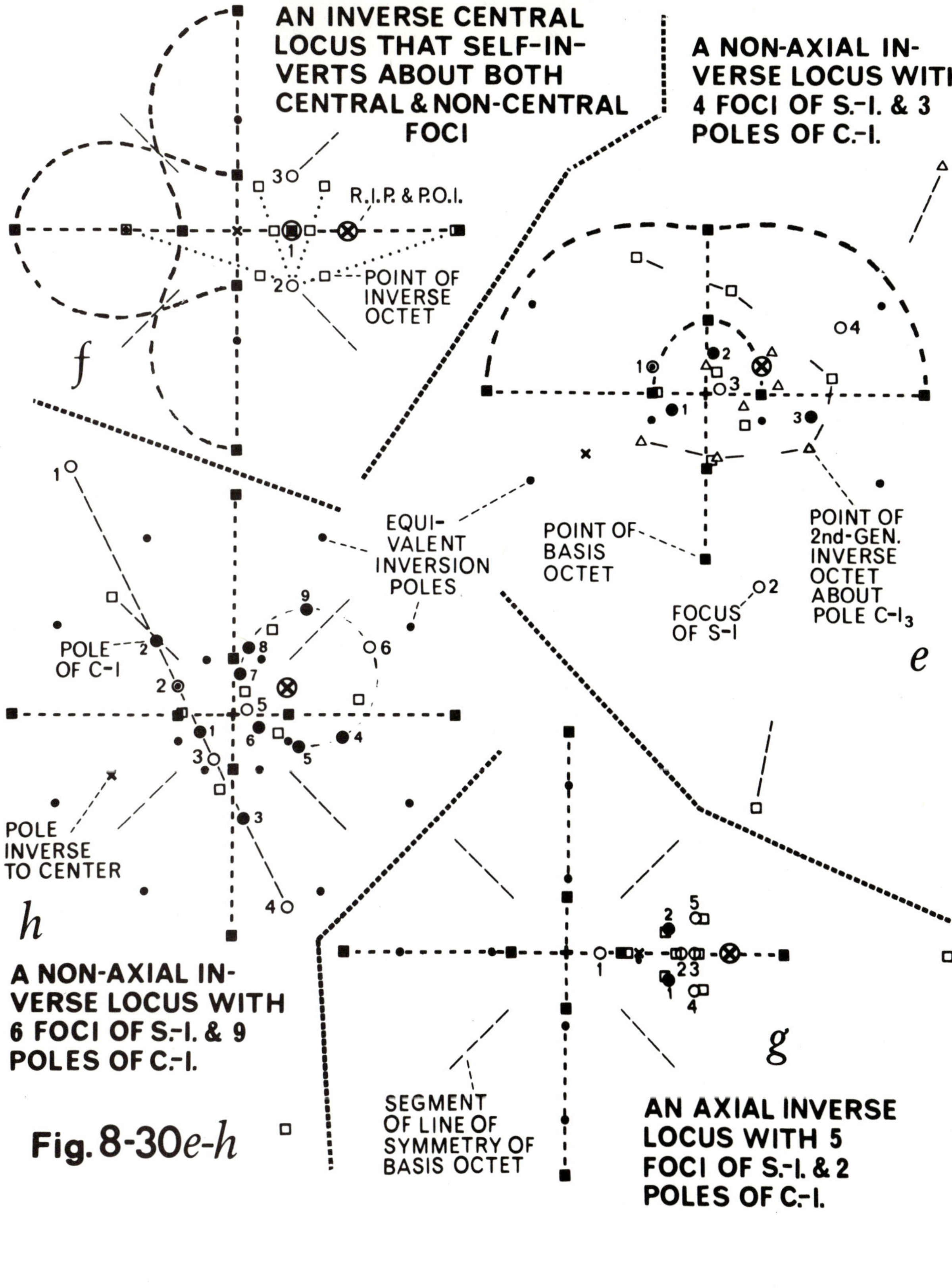

Fig. 8-30*e*-*h*

CIRCUMPOLAR
MAXIM 8-30-6: *In a locus B', the inverse about pole P of a central basis curve B that self-inverts about the center at 0° and 180°, the pole P*' inverse to pole P*, modularly inverse to P, is the 0° focus of self-inversion; the pole $P_R^{*\prime}$, inverse to P_R^*, the reflection of P* in the center, is the 180° focus of self-inversion* (see also Maxim 8-22-11).

The positions of the points of the 1st-generation inverse octet are marked by the open squares. Four of the points thereof (two of one inverse oval and two of the other) lie along a large peripheral arc, indicated by short lines and dashes emanating from them, while the other 4 lie proximal to one another in a roughly rectangular array. The foci of self-inversion 1 and 2 (open circles) "replace" the lines of symmetry, whereas foci of self-inversion 3 and 4, which are collinear with the pole of inversion and the center (central cross), "replace" the 0° and 180° center-foci of the basis curve. Focus 3 is a 180° focus, while focus 4 is a 0° focus.

At first sight, the two foci of self-inversion of the inverse locus that "replace" the coincident foci of self-inversion of the basis curve would seem to stand in a different relationship to poles inverse to the vertices of the basis curve than do the foci of self-inversion that "replace" lines of symmetry (Maxim 8-30-4). Each pair of coincident radii or chord segments from the former foci intersect two points of the inverse curve that are inverse to collinear modularly-inverse vertices of the basis curve.

In other words, these foci of self-inversion of the inverse curve stand in the same relationship to modularly-inverse points of the inverse curve as do the corresponding center-foci of self-inversion of the basis curve to the corresponding modularly-inverse points of the basis curve (a relationship that follows from Maxim 6-5-1). These relationships apply regardless of whether the inverse locus is composed of opposed or annular ovals or is an inversion cubic. For example, for inverse opposed ovals, pole 4 is a lateral 0° inter-oval-mode focus and pole 3 is an inter-oval 180° focus (i.e., it lies between the ovals), while for inverse annular ovals, poles 3 and 4 are interior 180° and 0° inter-oval-mode foci, respectively.

On the other hand, the two foci of self-inversion that replace lines of symmetry (foci 1 and 2) do not exist in the finite plane of the basis curve. The poles intersected by radii from these foci in the plane of the inverse curve are not modularly-inverse points of the basis curve, but are equivalent points reflected in lines of symmetry. But if one considers that these foci replace the two foci of the basis curve on the ideal line at infinity, all of the foci of self-inversion of the inverse curve preserve the relationships possessed by the corresponding foci of the basis curve (see Maxims 8-30-9 to 12 and Table 8-30-1).

CIRCUMPOLAR
MAXIM 8-30-7: *An annular central Cartesian B is inverted about a pole P to a non-axial bicircular quartic B'. Modularly-inverse point-pairs (p and p*)*$_{0°}$ *and (p and p*)*$_{180°}$ *of B are intersected by coincident radii and chord segments, respectively, from its center. In the plane of the inverse locus B', coincident radii from the 0° inter-oval-mode focus that "replaces" the 0° center-focus of B, intersect all point-pairs (p' and p*')*$_{0°}$ *inverse to the point-pairs (p and p*)*$_{0°}$*, while chord segments from the 180° focus of self-inversion that "replaces" the 180° center-focus of B intersect all point-pairs (p' and p*')*$_{180°}$ *inverse to point-pairs (p and p*)*$_{180°}$*.*

The open triangles of Fig. e mark the positions of an octet of points that comprise the 2nd-generation inverse locus of the 1st-generation octet (open squares) about pole 3 of congruent-inversion (large solid circle 3), shown at smaller than congruent size. Short line-segments extending from the triangles marking the peripheral positions help to indicate the positions of the points, for comparison with the 1st-generation figure.

The inverse loci of the 1st-generation figure about congruent-inversion poles 1 and 2 also are congruent with this figure (for the appropriate values of the units of linear dimension). Since the primary objective of Figs. e-h is to illustrate the numbers, locations, and types of poles of identical sub-specific-inversion, a congruent inverse locus is illustrated only in this one instance.

A Central Curve That Self-Inverts About Both Central And Non-Central Foci

Since the inversion properties of curves with more than 2 lines of symmetry have not yet been illustrated (for analytical treatments see pages XIV-77 to 86), and since the inverse loci of these curves possess more poles of identical subspecific-inversion than members of the QBI and central Cartesian-based inversion superfamilies, several inverse loci of a curve with 4 lines of symmetry now are considered.

The basis locus for Figs. f-h consists of a point-octet analogue of two centered, orthogonal, intersecting congruent monovular Cassinians ($C = 17/60$; outlined in part at the left in Fig. f); specifically it represents the vertices of the latter ensemble at $x = y = \pm\frac{1}{2}, \pm 2$. Accordingly, the points of this ensemble self-invert about the center in the 4 inter-oval modes; 0°, 90°, 180°, and 270°. [The basis locus in question has an 8th-degree degenerate eq.]

In the case of any given annular central Cartesian, the basis curve uniquely defines two complementary dissimilar opposed-oval central curves. A curve with 4 lines of symmetry that self-inverts about the center at angles of 0°, 90°, 180°, and 270° also uniquely defines two dissimilar inverse central curves, each of which is its inverse about 4 modular poles on sets of orthogonal axes. The inverse central octet of Fig. f was obtained by inverting the basis octet about an x-axis pole at modular distance from center; it possesses the highly unusual property that it self-inverts about both central and non-central foci. The locus obtained by inverting about a modular pole on one of the $\pm 45^\circ$ lines of symmetry also possesses this property (see section below).

[The two intersecting curves that make up the inverse locus in question consist of two congruent opposed ovals, each almost identical with the locus of Fig. 8-16g, with the lateral lobes overlapping to a depth beyond the level of the focus of self-inversion. Only the inverse loci about *modular* poles are congruent. Monovular parabolic Cartesians obtained, for example, by inverting monovular Cassinians about poles at $\pm\frac{1}{2}d$ on the transverse axis (x-axis of Fig. 8-19b) and $\pm 2d\sqrt{(C^2-1/16)}$ on the orthogonal axis (y-axis of Fig. 8-19b) belong to the same subspecies but are incongruent.

The rightmost poles of the basis and inverse loci are in coincidence but are illustrated as if merely in close juxtaposition, to show the presence of both. Similarly, the center-focus of the inverse locus is in coincidence with a point of the basis octet, and a point of the inverse octet is in coincidence with an equivalent inversion pole of the pole of inversion.]

To locate the poles of identical subspecific-inversion of the inverse locus, one finds the poles inverse to the equivalent inversion poles of the basis figure. Since the pole of inversion is both axial and modular, only 3 equivalent (inversion) poles exist (small solid circles). Moreover, each of these is a reflection of the pole of inversion in a line of symmetry, which means that each inverse pole is a focus of self-inversion (Maxim 8-30-3).

The pole inverse to the reflection of the pole of inversion in the orthogonal line of symmetry would achieve focal rank even if the basis locus self-inverted at only 90° and 270°, but the other two poles (foci of self-inversion 2 and 3) would have been poles of congruent-inversion only. Accordingly, the latter poles achieve focal rank by virtue of the additional modes (0° and 180°) of self-inversion and lines of symmetry.

Although all 3 equivalent poles in the plane of the basis curve become foci of self-inversion in the plane of the inverse locus, the latter gains only two foci of self-inversion and loses only two lines of symmetry. The factors limiting the gain and loss are as follows. Since the pole of inversion of the basis octet is modular, it is coincident with its inverse at 0° in the center-focus. By virtue of this factor alone, an orthogonal line of symmetry (inverse to the modular circle) is gained and a 0° inter-oval-mode focus of self-inversion is transferred from the finite plane to the ideal line at infinity in the orthogonal direction. But since the pole of inversion also is axial, an orthogonal line of symmetry is lost and an intra-oval-mode focus of self-inversion is transferred from the ideal line at infinity to the finite plane. The net result of the operation of both factors is that there is an exchange of modes of self-inversion between 0° foci in the finite plane and at infinity but neither a loss of the orthogonal line of symmetry nor a gain of a 3rd focus of self-inversion.

Radii emanating from the 0° inter-oval-mode foci 2 and 3, as expected, intersect the inverse locus at all point-pairs inverse to pairs reflected in the $\pm 45^{\circ}$ lines of symmetry of the basis curve. Since no points of the locus are illustrated on the $\pm 45^{\circ}$ lines of symmetry, there are no single points of contact of radii from foci 2 and 3 with inverse points. If 4 points are added to the basis octet at modular distance from center on the $\pm 45^{\circ}$ lines of symmetry (the points of intersection of the orthogonal basis monovals), the inverse points assume the expected relationships. Thus, they become modularly inverse in the center at 0° (intra-oval mode) and 180° ($j = \frac{1}{2}$), and modularly inverse in foci 2 and 3 in the 0° inter-oval mode ($j = \sqrt{2}/2$).

In Fig. g, the basis octet (solid squares) is unaltered from Fig. f and the pole of inversion is axial, but with the crucial difference that it is not at modular distance from center. As a result of this change, the otherwise somewhat similar inverse locus (open squares) is axial, rather than central, and its symmetry properties are altered markedly.

[The first axial point of the inverse octet to the right of center is coincident with a point of the basis octet but is shown slightly to the right of it. In addition, the position of the axial equivalent inversion pole modular to the pole of inversion has been offset slightly below the x-axis, since it is coincident with the point inverse to the center (cross).]

Since the pole of inversion is not modular, a rectangular quartet of equivalent inversion poles complements the 3 poles equivalent to the pole of inversion. This brings the total number of equivalent inversion poles of the basis locus to 7, and the total number of poles of identical subspecific-inversion of the inverse locus in the finite plane to 7 (the same as for the 1st-generation inverse locus of Fig. e).

Each of the 3 equivalent inversion poles is a reflection of the pole of inversion in a line of symmetry. In consequence, since these lines of symmetry are lost in the inversion, the inverse of each of these poles becomes a focus of self-inversion in the inverse locus, possessing the mode of the corresponding focus on the ideal line at infinity. Thus, focus 3, which replaces the y-axis line of symmetry, is an intra-oval-mode focus, while foci 4 and 5, which replace the $\pm 45^{\circ}$ lines of symmetry, are inter-oval-mode foci.

Radii emanating from foci 4 and 5 intersect all points of the inverse octet in modularly-inverse pairs, because all of these points are inverse to point-pairs that are reflected in the lost $\pm 45^{\circ}$ lines of symmetry. Radii emanating from focus 3, however, intersect the *axial* quartet of points in modularly-inverse pairs but intersect the *non-axial* points individually at modular distance. The latter intersections are at poles inverse to the vertices of the lost y-axis line of symmetry (Maxim 8-30-4).

The two equivalent inversion poles on the x-axis are modularly inverse to the pole of inversion at 0° and 180°. The inverses of these poles, foci 1 and 2, are the 0° and 180° foci, respectively, of the inverse locus, that replace the corresponding center-foci of the basis curve, and that retain the inter-oval-mode possessed by the latter foci. All the radii and chord segments from these foci that intersect one point of the inverse octet also intersect a corresponding modularly-inverse point. Thus, these 5 poles of the basis curve account for the 5 foci of self-inversion of the inverse locus.

The two remaining equivalent inversion poles lie on the y-axis of the basis curve and are modularly inverse (at both 0° and 180°) to the corresponding two equivalent poles on the y-axis (i.e., to the two reflections of the pole of inversion in the $\pm 45^{\circ}$ lines of symmetry). The inverses of these poles are the two poles (1 and 2) of congruent-inversion. All radii that emanate from these poles have only single intersections with points of the inverse octet (which are points inverse to the axial vertices of the basis octet).

CIRCUMPOLAR
MAXIM 8-30-8: *A given line through a pole of congruent-inversion of an inverse locus intersects the locus at no more than one point inverse to an axial vertex of the basis curve.*

A Non-Axial Inverse Locus With 15 Poles of Identical Subpecific-Inversion In The Finite Plane

Fig. h illustrates the most general case of an inverse locus of the basis Cassinian monovals (partially outlined in Fig. f). In the case of Fig. f, the inverse locus is central and the 3 foci of self-inversion are collinear because the pole of inversion is both modular and axial. In the case of Fig. g, the inverse locus is axial and possesses 4 concyclic and 3 collinear poles of identical subspecific-inversion, because the pole of inversion is axial but not modular.

In the present instance, since the pole of inversion is neither axial nor modular, the basis curve possesses 7 poles equivalent to the pole of inversion and 8 modularly-inverse poles, giving a total of 15 equivalent inversion poles and 15 poles of identical subspecific-inversion of the inverse non-axial locus. Since the basis curve possesses 4 lines of symmetry and self-inverts about the center at 0°, 90°, 180°, and 270°, the non-axial inverse locus possesses 6 foci of self-inversion. Four of these replace the lines of symmetry and the other two replace the coincident central 0° and 180° foci of self-inversion of the basis curve.

Since 7 equivalent inversion poles of the basis octet are concyclic with the pole of inversion, these become collinear poles of identical subspecific-inversion in the inverse locus. Four of these 7 poles are foci of self-inversion by virtue of being inverse to the 4 reflections of the pole of inversion in the 4 lines of symmetry. The other 8 equivalent inversion poles of the basis octet are modularly inverse to the 7 equivalent poles and the pole of inversion. Since these are concyclic, and the pole of inversion is interior to them, they also are concyclic in the inverse locus, for which they also become poles of identical subspecific-inversion. The latter include among their number the 0° and 180° foci of self-inversion (foci 6 and 5, respectively) that replace the corresponding center-foci of the basis curve. These are inverse to the two modularly-inverse poles that are collinear with the pole of inversion and the center of the basis curve (Maxim 8-30-6).

Of the latter two foci, focus 5 possesses the 180° inter-oval mode and focus 6 the 0° inter-oval mode, the same as for the center focus of the basis curve. The characteristics of foci 2 and 4, which replace the 0° and 90° lines of symmetry, and foci 1 and 3, which replace the $\pm 45^{\circ}$ lines of symmetry, are essentially unaltered from those described for the corresponding foci of the inverse loci of Figs. f and g. Radii from each of foci 2 and 4 intersect the points of the octet in two modularly-inverse pairs and in 4 modular singlets (the latter corresponding to the x and y-axis vertices, in two concyclic arrays), while radii from foci 1 and 3 intersect all 8 points of the inverse locus in modularly-inverse pairs.

Radii from the 9 poles of congruent-inversion intersect the points of the inverse octet (corresponding to vertices) only singly (Maxim 8-30-8). The orientations and sizes of the various congruent-inversion loci (for any given fixed value of j) are related to one another in various complex ways, a discussion of which is deferred.

[In this connection, recall that in the examples of Fig. d, locus B_3'' is simply a reflection of of the quartet B' in line segment P_B-P_B*, while loci B_1'' and B_2'' are congruent for a common value of j and are simply rotated $180°$ about poles c-i_1 and c-i_2 and translated along the line of collinear poles of identical subspecific-inversion by amounts equal to the distance between poles c-i_1 and c-i_3.]

Conservation of 0° Inter-Oval And of 0° Intra-Oval Modes of Self-Inversion In The Extended Plane

It already has been noted in the above treatment and previous figures (see Maxim 8-15-4b) that 180° foci are neither lost nor gained in an inverse locus (although we have dealt with only the inter-oval or inter-segment type, the intra-oval type possesses the same conservation properties). Because of their exchange properties with lines of symmetry, neither individual 0° foci nor 0° inter-oval or intra-oval modes of self-inversion in the finite plane are conserved, though, if a curve possesses both a 0° focus and a 180° focus of self-inversion in coincidence, all its inverse loci will possess a corresponding pair of 0° and 180° foci.

The above treatment of curves with 4 lines of symmetry and as many as 6 foci of self-inversion provides excellent illustrative material relating to conservation and exchange properties of 0° foci of self-inversion, and it will be evident from the accompanying Table 8-30-1 that, in the union of the ideal line at infinity and the finite plane--the extended plane--both 0° inter-oval and 0° intra-oval-mode foci of self-inversion are conserved (where a focus possessing both 0° and 180° modes of self-inversion is reckoned as coincident 0° and 180° foci). Thus, all types of foci of self-inversion at 0° and 180° are conserved in the extended plane; only their locations change.

CIRCUMPOLAR MAXIMS 8-30-

9: *A focus of self-inversion of an inverse curve that replaces a line of symmetry of a basis curve acquires the mode of self-inversion of the focus of self-inversion of the basis curve that lies on the ideal line at infinity in the direction orthogonal to the lost line of symmetry.*

10: *In modular axial inversions of a central curve that self-inverts about the center at 0° and 180°, to dissimilar central curves that also self-invert about their centers at 0° and 180°, the basis 0° center-focus "exchanges positions" (modes of self-inversion) with the basis 0° focus of self-inversion on the ideal line at infinity in the direction of the axis in question. In non-modular inversions of the same central curve to non-central curves, the inverses of the basis 0° center-focus possess the mode of self-inversion of the latter.*

11: *Inter-oval-mode and intra-oval-mode (and inter-segment-mode and intra-segment-mode) 0° foci of self-inversion in the extended plane are conserved by the inversion transformation.*

12: *The number of foci in the extended plane in each category of self-inversion, i.e., 0° and 180° inter-oval mode (and inter-segment mode) and 0° and 180° intra-oval mode (and intra-segment mode) is conserved by the inversion transformation; only the locations of the foci change.*

13: *If a central basis curve with n lines of symmetry that does not self-invert about the center is inverted about a non-central axial pole, the inverse locus is axial and possesses n-1 foci of self-inversion in the finite plane.*

As examples of Maxim 8-30-9 one may take any case of Figs. d-h in which a line of symmetry is lost. In each case, the focus of self-inversion that replaces the line of symmetry possesses the 0° mode of the corresponding basis focus of self-inversion on the ideal line at infinity.

Fig. f provides an example of Maxim 8-30-10 (see also Figs. 8-22a,c,f). Since the pole of inversion lies on the x-axis, the corresponding focus of self-inversion on the ideal line at infinity is an intra-oval-mode focus. Accordingly, the center-focus of the inverse locus possesses the 0° intra-oval mode, and the focus on the ideal line at infinity (in the direction of the x-axis) possesses the 0° inter-oval mode (the inverse locus consists of opposed overlapping curves, each similar to Fig. 8-16g, inverse to one another about the center-focus). On the other hand, for a modular pole on a

Table 8-30-1. Numbers of 0° Inter-Oval And Intra-Oval-Mode Foci of Self-Inversion of Dual, Concentric, Orthogonal, Congruent Monovular Cassinians

	Numbers of Foci					
	FINITE PLANE		IDEAL LINE AT ∞		TOTALS	
LOCUS	Inter-oval mode	Intra-oval mode	Inter-oval mode	Intra-oval mode	Inter-oval mode	Intra-oval mode
Basis	1 (center)	0	2	2	3	2
Central inverse locus (Fig. f)	2 (foci 2 and 3)	1 (center, focus 1)	1	1	3	2
Central inverse locus of Cassinian modular serpentine cubics	1 (center)	2 (double points)	2	0	3	2
Axial inverse locus (Fig. g)	3 (foci 1, 4, & 5)	1 (focus 3)	0	1	3	2
Non-axial inverse locus (Fig. h)	3 (foci 1, 3, & 6)	2 (foci 2 and 4)	0	0	3	2

$\pm 45^{\circ}$ line of symmetry, the corresponding focus of self-inversion on the ideal line at infinity is an inter-oval-mode focus. As a result, the center-focus, since it also possesses this 0° mode in the basis curve, remains a 0° inter-oval-mode focus of self-inversion in the inverse curve.

The rightmost two columns of Table 8-30-1 provide excellent examples of Maxim 8-30-12. Maxim 8-30-13 emphasizes the fact that the inverse locus of a central basis curve about an axial but non-central pole gains one focus of self-inversion in the finite plane for each line of symmetry that is lost, and possesses but one line of symmetry. The above sections have illustrated that, for central curves with more than two lines of symmetry that self-invert about the center in multiple modes, the situation is more complex.

A Central Ensemble of Cassinian Modular Serpentine Cubics

The second central locus obtained by a modular inversion of the basis concentric, orthogonal, congruent monovular Cassinians of Figs. f-h is dealt with briefly here (see further treatment and illustrations in the figures for Chapter IX in *Curves And Symmetry, Vol. II*, in preparation).

A modular incident inversion of the $C = 17/60$ Cassinian monoval produces an unusual serpentine cubic with equidistant reflective symmetry in a point. Of all monovular central Cartesian species, only monovular Cassinians possess modular serpentine cubics (since only Cassinian monovals self-invert about the center). Accordingly, the equidistant reflective symmetry of these curves in a point is another symmetry property conferred upon the inverse loci of Cassinian monovals by virtue of the property of central self-inversion at 90° and 270°.

Since the modular serpentine cubic has a linear 180° circumpolar intercept transform about the pole of reflective symmetry, it has the highest degree of circumpolar symmetry of any incident inversion locus. Furthermore, monovular Cassinians are the only curves in which this relationship is known to apply. [But the *circumlinear* symmetry of the axial serpentine cubic exceeds that of the modular serpentine cubic.]

When the orthogonal Cassinian monovals of Figs. f-h are inverted about a point of intersection, the orientations of the resulting individual modular serpentine cubics (which intersect at 3 points) are such that the composite locus has two orthogonal lines of symmetry. The locus somewhat resembles two equilateral strophoids (Fig. 6-5a) tangent at their loop vertices but with diagonal asymptotes through the point of "tangency."

An individual modular serpentine cubic self-inverts about two incident foci that are reflected in the pole of reflective symmetry (but it does not self-invert in the pole of reflective symmetry itself). The composite locus retains the property of self-inversion in the two pairs of reflected poles (which become coincident pairwise). In addition, it acquires two modes of self-inversion about a true center, which is formed by the coming into coincidence of the two individual poles of equidistant reflective symmetry.

The 0° mode about the true center, of course, is inter-segment; otherwise it would exist in each locus alone. The same mode (0° inter-oval) is possessed by the dual Cassinian basis locus, and it does not become altered upon exchange with the corresponding focus on the ideal line at infinity (in the direction of the line of symmetry upon which the pole of inversion lies) because the latter also is an inter-oval-mode focus.

Unlike the situation in the central inversion locus of Fig. f, which possesses both an inter-oval and an intra-oval-mode focus of self-inversion on the ideal line at infinity, both foci of the present central ensemble of serpentine cubics on the ideal line at infinity are of the inter-oval type. This property is a consequence of the fact that the individual loci have no line of symmetry.

Poles of Identical Subspecific-Inversion of The Inverse Loci of A Non-Central Figure With 3 Lines of Symmetry

To round out this brief discussion of poles of identical subspecific-inversion, a figure with 3 lines of symmetry and 3 modes of self-inversion about a single pole is treated here, and a summarizing tabulation is given (Table 8-30-2). Consider a figure for which one line of symmetry is the y-axis and the other two lines of symmetry are lines through the origin at $\pm 30^{\circ}$ to the x-axis. Let the figure consist of 6 points incident upon these lines, 3 points at a distance of ½ unit and 3 points at a distance of 2 units from the origin, one pair of points on each line, as in Fig. i.

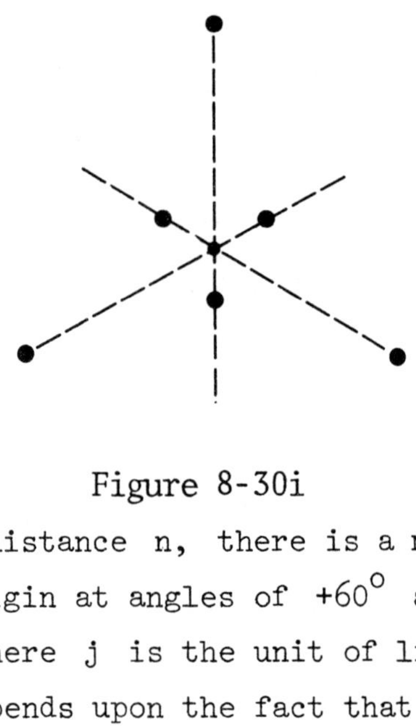

Figure 8-30i

This figure self-inverts at the 3 angles, 60°, 180°, and 300°. The 60° and 300° modes of self-inversion are consequent upon the fact that, for each point on a line of symmetry at a distance n, there is a modularly-inverse reflection of the point in the origin at angles of $+60^{\circ}$ and $+300^{\circ}$ (or -60°) at a distance of j^2/n (where j is the unit of linear dimension). The 180° mode of self-inversion depends upon the fact that a 180° modularly-inverse reflection of each point in the origin also exists on the same line of symmetry. The locus is not a *central figure* in the sense that this term is employed here, i.e., it does not possess two orthogonal lines of symmetry, nor even does it have simple (equidistant) reflective symmetry in a point.

Since the figure possesses a 180° focus of self-inversion, all of its inverse loci also self-invert at 180° (Maxims 8-15-4b and 8-30-12). For each line of symmetry that is lost in an inversion, the inverse locus gains a 0° focus of self-inversion; the focus that is gained possesses the mode of the focus of self-inversion on the ideal line at infinity that becomes transferred to the finite plane (Maxim 8-30-9).

In the case of a modular axial inversion, there are two poles equivalent to the pole of inversion and 3 additional equivalent inversion poles. In the inverse locus, only the poles inverse to the equivalent poles are 0° foci of self-inversion (each replacing a line of symmetry). Two equivalent inversion poles invert to poles of congruent-inversion; the third (the 180° reflection of the pole of inversion in the basis focus of self-inversion) inverts to the 180° focus of self-inversion. All 5 inverse poles are collinear.

In a non-modular axial inversion, there also are two equivalent poles and 3 equivalent inversion poles. Only the poles inverse to the two equivalent poles are 0° foci of self-inversion in the inverse figure. The 180° focus of self-inversion is inverse to the equivalent inversion pole of the basis figure that is modularly inverse to the pole of inversion at 180°; the other two equivalent inversion poles invert to poles of congruent-inversion.

In a non-axial non-modular inversion, there are 5 equivalent poles to the pole of inversion and 6 additional equivalent inversion poles. Of the former, only the 3 that are reflections of the pole of inversion in the lines of symmetry invert to 0° foci of self-inversion. The other two equivalent poles and 5 of the 6 equivalent inversion poles invert to poles of congruent-inversion, while the sixth equivalent inversion pole inverts to the 180° focus of self-inversion. Table 8-30-2 lists the poles of identical subspecific-inversion and lines of symmetry of the foregoing inverse loci of figures with two or more lines of symmetry and two or more modes of self-inversion about a single pole (in the finite plane).

Curves With Infinite Numbers of Foci of Self-Inversion

Curves of numerous periodic functions, such as a sine wave, possess an infinite number of (parallel) lines of symmetry. In consequence (Maxim 8-30-13), an inverse locus of such a curve about a point in the plane possesses an infinite number of collinear foci of self-inversion, one at each point inverse to the reflections of the pole of inversion in the lines of sym-

Table 8-30-2. Poles of Identical Subspecific-Inversion of Loci Inverse To Figures With Two Or More Lines of Symmetry And Two Or More Modes of Self-Inversion About A Single Pole

Basis figures	Modes of self-inversion, lines of symmetry, and poles of congruent-inversion			
	basis figure	Inverse loci		
		axial modular inversion	axial non-modular inversion	non-axial non-modular inversion
	S L C	S L C	S L C	S L C
Cassinian monoval, 2 lines of symmetry, 2 modes of self-inversion (90°, 270°)	0 2 0	1 1 2 (collinear)	1 1 2	2 0 5 (Fig. d, collinear)*
biovular central Cartesian, 2 lines of symmetry, 2 modes of self-inversion (0°, 180°)	2 2 0 (central Cartesian)	2 2 0 (central Cartesian, collinear)	3 1 0 (axial bicircular quartic)	4 0 3 (Fig. e)
point sextet with 3 lines of symmetry, 3 modes of self-inversion (60°, 180°, and 300°)	1 3 0	3 1 2 (collinear)	3 1 2	4 0 7
dual Cassinian monovals, 4 lines of symmetry, 4 modes of self-inversion (0°, 90°, 180°, and 270°)	2 4 0 (Fig. f)	4 2 0 (Fig. f, collinear)	5 1 2 (Fig. g)	6 0 9 (Fig. h)

S, modes of self-inversion at 0° or 180° in the finite plane (or foci of self-inversion, allowing for multiplicity)
L, lines of symmetry (or modes of 0° self-inversion about foci on the ideal line at infinity)
C, poles of congruent-inversion in the finite plane
* Non-axial, modular

metry. In addition, such an inverse locus always possesses an infinite number of poles of congruent-inversion that are collinear with the foci of self-inversion and may also possess an infinite number of concyclic poles of congruent-inversion.

Fig. 8-31. Representative axial subspecies of the Cassinian-based inversion superfamily for $C = 0.2$ and 0.3. The two basis curves are shown in confocal arrangement at the center (see also Fig. 8-16b). Eight axial inverse loci of the monovular $C = 0.3$ central subspecies are shown for poles from $\frac{1}{4}d$ to $2\frac{1}{2}d$, including a vertex inversion for $h = \sqrt{2}d/2$ (Fig. d).

Where feasible, the reciprocal inversion poles are shown as solid circles (the pole for curve g lies to the right of the axial-vertex cubic d). The locations of the poles of inversion are indicated by open circles along the axes of the basis curves (two of these overlap curve f). In the cases of the inverse loci, c and e, magnified views are shown for the regions of the reciprocal inversion poles.

For the biovular species, the central basis curve is taken to be the $C = 0.2$ opposed-oval subspecies. The annular central curve of Fig. 8-20d, or any axial inverse locus thereof, also could have been taken to be the basis subspecies, since the ensemble of axial inverse loci of any specific axial inverse curve contains representatives of all the axial-inversion subspecies.

While a total of only 12 representatives are shown, these represent the inverse loci about 24 different axial poles, since the same 12 subspecies as those illustrated are obtained by inverting about poles modularly inverse to each of the 12 poles for which an inverse locus is given. The two identical subspecies obtained by inverting about modularly-inverse poles, however, are of different size and have opposed orientations (see Maxim 8-22-2). In the cases of the 8 inverse loci of the monovals, the poles modularly inverse to those employed lie on the orthogonal axis (the y-axis) of the curve (since the monovals self-invert at 90°). For example, the axial-vertex serpentine cubic d represents the inverse loci obtained by inverting about all 4 axial vertices.

Since $j^2 = 0.15d^2$ for central self-inversion of the biovals, the inverse locus for a pole at $h = \frac{1}{2}d$, which is the $C = 5$ parabolic Cartesian, also is the inverse locus for a pole at $h = 0.3d$ (these being one pair of the modularly-inverse poles of the Cartesian condition; see Fig. 8-20). The interior oval of this curve also is shown magnified by a factor of 5 (the same locus is shown in Figs. 20a_2,b_2, and d_2). Similarly, the locus for $h = 0.1d$ also represents the locus for $h = 1\frac{1}{2}d$, the locus for $h = 2d$ also represents the locus for $h = 0.075d$, and the locus for the inner vertex also represents the locus for the outer vertex (see also Fig. 8-21).

The far vertex of the small oval of the serpentine circular cubic j is a generic triple-focus, since the focus defined by the $\cos^2\theta$-condition and the 0° inter-segment focus of self-inversion are in coincidence with this vertex. The serpentine arm is shown both in location to scale, and offset and reduced by a factor of 10 (in the dashed tracing).

As already noted in the preceding, monovular Cassinians possess a property not known to be characteristic of any other species of monovular Cartesians, namely, the property that they self-invert about the center at 90° and 270°. This means that if the curves are inverted about the center at 0°, the sub-specifically-identical inverse loci are rotated 90°, bringing the $\cos^2\theta$-condition foci into coincidence with the y-axis. This property has a parallel in the inversion properties of the biovular Cassinians about poles at modular distance from center. Thus, if a biovular Cassinian in standard position (with its transverse axis coincident with the x-axis) is inverted to an annular central Cartesian [about a pole at $h = \pm D = \pm(1/16 - C^2)^{1/4}d$], the $\cos^2\theta$-condition foci of the latter locus lie on the y-axis, whereas those of the basis curve lie on the x-axis.

The standard eq. of central Cartesians (eq. 8-20-1) is repeated here for convenience as eq. 1. This eq. represents both the monovular and biovular species, including, in the latter category, both annular and opposed-oval types. For every species of biovular central Cartesian belonging to a given genus, there exist two other biovular species belonging to the same genus, namely the two ensembles of inverse biovular loci obtained by inverting members of the biovular species in question about the modular poles on the two lines of symmetry. Of the 3 congeneric (belonging to the same genus) species, two possess opposed ovals and the third annular ovals.

$$(x^2+y^2)^2 + B^2x^2 + C^2y^2 + D^4 = 0 \qquad \text{central Cartesians} \qquad (8\text{-}31\text{-}1)$$

For every known species of monovular central Cartesian, except Cassinian monovals, there exists a second congeneric monovular species, namely the ensemble of inverse central loci obtained by inverting members of the basis species about the center. This is the simplest method of obtaining the eq. of the congeneric monovular species. Given that eq. 1 represents some particular one of the central biovular species (for given specific ranges of values

Representative Axial Subspecies of The Cassinian-Based Inversion Superfamily For C=0.2 & 0.3
(Monovular & Opposed-Oval Central Basis Curves)

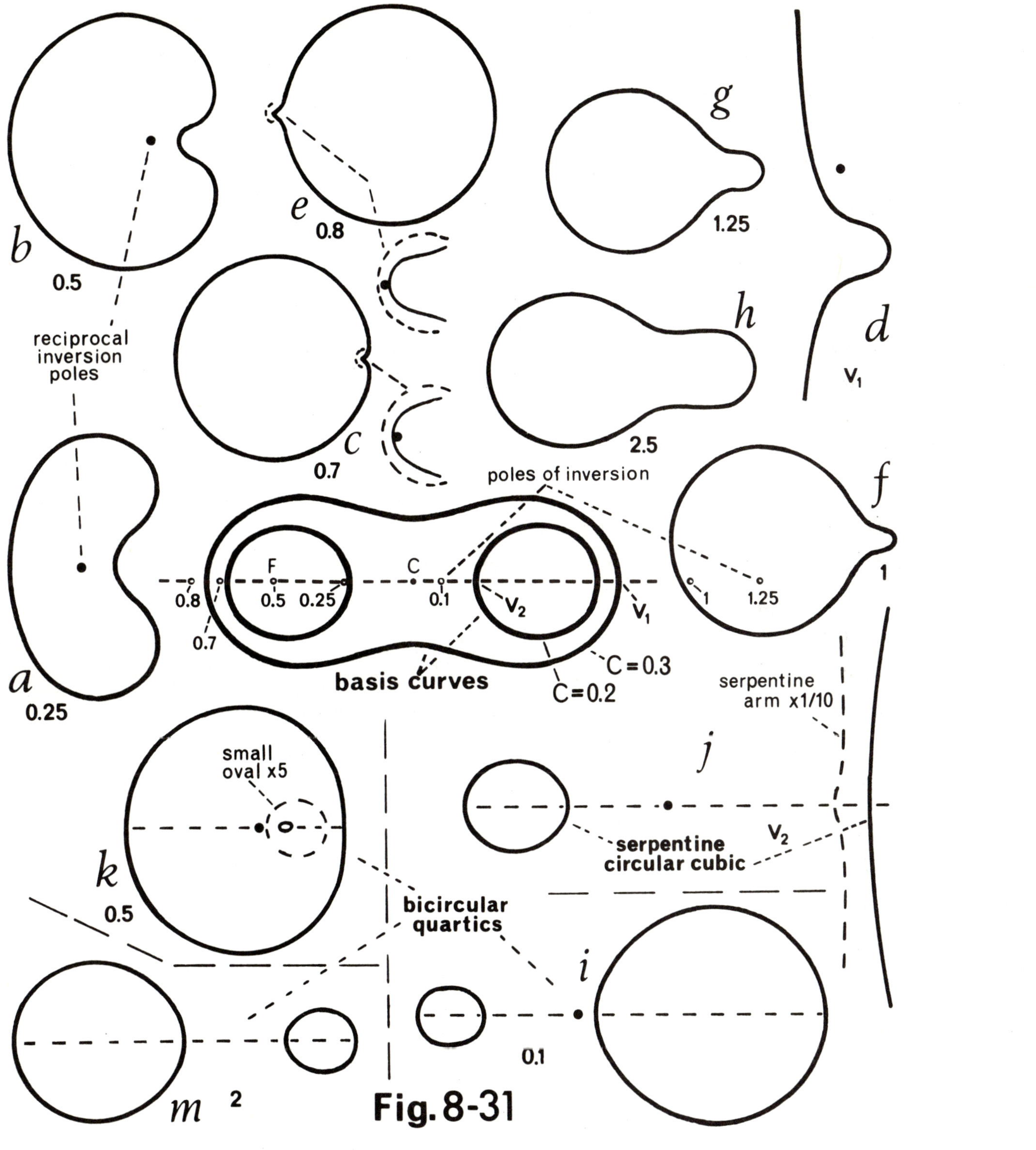

Fig. 8-31

of the parameters B^2, C^2, and D^4), the eqs. for the other two congeneric biovular species cast about their centers are obtained readily as follows.

Eqs. 8-20-9 and 8-20-10 represent all the inverse loci of eq. 1 obtained by inverting about a pole at $x = h$ and translating the origin of the inverse loci to the pole $x = H$. For $h = H = 0$ the curves in question are simply the inverse loci of the basis central curves about their centers. In the cases of the biovular species, these inverse loci belong to the basis species, since all biovular central Cartesians self-invert about the center. Accordingly, the pole of inversion $h = 0$ is excluded in the following considerations.

Among the ensembles of inverse loci of members of a given biovular species about the pole $x = h$, there exists one ensemble--for a particular value of h--which consists of central curves of one of the other two congeneric species, with their eq. cast about the pole of inversion at $x = h$. The purpose of the translation of the pole of this eq. to $x = H$ (where H is the abscissa relative to the pole h as origin) is to obtain the eq. of the species cast about the center.

In the polar eq. of central curves cast about their centers, there occurs no term in an odd power of $\cos\theta$ or $\sin\theta$. However, the simple intercept eq. for any inverse locus of a central Cartesian about a pole h translated to a pole H as origin (eq. 8-20-10) contains terms in both $x^3\cos\theta$ and $x\cos\theta$. Accordingly, the conditions for the inverse locus to be a centered central curve are that the coefficients of both of these terms vanish.

If one sets the coefficient of the $x^3\cos\theta$ term of eq. 8-20-10 equal to 0, one obtains the condition of eq. 2a for H, the pole of translation for centering the central loci inverse about pole h. If this expression for H is substituted in the coefficient of the $x\cos\theta$ term and the latter equated to 0, one obtains the degenerate eq. 2b. The factor (B^4-4D^4) equated to 0 is the familiar condition for congruent non-concentric circles, which is dis-

(a) $H = \dfrac{-hj^2(2h^2+B^2)}{2(h^4+h^2B^2+D^4)}$ condition for vanishing of the coefficient of the $x^3\cos\theta$ term of eq. 8-20-10 (8-31-2)

(b) $(h^4 - D^4)(B^4-4D^4) = 0$ condition for vanishing of the coefficient of the $x\cos\theta$ term of eq. 8-20-10

(c) $H = \mp j^2/2D$ 2nd condition for the vanishing of the coefficient of the $x\cos\theta$ term of eq. 8-20-10

regarded. Setting the factor (h^4-D^4) equal to 0, however, defines the poles $h=\pm D$, for $D^4>0$ (i.e., for biovular basis curves), as being the poles for inversion to central curves. Since the modular value for self-inversion of a biovular central Cartesian about the center is $j=|D|$, these poles are the *modular* poles.

Substituting $h=\pm D$ in eq. 2a leads to eq. 2c for the location of the pole of translation, H, for centering the central loci inverse about pole h. This is the familiar formula for translating the pole of an inverse locus to the position of a 0° focus of self-inversion that replaces a line of symmetry, when a curve is inverted about a pole at a distance $\pm D$ from the line. In the present instance, even though a line of symmetry is not lost, the formula for the location of the pole is unchanged. Accordingly, the poles $H=\pm D$ also could have been located by substituting $h=\mp\frac{1}{2}j^2/h$ in the cosθ-constant condition of eq. 8-20-10 (rather than by making the substitution of eq. 2a). [The cosθ-constant condition is the condition obtained by setting the coefficient of the xcosθ term equal to 0.]

Making the substitutions $h=D$ and $H=-\frac{1}{2}j^2/D$ (or $h=-D$ and $H=\frac{1}{2}j^2/D$) in eq. 8-20-10 leads to eq. 3 for a species of central Cartesian congeneric with any biovular species represented by eq. 1 (i.e., for the same values of B^2, C^2, and D^4). The corresponding poles on the y-axis for inversion to the third congeneric biovular central Cartesian species are the modular poles at $k=\pm D$. Since a 90° rotation of a central Cartesian merely exchanges the coefficients B^2 and C^2 of the basis eq., the eq. for the centered central curves of the third congeneric species is the same as eq. 3 but with the parameters B^2 and C^2 exchanged at every occurrence.

equation of a species of biovular central Cartesian congeneric with any biovular species represented by eq. 1, in terms of the parameters of eq. 1

$$(x^2+y^2)^2+\frac{j^4(3D^2-\frac{1}{2}B^2)x^2}{D^2(B^2+2D^2)}+\frac{j^4(C^2+D^2-\frac{1}{2}B^2)y^2}{D^2(B^2+2D^2)}+\frac{j^8}{16D^4}=0 \qquad (8\text{-}31\text{-}3)$$

locations of the foci defined by $\cos^2\theta$-condition of eq. 3

$$4h^2=\frac{j^4[(C^2+D^2-\frac{1}{2}B^2)-(3D^2-\frac{1}{2}B^2)]}{D^2(B^2+2D^2)}=\frac{j^4(C^2-2D^2)}{D^2(B^2+2D^2)} \qquad (8\text{-}31\text{-}4)$$

All of the prior findings regarding properties of the loci of eq. 1 apply with equal force to the loci of eq. 3, when one employs the compound coefficients of the terms of eq. 3 in place of the simple coefficients of eq. 1. For example, to locate the foci of the $\cos^2\theta$-condition of eq. 3 by applying the condition, $C^2 - B^2 = 4h^2$, of eq. 8-20-6a, one subtracts the coefficient of the x^2 term from the coefficient of the y^2 term and equates the difference to 4 times the square of the abscissa of the pole, as in eq. 4.

If the biovular species represented by eq. 1 possesses annular members (which requires that $D^4 > 0$, $B^4, C^4 > 4D^4$, $B^2, C^2 < 0$), the members of both of the other congeneric species--one represented by eq. 3 and the other by eq. 3 with the parameters B^2 and C^2 exchanged--possess opposed ovals. If the biovular species represented by eq. 1 possesses members with opposed ovals (see conditions on the parameters following eq. 8-20-3), one of the congeneric species represented by eq. 3 (and by eq. 3 with the parameters B^2 and C^2 exchanged) also possesses members with opposed ovals, and the other possesses members with annular ovals. The latter relationship (possession of annular ovals) applies when the modular poles of inversion lie on the line of symmetry that intersects the opposed ovals (and passes through the $\cos^2\theta$-condition foci), whereas members of the inverse congeneric species possess opposed ovals if the poles lie on the non-intersecting axis.

It remains to establish the identity of the non-central quartic loci of the Cassinian-based inversion superfamily that possess the highest degree of circumpolar symmetry. In preceding figure legends it has been asserted that these curves are parabolic Cartesians, but this remains to be proved. To establish this relationship the substance of Maxim 8-20-5 is employed. In other words, one must locate the axial poles about which inversions yield loci for which the $\cos^2\theta$-condition foci are in coincidence as a generic double-focus, and identify the inverse loci as parabolic Cartesians.

Accordingly, the Cassinian ovals of eq. 5 are translated along the x-axis to bring the origin to a pole h ($x+h$ translation) and inverted about this pole. The inverse locus then is cast about a pole at $x = H$ ($x+H$ translation relative to pole h) and the coefficient of the $\cos^2\theta$ term is equated to 0 yielding eq. 6 (the same result is achieved by substituting Cassinian parameters in eq. 8-20-11).

The condition for the existence of a $\cos^2\theta$-condition generic double-focus (which leads to the non-central quartic loci with the highest degree of

$(x^2+y^2)^2-\frac{1}{2}d^2(x^2-y^2)+d^4(1/16-C^2) = 0$ Cassinian ovals (8-31-5)

$4H^2[(h^2-\frac{1}{4}d^2)^2-d^4C^2]+8hHj^2(h^2-\frac{1}{4}d^2) + 4j^4(h^2-\frac{1}{4}d^2) = 0$ $\cos^2\theta$-condition for axial inverse loci of Cassinians about a pole $x = h$ (8-31-6)

the Cartesian condition for Cassinians (radicand of the quadratic-root solution of eq. 6 equated to 0)

$$h^2(h^2-\tfrac{1}{4}d^2)^2-(h^2-\tfrac{1}{4}d^2)[(h^2-\tfrac{1}{4}d^2)^2-d^4C^2] = 0 \qquad (8\text{-}31\text{-}7)$$

(a) $h = \pm\frac{1}{2}d$ pair of roots of eq. 7 identical with $\cos^2\theta$-condition (8-31-8)

(b) $h = \pm 2d\sqrt{(1/16-C^2)}$ 2nd pair of roots of eq. 7

circumpolar symmetry) is the vanishing of the radicand of the quadratic-root solution of eq. 6 for H. This condition--the Cartesian condition for Cassinians-- is eq. 7, the roots of which are given by eqs. 8. The two roots, 8a, of the Cartesian condition are identical with those defining the locations of the $\cos^2\theta$-condition foci. The other two roots, 8b, of the Cartesian condition are modularly inverse to those of 8a.

Accordingly, if Cassinian ovals are inverted about any of the poles defined by eqs. 8a,b, the loci obtained have the highest degree of circumpolar symmetry of any non-central quartic genus in the Cassinian-based inversion superfamily. To identify these inverse loci--all 4 of which, when real, belong to the same subspecies--those obtained by inverting about the pole $h = -\frac{1}{2}d$ are taken as examples.

Inversion of the loci of eq. 5 about the pole $h = -\frac{1}{2}d$, for $j = d$, leads to eq. 9. An $x+d$ translation of the loci of eq. 9 brings the origin to the focus of self-inversion that replaces the line of symmetry, yielding eq. 10.

polar equation of the inverse loci of Cassinians about a pole at $h = -\frac{1}{2}d$

$$C^2r^4 - d^2r^2 + 2d^3r\cos\theta - d^4 = 0 \qquad (8\text{-}31\text{-}9)$$

loci of eq. 9 after an $x+d$ translation, cast in the form of a simple intercept eq. (x is the intercept at the angle θ)

$$4d^2x^2\cos^2\theta+4dx(d^2+x^2)\cos\theta + x^4 + (d^2/C^2)(2C^2-1)x^2 + d^4 = 0 \qquad (8\text{-}31\text{-}10)$$

(a) $\cos\theta = \dfrac{x^2 \mp dx/C + d^2}{-2dx}$ simple intercept format of eq. 10 (8-31-11)

(b) $\cos\theta = \dfrac{x^2 + Cdx + d^2}{2dx}$ simple intercept format of parabolic Cartesians

Casting eq. 10 in the form of a simple intercept format (solving for $\cos\theta$) gives eq. 11a, which is recognized as being that of a Cartesian self-inverter.

Since the differences in the signs of the terms in x in the numerator and denomonator of eqs. 11a and 11b merely relate to differences in the orientation and manner of tracing of the loci (see Fig. 8-13), the format of eq. 11a is identical with that for parabolic Cartesians, with the proviso that the parameter "C" of one eq. is the reciprocal of the parameter "C" of the other eq. [This result also was evident from inspection of eq. 8-17-1a.]

Thus, the curves obtained by inverting Cassinians about the poles of the bipolar eq. $uv = Cd^2$ (which are the $\cos^2\theta$-condition foci), are the same curves that are represented by the bipolar eq. $u^2 = Cdv$, if the parameter "C" of one eq. is taken to be the reciprocal of the parameter "C" of the other.

Fig. 8-32. Representative axial subspecies of the Cassinian-based inversion superfamily for $C = 0.2497$ (annular central basis curve). Although all representatives of the Cassinian-based inversion superfamily are obtained by inverting all subspecies of Cassinian ovals about all points in the plane, representatives of all axial genera (i.e., genera with subspecies that possess a single line of symmetry) are not obtained by transverse-axis inversions alone.

It is evident that axial and central loci obtained by inverting about poles on the orthogonal line of symmetry are not represented, and the present studies have shown that axial loci also are obtained by inverting about poles on the modular circle, which also are not represented (the latter category includes the axial-vertex serpentine circular cubics of opposed-oval central modular-focal quartics; Fig. 8-22f and curves c and h of the present figure).

Only ensembles of axial inverse loci of subspecies of the annular central species--in this case annular central parabolic Cartesians--possess representatives of both species of axial serpentine circular cubics (Figs. 8-22b,d). A subspecies of the latter group (curve a and eq. 1) is taken to be the basis curve for the present figure, and the inversions are carried out about poles

Representative Axial Subspecies of The Cassinian-Based Inversion Superfamily For C=0.2497
(Annular Central Basis Curve)

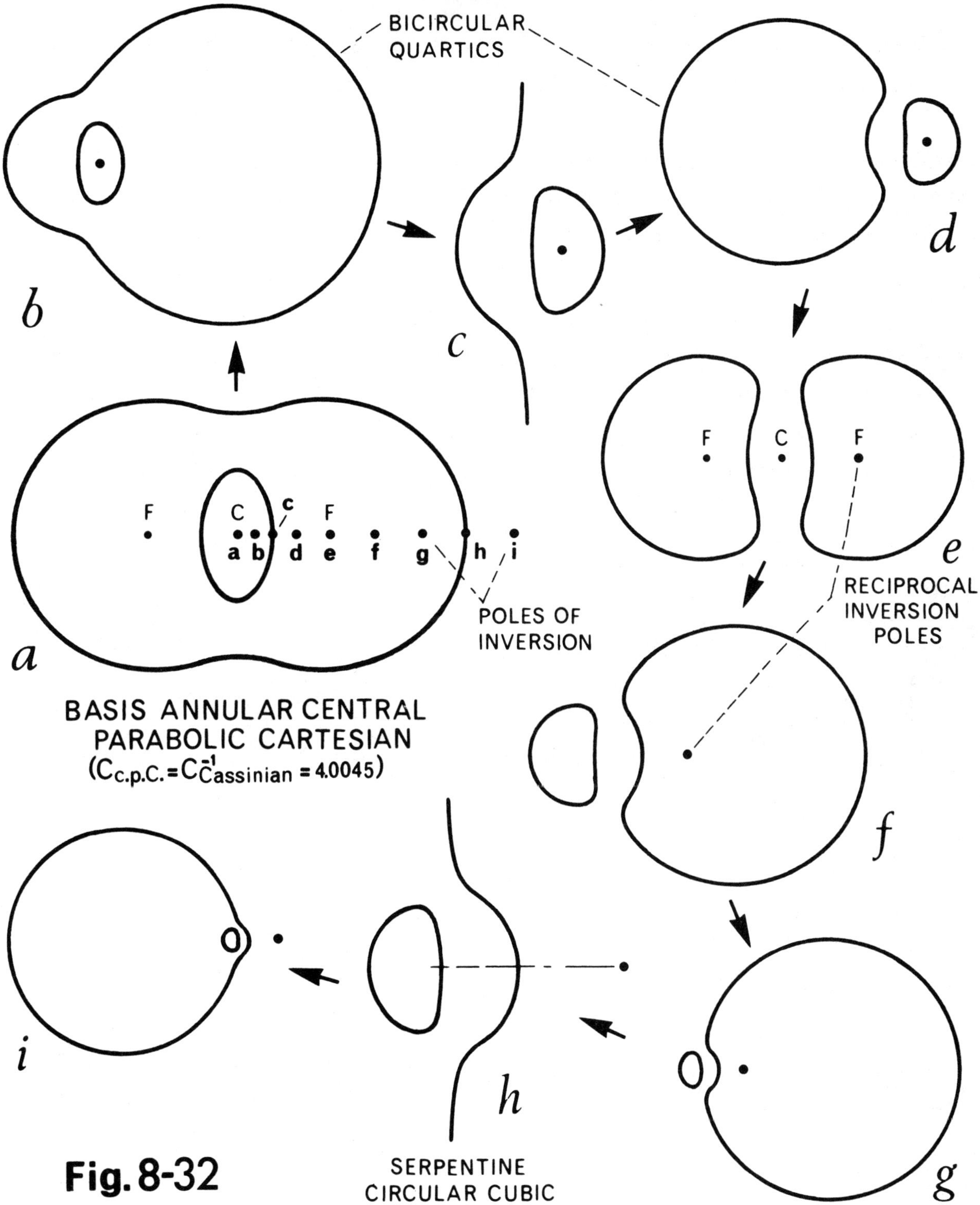

Fig. 8-32

equation of basis annular central parabolic Cartesians of Fig. a in terms of d, the distance between the $\cos^2\theta$-condition foci

$$(x^2+y^2)^2 - 1.6d^2x^2 - 0.6d^2y^2 + d^4/16 = 0 \qquad (8\text{-}32\text{-}1)$$

on the axis that passes through the $\cos^2\theta$-condition foci. Since this axis is inverse to the modular circle of Cassinian ovals, the inverse loci about the x-axis vertices are the serpentine circular cubics (curves c and h) of the corresponding opposed-oval central modular-focal quartic (curve e).

The vertices of the basis curve lie at $\pm 0.200164414d$ and $\pm 1.248973262d$ [modular distance $\tfrac{1}{2}d = (0.200164414 \cdot 1.248973262)^{\frac{1}{2}}d$] and the poles of inversion in order from the center (labelled a to i) are; (a) the center, (b) a point halfway to the inner vertex, (c) the inner vertex, (d) the point $h = 0.316227766d$, (e) the modular pole $h = \tfrac{1}{2}d$, (f) the point $h = 0.758946638d$, (g) the point $h = 1.01192885d$, (h) the outer vertex, and (i) the point $h = 1.517893276d$.

The arrows indicate the progression of the inverse loci from the center outward. The relative sizes of the inverse loci have been altered for the purpose of illustration. The reciprocal inversion poles of curves b to i are marked by the solid circles (in the case of curve e, the center and contralateral $\cos^2\theta$-condition focus also are marked by solid circles).

Inversion at 0° or 180° about the center-pole a, yields the basis curve itself. Inversion about any pole within the inner oval yields an annular inverse locus. For the axial pole b, this locus is curve b. It follows from principles of Inversion Taxonomy that all the inverse loci obtained by inverting about poles on the transverse axis of the basis curve a also are obtained by inverting about axial poles of any of the inverse loci. In other words, inversions about axial poles of curve b also yield all the other inverse loci of the figure, with the reciprocal inversion pole and the vertices of curve b marking the poles for inversion to curve a and the serpentine circular cubics c and h.

Inversion of curve a about the inner vertex, c, "opens" the inner oval to a serpentine arm, yielding the serpentine circular cubic, c. The reciprocal inversion pole lies within the oval of curve c, which is the inverse of the large oval of curve a. Proceeding outward from the inner vertex, all inverse loci about poles in the inter-oval region possess opposed ovals.

Inversion about the first axial inter-oval pole, d, yields the opposed-oval locus, d, for which the reciprocal inversion pole, as for curves b and c, is within the small oval.

Since pole e (coincident with the $\cos^2\theta$-condition focus F) is the modular pole, inversion of curve a about this pole produces another central curve (Maxim 8-20-3). Furthermore, the modular pole e also is the position on the right horizontal axis beyond which all of the inverse loci merely repeat--but with reversed orientation and in reverse sequence--the inverse loci obtained in inversions about the preceding poles (Maxim 8-20-6). For modularly-inverse pole-pairs, the inverse loci are subspecifically identical (Maxim 8-20-6).

If, for example, pole f or g had been selected to be modularly inverse to pole d, the corresponding inverse curve would belong to the same subspecies as curve d, and similarly for curve i with respect to curve b had pole i been selected to be modularly inverse to pole b. In point of fact, aside from the vertices, only pole f is approximately modularly inverse to a pre-modular pole (pole d), as a result of which curve f greatly resembles curve d, but with opposed orientation. Note that for intra-oval locations of the reciprocal inversion pole, the central curve e demarcates between curves for which this pole is in the small oval, as opposed to being within the large oval.

As the location of the pole of inversion progresses toward the vertex of the exterior basis oval (pole h), the large oval of the inverse locus progressively enlarges relative to the size of the small oval (see curves f and g) until, when the pole is at the vertex, the large oval "opens" at the right, yielding the serpentine circular cubic h, which is subspecifically identical to curve c but with opposed orientation. For a pole just beyond vertex h, the serpentine arm "closes to the left," once again yielding annular inverse loci, that duplicate those obtained about axial poles interior to both ovals, of which curve b is the sole representative.

Fig. 8-33. Representative axial subspecies of the Cassinian-based inversion superfamily for $C = 0.2497$ (non-intersecting line of symmetry of opposed-oval central basis curve). A consideration of the line of symmetry inverse loci of curves belonging to the Cassinian-based inversion superfamily would not be complete without an examination of the loci obtained by inverting about points on axes that do not intersect the curves. These loci are illustrated for Cassinian-based inversion subspecies which, like those of Figs. 8-22 and 8-32, have C-values only 3 ten-thousandths removed from the value for their curves of demarcation (the latter are equilateral QBI curves obtained by inverting an equilateral lemniscate about points on the line of symmetry that intersects the curve only at the DP).

The basis opposed-oval curve a--with ovals of the monolobed type--is oriented with its non-intersecting axis in coincidence with the x-axis, and the poles of inversion are spaced to the right from center in geometrical progression beginning at a pole a 0.025d; thereafter the distance from center is doubled for each of 8 successive additional poles out to a distance of 6.4d. The locations of the poles out to 0.8d are shown on the horizontal axis of the basis curve as small solid circles, with those at 0.2d, 0.4d, and 0.8d labelled.

The value of the abscissa of the pole of inversion for each inverse locus is given beneath the corresponding curve. The poles marked on the vertical axis of the basis curve are the two $\cos^2\theta$-condition foci (large solid circles labelled F) and the two poles at modular distance from center (small solid circles labelled M); the inverse locus about either of the latter poles is the corresponding annular central parabolic Cartesian. The 0° focus that replaces the line of symmetry is marked by the solid circles of intermediate size labelled "0°." In cases in which they lie within the confines of the figure, the two poles at modular distance on each side of this focus are marked by small solid circles and labelled M (the modular pole to its left) and R (the reciprocal inversion pole at modular distance to its right).

Inasmuch as the modular value for self-inversion of the basis central subspecies is $j = 0.109108877d$, the subspecies d (for $h = 0.1d$) represents the inverse locus about a pole only 9 thousandths of a unit from a modular pole, the inverse locus about which is the bilobed opposed-oval central curve of Figs. 8-22f and 8-32e. Accordingly, the subspecies d departs only slightly from having a second line of symmetry and ovals of the bilobed type.

Representative Axial Subspecies of The Cassinian-Based Inversion Superfamily For C=0.2497 (Non-Intersecting Line of Symmetry, Opposed-Oval Central Basis Curve)

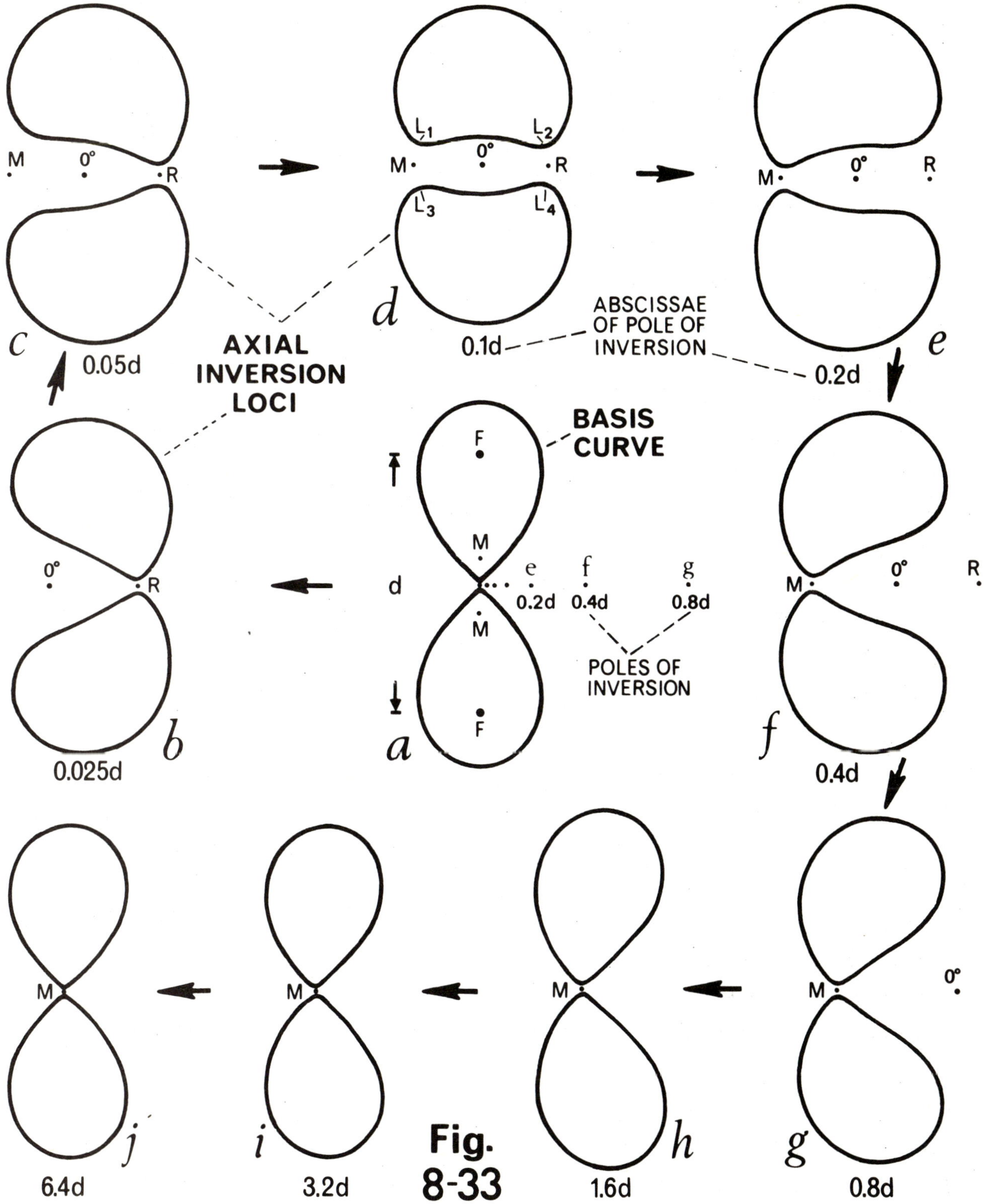

Fig. 8-33

It will be recalled from the discussion above (legend of Fig. 20 and Maxim 8-20-7) that, for a given value of C, representatives of every inversion subspecies about poles on a given line of symmetry of a central basis curve are obtained by inverting about all axial poles from the center to either pole at modular distance from center. Accordingly, the ensemble of inversion subspecies for poles proceeding in geometrical progression from 0.025d to 0.05d to 0.1d, etc., includes only 3 subspecies of the complete pre-modular axial inversion ensemble--curves b, c, and d--(or 4 subspecies, if the basis curve is included) but 6 from the complete post-modular ensemble.

In this connection, recall that identical subspecies are obtained by inverting about poles modularly inverse to one another (Maxim 8-20-6), say ½j and 2j, ¼j and 4j, etc., and note that the poles of inversion for curves c and e, as well as the poles for curves b and f roughly approximate being modularly inverse to one another. It follows that curves c and e represent two very similar subspecies, and that curves b and f represent two other very similar subspecies (since curves c and e are for poles of inversion approximating distances of ½j and 2j from center, and curves b and f are for poles approximating distances of ¼j and 4j from center). These similarities are readily evident.

An unillustrated case for a minute axial displacement (0.001d) of the pole of inversion to the right of center is considered first. For inversion about this pole (for $j = d$), the 0° intra-oval-mode focus that replaces the line of symmetry lies 500d units to the left of the reciprocal inversion pole, the other 0° intra-oval-mode focus lies only 0.084007162d units to its right, and the 180° inter-oval-mode focus lies only 0.083999495 units to its left.

In other words, for minute displacements of the pole of inversion to the right of center, the 0° intra-oval-mode focus that replaces the line of symmetry becomes lateral and lies very far to the left of the curve (for the basis curve itself this focus lies on the ideal line at infinity in the same direction), whereas the other two foci of self-inversion are very close to the reciprocal inversion pole in an inter-oval location (an analogous situation already has been illustrated in Fig. 8-21; see also Maxim 8-30-6).

The pole at $h = 0.025d$ is the closest of the illustrated poles of inversion to the center-pole. Curve b is the inverse locus about this pole. As is evident, the 0° intra-oval-mode focus that replaces the line of symmetry has assumed an inter-oval location (which is 20d units to the left of R). The 180° inter-oval-mode focus, which always has an inter-oval location, also lies to the left of R, but at a distance of only 1.995251773d units (not shown). The other 0° intra-oval-mode focus retains an inter-oval location, at 2.21636176d units to the right of R (not shown).

The latter two foci of self-inversion are located readily using Maxim 8-30-6. These foci also can be located purely geometrically, as follows. Using curve d as an example, if cross tangents are drawn from the tip of the upper right lobe (labelled L_2) to the tip of the lower left lobe (labelled L_3), and from the tip of the upper left lobe (labelled L_1) to the tip of the lower right lobe (labelled L_4), these are concurrent with (intersect) each other and the x-axis at the 180° inter-oval-mode focus of self-inversion (0.434782609d units to the left of the labelled 0° focus), which, of course, always lies between the two ovals.

Similarly, bitangents drawn between the tips of the two upper lobes (L_1 and L_2) and the tips of the two lower lobes (L_3 and L_4) are concurrent with each other and the x-axis at the other 0° intra-oval-mode focus. For curve d, this focus has become lateral, lying far to the right of the two ovals ($57\frac{1}{2}$d units to the right of the inter-oval 0° intra-oval-mode focus).

As the pole of inversion progresses to the right of the pole at 0.025d and toward the modular pole of the basis curve, the changes that take place in the inverse loci can be described grossly in terms of (a) a "rocking" of the ovals about the 0° inter-oval intra-oval-mode focus (this is the 0° focus of self-inversion that replaces the line of symmetry) toward the pole, M, modular to this focus, and away from the reciprocal inversion pole (R), and (b) a "bi-lobed bilateralization" of the shape of each oval (i.e., changes in shape toward bilateral symmetry of a bilobed type).

At $h = 0.05$d (curve c), the manner by which bilobed bilateral symmetry is assumed has become evident. The locations of the two unmarked foci of self-inversion readily can be estimated visually from the geometry of the figure by a consideration of the points at which cross-tangents (180° focus of self-inversion) and bitangents (0° intra-oval-mode focus of self-inversion) become concurrent with each other and the x-axis. It can be seen from such visual inspection that the 0° intra-oval-mode focus defined in this manner has assumed a slightly lateral position, i.e., that it no longer lies between the ovals.

At $h = 0.1d$ (curve d), bilobed bilateral symmetry of each oval almost is achieved. At the modular distance, $h = 0.109108877d$, the inverse locus for which is not illustrated here (but see Figs. 8-22f and 8-32e), bilobed bilateral symmetry of each oval is achieved, with the result that the opposed ovals have two lines of symmetry, and the lateral 0° intra-oval-mode focus has receded to the axial point at infinity.

At progressively more distant locations of the pole of inversion to the right, beyond the right modular pole of the basis curve, "rocking" of the ovals of the inverse locus about the 0° inter-oval intra-oval-mode focus toward pole M continues and the lateral 0° intra-oval-mode focus approaches from the axial point at infinity at the left. The shape of the ovals, however, now changes from the bilobed bilateral type in the direction of the monolobed bilateral type of the basis curve. Although the latter shape is achieved for inversion about a pole at finite distance to the left of the right modular pole, namely at the center, it is achieved only for a pole at infinite distance to the right of the right modular pole (Maxim 6-7-3).

In this connection, one notes that the form changes that occurred between curves b and c for a translation of the pole of inversion by only 0.025d units require a translation of 0.2d units to occur in the reverse direction in passing from curve e to curve f. Notwithstanding the different pace of form changes for pre- and post-modular poles, the shape of the inverse locus for a pole of inversion at only $h = 6.4d$ already greatly resembles that of the basis curve (as noted above, congruence is achieved only for an infinitely distant pole).

Fig. 8-34. Parabola inversion cubics, with detail of inverse vertex regions for inversions about remote poles. Inverse loci of the parabola are of interest for several reasons: they elucidate the form and properties of QBI curves of unit eccentricity and the changes thereof with increasing distance of the pole of inversion from the vertex, and they illustrate aspects of the inversion of certain open curves about remote incident poles, including the fact that *portions* of the inverse loci of these curves merely approach congruence, rather than coincidence, with *portions* of the basis curves (Maxim 6-7-2).

The present figure illustrates the inverse loci of a basis parabola (lower left and dashed segment of Fig. b) about incident poles having abscissae of 0, ½a, a, 2a, 8a, 16a, 64a, 256a, 1024a, and 4096a. The locations of the first 7 of these poles are indicated by the triangular arrowheads peripheral to the basis curve (the remainder lie off the figure). The sequence of inverse loci for increasingly distant poles of inversion is indicated both by the alphabetized letter-labels and the heavy large arrows.

Each individual inverse locus is shown in proper orientation relative to the basis curve (lower left and dashed segment of Fig. b) and the abscissa of the pole of inversion is given adjacent to each of these loci. In every case the reciprocal inversion pole is at the tip of the cusp. For each of the loci b-i, the focus of self-inversion is marked by a solid circle that is incident upon the curve at the point directly opposite (below) the tip of the cusp (see dashed line of Fig. b). Each of these foci lies at the point inverse to the reflection of the pole of inversion in the line of symmetry.

Since all of the poles of inversion are incident, all of the inverse loci are (asymptotic) cubics. Of these, only curve a, for the pole lying on the line of symmetry (the vertex), retains the line of symmetry. In all other cases, the line of symmetry is lost but is replaced by a *point of 0° reciprocal reflective symmetry*. This is the focus of self-inversion referred to above.

Curves h' and i' of the inset show greatly magnified regions in the neighborhood of the pole inverse to the vertex of the basis curve (marked $V_{h'}$ and $V_{i'}$ in Figs. h and i). In addition, a magnification of the corresponding region of the inverse locus for a pole of inversion with an abscissa of 4096a is illustrated. The corresponding curve is not represented; its appearance differs but little from that of curve i (at the scale employed).

An understanding of the appearance of parabola inversion cubics and the progressive changes in their form with increasing distance of the pole of inversion from the vertex is facilitated by the realization that, with increasing distance from the vertex, the arms of the parabola come more and more to resemble parallel lines. Since the pole of inversion is incident upon one of these "lines," the inverse of the contralateral arm comes increasingly to resemble a circle, and the inverse of the basis arm to resemble a tangent line (although the "circle" is open and the opposing "line" segments of the "tangent line" are extensions from the "circle" that do not join).

Curve a, the axial-vertex inversion, is the well-known cissoid of Diocles (see pages IX-1 to 4). Being an inverse locus of the basis parabola about a point on the line of symmetry, this curve does not self-invert about any pole in the finite plane [for the appropriate (infinite) value of the unit of linear dimension, it self-inverts about a point on the ideal line at infinity in the direction of a line orthogonal to the line of symmetry; see Maxim 6-6-3].

Curve b is for a pole of inversion with abscissa $\frac{1}{2}a$, i.e., halfway from the vertex to the traditional focus. The generation of this inverse locus is elucidated by the accompanying dashed representation of a segment of the basis curve. The focus of self-inversion is incident upon the inversion cubic at a point directly beneath the cusp at a distance of $\sqrt{2}a/4$, this being the position to which the contralateral incident pole of the basis curve inverts. The small arrow indicates the location of the point inverse to the vertex of the basis curve. The approximate linearity of the $\frac{1}{2}a$ inversion cubic in this region is an indication of the fact that an incident pole of inversion with abscissa $\frac{1}{2}a$ lies close to being incident upon the vertex circle-of-curvature of the parabola ($R = 2a$).

With reference to the upper-arm segment of the basis curve shown in dashed outline, it is evident that: (a) the upper arm of the inversion cubic from the cusp-point to infinity is the inverse of the upper arm of the basis curve from infinity to the pole of inversion; (b) the segment of the lower arm of the inversion cubic from the cusp-point to the focus of self-inversion is the inverse of the segment of the lower arm of the basis curve from infinity to the point contralateral to the pole of inversion; (c) the segment of the lower arm of the inversion cubic from the focus of self-inversion to the pole in-

Parabola Inversion Cubics, With Detail of Inverse Vertex Regions For Inversions About Remote Poles

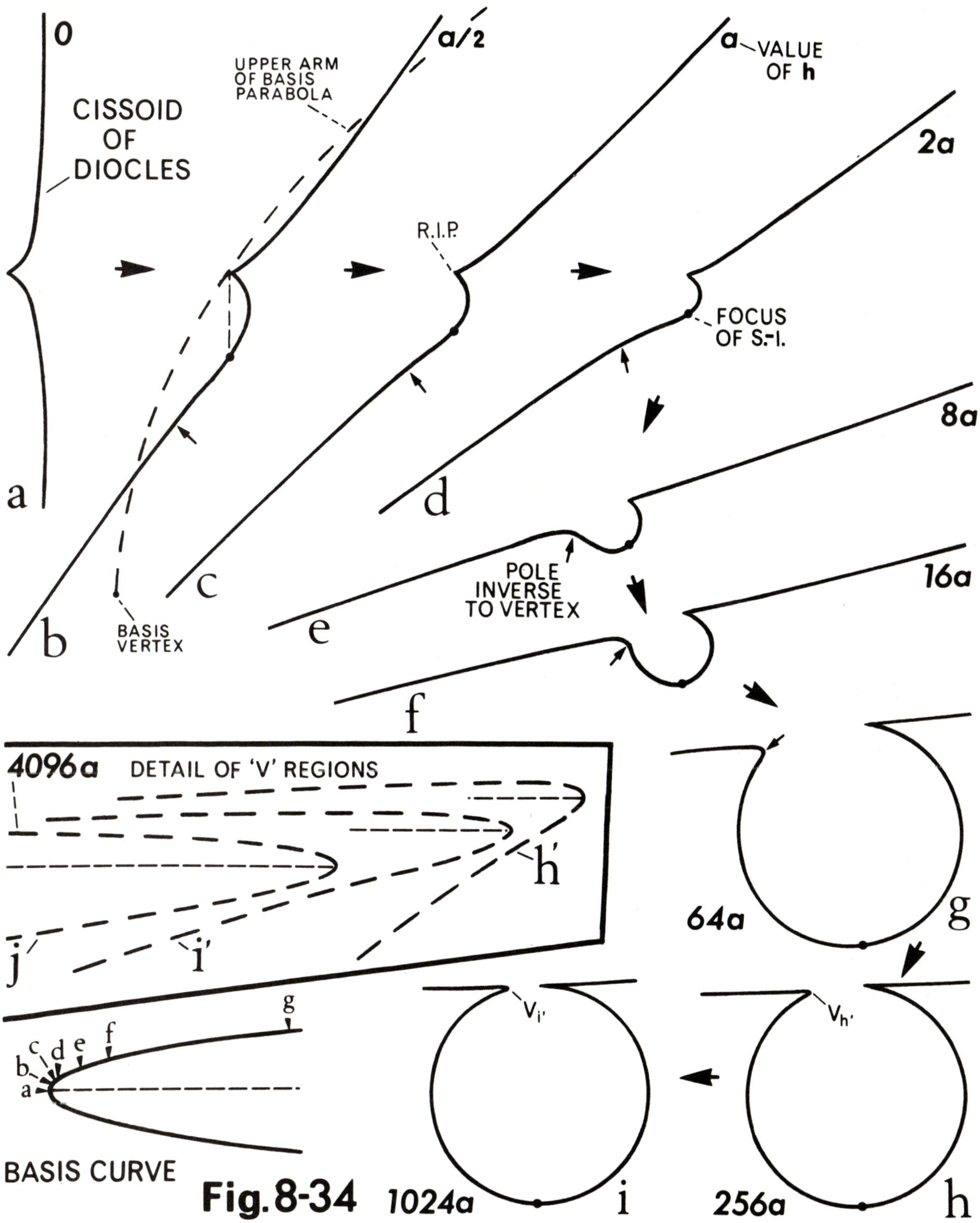

Fig. 8-34

verse to the basis vertex (small arrow) is the inverse of the segment of the lower arm of the basis parabola from the point contralateral to the pole of inversion to its vertex; and (d) the segment of the inversion cubic from the pole inverse to the basis vertex to infinity, is the inverse of the segment of the basis parabola from its vertex to the pole of inversion.

For poles of inversion progressively more distant on the upper arm of the basis curve, there occur several notable changes in the form and orientation of the inversion cubics. First, there is a CW rotation of the inverse locus reflecting the change in the slope of the upper arm of the basis curve. Second there is an increase in the approximation of the distal arms to linearity, reflecting a corresponding approximation to linearity of the upper arm of the basis curve. Third, the form of the inverse locus in the region of the focus of self-inversion takes on an increasingly circular appearance, and the focus of self-inversion comes increasingly to approximate the point of bisection of the "circular" segment.

These latter changes are an expression of the fact that the lower arm of the basis curve comes increasingly to approximate a parallel line with respect to the upper arm, and the contralateral pole comes increasingly to approximate the opposite point on this parallel line (note that the regions of curves b to i to the right of lines through the cusp-points and the foci of self-inversion are the regions inverse to the peripheral segments of the basis arms, and the regions to the left are inverse to the vertex-containing segments).

Fourth, the pole inverse to the vertex of the basis parabola occupies a region of increasing curvature that comes increasingly to resemble a reflection of the cusp region, giving the inverse locus an increasing appearance of bilateral symmetry. These changes reflect the fact that, with the vertex becoming increasingly distant, the vertex-containing segment of the basis curve comes increasingly to resemble two parallel lines extending infinitely to the left of the pole of inversion.

In essence, in the inverse locus, the differences between the peripheral and vertex-containing segments of the basis curve are expressed primarily as differences of the inverse vertex regions from the cusp regions (for example, $v_{h'}$ and $v_{i'}$ of curves h and i). The former, though they come increasingly to resemble cusps at the scale used in curves g-i, do not approximate the form of a cusp in the limit, but rather, they approach the form of a reflection of the vertex region of the basis parabola (Maxim 6-7-2).

The approach of the inverse vertex regions of the inversion cubic to a reflection of (congruence with) the vertex region of the basis parabola, with increasing distance of the pole of inversion, is illustrated in the insert at the lower left. Two of the illustrations are magnifications of the corresponding regions of curves h and i (for poles of inversion with abscissae of 256a and 1024a), the third is a magnification of the corresponding region of the inverse locus for a pole of inversion with the abscissa 4096a.

To facilitate comparisons, dashed lines are drawn through the poles inverse to the basis-curve vertex and parallel to the line of symmetry of the basis curve for curves h', i', and j. It is evident from the positions of these dashed lines, from the form of the curves (particularly curve j), and from a comparison with the form and orientation of the basis curve immediately below, that with increasing distance of the pole of inversion, these portions of the inverse loci come increasingly to take on the form of a reflection of the vertex region of the basis parabola (for the appropriate value of the unit of linear dimension).

As pointed out in the legend to Fig. 6-7, the complete loci of Figs. g, h, and i do not resemble parabolas, nor do they come increasingly to resemble parabolas for more distant poles of inversion. Rather, they come increasingly to resemble joined segments of lines and circles. But the region of the inverse locus that is inverse to the vertex region of the basis curve does come increasingly to resemble a reflection of the latter region. Moreover, the extent of the "vertex region" of the parabola that becomes congruent with the corresponding region of the inverse locus also increases progressively (and, in this sense, one could say that the inverse loci come increasingly to resemble parabolas for more distant poles of inversion).

In the limit, for an infinitely distant incident pole of inversion (and the appropriate infinite value of the unit of linear dimension), the inverse locus "opens" and it becomes congruent with the basis curve, but with an opposed orientation (Maxim 6-7-5).

Fig. 8-35. Parabola inversion quartics; for poles of inversion on the line of symmetry, the vertex tangent, and the line $y = 2x$ ($j = a$). Representative parabola inversion quartics are obtained by taking sequences of poles of inversion that lie on the line of symmetry (curves a-m), on the vertex tangent (curves s-x), and on a line concurrent with both of these lines at the vertex (curves n-r). The present figure illustrates the inverse loci for such pole locations. The locations of the poles lying relatively close to the vertex of the basis curve are shown by the large solid circles in the plane of the basis curve at left center (only the upper half of the basis curve is shown). The remaining poles of inversion lie on extensions of the dashed linear segments, mostly far off the figure.

Each inverse locus is identified by both a letter-label, matching the labelling of the selected poles in the plane of the basis curve, and a coordinate label of the abscissa (curves a-m), the ordinate (curves s-x), or both (curves n-r). The poles of self-inversion of curves p, q, and r are marked by small solid circles. In all cases, curves lie in proper orientation relative to the illustrated upper arm of the basis curve, but relative sizes and positioning have been adjusted to suit the purpose of illustration. The reciprocal inversion pole in every case lies at the tip of the cusp, which points in the same direction as the direction of the basis vertex.

The ensemble of curves a-d is comprised of inverse loci about interior poles on the line of symmetry. In essence, when the pole of inversion leaves coincidence with the vertex of the basis curve and enters the interior, the arms of the cissoid of Diocles "bend over to the left" and close, yielding a closed curve with an inward-pointing cusp (curve a). For a pole at the traditional focus this becomes the cardioid, b. As poles become more distant from the vertex, the curves come increasingly to resemble two joined congruent circles, reflecting the fact that the two arms increasingly take on the appearance of two equidistant parallel lines.

Although the region of the inverse curve opposite the cusp appears to come increasingly to resemble a cusp at low magnifications, this region, in fact, comes increasingly to resemble the vertex region of the basis curve (Maxim 6-7-2). For poles more distant than $h = 1024a$, the inverse loci greatly resemble curve d in shape, with the difference that the gap between the cusp and the region inverse to the vertex continually narrows. In the limit, of course, [for the appropriate (infinite) value of the unit of

Parabola Inversion Quartics: For Poles of Inversion On Line of Symmetry, Vertex Tangent, & Line, y=2x (j=a)

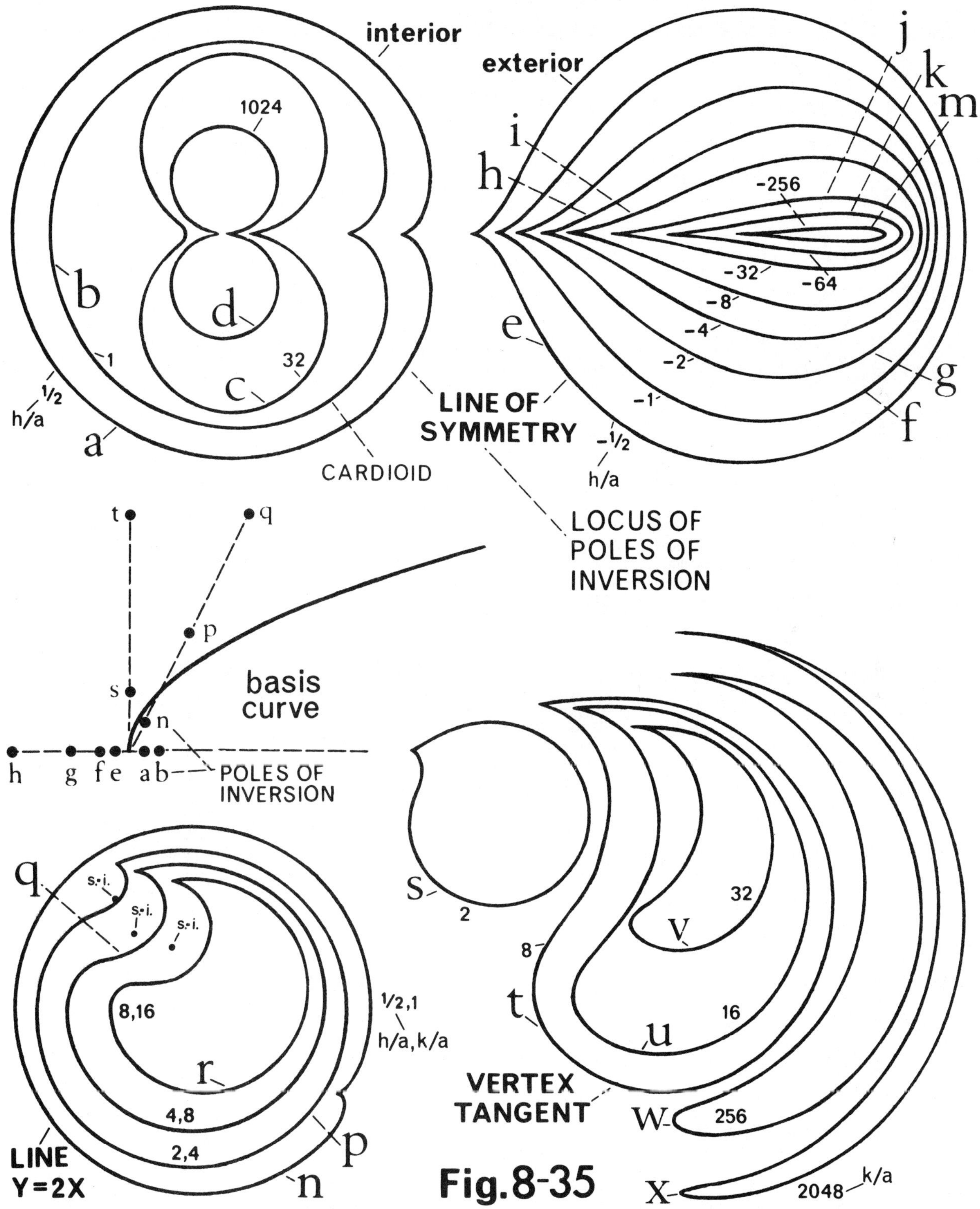

Fig. 8-35

linear dimension], the locus becomes congruent with the basis curve (Maxim 6-7-5).

The inverse loci about exterior poles on the line of symmetry are quite different in form, inasmuch as the arms of the cissoid now "bend over" to the right and close (as the pole of inversion leaves coincidence with the vertex of the basis curve and enters the exterior region); the poles of inversion no longer lie between the arms, with the result that the upper and lower portions of the curve no longer come increasingly to resemble two congruent joined circles. Instead, the curves take on an increasingly elongate appearance.

The regions inverse to the vertex region, as in curves a-d (and all the other curves represented), come increasingly to resemble a reflection of the vertex region. The extent of the region for which this reflective congruence applies increases progressively with increasing remoteness of the pole of inversion. In the limit, of course, the inverse locus becomes congruent with the basis curve (with opposed orientation).

In the case of the ensemble for poles on the line $y = 2x$, only curve n is for an interior pole. The abscissa for pole n is the same as for pole a, but pole n is displaced one unit upward. Curve n resembles curve a but its cusp region is smaller, distorted and displaced downward. For each curve, the nearly circular-appearing region to the left of a vertical line through the cusp point is the inverse of the same vertex region of the parabola. Both of these regions are nearly circular for the same reason, namely, that the vertex region of the parabola from the vertex itself to $h = \frac{1}{2}a$ is nearly circular ($R = 2a$) [which is the same reason that the lower arm of the $\frac{1}{2}a$ inversion cubic is nearly linear from the focus of self-inversion to infinity].

For poles exterior to the curve on the line $y = 2x$, the inverse loci at first resemble a portion of the figure formed by two intersecting circles of greatly disparate size. With increasing remoteness of the pole, the circles approach congruence and the shape of the inverse locus becomes ever more elongate. The changes in form are similar to those undergone by the inverse loci for poles on the vertex tangent (curves s-x), for which much more remote poles are represented.

The principal difference between the inverse curves for poles incident upon the vertex tangent and poles to one side or the other of it, lies in the orientation and position of the region inverse to the vertex region of the basis curve. For poles on the vertex tangent, the latter region lies directly beneath the cusp, and it increasingly approaches coincidence with the vertex region of the basis curve (for the appropriate value of the unit of linear dimension).

For poles to the left or right of the vertex tangent, the region inverse to the vertex region of the basis curve lies below to the left or right, respectively, of the cusp (rather than directly below) and is pointing upward or downward, respectively, rather than straight to the left. In other words, for increasingly distant poles on the vertex tangent and lines parallel to it, the inverse vertex region approaches coincidence with the vertex region of the basis curve, whereas for increasingly distant poles on lines at angle of less than or greater than 90° to the x-axis, the corresponding regions merely approach congruence with the vertex region of the basis curve.

[The difference between curve x and the inverse curve for a pole with the same ordinate but lying to the right or left of the vertex tangent can be visualized simply by rotating the page about center CW or CCW, respectively.]

Curve x of the vertex-tangent ensemble provides a convenient example for describing the changes in inverse loci, with increasing distance of the pole of inversion, in terms of *angle-doubling inversion rotation and reflection in a line orthogonal to a radius vector from the pole to the curve* (see Fig. 6-8 c-e). In the case of the vertex region of the curve for a pole at $k = 2048a$, the radius vector from the pole to the vertex is at 90° to the line of symmetry and the line orthogonal thereto is parallel to the line of symmetry, whereupon the vertex region essentially reflects onto itself.

For a radius vector to a portion of the curve to the right of the vertex, the angle, θ, with the positive x-axis, increases from 90°, while the direction of the *orthogonal line* increases from 0° (or 180°). For simplicity, we consider that the radius vector in question passes through the left member of a pair of incident points (see Fig. 6-8e). For a sufficiently distant pole, these points can be regarded as lying on a line parallel to the x-axis. In these circumstances, the reflection of the right member of each point-pair in the orthogonal line leads to an angle for the line connecting the two points of approximately double that of the *orthogonal line* itself (with the x-axis). Thus, the inverse locus (curve x) bends up in an almost circular arc.

For $\theta = 135^{\circ}$, the orthogonal line is at 45° and the line segment connecting the reflected points of a pair is at 90°. In other words, the segment of curve X with a tangent at 90° to the x-axis is inverse to a region of the basis curve with abscissa of roughly $h = 2048a$, the same value as for the ordinate of the pole of inversion (to account for an angle of the radius vector from the pole of inversion of roughly 135°).

As θ increases from 135°, the angle of the orthogonal line increases from 45°, and the position of the reflection of the right point of the pair under consideration comes to lie to the left of a normal to the x-axis from the left point, with the consequence that the connecting segment between the two points of the pair makes an angle of greater than 90° with the x-axis and the inverse locus bends back toward the pole of inversion.

As θ approaches 180°, the orthogonal line approaches 90° and the line segment connecting the pair of points also approaches 180°. The more distant the regions of the curve, the closer the corresponding inverse regions to the pole of inversion, and the closer the tangent angle to 180° (these inverse regions form the cusp).

This description accounts for the appearance of curve X for the fixed pole of inversion at $k = 2048a$. However, as the ordinate k increases from 2048a, the angle θ of the radius vector to any fixed point on the curve decreases. By choosing k sufficiently large, the angle θ for any fixed point can be made to approximate 90° (and the orthogonal line to approach coincidence with the x-axis) to any degree desired, with the result that the inverse curve approximates to being a reflection of itself in the x-axis.

Accordingly, with increasing k, the extent of the curve that inverts into coincidence with itself increases and can be made as great as desired. Only in the limit, however, with the pole of inversion on the ideal line at infinity in the direction of the vertex tangent, and for an infinite value of the unit of linear dimension, does the inverse locus also "open" to become a self-inversion in toto.

It is evident from this figure and Fig. 8-34, and the discussions thereof, that there are only two directions of increasingly distant poles in the plane of the parabola for which the inverse loci increasingly approach coincidence with the basis curve (for the appropriate value of the unit of linear dimention), namely, in one direction or the other orthogonal to the line of symmetry. For increasingly distant poles on radii in all other directions, the inverse loci merely approach congruence.

Fig. 8-36. Inversion cubics of the $e = 0.866$ ellipse ($a = 2b$). The present figure illustrates the $e = 0.866$ subspecies of the species *ellipse a-vertex inversion cubics* (curve a), the species *ellipse b-vertex inversion cubics* (curve b), the species *ellipse LR-vertex inversion cubics* (curve e), as well as several other representative subspecies of species in the superspecies *ellipse non-axial inversion cubics* (see Table 8-36-1).

Table 8-36-1. Conic Inversion Cubics

subfamily	Conic Inversion Cubics						
supergenus					Conic Non-Axial Inversion Cubics		
genus	Conic Line-of-Symmetry Inversion Cubics				LR-vertex cubics		other genera
species	ellipse a-vertex cubics	ellipse b-vertex cubics	hyperbola vertex cubics	parabola vertex cubic	ellipse LR-vertex cubics	hyperbola LR-vertex cubics	parabola LR-vertex cubic
sub-species	$e = 0.866$ ellipse a-vertex cubic	$e = 0.866$ ellipse b-vertex cubic	$e = 2$ hyperbola vertex cubic	parabola vertex cubic	$e = 0.866$ ellipse LR-vertex cubic	$e = 2$ hyperbola LR-vertex cubic	parabola LR-vertex cubic

Each representative inversion cubic is illustrated in proper orientation relative to the basis curve (at upper right) and to the same scale. The poles of inversion are marked by the large solid circles incident upon the basis curve; the reciprocal inversion poles (DP homologues) are marked by the small solid circles in the plane of (but not incident upon) the inverse loci a-i. The large solid circles incident upon curves e, g, and h mark the locations of the foci of self-inversion, while the small solid circles incident upon the same curves mark the locations of the *poles of congruent-inversion.*

The asymptote of curve e is illustrated along the curve's entire illustrated length by the tracing of long dashes, while shorter segments of the asymptotes of curves a and i are indicated by the tracings composed of short dashes. The poles of inversion follow the same progression from the a vertex to the b vertex as the sequence of letter-labels, and the ordinate of each curve (in units of a) is indicated by the number adjacent to the curve.

Among ellipse inversion cubics, the a-vertex cubic, a, is characterized uniquely by the fact that its asymptote lies entirely on the opposite side of the curve from the DP homologue (the reciprocal inversion pole), with the vertex between them. While curve b greatly resembles curve a, it and the inverse loci at all other incident points between the a and b vertices are characterized by the lack of a line of symmetry and the fact that the asymptote intersects the curve (see curve e).

In the progression from curve a to curve i, the asymptotes and the curves progressively rotate CCW from an initially vertical orientation to a terminally horizontal position. Concurrently, the shapes of the loci change from one in which the near region of the curve is convex toward the DP to one in which it is concave toward the DP. Accompanying these shape changes, the asymptote "migrates through the curve" to its eventual unique position (among ellipse line-of-symmetry cubics) between the curve and the DP--in the b-vertex inversion cubic, i. While it does not intersect the curve in this location, it nonetheless divides the curve into two segments, inasmuch as the DP homologue is an isolated DP of the locus (i.e., it is a double-root of the eq. of the curve; see eq. VIII-27b and note in legend to Fig. 8-24).

Since ellipses are central curves, all of their non-central inverse loci possess at least one focus of self-inversion. For the line-of-symmetry inverse loci a and i, only a single focus of self-inversion is present. This lies at the point inverse to the reflection of the pole of inversion in the line of symmetry (see page XI-37 and Maxim 6-4-2), In other words, at the point inverse to the contralateral vertex, namely at the vertices of curves a and i (since the unit of linear dimension of the basis curve was taken to be a, this means that lines tangent to the curve and passing through the vertex of curve a or i are of length a).

In the cases of inversions about poles that do not lie on a line of symmetry, the inverse curves possess two foci of self-inversion as well as a pole (which does not possess circumpolar focal rank) of congruent-inversion (see Maxims 6-4-2,12, 6-6-4,5). These poles are relatively simple to locate for cubics, inasmuch as all 3 must be incident upon the curve itself (Maxims 6-4-8,9 and 6-8-11). The two foci of self-inversion lie at the points of the

Inversion Cubics of The e=0.866 Ellipse

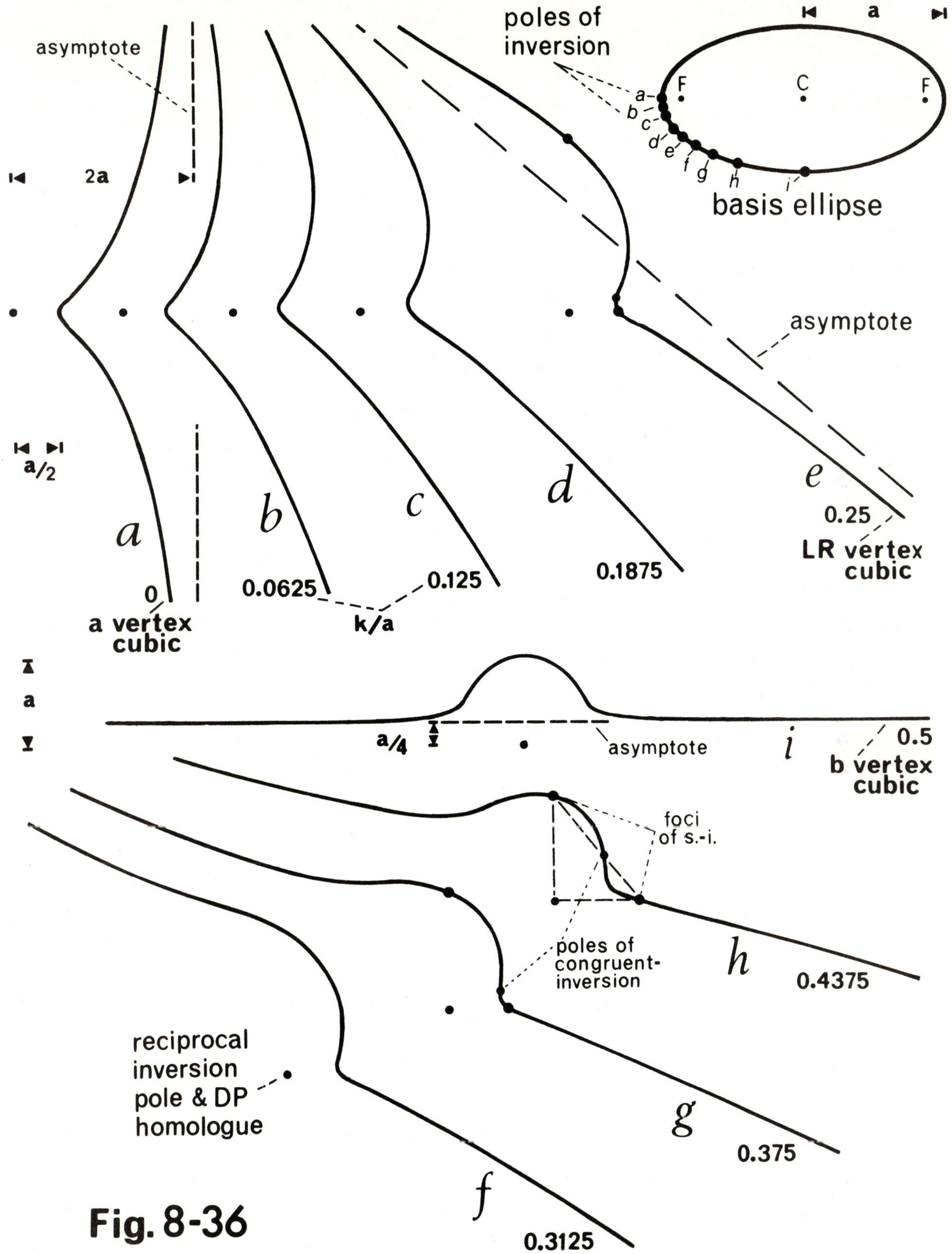

Fig. 8-36

curve that are intersected by lines through the reciprocal inversion pole that parallel the lines of symmetry of the basis curve (Maxim 6-6-5). The pole of congruent-inversion lies at the point where a line segment connecting the foci of self-inversion intersects the curve (Maxims 6-4-12 and 6-6-4,5).

The manner of locating these poles by means of the corresponding lines and intersections is illustrated in Fig. h, but the locations of the poles merely are marked for curves e and g. Whereas the inverse loci about the foci of self-inversion are coincident with the basis curve (by definition), the inverse locus about the pole of congruent-inversion is the reflection of the basis curve in a line passing through the reciprocal inversion pole and the pole of congruent-inversion, which is coincident with a line passing through the points of intersection of the circles of inversion of the two foci of self-inversion (see Maxim 6-6-6 and Fig. 8-38).

Of the several conventional analytical techniques for locating the asymptotes of curves, the following probably is the most powerful. If the eq. of the curve in question is $y = f(x)$, then the asymptote is the line $y = mx+b$, where m is the limit of the function $y = f'(x)$ for x approaching plus or minus infinity, and b is the corresponding limit of the function $[f(x) - xf'(x)]$. An elegant new direct method (though for inverse curves) devised by Christopher Kavanau, leads to the following Maxims.

CIRCUMPOLAR MAXIMS 8-36-

1: *The inverse locus of a curve about an incident pole at which the curvature is finite possesses an asymptote that is inverse to the circle of curvature at that pole.*

2: *The inverse locus of a curve about an incident pole at which the curvature is infinite possesses the ideal line at infinity as asymptote.*

3: *When an open curve with an asymptote at a finite distance from its nucleus is inverted about a non-incident pole, one obtains a closed curve with finite curvature at the reciprocal inversion pole.*

4: *When an open curve with the ideal line at infinity as asymptote is inverted about a non-incident pole, one obtains a closed curve with infinite curvature at the reciprocal inversion pole.*

The radius of curvature at the a vertex of an ellipse is b^2/a (the same length as that of the hemi-LR), whence, since $a = 2b$ for the basis curve, the asymptote of the a-vertex cubic lies at a distance $a^2/(\frac{1}{2}a) = 2a$ from the reciprocal inversion pole (as indicated for curve a). The radius of curvature of the b vertex is a^2/b, whence the asymptote of the b-vertex cubic lies at a distance of $a^2/4a = a/4$ from the reciprocal inversion pole (as indicated for curve i).

[For the cissoid of Diocles in standard position, the curvature at the cusp point is infinite, whence the asymptote of the parabola is the ideal line at infinity. For the parabola, the radius of curvature at the vertex is 2a, whence the asymptote of the cissoid of Diocles lies at $x = \frac{1}{4}a$ (where a is taken to be the unit of linear dimension).]

Fig. 8-37. Inversion quartics of the $e = 0.866$ ellipse: for interior poles of inversion on lines of symmetry, the line $x = 3a/4$, and the line $y = \frac{1}{4}a$. Since some representative axial and non-axial inversion quartics of the $e = 0.866$ ellipse already have been illustrated in Figs. 6-7,8, the present treatment is confined to some representative interior inversion quartics.

Each inversion quartic is illustrated in proper orientation relative to the basis curve (at the center) but the size of each curve has been adjusted for the purpose of illustration (to allow representation as nestling ensembles). The poles of inversion are marked by the solid circles in the interior of the basis curve, each of which has an adjacent letter-label which serves to identify the corresponding inverse locus. The poles of inversion are distributed along the lines of symmetry and a line parallel to each line of symmetry at $x = 3a/4$ and $y = \frac{1}{4}a$ (the traditional focus lies between poles a and b).

In addition to the letter-label that identifies the pole of inversion of the basis curve, each inverse locus is accompanied by an abscissa (major-axis ensemble), ordinate (minor-axis ensemble), or both coordinates (line ensembles) in units of a/16; these specify the location of the corresponding pole in the plane of the basis curve. Only the reciprocal inversion pole for the innermost locus of each ensemble is marked (by the solid circles). Foci of self-inversion and poles of congruent-inversion are not identified, but are shown in detail, together with other pertinent loci, in Fig. 8-38, for locus s of the line ensemble for $y = \frac{1}{4}a$.

The inverse locus of the basis curve about the center (curve h of the major and minor-axis ensembles) is the subspecies *e = 0.866 central quartic*. It is the only inverse locus to possess two lines of symmetry. All of the other curves in the major and minor-axis ensembles possess only one line of symmetry, but they also possess a *point of 0° reciprocal reflective symmetry*, namely a 0° focus of self-inversion.

In the cases of inversions about poles that do not lie on a line of symmetry, no inverse locus possesses a line of symmetry but, in place of lines of symmetry, each inverse locus possesses two points of 0° reciprocal reflective symmetry, namely two 0° foci of self-inversion. As shown in the following illustration (Fig. 8-38), each of these curves also possesses a single pole of congruent-inversion.

Fig. 8-38. A non-axial inversion quartic (3a/8,¼a) of the $e = 0.866$ ellipse, its foci of self-inversion, pole of congruent-inversion, and circles of inversion. The present figure illustrates the geometrical relationships that apply in the inverse loci of central QBI curves when the latter are inverted about a pole that does not lie on a line of symmetry--in this case a non-incident pole of the $e = 0.866$ ellipse at (3a/8,¼a), corresponding to pole s (yielding curve s) of Fig. 8-37.

All of the exceptional points and loci in the plane of the non-axial inversion quartic are labelled, and their exceptional properties are identified --in terms of the loci obtained by inverting the non-axial inversion quartic about the exceptional points.

The two foci of self-inversion, p_2' and p_4', are inverse to the equivalent poles p_2 and p_4 of the basis curve, the latter poles being the reflections of the pole of inversion p_1 in the two lines of symmetry. These foci of self-inversion, of course, lie on lines (not illustrated) parallel to the lines of symmetry of the basis curve. Since the lines of symmetry of the basis curve lie midway between the pole of inversion p_1 and the equivalent poles, p_2 and p_4, they invert to circles centered on the foci of self-inversion, p_2' and p_4' and, thus, are the circles of inversion of these foci.

Accordingly, since points on each of the circles of inversion are poles of equivalent inversion relative to corresponding points on the basis lines of symmetry (Maxim 6-4-5,6), the inverse loci about poles on the inverse minor-axis circle of inversion comprise an ensemble possessing as members the same axial curves possessed by ensembles of inverse loci about poles on the minor axis itself; a corresponding relationship holds for poles on the major-axis circle of inversion and the major axis.

It follows from this relationship that the inverse loci about points of intersection of the two circles of inversion are central curves. Of course, this conclusion also follows from the fact that one of the points of inter-

Inversion Quartics of The e = 0.866 Ellipse: For Interior Poles of Inversion On Lines of Symmetry, The Line x = 3a/4, & The Line y = a/4

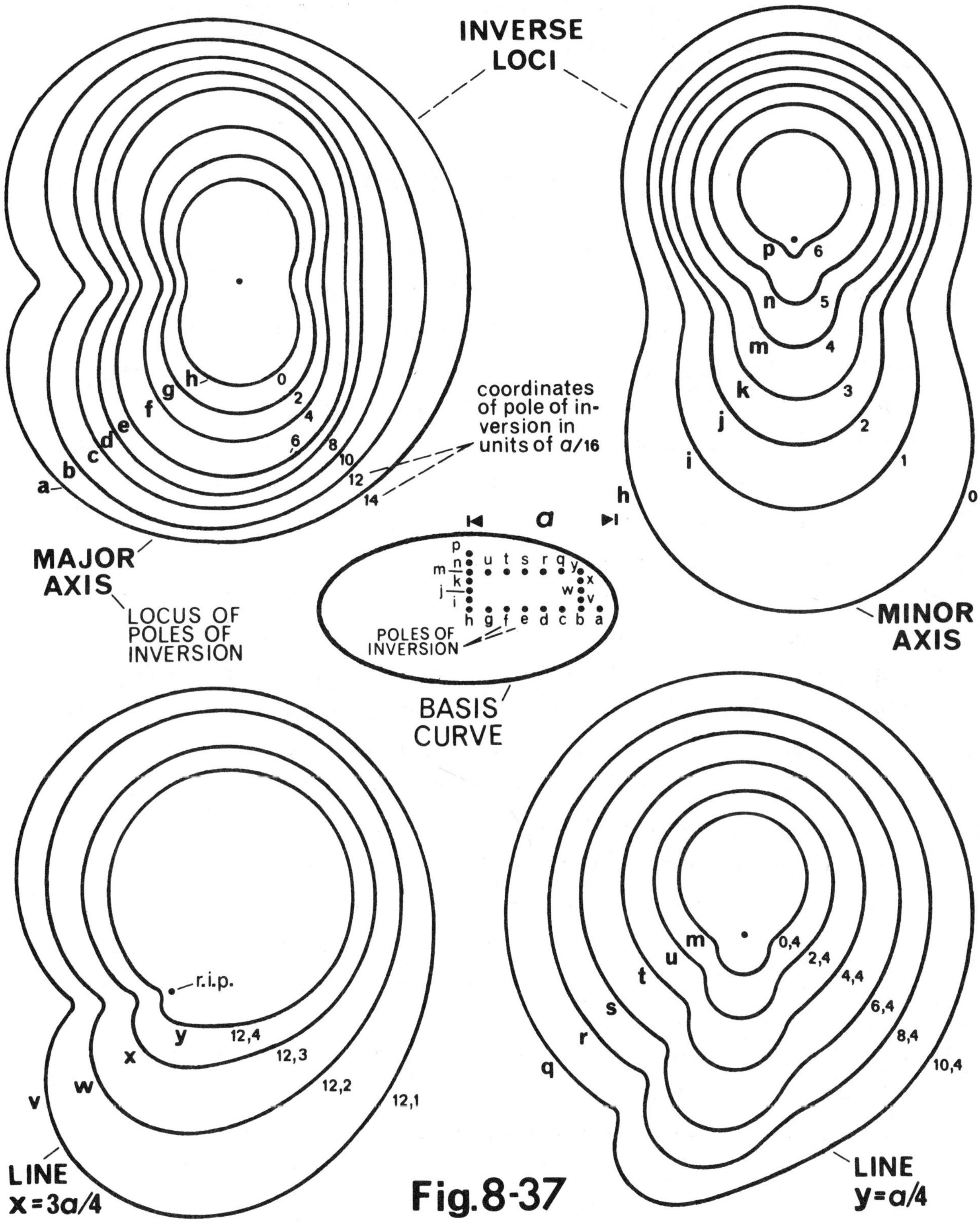

Fig.8-37

section is the reciprocal inversion pole and the other is the pole inverse to, and, hence, an equivalent inversion pole to, the center of the basis curve. It also follows from earlier Maxims that the points where the circles of inversion intersect the inversion quartic itself are poles of inversion to axial-vertex inversion cubics, and that these poles also are points of tangency of certain radii from the foci of self-inversion.

In this connection, there are 4 radii from the focus of self-inversion p_2' that are tangent to the inversion quartic. Two of these possess single points of tangency at the poles of inversion to the a-vertex cubic and also intersect the curve at two unexceptional poles of equivalent inversion. The other two are bitangents, for which the tangent points are unexceptional poles of equivalent inversion (in the present case these poles are not far from the poles of inversion to the b-vertex cubic and the LR-vertex cubic). There are only two radii from the focus of self-inversion p_4' that are tangent to the curve, and the tangent points are at the two poles of inversion to the b-vertex cubic. The radius from this focus to the poles of inversion to the a-vertex cubic intersects the curve at the latter poles.

The pole of congruent-inversion lies at the midpoint of a line segment between the two points of intersection of the two circles of inversion (Maxim 6-4-13), which is coincident with the line segment connecting the reciprocal inversion pole and the center of the basis curve. It also lies on a line segment between the two foci of self-inversion (Maxim 6-6-4) at the point where this line segment intersects the former coincident line segments (Maxim 6-6-5). With the exception of the poles of inversion to central curves, the points on the circle of inversion (not illustrated) of the pole of congruent-inversion appear to be unexceptional.

The congruent inverse locus of the inversion quartic is represented by the heavy dashed tracing, which is seen to be a reflection of the inversion quartic in a line passing through the reciprocal inversion pole and the pole of congruent-inversion. Since the inverse locus of the non-axial inversion quartic about the pole inverse to pole p_1, namely pole p_1' (points on the ideal line at infinity) also is a congruent reflection of the curve in a line (Maxim 6-8-2), these results suggest Maxim 8-38-4, below.

A Non-Axial Inversion Quartic (3a/8, a/4) of The e = 0.866 Ellipse, Its Foci Of Self-Inversion, Pole of Congruent Inversion, & Circles of Inversion

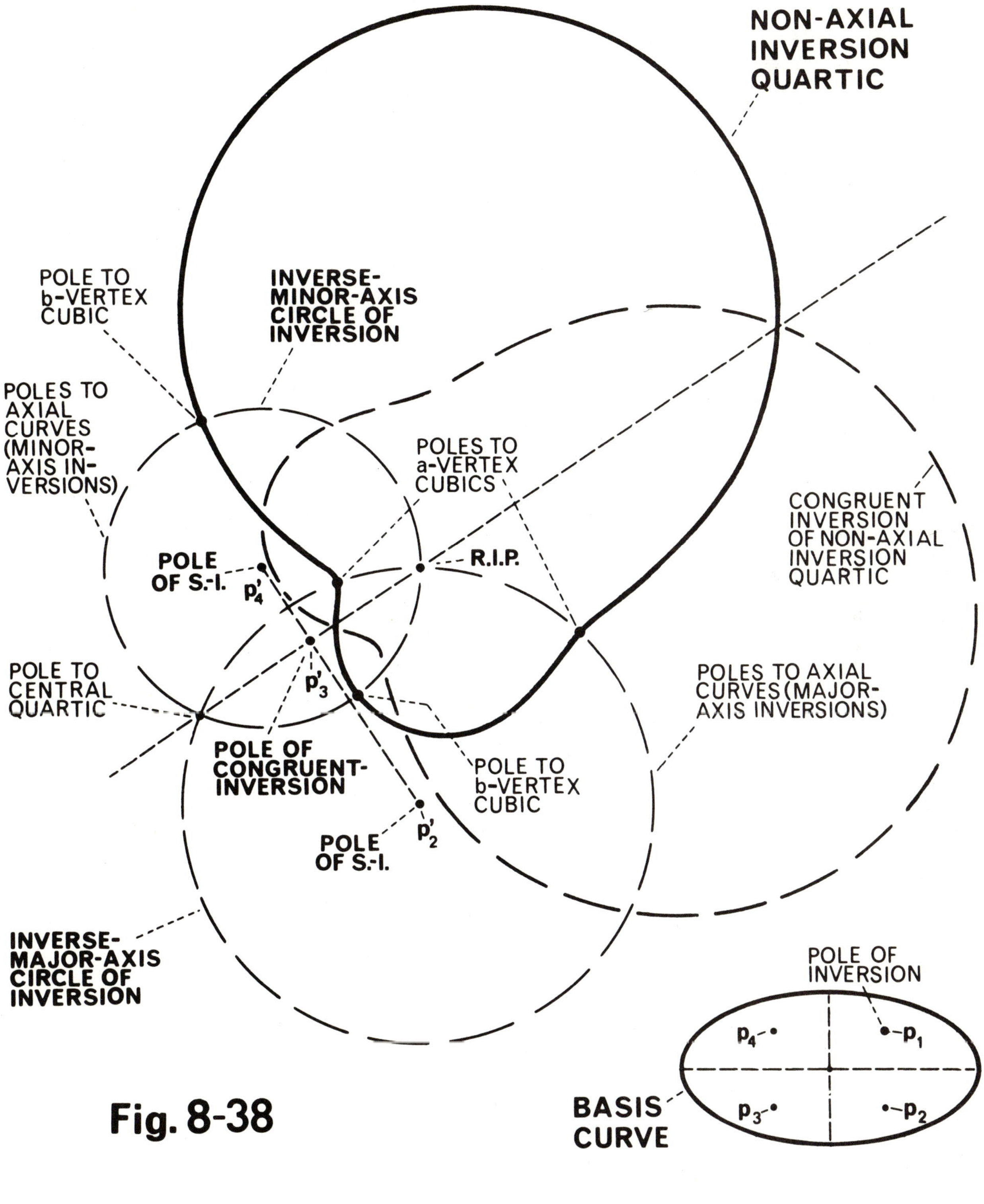

Fig. 8-38

CIRCUMPOLAR MAXIMS 8-38-

1: *Single points of tangency (i.e., excluding bitangents) of radii from foci of self-inversion of non-axial QBI cubics and quartics of non-unit eccentricity are poles of inversion to axial-vertex inversion cubics.*

2: *The 4 points of intersection of a non-axial elliptical QBI curve with its two circles of inversion are poles of inversion to axial-vertex inversion cubics.*

3: *The two points of intersection of a non-axial hyperbolic QBI curve with its inverse transverse-axis circle of inversion (excluding the DP) are poles of inversion to axial-vertex inversion cubics (axial strophoids).*

4: *In the extended plane, the pole of inversion of an inverse locus B' to a congruent inverse locus B", where B" also is a reflection of B' in a line, is either:*

(a) the inverse of the pole of inversion of B (points of the ideal line at infinity); or
(b) the pole inverse to the equidistant reflection of the pole of inversion of B in a true center.

Fig. 8-39. Approach to self-inversion of the $e = \sqrt{3}$ hyperbola about increasingly distant poles on lines parallel to the lines of symmetry but not about incident poles. The forms of incident inversions of hyperbolas already have been illustrated in Figs. 6-4 and 6-6, which dealt with incident inversions of the equilateral limacon and equilateral 10a cubic. Thus, the ensemble of cubics obtained by incident inversions of these other equilateral QBI curves is identical with the ensemble obtained by inverting the equilateral hyperbola itself, or any other equilateral subspecies of QBI curve, about all incident points. The loci of Fig. 6-3 complement the incident inverse loci and contribute 7 non-incident line-of-symmetry inverse loci of the equilateral hyperbola itself, together with two non-axial non-incident inverse loci.

The present figure complements the earlier figures with illustrations of the inverse loci of the hyperbola $e = 2$ subspecies, concentrating on inversions about remote poles lying upon 3 loci, namely incident poles, poles on the line $y = 17.23a$, and poles on the line $x = 10a$. Further, it illustrates those regions of the inverse loci that are inverse to the vertex regions of the basis curves. In so doing, it depicts the relationships expressed in Maxims 6-8-5,6. These concern the forms of the inverse loci of open central basis curves or finite segments thereof about increasingly distant poles of inversion on lines orthogonal to a line of symmetry, as opposed to their forms

Approach To Self-Inversion of The e = √3 Hyperbola About Increasingly Distant Poles On Lines Parallel To The Lines of Symmetry But Not About Incident Poles

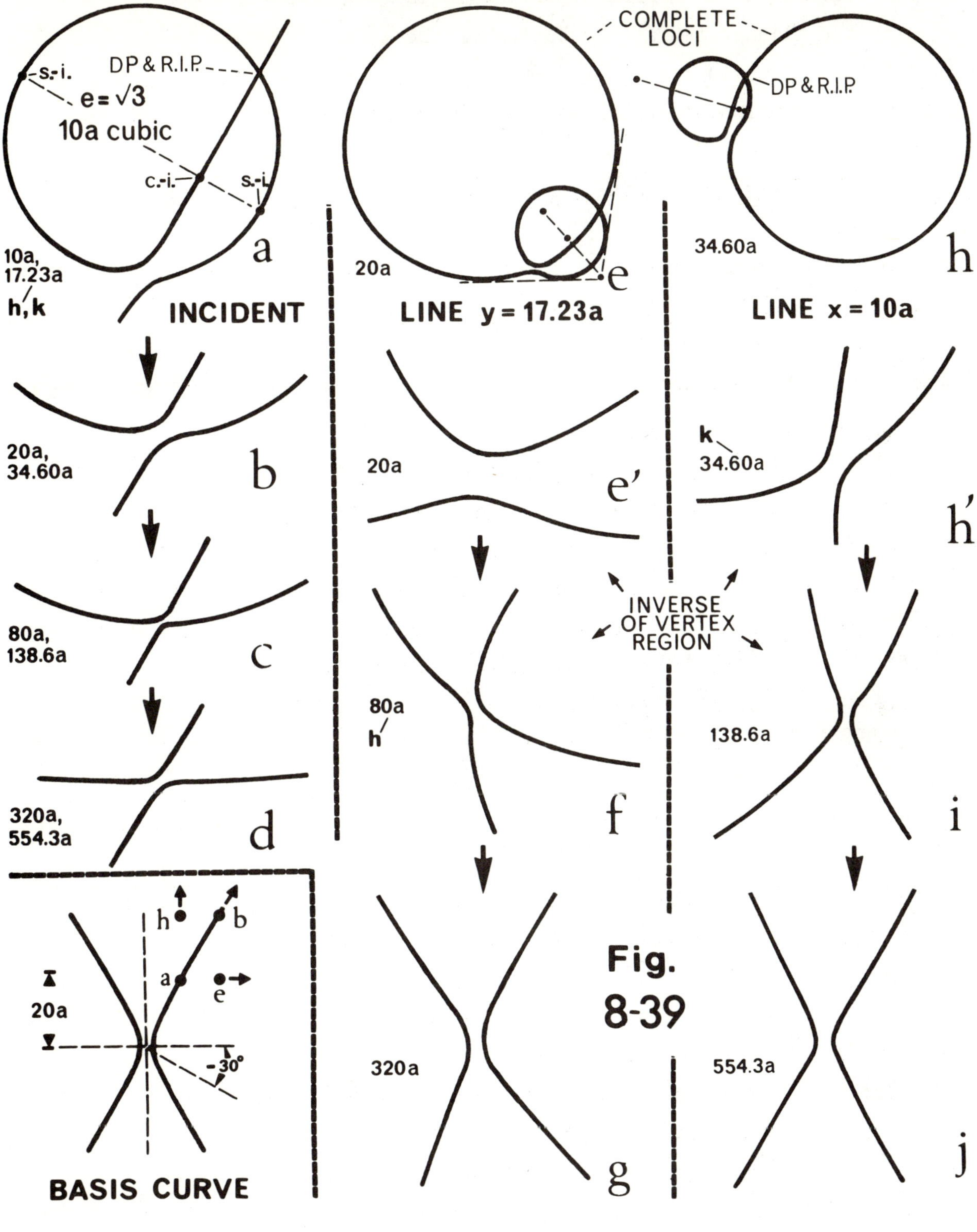

Fig. 8-39

about increasingly distant poles of inversion on lines not orthogonal to either line of symmetry, as described below.

The illustrations are presented in 3 series, progressing from the top of the figure downward in 3 columns. The basis curve and the nearest poles of inversion are illustrated at the bottom of the leftmost column. Each individual illustration is accompanied by a letter-label and the coordinate(s) of the pole of inversion. The figures are not to the same scale, but have been adjusted in size for the purpose of illustrating comparisons, particularly comparisons between the inverse vertex regions and the vertex region of the basis curve.

The poles of inversion were selected as follows. The pole nearest the vertex region yields the $e = \sqrt{3}$ 10a cubic, a (compare with the $e = \sqrt{2}$ 10a cubic of Figs. 6-3 and 6-6). The nextmost distant incident pole yields the corresponding 20a cubic, b. Non-incident poles next were selected, one with the same abscissa as the 10a cubic but with the ordinate of the 20a cubic and the other with the same ordinate as the 10a cubic but with the abscissa of the 20a cubic. The same format was followed for more distant poles, always maintaining the abscissa of the 10a cubic (poles on the line $x = 10a$) and the ordinate of the 20a cubic (poles on the line $y = 17.23a$) in inversions about more distant poles.

The series of curves a-d represent incident inverse loci with abscissae ranging from 10a to 320a. The nuclear region of the 10a cubic is shown in its entirety and presents the familiar appearance of a circle segment and two line segments. The 3 collinear incident poles of identical subspecific-inversion are marked and labelled (Maxim 6-6-4). Even for this very near incident inversion, the region inverse to the vertex region of the basis curve (lower region just above the arrow) resembles a reflection of the vertex region of the basis curve in a line orthogonal to the asymptote that lies in the 1st and 3rd quadrants (the orthogonal line in question is shown in part by the light dashed tracing at -30° to the x-axis in the 4th quadrant of the basis curve).

In the progression to the 20a cubic, the 80a cubic, and the 320a cubic, it is evident that there is an increasingly close approach of the form of the inverse vertex region of the inversion cubic to congruence with the vertex region of the basis curve. Also evident is an increasingly close approximation of the orientation of the former to a reflection of the latter in the line at

-30° to the x-axis (Maxim 6-8-7).

The series of curves e-g represent inverse loci about increasingly distant poles on a line ($y = 17.23a$) orthogonal to the y-axis, the poles having abscissae of 20a, 80a, and 320a. The appearance of the locus e is typical of distant non-axial inverse loci about poles interior to an arm of a hyperbola, i.e., it resembles two intersecting circles with an internalized loop. Here again, the 3 collinear poles of identical subspecific-inversion are marked (the pole of congruent-inversion, as for curves a and h, lies between the foci of self-inversion).

Curve e' is an enlargement of the region of curve e that is inverse to the vertex region of the basis curve. Clearly, for a pole of inversion at this distance and position relative to the vertex region, the orientation of the arms of e' is far from approximating the orientation of the arms of the basis curve, as occurs for increasingly distant poles (Maxim 6-8-5); the form of the arms (curve e' versus the basis curve) also is quite different.

At $h = 80a$, the orientation of the arms of curve f much more closely approximates that of the basis arms, although the individual arms still have distinctly different form from each other and the basis arms. At $h = 320a$, however, both the form and the orientation of the region inverse to the basis vertex region closely resemble the latter and, as h increases beyond 320a, coincidence with the latter (i.e., with a reflection of the basis curve in the y-axis) increasingly is approximated (Maxim 6-8-5).

The series of curves h-j represent inversions about increasingly distant poles on a line ($x = 10a$) orthogonal to the x-axis, the poles having ordinates of 34.60a, 138.6a, 554.3a. The appearance of the illustration of Fig. h is typical of non-axial inverse loci of hyperbolas about distant poles between the two arms, i.e., these loci resemble two intersecting circles with opposed ovals. Again the 3 collinear poles of identical subspecific-inversion are marked, but now two of the poles are between the loops rather than within the small loop (as in curve e). Note that in both curves e and h the "exterior" pole (a focus of self-inversion) lies at the point of concurrence of two bitangents to the curves (illustrated only in Fig. e).

Curve h' is an enlargement of the region of curve h that is inverse to the vertex region of the basis curve. As in the series e'-g, the progression of series h'-j shows increasing similarity of both the form and the orientation of the arms to the form and orientation of the arms of the basis curve (in this case reflected in the x-axis); as k increases beyond 554.3a, this increasing similarity approaches a condition of coincidence.

In inversions of hyperbolas about points in the plane, the reciprocal inversion poles are incident upon the inverse loci at the double points (see Maxim 13). These poles are indicated for curves a and h. The fact that, in the reciprocal inversions, the original pole of inversion of the basis hyperbola becomes incident in the former case but non-incident in the latter case--even though a DP is the pole of inversion in both cases--depends upon the fact that curve a is a cubic but curve h is a quartic. Thus, in the inversion of curve a about the DP, some of the neighboring points of the original pole of inversion of the basis hyperbola are the inverses of distant points on the asymptotes, whereupon this pole becomes an incident pole; contrariwise, in the inversion of curve h about the DP, all of the points neighboring the original pole of inversion of the basis hyperbola are the inverses of distant non-incident points in the plane of curve h.

Fig. 8-40. Incident inversions of the folium of Descartes. The folium of Descartes (eq. 1) is a *single-parameter curve*, i.e., a curve whose rectangular (or polar) eq. contains but a single parameter. It can be suggested that, in the inversion superfamily to which it belongs, the folium of Descartes is akin to the equilateral strophoid in the QBI superfamily, whereupon the folium, and all its inverse loci, would be among the most highly symmetrical subspecies in its inversion superfamily.

(a) $y^2(3x+b) = x^2(b-x)$ — folium of Descartes (8-40-1)

(b) $y^2(3x\pm b) = x^2(a-x)$ — eq. of possible genus of axial cubics of the corresponding inversion superfamily

(a) $bx(x^2+3y^2) = (x^4-y^4)$ — DP-inversion quartic of the folium of Descartes (8-40-2)

(b) $j^2x(x^2+3y^2) = (ax^2\mp by^2)(x^2+y^2)$ — eq. of possible basis axial quartic genus of the corresponding inversion superfamily

Unlike the equilateral strophoid, the folium of Descartes does not self-invert. As a consequence, no inverse locus of it about an axial pole possesses a

Incident Inversions of The Folium of Descartes

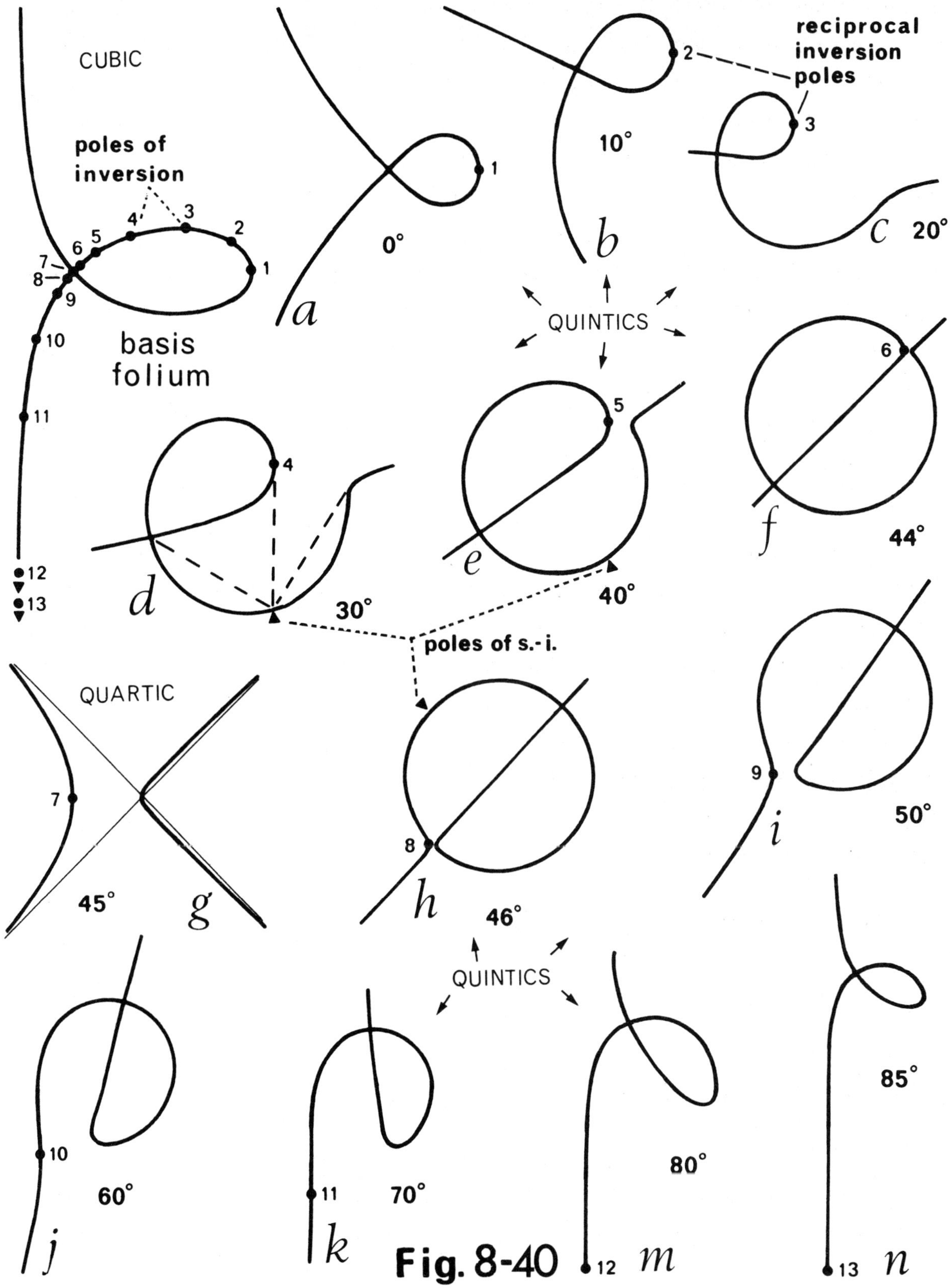

Fig. 8-40

second line of symmetry. Since the lowest-degree inverse locus of the folium of Descartes is the DP-inversion quartic (curve g and eq. 2a), this curve is a member of the basis genus for the inversion superfamily (see page VIII-18). One possible eq. for the basis genus is 2b, the inverse axial cubics of which are represented by eq. 1b (see pages IX-18,19).

In this view, the basis genus does not consist of central curves. Since all incident inverse loci of the folium of Descartes except the DP-inversion quartic are quintics, one also could take the view that the folium of Descartes is the only *cubic* locus in the inversion superfamily, which would lead to a basis curve for the superfamily different from eq. 2b. This matter is not pursued further here.

At any rate, it follows from principles of Inversion Taxonomy that the inverse loci of the folium of Descartes about all points in the plane comprise all of the corresponding subspecies in the inversion superfamily (just as the inverse loci of, say, the equilateral hyperbola about all points in the plane comprise all the equilateral subspecies in the QBI superfamily). Accordingly, all of the quartic and quintic inverse loci of the present figure about incident poles also are obtained by inverting any one of the curves a-n about the appropriate incident points.

Thirteen inverse loci about incident poles are depicted for poles lying at the positions of the solid black circles on the basis curve (upper left). The number of the pole position, the value of θ corresponding to that position for the eq., $r = b(1-2\cos^2\theta-3)\cos\theta$, cast about the DP, and an identifying letter-label are given with each inverse curve.

As mentioned above, if the basis folium is inverted about the DP (pole 7), the inverse curve is the asymmetrical hyperbola-like quartic, g (eq. 2a). As shown by the light continuous lines, the asymptotes of this curve are orthogonal to one another at $\pm 45^\circ$ to the x-axis. They are internal to the right arm, external to the left arm, and concurrent at the axial vertex of the right arm, which is the pole reciprocal to the loop vertex (pole 1).

[The position of the asymptotes is ascertained readily following Maxim 8-36-1. The slopes, $y' = dy/dx$, at the DP (the pole of inversion) are ± 1, with second derivative, $y'' = d^2y/dx^2 = 4/b$, whence the radius of curvature $\rho = \{1+(y')^2\}^{3/2}/y'' = \sqrt{2}b/2$. This places the asymptotes at a distance $b^2/(2\sqrt{2}b/2) = \sqrt{2}b/2$, from the DP (the vertex of the left arm of the inverse locus) and parallel to the DP tangents. Since circles of curvature of radius $\sqrt{2}b/2$, tangent at the DP (pole 7), pass through the loop vertex (pole 1), the asymptotes pass through the pole inverse to pole 1, namely the vertex of the right arm. Accordingly, they are tangent to this arm internally (the *internal-asymptote arm*). Note, in this connection, that the *internal* and *external-asymptote arms* are misidentified (reversed identities) on pages IX-19 and 22.]

Inasmuch as the arms possess asymptotes and the asymptotes are orthogonal, the inverse loci of the *folium of descartes DP-inversion quartic* about distant poles on the arms, such as curves f and h, resemble circles with extended diameters. These are almost indistinguishable from the corresponding inverse loci of the equilateral hyperbola (Fig. 6-3b-f). They differ essentially in that the "vertex" regions of the quintics are incongruent; one "vertex" region is more sharply curved and the other less sharply curved than the corresponding regions of equilateral QBI cubics and the equilateral hyperbola.

In the present instance of inverse loci of the folium, it is evident that the more-sharply-curved "vertex" regions are inverse to the loop-vertex region of the folium, whereas the less-sharply-curved "vertex" regions--which always are in the neighborhood of the reciprocal inversion pole--originate as the inverse of the distant regions of the basis arms. The *internal-asymptote arm* of the folium of Descartes DP-inversion quartic originates as the inverse locus of the loop of the folium, whereas the *external-asymptote arm* originates as the inverse locus of the asymptotic arms of the folium. The regions near the DP of the folium, of course, invert to distant regions on the arms of the DP-inversion quartic.

The inverse locus about the loop vertex (pole 1 of the basis curve) is the only inverse quintic with a line of symmetry, namely the *folium of Descartes vertex-inversion quintic* or *folium vertex quintic*. This curve also is obtained by inverting the DP-inversion quartic about the axial vertex of the external-asymptote arm. In fact, this curve and the folium of Descartes are obtained by inverting all the line-of-symmetry inverse sextics of the folium of Descartes about their axial vertices (see page IX-20) and by inverting all its point-in-the-plane inverse sextics about the homologous poles.

For example, in the case of Fig. f, the folium is obtained by inverting about pole 6 in the less-sharply-curved "vertex" region and the folium vertex quintic by inverting about the pole inverse to pole 1 in the more-sharply-curved "vertex" region. Inversion about the DP, of course, gives the folium DP-inversion quartic. The same relations hold for Fig. h, which is nearly congruent with Fig. f but rotated 180°.

For poles of inversion progressing from the loop vertex toward the DP, the forms of the inverse quintics change toward that of the circle-and-extended-diameter loci, f and h. For poles of inversion progressing outward along the lower arm, the forms of the inverse quintics undergo the reverse changes, except that these changes are toward the form of the folium itself, whereas the changes on the loop side of the DP (progressing toward the vertex) are toward the form of the folium vertex quintic.

As the pole of inversion becomes increasingly distant from the loop of the folium, the inverse quintics approach coincidence with the basis cubic (for the appropriate value of the unit of linear dimension), illustrating Maxim 6-6-2. For a pole of inversion for $\theta = 89.5^\circ$ (not illustrated), the inclination of the loop axis to the line of symmetry of the basis curve is only about -1.8°.

Since the folium of Descartes has a line of symmetry, all of its non-axial inverse loci self-invert (Maxim 18). As noted in earlier figures, the corresponding focus of self-inversion is the point inverse to the reflection of the pole of inversion in the line of symmetry. For incident inversions, of course, this pole is incident upon the inverse curve. It is located readily by dropping a line perpendicular to the line of symmetry of the basis curve from the reciprocal inversion pole; the focus lies at the intersection of this line with the inverse quintic. Its position is indicated by the triangular arrows of curves d, e, and h.

Self-inversion, of course, requires that the focus of self-inversion be equidistant from the DP, the tangent-point on the loop, and the tangent-point on the region of the arm inverse to the basis vertex, as indicated for curve d by the 3 dashed line-segments. All 3 of these points are poles of inversion to axial curves, namely, curve g, the basis curve, and curve a, respectively (see also Fig. 8-41, curve f). The situation is similar for parabola inversion cubics (Fig. 8-34). For the latter, however, there are only two poles of contact for inversion to axial curves (the cusp-point and the pole inverse to the basis vertex).

CIRCUMPOLAR
MAXIM 8-40-1: *Single points of tangency of radii from foci of self-inversion of cubics and quintics are poles of inversion to axial curves.*

8-40-2: *If a basis curve B is inverted about a pole P and the circle of curvature of B at P intersects B at points C, then the asymptote of the inverse curve B' intersects B' at points C' inverse to C.*

8-40-3: *An inverse curve intersects an asymptote at points of the ideal line at infinity but at no finite point only if the circle of curvature of the basis curve at the pole of inversion, P, has no other finite point of contact with the basis curve than its point of tangency at P.*

Fig. 8-41. Incident inversions of Tschirnhausen's cubic. Tschirnhausen's cubic and its inverse loci exhibit many parallels with the folium of Descartes: (a) it is a single parameter curve (and it can be suggested that it, and all its inverse loci are among the most symmetrical subspecies in its inversion superfamily); (b) it does not self-invert and, accordingly, does not invert to a central curve; (c) it inverts to sextics about points in the plane; (d) it inverts to an asymmetrical hyperbola-like quartic (see pages IX-28 to 30) about the DP (the latter is a member of the basis genus but the general eq. of the genus is not known); (e) it generally inverts to quintics about incident poles; and (f) it inverts to an axial looped quintic about the loop vertex (see pages IX-25 to 30).

It differs from the folium of Descartes in at least 3 notable respects; its DP tangents intersect at 60° (and 120°) rather than 90°, its asymptote is the ideal line at infinity, and its circles of curvature at the DP do not pass through the loop vertex (as a consequence of which the asymptotes of the DP-inversion quartic are not concurrent at the vertex of its right arm).

Differences in the appearance of the inverse loci about incident poles are traceable to these differences in the basis cubic. For example, since the asymptotes of the hyperbola-like *Tschirnhausen's cubic DP-inversion quartic* intersect at 60°, inverse loci about poles in the neighborhood of the DP resemble circles intersected by a secant (see curves g and i) at a distance ½R from the center, rather than by an extended diameter.

The fact that the asymptote is the ideal line at infinity means that the curvature of all inverse loci of Tschirnhausen's cubic is infinite at the reciprocal inversion pole (Maxim 8-36-4). The basis for this relatively uncommon property can be understood from the fact that Tschirnhausen's cubic is inverse to Cayley's sextic about the small-loop vertex (see pages IX-25 to 30). Cayley's sextic, in turn, is the roulette traced by the cusp of one cardioid rolling on the perimeter of a second congruent cardioid, with corresponding points in contact. The small-loop vertex is generated at the position where the cusps of the two cardioids are in contact. As a result, the small-loop vertex has infinite curvature.

[This is confirmed by calculating $y' = dy/dx$ and $y'' = d^2y/dx^2$, and substituting, as above, in the eq. for the radius of curvature, $\rho = [1+(y')^2]^{3/2}/y''$. Since the eq. of Cayley's sextic is $r = 4a\cos^3(\theta/3)$, and $\theta = 270°$ at the vertex of the small loop, it is evident that the property in question is a property of the $\cos^3\theta$ function as its value approaches 0.]

Although the tangent at the loop vertex is orthogonal to the line of symmetry, the 2nd derivative increases very rapidly as this point is approached, with the result that the radius of curvature tends to 0 and the curvature becomes infinite. As a consequence of this property of the small-loop vertex of Cayley's sextic, curves inverse to Cayley's sextic possess a corresponding property at the point inverse to the small-loop vertex. In the sole case in which Cayley's sextic is inverted about the small-loop vertex itself--to Tschirnhausen's cubic--the inverse pole is the ideal line at infinity, and the corresponding inverse property is the possession of this line as asymptote.

However, in every inverse locus of Tschirnhausen's cubic about poles in the finite plane, the pole inverse to the small-loop vertex of Cayley's sextic appears in the finite plane as an incident reciprocal inversion pole and point of infinite curvature (marked for each curve a-p by the solid circle). In the inversion of any of the loci a-p about any point in the plane except the reciprocal inversion pole itself, the pole inverse to the reciprocal inversion pole possesses infinite curvature. In the case where the inverse curve is Cayley's sextic (such as when curve h is inverted about the pole marked by the small x), this pole is the small-loop vertex.

The locations of the poles of inversion of the basis curve for each inverse locus are marked by the solid circles labelled and identified by the corresponding values of θ of the eq., $r = 9a(\cos^2\theta - 3/4)/\cos^3\theta$, which represents the curve cast about the DP. These values of θ appear with each locus together with a letter-label. For each inverse locus, the reciprocal inversion pole is marked by a solid circle. Inversion of the inverse locus about this pole restores the basis curve, with the ideal line at infinity as asymptote, because the curvature at the reciprocal inversion pole is infinite. For the basis curve and the quartic inverse locus h, the pole for inversion to Cayley's sextic also is indicated (by an x), as are the poles of self-inversion for curves e, f, and i.

Inversion of Tschirnhausen's cubic about the loop vertex yields *Tschirnhausen's cubic loop-vertex inversion quintic*, a. Its loop is relatively

Incident Inversions of Tschirnhausen's Cubic

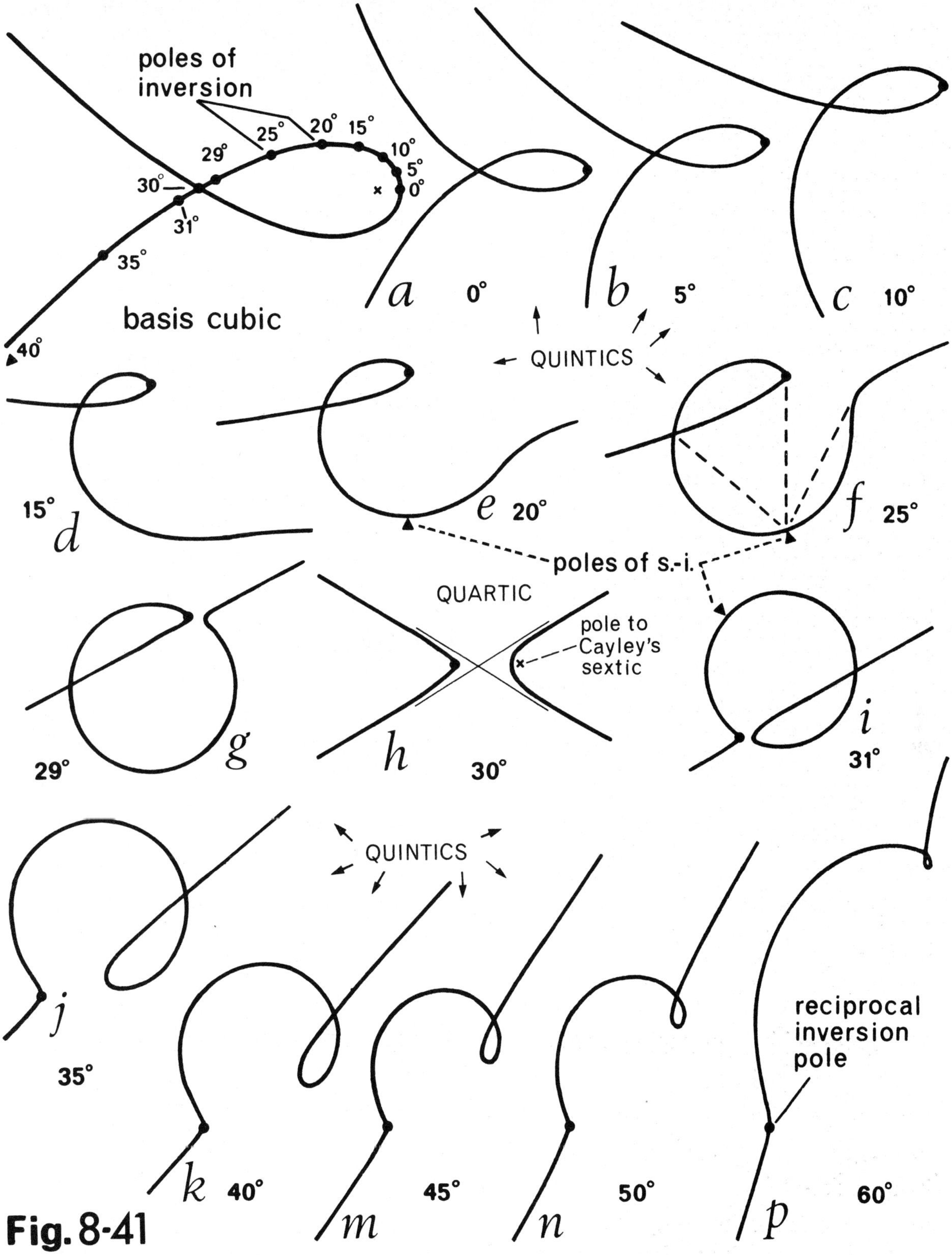

Fig. 8-41

narrower than the loop of the basis curve and derives from inversion of the latter's arms. The DP inversion yields an asymmetrical hyperbola-like quartic superficially similar to the folium of Descartes DP-inversion quartic but with the essential differences that the asymptotes, in the present instance, are exterior to both arms (light continuous lines that intersect at $h = a/6$, which is 3/8ths of the distance from the left vertex to the right vertex) and the curvature, as mentioned, is infinite at the reciprocal inversion pole.

For incident poles of inversion progressing from the loop vertex of Tschirnhausen's cubic toward the DP, the forms of the inverse quintics change from that of the loop-vertex quintic, a, in the direction of curves resembling a circle with a secant at a distance ½R from the center (see curve g). For poles of inversion progressing from the DP outward on one of the arms, the form progression of the inverse loci is from that of the latter type of curves to that of curves in which a portion of the locus comes increasingly to resemble a portion of the basis curve.

Maxims 6-6-2 and 6-6-3 do not apply to Tschirnhausen's cubic because the asymptote is not at a finite distance from the nucleus of the curve. Instead, Maxim 6-8-3 applies. According to the latter Maxim, self-inversion is achieved only in the limit, and only "finite segments" of the inverse curve increasingly resemble a reflection of the corresponding segments of the basis curve for finite incident poles of inversion. Approach of the loop regions of the inverse loci to coincidence with a reflection of the basis loop region in the line of symmetry is evidenced by the fact that the inclination of the inverse loop to the line of symmetry (which is about -80° for curve p) is only -8.3° for a pole of inversion at $\theta = 85^\circ$.

[An incident receding pole is equivalent to a pole receding on a line orthogonal to the line of symmetry, because the angles of lines from such poles to the nucleus increasingly approaches 90° with increasing distance.]

Three examples of foci of self-inversion are indicated by triangular arrows for curves e, f, and i. The dashed lines of curve f indicate equidistance of the DP, the tangent-point on the loop, and the inverse-vertex tangent-point from the focus of self-inversion. The inverse locus about the DP is curve h, that about the loop tangent-point (the reciprocal inversion pole) is the basis curve, while that about the inverse-vertex tangent-point is curve a (Maxim 8-40-1).

Fig. 8-42. Representative axial subspecies ($C = \frac{1}{4}$) of the inversion superfamily based upon equilateral hyperbolic polar-linear conchoids, $uv = Cd^2$, employing as basis curve the LCQ sextic curve of demarcation. The relatively high circumpolar symmetry of LCQ sextics about the point of tangency of the small loop and small oval, and the equational and other relationships between these curves and limacons, true exchange-limacons, and Cayley's sextic already have been dealt with in the legend to Fig. 8-28.

Inversion of LCQ sextics (eq. 1) about a point in the plane (h,k) yields the general eq. 2. It follows from principles of Inversion Taxonomy that, when the loci represented by eq. 2 are joined with the loci represented by the genus *LCQ sextics* (eqs. 1 and 8-28-3), the resulting ensemble contains all the members of the inversion superfamily of which LCQ sextics are members (see pages VIII-18 and 19, *Inversion Taxonomy*, and pertinent circumpolar Maxims). [Of course, if points on the ideal line at infinity are included among the poles of inversion (h,k), then eq. 2, alone, represents all members of the inversion superfamily (since eq. 2 then also represents congruent inverse loci of the curves of eq. 1).]

equation of LCQ sextics with the origin at (h,k) relative to the points of tangency of the small and medium-sized ovals

$$[x^2+y^2+2(hx+ky)+(h^2+k^2)]^3 = a^2[(x^2+y^2+(2h+b^2/a)x+2ky+(h^2+k^2+b^2h/a)]^2 \qquad (8\text{-}42\text{-}1)$$

curves inverse to the loci of eq. 1

$$[j^4+2j^2(hx+ky)+(h^2+k^2)(x^2+y^2)]^3 = a^2(x^2+y^2)[j^4+(2h+b^2/a)j^2x+2j^2ky+(h^2+k^2+b^2h/a)(x^2+y^2)]^2 \qquad (8\text{-}42\text{-}2)$$

(a) $(h^2+k^2)^3 = a^2(h^2+k^2+b^2h/a)^2$ — condition for vanishing of the sextic terms of eq. 2 (8-42-3)

(b) $h^2[h^4-a^2(h+b^2/a)^2] = 0$ — condition for vanishing of sextic terms of axial inversions of eq. 8-28-3 (eq. 2, $k = 0$)

(c) $k = 0,\quad 6h^5j^2 = 0$ — conditions for vanishing of quintic terms of eq. 2 with $k = 0$

The basis genus for the superfamily in question is the member genus of lowest even degree (or, if more than one such member genus exists, the genus with the highest degree of circumpolar symmetry). The basis genus for the inversion superfamily to which LCQ sextics belong is identified as follows.

First the origin of eq. 8-28-3 for LCQ sextics is translated to the finite pole (h,k), yielding eq. 1. Eq. 1 then is expressed in polar coordinates and inverted by replacing r with j^2/r. Restoring rectangular coordinates yields eq. 2, which represents all members of the inversion superfamily in question except members of the genus *LCQ sextics* (if LCQ sextics were self-inverting, eq. 2 would represent *all* members of the superfamily).

The first condition for the reduction of eq. 2 is that the sextic terms vanish, namely eq. 3a. But eq. 3a will be recognized as the eq. of the basis LCQ sextics (8-28-3) with x and y replaced by h and k. In other words, if LCQ sextics are inverted about an incident pole, the resulting inverse locus is of rectangular degree no greater than 5. The conditions for vanishing of the 5th-degree terms are given by eq. 3c, namely, $h = k = 0$. Accordingly, if the LCQ sextics of eq. 8-28-3 (or eq. 1 with $h = k = 0$) are inverted about the origin, the eqs. of the resulting loci will possess neither 6th-degree terms nor 5th-degree terms.

inverse loci of LCQ sextics of eq. 8-28-3 about the origin, i.e., about the point of tangency of the small and medium-sized ovals

$$(x^2+y^2)(x+aj^2/b^2)^2 = j^8/b^4 \qquad (8\text{-}42\text{-}4)$$

equilateral hyperbolic polar-linear conchoids cast about the point of intersection of the line of symmetry and the asymptote (line pole)

(a) $$[(x-d)^2+y^2]x^2 = C^2d^4 \qquad (8\text{-}42\text{-}5)$$

(b) $(x^2+y^2)(x+d)^2 = C^2d^4$ — eq. 5a with the origin translated to the point-pole (x+d translation)

$C = b^2/a^2, \quad d = aj^2/b^2$ — correspondences between the parameters of eqs. 4 and 5b (8-42-6)

[It is of great interest to note that, unlike the situation for QBI curves, inversions about the double points of LCQ sextics do not yield the inverse loci of lowest degree. The same relationship has been noted in the superfamily based upon the genus possessing the DP-inversion quartic of Cayley's sextic as a subspecies (see pages IX-28 to 30, *Cayley's Sextic*). The curves of the latter superfamily superficially resemble those of the present one, except for the absence of the small oval or unlooped open arm. In the case of Cayley's sextic, inversion about the DP yields a quartic, whereas inversion about the small-loop vertex (the homologue of the point of tangency of the small and medium-sized ovals) yields Tschirnhausen's cubic.]

Representative Axial Subspecies (C = 1/4) of The Inversion Superfamily Based On Equilateral Hyperbolic Polar-Linear Conchoids, $uv = Cd^2$, Employing As Basis Curve The LCQ Sextic Curve of Demarcation

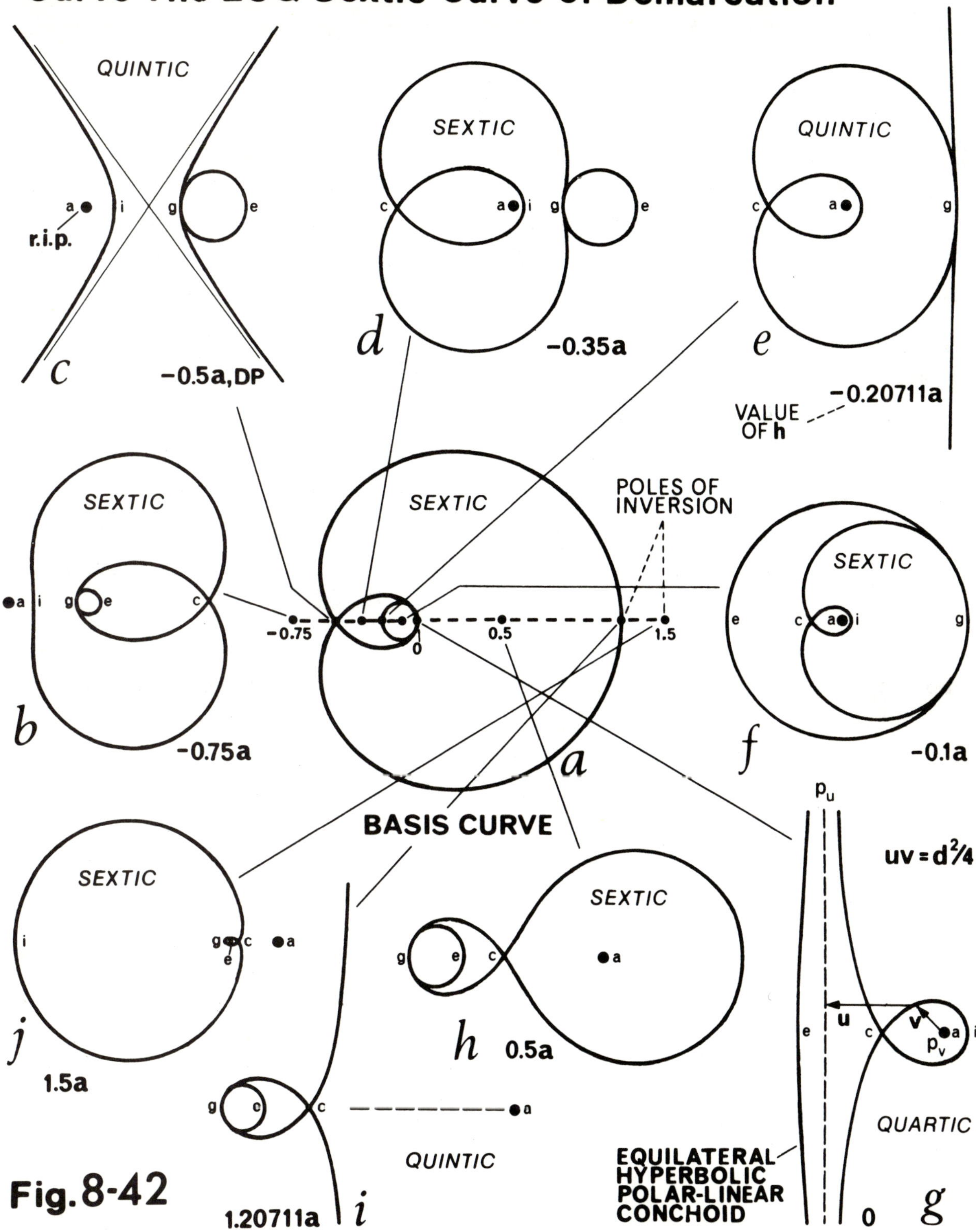

Fig. 8-42

Inversions of LCQ sextics about the origin yield the quartic loci of eq. 4. In the case of the LCQ sextic curve of demarcation, the resulting locus is the asymptotic conchoid-like locus, g. The latter is identified readily by reference to Chapter III (pages 14 and 15, *Polar Exchange Symmetry and Conchoids*). Eq. 4 differs from eq. III-30 (reproduced here for convenience as eq. 5a) for equilateral hyperbolic polar-linear conchoids only in that the origin of the latter curves is at the point of intersection of the line-pole or asymptote (p_u) with the line of symmetry, whereas that of the former curves is at the point-pole (p_v). Translating the origin of eq. 5a to the point-pole (x+d translation) yields eq. 5b. Comparison of eq. 5b with eq. 4 leads to the correspondences of eq. 6 between the parameters a and b of LCQ sextics and the parameters C and d of equilateral hyperbolic polar-linear conchoids.

Inasmuch as LCQ sextics are axial members of the inversion superfamily based upon equilateral hyperbolic polar-linear conchoids, it follows from principles of Inversion Taxonomy that for any given value of $C = b^2/a^2$, the ensemble of axial inverse loci (in the extended plane) of the LCQ sextic subspecies possess the same members (same subspecies) as the ensemble of the axial inverse loci of the corresponding conchoid. Accordingly, it suffices to portray the axial inverse loci of a representative curve from either genus in order to illustrate the general appearance of the axial members of the superfamily (since the axial curves do not self-invert, no central curve exists).

Furthermore, if an inverse locus is given for each axial region of the selected basis curve, all the different types of inverse loci are represented. For the present figure, the LCQ sextic curve of demarcation, a, has been selected as the basis curve, and representatives are given for all incident axial inverse loci and axial inverse loci about non-incident poles in each dissimilar region of the basis curve.

To emphasize the fact that any of the represented curves could serve as the basis curve, a letter adjacent to each point of intersection of the curves with their lines of symmetry (and a letter adjacent to the reciprocal inversion pole) identifies the locus obtained by inverting the curves in question about each labelled pole. For example, the inverse loci of curve d about the axial poles labelled c, a, i, g, and e yield, respectively, curves c, a, i, g, and e.

Beginning with the sextic curves b and j, these inverse loci about exterior poles to the left and right of the basis curve are merely distorted images of the basis curve and of one another. Since the small loop and small oval are at relatively great distance from the pole of inversion to curve j, as compared to their distance from the pole of inversion to curve b, these regions are comparatively very small in curve j, as compared to their size in curve b. However, as exterior poles of inversion recede along the line of symmetry to the right and left, this disparity gradually lessens and the corresponding inverse loci come increasingly to resemble one another and the basis curve; in the limit, congruence of all 3 loci is achieved (Maxim 6-7-1)

Curve c is the DP-inversion quintic. Inasmuch as the basis curve greatly resembles a limacon with a tangent circle within the small loop, curve c greatly resembles a hyperbola cast about a traditional focus with an interior tangent circle at one vertex. The two arms of the hyperbola-like portion of this locus are neither individual arms of hyperbolas nor congruent to one another. Notwithstanding the lack of congruence, however, the two arms are much more similar to one another in appearance than the arms of the quartic hyperbola-like curves obtained by inverting Tschirnhausen's cubic (Fig. 8-41h) and the folium of Descartes (Fig. 8-40g) about the DP.

The similarity of the two arms of curves c finds its counterpart in the fact that the asymptotes (light continuous lines), which intersect at angles of 109.47° and 70.53° [$\arctan(-2\sqrt{2})$ and $180^{\circ}-\arctan(-2\sqrt{2})$] are concurrent very near to the midpoint of the inter-vertex segment. The point of concurrence is at roughly 52.35% of the distance from the left to the right vertices. The corresponding radius of curvature at the DP of curve a is 0.4618a, which means that the DP circles of curvature lie within the large loop and enclose the small loop and its interior tangent oval. Consequently, the asymptotes intersect the curve only at points of the ideal line at infinity (Maxim 8-40-3).

Proceeding to the right from the DP, the next pole of inversion lies within the small loop about halfway to the left vertex of the small oval. Since an axial pole within the small loop is the location of the pole of self-inversion of hyperbolic limacons, all axial inversions within this region preserve the general appearance of the looped portions of curve d. For this particular location of the pole of inversion, the small oval comes to lie exterior to the remainder of the curve, tangent to it at the large-loop vertex.

For the next location of the pole of inversion--at the vertex of the small oval--the looped portions of the inverse locus are relatively little changed from curve d but the small oval inverts to an open, almost linear, exterior segment tangent to the remainder of the locus at the large-loop vertex.

For the next location of the pole of inversion--within the small oval--the portions of the inverse sextic locus f that are inverse to the looped portions of the basis curve continue to resemble these looped portions, but the inverse of the small oval now is an exterior oval that encloses the remainder of the locus and is tangent to it at the large-loop vertex.
[Since the small oval of the basis curve is tangent to the small loop at the latter's vertex, its inverse in all inverse loci is tangent to the inverse of the small loop at the pole inverse to the small-loop vertex.]

The next pole of inversion lies at the point of tangency of the small oval and the small-loop vertex. This is the pole of inversion to the corresponding ($C = \frac{1}{4}$) quartic equilateral hyperbolic polar-linear conchoid, g (eq. 4 with $a^2 = 4b^2$). The looped portions of the basis curve invert to the looped strophoid-like segment, while the small oval inverts to the open segment. Since the point of tangency now is the pole of inversion, the point of tangency in the inverse locus lies on the ideal line at infinity (where the arms of the curve meet the asymptote).

It is evident from eq. 4, for $a^2 = 4b^2$ and $j = a$, that curve g possesses two coincident asymptotes at $x = -4j^2/a = -4a$ (and that the possession of coincident asymptotes is a property characteristic of all the quartic subspecies). These asymptotes are at the position of the vertical line-pole p_u (shown by the heavy dashed line). Since these asymptotes are inverse to the circles of curvature at the tangent point between the small oval and small loop, both of these segments have the same radius of curvature at this point, namely a/8. Accordingly, their coincident circles of curvature, of diameter $\frac{1}{4}a$, enclose the small oval, lie within the small loop (diameter $\frac{1}{2}a$), and make no contact with these segments at any point but their tangent point.

For all axial locations of the pole of inversion within the large loop between the vertices of the two loops, the inverse loci are sextics. These possess opposed loops of varying relative proportions, within one of which--and tangent to it at its vertex--lies the oval inverse to the small oval of the basis curve h. The closer the pole of inversion approaches to the large-loop vertex, the larger the right loop of the inverse curve becomes, and the smaller the corresponding left loop, until, when the pole is at the vertex, the right loop opens to form the asymptotic quintic locus i. The latter consists of a strophoid-like loop very similar to that of curve g but with an interior oval tangent to it at its vertex.

Lastly, when the pole of inversion is displaced to the right of the large-loop vertex, the arms of the strophoid-like curve i "bend around" to the left and close to form the sextic curve j. For the pole of inversion at $h = 1\frac{1}{2}a$, the small loop and associated oval of curve j are very similar in shape to the corresponding portion of locus i, which, by comparison, is magnified by a factor of 5.

In summary, the axial inverse loci of the LCQ sextic curve of demarcation have characteristic appearances that depend upon the location of the pole of inversion. Since there exists a point of tangency of two segments of the basis curve, the corresponding segments of the inverse loci also are tangent at the poles inverse to this point. Incident axial inversions are reduced in degree to quintics or a quartic, depending upon whether the point of tangency remains in the finite plane (quintics) or inverts to a location on the ideal line at infinity (the quartic). Inasmuch as the basis curve does not self-invert and no line of symmetry is lost in inversions about axial poles, none of the axial inverse loci self-inverts.

APPENDIX 1. CIRCUMPOLAR MAXIMS

6-4-1 : One obtains members of a single subspecies by inverting a basis curve about a given incident pole and by inverting any inverse curve about the pole inverse to the given incident pole.

6-4-2 : One obtains two or 4 members of the same subspecies by inverting an inverse curve about poles inverse to equivalent poles of a basis curve with one or two lines of symmetry, respectively.

6-4-3 : One obtains members of only a single subspecies of curve from inversions of a basis curve about equivalent poles and from inversions of any inverse curve about the poles inverse to the equivalent poles.

6-4-4 : One obtains members of only a single subspecies of curve by inverting a basis curve (or any figure) about a given pole and by inverting any inverse curve or figure about the pole inverse to the given pole.

6-4-5 : The equivalent inversion poles for an ensemble of inverse curves obtained by inverting a basis curve B about all points on a line L comprise the circle L' inverse to L in the plane of any curve B' inverse to B about a pole not incident upon L.

6-4-6 : The equivalent inversion poles for an ensemble of inverse curves obtained by inverting a basis curve B about all points on a given locus L comprise the locus L' inverse to L in the plane of any curve B' inverse to B.

6-4-7 : Inversion of a given basis curve B about all points in the plane generates an ensemble of inverse curves B', each member of which inverts to every other member but to no non-member. This is the property of *immediate closure* or *one-step closure*.

6-4-8 : Inverse loci of QBI quartics about unexceptional incident poles are of 3rd degree (exceptional incident poles are the DP and DP homologue, about which all QBI quartics invert to quadratics).

6-4-9 : Inverse loci of QBI cubics about unexceptional incident poles also are cubics (inverse loci about exceptional incident poles are quadratics).

6-4-10: Inverse loci of cubics about non-incident poles in the finite plane are of higher degree than 3.

6-4-11: A non-axial QBI curve of non-unit eccentricity inverts to axial QBI curves about points incident upon its two (orthogonal) circles of inversion (circles of modular radius centered upon its foci of self-inversion), excepting the two points of intersection of these circles. In these transformations, the circle upon which the pole of inversion is incident inverts to the line of symmetry of the inverse curve.

6-4-12: A non-axial QBI curve of non-unit eccentricity inverts to two subspecifically different central QBI curves about the two points of intersection of its two circles of inversion; one of these curves is a central conic, the other is a central quartic. In these transformations, the two circles of inversion invert to the two lines of symmetry of the central curves.

6-4-13: A non-axial QBI curve of non-unit eccentricity undergoes identical subspecific-inversion about 3 finite poles. Two of these poles are foci of self-inversion, the 3rd is a pole of congruent-inversion. The latter lies on the line segment between the two foci of self-inversion.

6-4-14: The pole of congruent-inversion of a non-axial QBI curve of non-unit eccentricity lies midway between the two poles for inversion to central curves.

6-4-15: An axial QBI curve of non-unit eccentricity inverts to central QBI curves about the two points of intersection of its circle of inversion with its line of symmetry. In these transformations, the circle of inversion inverts to the orthogonal line of symmetry of the inverse central curve.

6-4-16: The two inverse central loci of an axial QBI curve of non-unit eccentricity consist of a central conic (the inverse locus about the DP or DP homologue) and a central quartic.

6-4-17: A non-axial QBI curve of unit eccentricity inverts to the entire ensemble of axial QBI curves of unit eccentricity about points on its circle of inversion.

6-4-18: About the pole on its circle of inversion diametrically opposite the pole for inversion to a parabola, a non-axial QBI curve of unit eccentricity inverts to the $x = h$ parabola axial-inversion quartic, where h is the abscissa of the reciprocal inversion pole in the plane of the basis parabola.

6-4-19: Equivalent inversion poles on the line of symmetry of axial inversions of central conics lie in modularly-inverse pairs, both members of a given pair lying either to the right or the left of the focus of self-inversion.

6-5-1 : The two inverse poles at which a given radius through a 0° focus of self-inversion (or given chord segments through a 180° focus of self-inversion) intersects a curve are equivalent inversion poles.

6-6-1 : For the appropriate value of the unit of linear dimension, the inverse curves obtained by inverting a non-axial asymptotic looped cubic about incident poles increasingly distant from the loop approach congruence with the basis curve. In the limit, congruence is achieved (a congruent-inversion).

6-6-2 : For the appropriate value of the unit of linear dimension, the inverse curves obtained by inverting an axial asymptotic looped cubic about incident poles increasingly distant from the loop approach coincidence with the basis curve. In the limit coincidence is achieved (a self-inversion).

6-6-3 : For the appropriate value of the unit of linear dimension, the inverse of a basis asymptotic arm of an axial curve, obtained by inversion about a pole incident upon the arm at points increasingly distant from the nucleus, approaches coincidence with the arm. In the limit, coincidence is achieved between the inverse arm and the basis arm.

6-6-4 : A non-axial inversion cubic of a central conic or central quartic undergoes identical subspecific-inversion about 3 finite collinear poles. Two of the identical subspecific-inversions are self-inversions, the 3rd is a congruent-inversion (for the appropriate value of the unit of linear dimension). The pole for a congruent-inversion lies on the line segment connecting the foci of self-inversion.

6-6-5 : A non-axial inverse curve of a central conic or a central quartic possesses 3 finite collinear poles of identical subspecific-inversion. Two of these poles are foci of self-inversion. They lie on lines parallel to the lines of symmetry of the basis curve that pass through the pole of inversion. The 3rd is a pole of congruent-inversion (for the appropriate value of the unit of linear dimension). It lies at the point of concurrence of the line segment from the pole of inversion to the center of the basis curve, and the line segment connecting the foci of self-inversion.

6-6-6 : The congruent inverse locus of a non-axial QBI basis curve of non-unit eccentricity (obtained by inverting about the pole of congruent-inversion) is the reflection of the basis curve in a line through the points of intersection of its circles of inversion. The unit of linear dimension for the congruent-inversion is one-half the length of the line segment connecting these points (the distance from the pole of congruent-inversion to either pole for inversion to a central curve).

6-7-1 : For the appropriate value of the unit of linear dimension, inversions of a closed basis curve or finite segments of an open basis curve about distant poles yield inverse curves that increasingly approach congruence with the basis curve, the more distant the pole.

6-7-2 : For the appropriate value of the unit of linear dimension, inversions of a basis curve about distant poles yield inverse curves that, in part or in their entirety, increasingly approach congruence with the basis curve (in part or in its entirety) the more distant the pole.

6-7-3 : For the appropriate value of the unit of linear dimension, inversions of a closed central basis curve or finite segments of an open central basis curve about distant poles on either of the lines of symmetry yield inverse curves that increasingly approach coincidence with the basis curve (self-inversion), the more distant the pole.

6-7-4 : For the appropriate value of the unit of linear dimension, inversions of an open central basis curve about distant poles on either of the lines of symmetry yield inverse curves, portions of which increasingly approach coincidence with portions of the basis curve, the more distant the pole.

6-7-5 : For the appropriate (infinite) value of the unit of linear dimension, inversion of a basis curve or figure about an infinitely distant pole yields a congruent curve or figure.

6-8-1 : When two points P_1 and P_2 on the x-axis are inverted about increasingly distant points on a line L in the 1st quadrant that emanates from the origin at an angle θ to the x-axis, the line $P_1'P_2'$, connecting the inverse points P_1' and P_2', forms an angle with the x-axis whose limiting value is 2θ.

6-8-2 : For the appropriate (infinite) value of the unit of linear dimension, the inverse locus of a basis curve about an infinitely distant pole is a reflection of the curve in a line (specifically, the circle of inversion of infinite radius).

6-8-3 : For the appropriate value of the unit of linear dimension, the inverse locus B' of an axial closed basis curve, or of finite segments of an axial open basis curve B about increasingly distant poles on a line orthogonal to the line of symmetry, increasingly resemble the reflection of B, or the reflection of finite segments of B, in the line of symmetry. In the limit, B' is a coincident reflection of B, whereupon B self-inverts.

6-8-4 : For the appropriate value of the unit of linear dimension, the inverse locus B' of an axial closed basis curve, or of finite segments of an axial open basis curve B about increasingly distant poles on a line L not orthogonal to the line of symmetry, increasingly resemble a reflection of B, or a reflection of finite segments of B, in a line L* orthogonal to L. In the limit, B' is a congruent reflection of B in L* (a congruent-inversion).

6-8-5 : For the appropriate value of the unit of linear dimension, the inverse locus B' of a central closed basis curve, or of finite segments of a central open basis curve B about increasingly distant poles on a line parallel to one of the lines of symmetry, increasingly resemble a reflection of B, or a reflection of finite segments of B, in the other line of symmetry. In the limit, B' is a coincident reflection of B, whereupon B self-inverts.

6-8-6 : For the appropriate value of the unit of linear dimension, the inverse locus B' of a central closed basis curve, or of finite segments of a central open basis curve B about increasingly distant poles on a line L not orthogonal to either line of symmetry, increasingly resemble a reflection of B, or a reflection of finite segments of B, in a line L* orthogonal to L. In the limit, B' is a congruent reflection of B in L*.

6-8-7 : For the appropriate value of the unit of linear dimension, the inverse locus B' of a non-axial closed basis curve, or of finite segments of a non-axial open basis curve B about increasingly distant poles on a line L, increasingly resemble a reflection of B, or a reflection of finite segments of B, in a line L* orthogonal to L. In the limit, B' is a congruent reflection of B in L*.

6-8-8 : The possession of a line of symmetry by a curve is equivalent to the possession of a 0° focus of self-inversion on the ideal line at infinity in the direction orthogonal to the line of symmetry.

6-8-9 : Central curves with two lines of symmetry possess two 0° foci of self-inversion on the ideal line at infinity, one in the direction of each line of symmetry.

6-8-10: The gain (or loss) of a line of symmetry in an inversion is accompanied by the transfer of a 0° focus of self-inversion to (or from) the ideal line at infinity from (or to) the finite plane (see also Maxim 8-15-4a).

6-8-11: In the finite plane, curves of odd degree self-invert only about incident foci; contrariwise, curves of even degree self-invert only about non-incident foci.

8-10-1: There are two axial poles about which a self-inverting monovular basis curve with a line of symmetry inverts to curves with true centers. These poles are reflections of one another in the focus of self-inversion at the modular distance.

8-10-2: For the appropriate value of the unit of linear dimension, the two curves with true centers, derived by inverting monovular self-inverting curves with a line of symmetry, are central inversions of one another. These curves may be subspecifically identical or dissimilar.

8-10-3: If a self-inverting QBI curve with a line of symmetry is inverted about the reflection of the DP or DP homologue in the focus of self-inversion, a central quartic is obtained.

8-10-4: There are two pairs of axial poles about which axial bicircular quartics invert to central curves. The members of one pair of poles are modular reflections of one another in one 0° focus of self-inversion, and the members of the other pair are modular reflections of one another in the other 0° focus of self-inversion.

8-11-1: All the non-focal self-inversion loci of a basis curve that self-invert about a focus of self-inversion of the latter, taken as reference pole, also are non-focal self-inversion loci of one another.

8-15-1: Upon inversion of points in the plane about a pole P, all point-pairs inverse at 0° in a circle of inversion C_1 become inverse at 0° in a circle of inversion C_2 centered on pole P', the inverse at 0° of pole P in circle C_1, provided that poles P and P' are not coincident (i.e., provided that pole P is not incident upon circle C_1).

8-15-2: Upon inversion of points in the plane about a pole P incident upon a circle of inversion C, all points inverse at 0° in C become reflected in a common line.

8-15-3: Upon inversion of points in the plane about a pole P, all point-pairs inverse at 180° in a circle of inversion C_1 become inverse at 180° in a circle of inversion C_2 centered on pole P', the inverse at 180° of pole P in circle C_1.

8-15-4: In the inversion of curves to dissimilar inverse loci:

(a) the gain (or loss) of a 0° focus of self-inversion in the finite plane is accompanied by the loss (or gain) of a line of symmetry and a focus of self-inversion on the ideal line at infinity in the othogonal direction (see also Maxim 6-8-10);

(b) 180° foci of self-inversion are conserved, i.e., they are neither gained nor lost;

(c) 0° and 180° foci of self-inversion that are coincident are conserved but are non-coincident in the inverse locus.

8-15-4 : (d) foci of self-inversion at 90° and 270° are not conserved in any dissimilar inverse locus. Those of quartics are replaced by two poles of congruent-inversion in axial inverse loci and by 4 poles of congruent-inversion in non-axial inverse loci. Together with one pole of congruent-inversion possessed by virtue of the basis curve having two lines of symmetry, this makes a total of 5 poles of congruent-inversion for the non-axial loci (for the appropriate values of the units of linear dimension).

8-15-5 : Conservation of a focus of self-inversion by the inversion transformation implies conservation of the corresponding angle of self-inversion.

8-15-6 : Modularly-inverse pole-pairs about foci of self-inversion at angles of 0°, 180°, 0° and 180°, and θ and 180°+θ (θ ≠ 90° or 270°, except for central curves) are equivalent inversion poles.

8-20-1 : Axial inversions to and from central Cartesians conserve the sum of the number of lines of symmetry and the number of 0° modes of self-inversion for finite moduli.

8-20-2 : The inversion transformation conserves the sum of the number of lines of symmetry and the number of 0° modes of self-inversion for finite moduli.

8-20-2': The inversion transformation conserves the sum of the numbers of 0° modes of self-inversion for finite and infinite moduli.

8-20-3 : On each line of symmetry there are two poles about which a biovular central Cartesian inverts to other subspecies of biovular central Cartesians; these poles lie at modular distance on each side of center, i.e., at the distance of the unit of linear dimension for self-inversion. In these inversions, annular curves invert to two subspecies of opposed-oval curves, whereas opposed-oval curves invert to one annular and one opposed-oval subspecies.

8-20-4 : Single radicand conditions for simple intercept formats of central curves of greater than 2nd degree do not define additional circumpolar focal loci beyond those specified by conditions on the corresponding simple intercept equations.

8-20-5 : The most symmetrical axial quartics in a central Cartesian-based inversion superfamily are subspecies of Cartesian self-inverters, i.e., they are subspecies of the genera in which the two foci defined by the $\cos^2\theta$-condition are in coincidence as a double-focus.

8-20-6 : On a given line of symmetry, the poles about which a given subspecies of central Cartesian inverts to a given subspecies of axial bicircular quartic or serpentine circular cubic occur in quartets. The poles in each quartet lie in modularly-inverse pairs on each side of center, i.e., if two poles lie at $\pm nj$, the other two lie at $\pm j/n$, where n can take on any real value except 0 or 1, and j is the modular value (the unit of linear dimension for self-inversion).

8-20-7 : All axial inverse loci of biovular central Cartesians are obtained by inverting about poles on 3 loci: all axial poles from $h = 0$ to $h = j$; all axial poles from $k = 0$ to $k = j$; and all poles incident upon the modular circle in one quadrant.

8-20-8 : Axial serpentine circular cubics and axial bicircular quartics invert to two subspecies of biovular central Cartesians about the members of two pole-pairs on their (single) lines of symmetry; they invert to one subspecies about the members of one pole-pair, and to the second subspecies about the members of the other pole-pair.

8-20-9 : An axial bicircular quartic with a non-intersecting line of symmetry possesses two axial 0° intra-oval-mode foci of self-inversion; the inverse loci about the axial modular poles of these foci belong to different subspecies of opposed-oval central Cartesians, one subspecies for the poles of each focus.

8-20-10: An axial serpentine circular cubic or axial bicircular quartic with an intersecting line of symmetry inverts to a single subspecies of axial-vertex serpentine circular cubic about its vertices, to a single subspecies of opposed-oval central Cartesian about the two axial poles at modular distance from the 0° inter-oval-mode focus of self-inversion, and to a single subspecies of annular central Cartesian about the two axial poles at modular distance from the 0° intra-oval-mode focus of self-inversion.

8-20-11: Axial inverse loci of a given biovular central Cartesian possess only two axial poles about which inversions yield curves that are congruent with one another for a common value of the unit of linear dimension. These are the poles at modular distance from the 180° inter-oval-mode focus of self-inversion.

8-20-12: An axial serpentine circular cubic and an axial bicircular quartic whose line of symmetry intersects the curve possess 5 axial multi-member sets of exceptional equivalent inversion poles. For any given basis curve, all of the inverse loci about the poles of any given set are members of the same subspecies. Three of these sets consist of two poles each, and the pole-pairs lie at modular distances from the foci of self-inversion; a fourth set consists of the 3 foci of self-inversion, and the fifth set consists of the 3 or 4 vertices. An axial bicircular quartic whose line of symmetry does not intersect the curve has only 4 such exceptional multi-member sets of equivalent inversion poles, since it possesses no axial vertices.

8-20-13: There are 4 generally non-focal poles about which biovular central Cartesians invert to Cartesian self-inverters. For annular central curves these lie on a line of symmetry in pairs interior and exterior to both ovals. For opposed-oval central curves they are located in pairs within each oval; in the members of one opposed-oval species they lie on the line of symmetry that intersects the curve, while in the members of the other species they are concyclic.

8-20-14: The inversion image obtained by first inverting a biovular central Cartesian about an axial pole for modular j (j for self-inversion about the center), and then inverting the resulting axial bicircular quartic or serpentine circular cubic about the pole inverse to the center of the basis curve, also for modular j (j for self-inversion about the focus of self-inversion that replaces the line of symmetry), is ¼ the linear size of the basis curve.

8-22-1 : Biovular central Cartesians invert to subspecifically different central Cartesians about modular poles on the lines of symmetry. In these inversions, the modular distance becomes halved, i.e., if j is modular for self-inversion of the basis curve, ½j is modular for self-inversion of the inverse central curve.

8-22-2 : The linear size ratio of subspecifically identical inverse loci about modularly-inverse poles on the lines of symmetry of a central Cartesian is the inverse ratio of the distances of these poles from the center.

8-22-3 : A given axial bicircular quartic or serpentine circular cubic inverts to a different subspecies of central Cartesian about each of 3 pairs of poles. One of these central subspecies is annular, the other two possess opposed ovals.

8-22-4 : A monovular central Cartesian inverts to axial monovals or axial serpentine cubics about poles on its lines of symmetry. It inverts to a generally dissimilar central curve about its center.

8-22-5 : An axial inverse locus of a monovular central Cartesian inverts to axial monovals about poles on its line of symmetry and on its inverse-axis circle (the circle inverse to the orthogonal line of symmetry of the basis curve). It inverts to central Cartesians about the two points of intersection of these loci, and self-inverts about the center of the inverse-axis circle. The central Cartesian subspecies obtained in this manner invert to one another through their centers and generally are dissimilar.

8-22-6 : Non-axial inverse loci of monovular central Cartesians invert to axial monovals about points on the inverse-axis circles (circles inverse to the lines of symmetry of the central basis curves); they invert to the basis central Cartesians about the two points of intersection of these circles. The basis central Cartesian subspecies obtained in this manner invert to one another through their centers and generally are dissimilar.

8-22-7 : A given biovular central Cartesian inverts to both subspecifically identical and different biovular central Cartesians about the 5 points of intersection of the lines of symmetry and the modular circle (the centered circle of inversion). The 0° and 180° inverse loci about the center are self-inversions, whereas the pairs of inverse loci about the two points of intersection of each line of symmetry with the modular circle are dissimilar to each other and to the basis curve. If the basis curve possesses opposed ovals, one pair of inverse loci also possess opposed ovals, whereas the other are annular; if the basis curve is annular, both pairs of inverse loci possess opposed ovals (see also Maxim 8-20-3).

8-22-8 : Biovular central Cartesians invert to axial bicircular quartics or serpentine circular cubics about points on one or the other line of symmetry or the modular circle, excluding the points of intersection of these 3 loci (they invert to central Cartesians about the points of intersection). About other points in the plane, they invert to non-axial curves.

8-22-9 : An axial bicircular quartic or serpentine circular cubic possesses one 180° and two 0° axial foci of self-inversion in the finite plane. The 3 circles of inversion centered on these foci (circles A and B at 0°; circle C at 180°) intersect in two common points at the axial level of the 180° focus, which lies midway between the points of intersection.

8-22-10: There are 3 loci in the plane of an axial bicircular quartic or serpentine circular cubic about which it inverts to other axial curves. These are the two circles of inversion, A and B, centered on the two 0° foci of self-inversion, and the line of symmetry, L. Inverse curves about points on the 3rd circle of inversion, C, centered on the 180° focus of self-inversion (except for the points where C intersects A, B, or L) are non-axial.

8-22-11: In the inversion of a central Cartesian about any member of a quartet of axial poles consisting of reflected modularly-inverse pole-pairs, the poles inverse to the other 3 members of the quartet are the foci of self-inversion in the finite plane of the inverse curve (see also Maxim 8-30-6).

8-22-12: The 180° circle of inversion, C, of an axial bicircular quartic or serpentine circular cubic inverts to an unexceptional (and uncentered) circle, C', in the plane of each of its 3 basis central curves. The inverse circle C' intersects both lines of symmetry and the modular circle of each basis central curve; the two points of intersection of circle C' with the modular circle are coincident with the two points where the modular circle intersects one of the lines of symmetry.

8-22-13: An axial bicircular quartic or serpentine circular cubic inverts to central Cartesians about the 6 points of intersection of the line of symmetry, L, and the two circles of inversion, A and B, centered on the two 0° foci of self-inversion (taking L, A, and B pairwise). The inverse central loci about the two points of intersection of circle A with line L have lines of symmetry inverse to A and L and a modular circle inverse to circle B; the inverse central loci about the two points of intersection of circle B with line L have lines of symmetry inverse to B and L, and a modular circle inverse to circle A; the inverse central loci about the two points of intersection of circles A and B have lines of symmetry inverse to A and B, and a modular circle inverse to L.

8-22-14: An axial bicircular quartic or serpentine circular cubic inverts to non-axial curves about points in the plane that are not incident upon its line of symmetry or either of its circles of inversion, A or B, centered on the two (axial) 0° foci of self-inversion.

8-22-15: A non-axial bicircular quartic or serpentine circular cubic inverts to 3 pairs of subspecifically different central Cartesians about the 6 points of intersection of 3 mutually orthogonal 0° circles of inversion, A, B, and D. In a given inversion to a central curve, if a point of intersection of circles A and B provides the pole of inversion, the circles A and B invert to the two lines of symmetry; the 3rd circle, D, inverts to the modular circle. If a point of intersection of circles B and D provides the pole of inversion, then circles B and D invert to the two lines of symmetry and circle A inverts to the modular circle, etc.

8-22-16: The line of symmetry of the inverse axial curve obtained by inverting a biovular central Cartesian about a non-axial pole on its modular circle is inverse to the modular circle and passes through the two poles inverse to the reflections of the pole of inversion in the lines of symmetry (these poles become the $0°$ foci of self-inversion in the finite plane).

8-22-17: The single points of tangency with the curve, of radii from $0°$ intra-oval-mode foci of self-inversion of axial bicircular quartics, are the poles of inversion to axial-vertex serpentine circular cubics.

8-22-18: The poles inverse to the 8 vertices of annular central Cartesians lie at the points where the inverse curve intersects the loci inverse to the basis lines of symmetry.

8-22-19: A pair of equivalent inversion poles of a given central QBI curve, that are reflected in a line of symmetry, are modularly inverse in a focus of self-inversion in the finite or extended plane in all of its inverse loci.

8-22-20: A pair of equivalent inversion poles that are modularly inverse in a focus of self-inversion of a given mode ($180°$ inter-oval, $0°$ inter-oval and intra-oval) in the finite or extended plane, in a given subspecies of curve, are modularly inverse in a focus of self-inversion of the same mode in all inverse loci of that subspecies.

8-22-21: A non-axial curve, B, that self-inverts at $0°$ and/or $180°$ about foci Q, R, etc., is inverted about a pole, P. The foci of self-inversion of the inverse locus, B', lie at poles, $P^*_Q{}'$, $P^*_R{}'$, etc., in the plane of B'. The latter poles are inverse to the modularly-inverse poles, P^*_Q, P^*_R, etc., of P in the foci of self-inversion, Q, R, etc., in the plane of B. $0°$ foci of self-inversion of B' may lie in the finite plane or on the ideal line at infinity; a $180°$ focus of self-inversion of B' lies only in the finite plane.

8-29-1 : Any unexceptional inversion of a locus C to a locus D can be regarded as a self-inversion; conversely, any self-inversion can be regarded as an unexceptional inversion of a locus A to a locus B. In the former case, one regards the union of C and D as the self-inverting curve; in the latter case, one regards the segments A and B of the self-inverting curve as the individual loci that are the inverse of one another.

8-30-1 : The equivalent inversion poles of a pole of inversion, P_h, of a basis curve that self-inverts about a single pole, C, at angles θ and $\theta+180°$ are:

(a) the reflection P_{-h} of P_h in C,

(b) poles P^*_h and P^*_{-h}, modularly inverse to poles P_h and P_{-h} (at distances $\pm j^2/h$ along lines through C at the angles θ and $\theta+180°$), and

(c) the reflections of poles P_h, P_{-h}, P^*_h, and P^*_{-h} in lines of symmetry.

8-30-2 : Poles Q', inverse to equivalent inversion poles Q, of the pole of inversion P, of a basis curve B, are either poles of congruent-inversion or foci of self-inversion of the inverse locus B'. Poles Q' are foci of self-inversion if the corresponding poles Q are:

(a) reflections of P in lines of symmetry,

(b) modularly inverse to P in 0° or 180° foci of self-inversion, or

(c) modularly inverse to P in a 0° and 180° focus of self-inversion.

8-30-3 : If a basis curve that self-inverts about a single pole in multiple modes is inverted about a pole on its modular circle, all of the poles of identical subspecific-inversion of the inverse locus will be collinear.

8-30-4 : If radii from a 0° intra-oval-mode focus of self-inversion that replaces a line of symmetry, L, of a central Cartesian are drawn to points p'_n of the inverse locus that are inverse to points p_n, incident upon L, these radii are tangent to the inverse locus at the points p'_n. Such points, p'_n, are poles of inversion to axial-vertex cubics.

8-30-5 : If a basis curve that self-inverts about a single pole in multiple modes is inverted about a non-modular pole, the poles of identical subspecific-inversion in the finite plane of the inverse locus comprise two ensembles, one concyclic and the other collinear. The concyclic ensemble possesses one more pole than the collinear ensemble.

8-30-6 : In a locus B', the inverse about pole P of a central basis curve B that self-inverts about the center at 0° and 180°, the pole P*' inverse to pole P*, modularly inverse to P, is the 0° focus of self-inversion; the pole $P^{*\prime}_R$, inverse to P^*_R, the reflection of P* in the center, is the 180° focus of self-inversion (see also Maxim 8-22-11).

8-30-7 : An annular central Cartesian B is inverted about a pole P to a non-axial bicircular quartic B'. Modularly-inverse point-pairs (p and p*)$_{0°}$ and (p and p*)$_{180°}$ of B are intersected by coincident radii and chord segments, respectively, from its center. In the plane of the inverse locus B', coincident radii from the 0° inter-oval-mode focus that "replaces" the 0° center-focus of B intersect all point-pairs (p' and p*')$_{0°}$ inverse to the point-pairs (p and p*)$_{0°}$, while chord segments from the 180° focus of self-inversion that "replaces" the 180° center-focus of B intersect all point-pairs (p' and p*')$_{180°}$ inverse to point-pairs (p and p*)$_{180°}$.

8-30-8 : A given line through a pole of congruent-inversion of an inverse locus intersects the locus at no more than one point inverse to an axial vertex of the basis curve.

8-30-9 : A focus of self-inversion of an inverse curve that replaces a line of symmetry of a basis curve acquires the mode of self-inversion of the focus of self-inversion of the basis curve that lies on the ideal line at infinity in the direction orthogonal to the lost line of symmetry.

8-30-10: In modular axial inversions of a central curve that self-inverts about the center at 0° and 180°, to dissimilar central curves that also self-invert about their centers at 0° and 180°, the basis 0° center-focus "exchanges positions" (modes of self-inversion) with the basis 0° focus of self-inversion on the ideal line at infinity in the direction of the axis in question. In non-modular inversions of the same central curve to non-central curves, the inverses of the basis 0° center-focus possess the mode of self-inversion of the latter.

8-30-11: Inter-oval-mode and intra-oval-mode (and inter-segment-mode and intra-segment-mode) 0° foci of self-inversion in the extended plane are conserved by the inversion transformation.

8-30-12: The number of foci in the extended plane in each category of self-inversion, i.e., 0° and 180° inter-oval mode (and inter-segment mode) and 0° and 180° intra-oval mode (and intra-segment mode) is conserved by the inversion transformation; only the locations of the foci change.

8-30-13: If a central basis curve with n lines of symmetry that does not self-invert about the center is inverted about a non-central axial pole, the inverse locus is axial and possesses n-1 foci of self-inversion in the finite plane.

8-36-1 : The inverse locus of a curve about an incident pole at which the curvature is finite possesses an asymptote that is inverse to the circle of curvature at that pole.

8-36-2 : The inverse locus of a curve about an incident pole at which the curvature is infinite possesses the ideal line at infinity as asymptote.

8-36-3 : When an open curve with an asymptote at a finite distance from its nucleus is inverted about a non-incident pole, one obtains a closed curve with finite curvature at the reciprocal inversion pole.

8-36-4 : When an open curve with the ideal line at infinity as asymptote is inverted about a non-incident pole, one obtains a closed curve with infinite curvature at the reciprocal inversion pole.

8-38-1 : Single points of tangency (i.e., excluding bitangents) of radii from foci of self-inversion of non-axial QBI cubics and quartics of non-unit eccentricity are poles of inversion to axial-vertex inversion cubics.

8-38-2 : The 4 points of intersection of a non-axial elliptical QBI curve with its two circles of inversion are poles of inversion to axial-vertex inversion cubics.

8-38-3 : The two points of intersection of a non-axial hyperbolic QBI curve with its inverse transverse-axis circle of inversion (excluding the DP) are poles of inversion to axial-vertex inversion cubics (axial strophoids).

8-38-4 : In the extended plane, the pole of inversion of an inverse locus B' to a congruent inverse locus B", where B" also is a reflection of B' in a line, is either:

(a) the inverse of the pole of inversion of B (points of the ideal line at infinity); or

(b) the pole inverse to the equidistant reflection of the pole of inversion of B in a true center.

8-40-1 : Single points of tangency of radii from foci of self-inversion of cubics and quintics are poles of inversion to axial curves.

8-40-2 : If a basis curve B is inverted about a pole P and the circle of curvature of B at P intersects B at points C, then the asymptote of the inverse curve B' intersects B' at points C' inverse to C.

8-40-3 : An inverse curve intersects an asymptote at points of the ideal line at infinity but at no finite point only if the circle of curvature of the basis curve at the pole of inversion, P, has no other finite point of contact with the basis curve than its point of tangency at P.

APPENDIX 2. GLOSSARY OF TERMS

Algebraically complete transform equation see Intercept transform

α-Transform (α-Circumpolar intercept transform) the general expression for the circumpolar intercept transform for any angle α between the two radii; see Intercept transform.

α-θ-Transform circumcurvilinear or circumlinear intercept transform; relates the x intercepts of a radius vector from a pole at the angle θ to the x-axis to the y intercepts of a radius vector from the same pole at the angle θ+α to the x-axis; the poles from which these vectors are measured are points on the locus to which the symmetry of the basis curve is being referred.

Asymptote the tangent to a curve whose points of contact lie on the ideal line at infinity. If the limit of a tangent line is the ideal line at infinity itself, the latter is said to be the asymptote.

- **point** point of intersection of an asymptote with a line of symmetry.

Axial pertaining to a line of symmetry.

- **curves** curves with a single line of symmetry.
- **equality** equality of the number of lines of symmetry for two or more different configurations of a given ensemble of reference elements.
- **focal locus** a line, all points of which have focal rank.
- **inverse loci** inverse curves with one line of symmetry.
- **inversions** inversions of curves about poles on a line of symmetry, generally the axis coincident with the x-axis (for curves in standard position).
- **locations** refers to incidence of points, poles, or foci upon a line of symmetry, generally the axis coincident with the x-axis (for curves in standard position).

Basis curve subject or "starting curve" for an operation or transformation.

Basis genus genus of an inversion superfamily with an equation of the lowest even exponential degree, member curves of which possess the highest degree of circumpolar symmetry (a genus of central curves, if the superfamily possesses central curves).

Bicircular relating to two circles.

- **quartic** a quartic locus consisting of two conjugate (self-inverting in multiple modes) non-intersecting ovals.

Bilinear relating to two lines.

Bipolar relating to two poles.

- **coordinates** a coordinate system based upon undirected distances from two fixed point poles.

Cartesian locus defined by certain simple bipolar, tripolar, and polar-circular equations.

- **condition** vanishing of the Cartesian term, the condition on an axial inverse locus of a central Cartesian for which its $\cos^2\theta$-condition foci are in coincidence--characterizes Cartesian self-inverters.
- **term** radicand of the $\cos^2\theta$-condition of the simple intercept equation of axial inverse loci of central Cartesians (vanishing of this term is the condition for coincident $\cos^2\theta$-condition foci and Cartesian self-inverters).

Cartesian self-inverters supergenus or subfamily of monovular and biovular self-inverting curves with polar equation, $Gr\cos\theta = r^2+Er+F^2$.

Cassinians (Cassinian ovals, Ovals of Cassini) loci defined by the equilateral hyperbolic bipolar equation, $uv = Cd^2$.

Catacaustic of a curve the envelope of light rays, emitted from a pole, after reflection at all points of the curve.

Caustic see Catacaustic, Diacaustic

Central Cartesian any central curve derived by an inversion of a Cartesian self-inverter [equation $(x^2+y^2)^2+B^2x^2+C^2y^2+D^4 = 0$]

Central conic conic section with two lines of symmetry.

Central curve curve with two (or more) orthogonal lines of symmetry.

Central modular-focal quartic contral Cartesian for which the foci of the $\cos^2\theta$-condition are at modular distance from center.

Central quartic curve obtained by inverting a central conic about its center.

Circle of inversion circle centered on a pole of inversion, of radius equal to the unit of linear dimension for the inversion.

Circumcurvilinear relative to a curve as reference element (includes the line).

Circumlinear relative to a line as reference element.

Circumpolar relative to a point as reference element.

point-focal rank a pole possesses circumpolar point-focal rank for a given angle and a given curve if the rectangular degree of its intercept transform for the curve is lower than that of transforms about all neighboring points.

Class of circumpolar symmetry the exponential degree of the intercept transform of a curve about a given pole at a given angle.

Closure, immediate see Immediate closure

Compound intercept equation equation for a specified angle α expressing the relationship between the x and y intercepts of radius vectors from the reference pole at angles θ and $\theta+\alpha$, respectively, as a function of θ.

Compound intercept format solution of the compound intercept equation for $\cos^n\theta$ or $\sin^n\theta$, but often for $\cos\theta$ or $\sin\theta$.

Conchoid in the classical conchoid construction equal distances along a line are marked off from its point of intersection with a curve for various positions of the line as it rotates about a point. The conchoid is the locus of the endpoints of the line segments.

Concentric having a common center.

Concurrent lines passing through a common point are said to be concurrent.

Congeneric belonging to the same genus.

Congruent having the same size and shape.

Congruent-inversion an inversion of a basis curve to a congruent but non-coincident locus (for the appropriate value of the unit of linear dimension).

Conic axial-vertex inversion cubic cubic obtained by inverting an axial or central QBI curve about a major or transverse-axis vertex.

Conjugate point or pole the reflection of a basis point or pole in a reference element.

Conservation of degree property of certain equations whereby degree does not increase under the inversion transformation, i.e., no inverse locus is of higher degree than the basis curve.

ultimate conservation of degree describes the property of the inversion transformation wherein there is an upper limit to the exponential degree of inverse loci obtained by a chain of successive inversions of a basis curve.

Conservation of eccentricity property of QBI inversion cycles whereby the image conic (see Inversion image) has the same eccentricity as the basis conic.

Constant condition a condition obtained by setting the constant term of a simple intercept equation equal to 0 (defines axial intersections--vertices, double points, etc.).

$\text{Cos}^2\theta$-condition condition for vanishing of the coefficient of the $\cos^2\theta$ term of a simple intercept equation.

focus or foci focus or foci defined by the $\cos^2\theta$-condition.

Covert linear focal locus a linear focal locus that is not a line of symmetry.

Cubic curve of 3rd degree in the variables.

Curve of demarcation typically a subspecies (single-membered species) that demarcates two species with differing numbers of ovals or segments, while itself possessing intersecting loops.

dual curves in cases in which two curves demarcate the members of 3 species of a genus, they are referred to as *dual curves of demarcation.*

Curve of intersection curve possessed in common by two or more inversion superfamilies. All known curves of intersection are curves of demarcation.

Cusp a variety of DP at which the tangents coincide.

Decic curve of 10th degree in the variables.

Diacaustic of a curve the envelope of light rays emitted from a pole, after refraction at all points of the curve.

Displaced parabolic Cartesians a biovular species of Cartesian self-inverters with the bipolar equation, $u^2 = Cdv+2d^2$.

Double point (DP) for a hyperbolic QBI curve, the point of self-intersection of the curve.

Double-point homologue for elliptical QBI curves, an isolated incident double point of the locus (double-root of the equation of the locus). For QBI curves of unit eccentricity, the cusp point (also a double-root of the equation).

Eccentricity for central conics, $b^2x^2 \mp a^2y^2 = a^2b^2$, and all their inverse loci, the eccentricity is $e = \sqrt{(a^2 \pm b^2)}/a$. For the parabola and all its inverse loci $e = 1$.

Eccentricity-correlated rank phenomenon whereby the α-transform of a hyperbola reduces, not for a specific angle of transformation, but for variable angles determined by eccentricity. For any given value of the eccentricity, the angles in question are the angles of the asymptotes.

Eliminant the equation obtained by eliminating one or more variables between two or more equations.

Elliptical limacon see Limacons

Equational pertaining to relationships between curves based upon similarities between their equations, i.e., the rules for their construction.

focal condition a focal condition on the entire simple intercept equation; for example, a condition for which the equation becomes a perfect square.

Equilateral QBI curve the equilateral hyperbola or any of its inverse loci.

Equivalent poles poles reflected in a line of symmetry or a point of symmetry.

Equivalent inversion poles to a given reference pole, points in the plane of a curve or figure about which inversion yields loci that are subspecifically identical (similar) to the locus obtained by inverting about the given reference pole.

Extremal-distance symmetry designation for the relationship between curves defined by a single equation in undirected-distance coordinates, one locus given by the greatest-distance construction, the other by the least-distance construction.

Focal condition condition on an intercept equation or format that leads to a reduction in the degree of the corresponding transforms.

Focal rank

circumpolar of a point with respect to a curve, the reciprocal of the rectangular degree of the circumpolar intercept transform about a point, or of the circumpolar intercept transforms about points on a focal locus for a given angle of transformation

point-focal see Circumpolar point-focal rank

Focus a locus about which a curve has greater symmetry than about neighboring points not incident upon the locus.

angle-independent circumpolar circumpolar focus defined by a condition on a simple intercept equation or format

axial a line, all points of which have focal rank.

compound a curvilinear focal locus upon which there also lie point foci.

eclipsable a pole or curvilinear locus that loses its focal rank in certain (highly symmetrical) subspecies.

loop a point focus that lies within a loop in all subspecies.

of self-inversion any pole about which a curve self-inverts(at any angle).

multiple two or more point foci in coincidence.

generic a focus that is multiple in all subspecies of a genus (or in all but a finite number of subspecies).

subspecific a focus that is multiple only in a certain subspecies.

simple a curvilinear focal locus upon which there lie no point foci.

variable a focus that has different relationships to a curve in different subspecies, for example, within a small loop in some, within a large loop in some, and outside of the curve in others.

Focus-vertex group ensemble of axial inversions of central conics about poles between the focus and the vertex.

4-Quadrant transform circumpolar intercept transform represented for all combinations of positive and negative intercepts.

Genus taxon one level above *subgenus* and one level below *supergenus*. Broadly speaking, all curves obtained from a rectangular or polar equation having two parameters, for all parameter values (excluding parameters that merely influence the location or orientation of a locus).

h or H axis any axis parallel to the x-axis.

Hyperbolic limacon see Limacons

Ideal line at infinity the set of ideal points at infinity, one for each direction in the plane.

Identical congruent to and coincident with.

Identical subspecific-inversion either a self-inversion or a congruent-inversion, i.e., an inversion to a curve of the same subspecies.

Immediate closure the property of the inversion transformation that in an ensemble of inverse loci of a given basis curve, each member inverts to every other member but to no non-member.

Individual bipolar focal rank the bipolar focal rank of single poles as opposed to the bipolar focal rank of pole-pairs.

Inner segment segment closest to the origin--of a circumpolar intercept transform or of a plot of $ds/d\theta$ versus θ for a circumpolar intercept transform.

Intercept product the expression for the product of intercepts from a point to a curve along a line.

Intercept product
  chord-segment the distance (in the positive sense) from the reference point to the points of intersection with the curve are measured in opposite directions along a chord through the reference point.
  coincident-radii the distances (in the positive sense) from the reference point to the points of intersection with the curve are measured in the same direction along a radius from the reference point.
  "pure" product of the intercepts of the form $xy = f(\theta,h,k)$ that is valid for all points in the plane; these exist in simply expressible form only for conic sections.

Intercept transform equation relating the intercepts with a curve of two radii emanating from a point at a fixed angle to one another, as these radii rotate about the point to which the symmetry is being referred, or translate or travel along the reference locus.
  algebraically complete transform equation transform equation, usually degenerate, that is valid for all possible combinations of positive and negative intercepts.
  circumcurvilinear the pole from which the radii emanate travels along a curve.
  circumlinear the pole from which the radii emanate travels along a line.
  circumpolar the two radii *rotate* about the pole from which they emanate.
  condition a condition on an intercept transform which, if satisfied, leads to a reduction in its degree in the variables.
  trivial a transform that merely expresses the equality of identical intercepts, i.e., intercepts for which the radius vectors are both equal and coincident, excluding cases for which the intercepts are 0 for all angles for which they are defined.

Inter-vertex group ensemble of axial inversions of central conics at poles lying between the vertices.

Inverse-axis circle circular locus in the plane of an inverse curve that is inverse to a line of symmetry of the basis curve.

Inverse $\cos^2\theta$-group curves defined by polar equation, $r = j^2/(a+b\cos^2\theta)$.

Inverse locus of a curve or figure, the locus obtained relative to a point reference pole by measuring off distances along radius vectors from the pole that are the reciprocals of the distances to the curve along the same radius vectors.

Inverse modular circle circular locus in the plane of an inverse curve that is inverse to the modular circle of the basis curve.

Inverse-root-$\cos^2\theta$ group curves defined by polar equation, $r^2 = j^4/(a^2 \pm b^2\cos^2\theta)$.

Inversion the transformation whereby a curve is inverted about a point-pole (see Inverse locus).
  centers points on a line in a given region of intersection with a curve, such that the product of the lengths of the delineated segments is constant at a given value.
  cycle closed inversion chain, i.e., an inversion chain in which the basis curve is recovered as an image in the terminal member of the chain (see Inversion image).
  image a curve belonging to the same subspecies as the basis curve, recovered through a chain of inversions.
  Taxonomy the classification or systematic organization of curves on the basis of their relatedness through the inversion transformation.

k or K axis any axis parallel to the y-axis.

Latus rectum chord of a conic section at the level of a traditional focus and orthogonal to the axis upon which the traditional focus lies.

LCQ Sextics see Sextics

Limacons inverse loci of conic sections about the traditional foci.
  elliptical inverse loci of ellipses about their traditional foci.
  hyperbolic inverse loci of hyperbolas about their traditional foci.
  parabolic inverse locus of the parabola about its traditional focus (the cardioid).

Limiting locus for some biovular species of curves, the two ovals achieve coincidence for certain values of the parameters of their equations. The locus of coincidence is the limiting locus or limiting curve. The same concept applies to an ensemble of curves for the case in which the curves of the ensemble achieve coincidence (the limiting locus commonly is a curve of demarcation).

Linear-circular relating to a line and a circle.

Line of Symmetry a line about which a curve possesses equidistant reflective symmetry, or a line of equidistant reflective symmetry of a coordinate system.

Loop a segment of a curve that is self-intersecting.

Maxim a rule based upon current knowledge.

Mimics see Symmetry mimics

Modular for the modular value of distance, i.e., for the unit of linear dimension for self-inversion.
  circle circle centered upon a focus of self-inversion of radius equal to the unit of linear dimension for self-inversion (the modular value).
  locus the non-focal self-inversion locus for modular j referred to a focus of self-inversion as reference pole.
  poles axial poles at modular distance from a focus of self-inversion.
  value the unit of linear dimension for self-inversion.

Monoconfocal the condition wherein two central conics have a common traditional focus.

Multiple focus see Focus

Multipolar relating to more than two poles.

Mutual non-focal self-inversion loci an ensemble of curves, each of which is, in one respect or another, a non-focal self-inversion locus of every other member of the ensemble.

Near-circles see Near-conics

Near-conics near-circle, near-ellipse, and near-line transforms about the $-\frac{1}{2}b$ focus of limacons and its homologue in linear Cartesians.

Near-ellipses see Near-conics

Near-lines see Near-conics

Non-axial curve curve lacking a line of symmetry.

Non-concentric, congruent, bicircular coordinates undirected-distance coordinate system in which the fixed-position reference elements consist of two non-concentric congruent circles.

Non-focal self-inversion self-inversion about points that lie on a curvilinear locus defined by the inversion centers on lines concurrent at a given reference pole for the inversion.

Non-polarized bipolar coordinates bipolar coordinate system in which the poles are unordered, preserving reflective redundancy in both the polar axis and the midline.

Nucleus of open curves, central or axial region, or region of oval, loop, or exceptional point(s).

Octic curve of 8th degree in the variables.
  LLC octic with linear 0° and 180° transforms and a circular 90° transform about a given pole.

Ordinal rank the degree in x and y of the ordinal rank equation for $\mathring{s}^2$ of the α-transform.

Orthotomic of a curve the locus of the points that are the reflections of the reference pole in the tangents to the basis curve (same subspecies as the tangent pedal).

Outer segment segment most distant from the origin--of a circumpolar intercept transform or of a plot of $ds/d\theta$ versus θ for a circumpolar intercept transform.

Oval a non-self-intersecting closed curve or closed segment of a curve (possessing continuous first derivatives at all points).

Parabolic Cartesian genus of Cartesian self-inverters with bipolar equation $u^2 = Cdv$.

Pedal

- **normal** the normal pedal P of a basis curve B with respect to a point O is the locus of the feet of the perpendiculars from O on the normals to B (also called the "contrapedal").
- **tangent** the tangent pedal P of a basis curve B with respect to a point O is the locus of the feet of the perpendiculars from O to the tangents to B.
- **Taxonomy** the classification or systematic organization of curves according to their pedals about all points in the plane, the relationships between these pedals, the pedals of these pedals, etc.

Peripheral vertices the intersections with the curve of normals to its major or transverse axis at the positions of the axial foci.

Phylum of circumpolar symmetry the lowest degree non-trivial class of circumpolar symmetry of a curve about any point at any angle determines its membership in a phylum of circumpolar symmetry.

Plane

- **extended** the union of the finite plane and the ideal line at infinity.
- **finite** all points in the plane at a finite distance from a reference point.

Polar-circular relating to a point and a circle

Polarized bipolar coordinates a bipolar coordinate system in which the poles bear fixed labels and there is no reflective redundancy in the midline of the system.

Polar-linear relating to a point and a line.

Pole of conditional focal rank a point that attains point-focal rank only in certain subspecies of a species or genus through coming into coincidence with a point focus.

Pole of concentricity pole about which two or more central curves are centered.

Post-focal group ensemble of axial inversions of central conics at points beyond the focus, as reckoned from the center.

Post-vertex group ensemble of axial inversions of central conics at poles external to the a vertices of ellipses, and beyond the vertices (in the directions of the foci) of hyperbolas.

Potential focal condition condition on an intercept equation for which a term in the variables vanishes (without leading to the factoring out of a term in a variable).

Pre-focal group ensemble of axial inversions of central conics at points between the center and the focus.

Pseudo-exchange-limacon the curves obtained by axial inversions of central conics about a pole at twice the distance of the a vertex from the center.

QBI curves quadratic-based inversion curves, i.e., the curves obtained by inverting conic sections about all points in the extended plane.

Quadratic — curve of second degree in the variables.
Quartic — curve of fourth degree in the variables.
Quintic — curve of fifth degree in the variables.
Radial of a curve — the locus of the endpoints of line segments drawn from a pole, where these are parallel and congruent to the radii of curvature for all points on the basis curve.
Radicand constant-condition — a condition resulting from setting the constant term of the radicand of a simple intercept format equal to 0.
Rank — usually a comparative quantitative measure of symmetry.
- focal — of a focal locus with respect to a curve, the reciprocal of the degree of the rank-determining transform.
- ordinal — degree of the ordinal rank equation.
- subfocal — based upon comparisons of the complexity of the rank-determining equations (no quantitative index).
- subordinal — degree of the subordinal rank equation.
- symmetry — non-specific characterization of either or both focal and subfocal rank.

Reciprocal inversion pole — the pole in the plane of an inverse curve coincident with the pole of inversion of the basis curve (inversion about this pole restores the basis curve).
Reciprocal reflective symmetry — of a locus about a point at the angle α. The condition wherein the product of the distances to the locus from the pole along two radii at the fixed angle, α, is independent of the variable angle, θ, of one of the radii to a reference axis.
Requirement for axial equality — a necessary condition for valid comparisons of the symmetry of different curves relative to different configurations of a given ensemble of reference elements (see Axial equality).
Root-$\cos^2\theta$ group — curves defined by the polar equation, $r^2 = a^2 \pm b^2\cos^2\theta$.
Self-inversion — an inversion in which the inverse locus is identical to (congruent and coincident with) the basis curve.
- α- (reciprocal reflective symmetry about a point at the angle, α) the intercepts of radii from the reference pole to the curve are measured for lines at and angle, α, to one another.
- focal — self-inversion of a curve about a point (the classical self-inversion).
- negative — the intercepts are measured in opposite directions from the reference pole (self-inversion at 180°).
- non-focal — see Non-focal self-inversion
- positive — the intercepts are measured in the same direction from the reference pole (self-inversion at 0°).
- θ loci — the loci defined by the inversion centers in a given region of a curve along lines at an angle θ to the x-axis.

Serpentine circular cubic — inverse locus of a biovular central Cartesian about an incident pole--consists of a self-inverting ensemble of an oval and a non-intersecting serpentine arm.
Serpentine cubic — inverse locus of a monovular central Cartesian about an incident pole--consists solely of a self-inverting serpentine arm.
Sextic — curve of sixth degree in the variables.
- LCQ — sextic with linear 0° transform; circular 180° transform, and quartic 90° transform about a given pole.

Simple intercept equation equation expressing the x intercept of a radius vector from the reference pole to a point on the curve as a function of θ, the angle of the radius vector to the x-axis.

mixed transcendental simple intercept equation containing more than one transcendental function of θ.

Simple intercept format solution of the simple intercept equation for $\cos^n\theta$ or $\sin^n\theta$, usually for $n = 1$.

Single-parameter curve subspecies of a species or genus of curves in which all the parameters of the rectangular and polar equations of the curve in standard position have the same value (only a single parameter occurs). Different values of this parameter produce similar curves of different size. Some examples of single-parameter QBI curves are the parabola, the cissoid of Diocles, the equilateral hyperbola, and the circle.

Species taxon one level below *superspecies* and one level above *subspecies*. Broadly speaking, all curves obtained from an equation for a certain range of parameter values.

Specific symmetry class a more precise categorization than merely the exponential degree, for the class of circumpolar symmetry defined by an intercept transform. The equation of the intercept transform about a given point at a given angle is the most precise specification of the class of circumpolar symmetry of a curve about the given point at the given angle.

Squared-$\cos^2\theta$ group curves defined by polar equation, $jr = (a+b\cos^2\theta)^2$.

Subfamily taxon one level below *family* and one level above *supergenus*.

Subfocal rank a comparative ranking within focal rank categories based upon measures of the simplicity or complexity (aside from exponential degree in the variables) of the equations or intercept transforms of two or more curves or of a given curve about two or more poles.

Subordinal rank the degree of the ordinal rank equation with y replaced by f(x) for the specific pole and angle of the transformation.

Subspecifically-identical curves curves belonging to the same subspecies, hence "similar."

Superfamily taxon one level below *suborder* and one level above *family*.

inversion for a given genus of basis curves, for example, conic sections, the inversion superfamily (the QBI superfamily) includes the inversion loci of all subspecies in the genus (i.e., all hyperbolas, all ellipses, and the parabola) about all points in the extended plane.

Supergenus taxon one level below *subfamily* and one level above *genus*.

Symmetric function a function that remains unchanged upon interchanging of the variables.

Symmetry the measure of the simplicity of construction rules for a curve or for the intercept transform of a curve relative to specific reference elements.

circumcurvilinear symmetry about a curve

circumlinear symmetry about a line

circumpolar symmetry about a point

greater symmetry see more symmetrical

more symmetrical having a simpler construction rule (relative to specific reference elements).

most symmetrical having the simplest construction rule (relative to specific reference elements).

Symmetry class see Class of circumpolar symmetry and Specific symmetry class.
Symmetry mimics curves of higher degree than a basis curve that have identical 0° and 180° intercept transforms to those of the basis curve about a homologous pole.
Taxonomy a paradigm of classification.
θ Self-inversion locus see Self-inversion, θ loci
Traditional focus term used primarily to refer to the only heretofore recognized real focus of a conic section, to distinguish it from the other real circumpolar foci (center, vertices, etc.).
Transform see Intercept transform, α-Transform, α-θ-Transform.
Transformation format general equations for α-transforms in terms of simple intercept formats expressed as the $\sin\theta$, $\cos\theta$, $\sin^2\theta$, $\cos^2\theta$, etc. (see Appendix I in *Symmetry, An Analytical Treatment*).
Tripolar relating to 3 points.
Trivial intercept transform see Intercept transform, trivial
True center point of intersection of two orthogonal lines of symmetry.
True two-arm parabola a parabola possessing two arms or branches (not necessarily congruent) represented by a non-degenerate equation.
Valid locus a locus for which non-trivial transforms exist at the angle in question.
Variable-angle transform α-transform of the hyperbola for which there exists a variable-angle focal condition dependent upon eccentricity (see Eccentricity-correlated rank).
Variable focus see Focus, variable
Vertex
  axial a point of intersection of a curve with a line of symmetry.
  peripheral see Peripheral vertices
Vertex-focus group see Focus-vertex group

BIBLIOGRAPHY

Daus, P. H., College Geometry, New York, Prentice-Hall, Inc., 1941.

Hilton, H., Plane Algebraic Curves, Oxford, Clarendon Press, 1920.

Kavanau, J. Lee, Symmetry, An Analytical Treatment, Los Angeles, Science Software Systems, 1980.

SUBJECT INDEX